Introduction to Modern Quantum Optics

Introduction to Modern Quantum Optics

Jin-Sheng Peng
Gao-Xiang Li
Huazhong Normal University, China

World Scientific
Singapore • New Jersey • London • Hong Kong

Published by

World Scientific Publishing Co. Pte. Ltd.

P O Box 128, Farrer Road, Singapore 912805

USA office: Suite 1B, 1060 Main Street, River Edge, NJ 07661

UK office: 57 Shelton Street, Covent Garden, London WC2H 9HE

British Library Cataloguing-in-Publication Data
A catalogue record for this book is available from the British Library.

INTRODUCTION TO MODERN QUANTUM OPTICS

ISBN 981-02-3448-1

This book is printed on acid-free paper.

Printed in Singapore by Uto-Print

Preface

Quantum optics is one of the main fields in modern physics. As we know, it has been only 30 years since quantum optics became a field of science. Although the birth of the quantum theory of light was due to the Planck's quantum postulate and Einstein's photons hypothesis, using quantum mechanics to study the effects of the interaction between matter (atom, molecular, etc.) and light comprehensively and systematically, as well as to investigate them by experiments, started after the development of the laser at the end of the 1950s. Many new optical effects have been revealed from several different subjects of quantum optics by using lasers. With the development of new experiments, additional optical effects will be revealed which will promote further development and understanding.

The key point of modern quantum optics is to investigate the effects of the interactions between atoms and laser fields. Both atoms and radiation fields display quantum properties in the interaction of one or more atoms with laser fields (single or multimode), which can not be discussed in classical terms. The theory of modern quantum optics demonstrates that one can understand many quantum properties of both the radiation field and the atomic behavior by means of quantum electrodynamics.

The development of quantum optics shows that it is necessary to make a specialized statement on the main contents of quantum optics according to the quantum properties of both atoms and radiation fields in the atom-field coupling system. This book is devoted to this purpose, and is based on our lectures in quantum optics for graduate students, on our research publications over the past few years, and on the latest achievements in the field of quantum optics. We hope that this book will be regarded as a useful book for graduate students, undergraduate students, teachers, as well as scientists who are interested in quantum optics and theoretical physics.

This book is divided into three parts. The first part is devoted to introducing the elementary theory of the interaction between atoms and light. It includes five chapters which present the basic theory of quantum optics. For better understanding, we have discussed some practical examples. The first part is also as an introduction to the next two parts. In the first chapter of this book, the three pictures of quantum mechanics and the theory of the density operator are outlined. The two-level atom and the optical Bloch equations are discussed in Chapter 2. Chapter 3 deals with the quantum description of the light. We present not the quantization of light but also a specialized discussion on three state functions of the radiation field. Chapter 4 introduces systematically the Dicke model and the Jaynes-Cummings (J-C) model, which are two typical theoretical models. These two models are of wide application. Chapter 5 discusses the quantum theory of a small system coupling to a thermal reservoir. Besides introducing the Langevin equation and the Fokker-Planck equation, we deal with the master equation of the quantum harmonic oscillator. Chapter 5 is about the fundamental theory of the quantum statistical properties of a system in quantum optics. Part II gives a concentrated discussion of the quantum properties of light fields. First we introduce the theory of coherent light in Chapter 6. Then, in Chapter 7 we discuss the theory of squeezed light, which has been recently the subject of extensive investigations. In the following three chapters, we respectively deal with the three important subjects, which are resonance fluorescence, superfluorescence, and optical bistability. The research on these three subjects has almost been completed both theoretically and experimentally. We have tried to keep our presentation simple and fundamental. Chapter 11 describes the effects of the virtual light field on the atom-field coupling system. From the theoretical point of view, it is of interest to investigate the influence of virtual photon processes on the radiation field. Here we give more details according to our recent publications on this subject. Part III deals with the quantum dynamic properties of the atoms interacting with the laser field. First we introduce the periodic collapse-and-revival effects of atomic behavior under the interaction of the laser field in Chapter 12. The squeezing of atomic operators is discussed in Chapter 13. As for the coherent trapping of atomic populations, we give a more detailed discussion in Chapter 14. Chapter 15 is devoted to presenting a theory of a two-atom coupling system under the interaction of the laser field. Here, the effects of the dipole-dipole interaction have been examined in detail. Chapter 16 introduces the autoionization of the atom induced by the laser field. Chapter 17 deals with the atomic motion in the laser field, the deflection and refraction of an atom moving in a standing-wave field and the force acting on the atom driven by the radiation field are discussed. In the last chapter, we concentrate on the recently developing subject of laser cooling. We introduce a quantum theoretical description of the laser cooling, and give some statements on the problems which are related to its applications. Some key references are listed at the end of each chapter, but the lists are not comprehensive.

Although the development of modern quantum optics continues, we believe that the basic theory and the main treatments presented in this book are useful. The presentation of the subject is novel, rigorous and as clear as possible. Readers may learn the simple physical ideas, the advanced theoretical description and the general methods from this book.

We would like to express our gratitude to Prof. F.Persico and Prof. C.Leonardi for very helpful suggestions and discussions, particularly in connection with Chapters 5, 8, 9 and 11. We also wish to thank Prof. M.U.Palma, Prof. S.Swain , Prof. L.S.Liu, Prof. W.Liu and Dr. P.Zhou for helpful discussions and encouragement. Our thanks are due to Prof. Bai-xiu Huang, who has given much encouragement and provided assistance in the preparation of the figures. The encouragement of Ms. Ning-jiang Yan for writing this book is acknowledged. We also would like to thank to Ms. Jian-feng Ba. Finally, we acknowledge financial supports from the Committee of the National Natural Science Foundation of China and Huazhong Normal University. Jin-sheng Peng is grateful to the International Atomic Energy and UNISCO for hospitality at the International Centre for Theoretical Physics (Trisete, Italy).

CONTENTS

PART II. The quantum properties of light

Chapter 6. Coherence of light

Chapter 7. Squeezed states of light

Chapter 8. Resonance fluorescence

PART I

THEORY OF THE INTERACTION BETWEEN ATOM AND RADIATION FIELD

It is assumed that the reader is familiar with the fundamental theory of quantum mechanics. In the first chapter of this book, we only give a short description of three pictures and a brief review of the basic principles in quantum mechanics. Since the density matrix theory is frequently used in quantum optics, here we also give a general discussion of the density matrix theory. In Chapter 2 we introduce the concept of a two-level atom. This is an idealized model of a real atom, routinely adopted in theoretical descriptions. The optical Bloch equations describing the two-level atom interacting with the radiation field are also discussed in detail. Chapter 3 is devoted to explaining how to represent the light field in quantum theory. After outlining the classical description of the electrodynamic field in the vacuum, the observables and operators which describe the light field are introduced by quantizing the radiation field. Then we introduce some state functions of light fields. Those state functions may revival different aspects of the light field. The Dicke model and the Jaynes-Cumming model are usually adopted to describe atoms interacting with the light field in quantum optics. Subsequently in Chapter 4 we give a specialized discussion on these two typical models which are extensively applied in theoretical calculations. In the last chapter (Chapter 5) of this volume, we introduce the basic theory of quantum statistics, as applied to quantum optics by treating a small system coupled with a reservoir. After introducing the Langevin equation and the Fokker-Plank equation, we give a specialized discussion on the master equation for a quantum harmonic oscillator and a

two-level atom. Finally the characteristic function and the quasi-probability distribution for the quantum harmonic oscillator are also discussed.

CHAPTER 1

THREE PICTURES IN QUANTUM MECHANICS

In quantum mechanics, we may adopt three different pictures, that is, the Schrödinger picture, the Heisenberg picture, and the interaction picture to describe the dynamics of a quantum system. In fact, in the frame of classical mechanics, the dynamics of a classical material point can be described by different methods. For example, a material point can be described by the time development of a state vector with fixed coordinate vectors, or by the time development of the coordinate vectors with the fixed state vector. The mathematical representations of these two descriptions are different, but these two different descriptions give the same physical results. In quantum mechanics, the three different pictures adopted to describe a quantum system indicate that there exist three different ways of description. No matter which picture is applied to depict a quantum system, the laws of motion of the micro-system must be identical. This means that unitary transformation relations among the three pictures must exist. Which picture we actually choose to describe a quantum system depends on the characteristics of the system. Generally, we would rather choose the picture in which the physical properties of a system are more evident and the calculation are simple. In the following, we describe these three pictures.

1.1 The Schrödinger picture

In quantum mechanics when we describe a micro-system such as an atom, a system of an atom coupling to a single-mode field or two atoms interacting with a bimodal field, we usually assume that a state of the system is described by a state function $|\Psi(t)\rangle$. If the system is a single micro-particle, it can be represented by a state function $|\Psi(\mathbf{r},t)\rangle$, here $\mathbf{r}$ is the space coordinate of the system (micro-particle) and t is the time coordinate. When the exact expression of the state function of the system is known at a given time, the time evolution of the system can be deduced. For example, we can learn the probability of the micro-particle at the position $\mathbf{r}$ at time t which is defined by $\langle \Psi(\mathbf{r},t)|\Psi(\mathbf{r},t)\rangle$. And the probability of finding the particle in a volume $d^3\mathbf{r} = dxdydz$ about the point $\mathbf{r}$ at time t is $\langle \Psi(\mathbf{r},t)|\Psi(\mathbf{r},t)\rangle d^3\mathbf{r}$. Since the probability of the particle over the space is equal to 1, the state function must obey the normalization condition, i.e.,

$$\int \langle \Psi(\mathbf{r},t)|\Psi(\mathbf{r},t)\rangle d^3\mathbf{r} = 1 \tag{1.1}$$

Another basic assumption in quantum mechanics is that the physical variables such as position, momentum or spin, are represented by operators. However, the property that any physical variable should be measurable requires that the eigenvalues of the corresponding operator must be restricted to a real number. Such operator is said to be Hermitian. An arbitrary Hermitian operator satisfies the following eigenvalue equation

$$A|u_n\rangle = \lambda_n|u_n\rangle \tag{1.2}$$

Here the symbol $|u_n\rangle$ is called the eigenfunction of the Hermitian operator A, and λ_n is called the corresponding eigenvalue. An Hermitian operator has three important properties: (1). its eigenvalues λ_n are real numbers, (2). its two eigenvectors $|u_n\rangle$ and $|u_m\rangle (n \neq m)$ belonging to different eigenvalues are orthogonal, (3). the eigenvectors of A form a complete set $\{|u_m\rangle\}$, this completeness property allows the expansion of any state $|\Psi(t)\rangle$ of the system by means of the eigenkets of A, namely

$$|\Psi(t)\rangle = \sum_n |u_n\rangle\langle u_n|\Psi(t)\rangle = \sum_n C_n(t)|u_n\rangle \tag{1.3}$$

here

$$C_n(t) = \langle u_n|\Psi(t)\rangle$$

is the probability amplitude of the system characterized by $|\Psi(t)\rangle$ in the eigenvector set $\{|u_n\rangle\}$. So we can utilize the linear superposition of the set of eigenvectors $\{|u_n\rangle\}$ to represent the state vector of the quantum system.

Inasmuch as the description of a micro-system (for example, a single Hydrogen atom) needs a lot of physical variables such as position, momentum, angular momentum, energy, spin, etc., the question arises of the relationships among the operators of these physical variables. The actual relations among physical variables are determined by the physical characteristics of the system. In view of theoretical sense, there are two kinds of relations among arbitrary operators. If two operators A and B have a common eigenvector set, they satisfy the following commutation relation

$$[A, B] = AB - BA = 0$$

we say that the operators A and B commutate. If operators A and B do not have a common set of eigenvectors, A and B do not commutate. In this case,

$$[A, B] = iC \tag{1.4}$$

where C may be a constant or another operator. The commutation relation eq.(1.4) reflects the physical correlation between the physical quantities A and B .

One of the main problems in quantum mechanics is how to determine the dynamic behavior of a quantum system. In quantum mechanics, the time development of the state vector $|\Psi(t)\rangle$ of the system is postulated to be determined by the Schrödinger equation

$$i\hbar\frac{\partial}{\partial t}|\Psi(t)\rangle = H|\Psi(t)\rangle \tag{1.5}$$

here H is an operator representing the energy of the system, which is usually interpreted as the Hamiltonian of the system. Generally an arbitrary system may be associated with a certain Hamiltonian. Thus the state vector $|\Psi(t)\rangle$ of the system can be determined by (1.5) and the initial condition. Consequently the time evolution of the system can be determined.

It is very important to choose an appropriate picture in studying the dynamic behavior of a quantum system. In quantum mechanics, one important picture is the Schrödinger picture. The key point of this picture is that the state vector $|\Psi(t)\rangle$ describing the dynamic behavior of the system changes continuously according to the Schrödinger equation from an initial state $|\Psi(t_0)\rangle$ to a final state $|\Psi(t)\rangle$ at time t, but the operators of physical variables (such as H, $\mathbf{P}, \mathbf{r}$) are time-independent. In order to distinguish this picture from others, the subscript (or superscript) S is used to indicate that the operators and the state vectors are in the Schrödinger picture. Thus we write H_S , q_S , p_S , $|\Psi_S(t)\rangle$ and the like. In general, if the subscript or superscript of picture is not explicitly written, we mean that the quantities are in the Schrödinger picture. Inasmuch as physical quantity A_S is time-independent in the Schrödinger picture, the eigenvectors of A_S are stationary (time-independent). Therefore, in the Schrödinger picture the eigenvectors of an arbitrary Hermitian operator can form a fixed basis vectors to describe the state vector of the system. That is to say, the basis vectors are stationary and the dynamical state vector changes in time in the Schrödinger picture. From the Schrödinger equation (1.5) and the state vector $|\Psi_S(t_0)\rangle$ of the system at time t_0 , the state vector $|\Psi_S(t)\rangle$ at time t can be obtained as

$$|\Psi_S(t)\rangle = U(t, t_0)|\Psi_S(t_0)\rangle \tag{1.6}$$

where $U(t, t_0)$ is a time evolution operator which depends on the Hamiltonian of the system. Substituting (1.6) into (1.5), then

$$i\hbar\frac{\partial}{\partial t}U(t, t_0)|\Psi_S(t_0)\rangle = H_S U(t, t_0)|\Psi_S(t_0)\rangle \tag{1.7}$$

Since $|\Psi_S(t_0)\rangle$ is arbitrary, the time evolution operator obeys

$$i\hbar\frac{\partial}{\partial t}U(t, t_0) = H_S U(t, t_0) \tag{1.8}$$

Integrating the above equation, gives

$$U(t, t_0) = \exp\left[-\frac{i}{\hbar}\int_{t_0}^{t} H_S(t')dt'\right] \tag{1.9}$$

Here H_S is a Hermitian operator, so $U(t, t_0)$ is a unitary operator. If the system is conservative and H_S is explicitly independent of time, then equation (1.9) reduces to

$$U(t, t_0) = \exp\left[-\frac{i}{\hbar}H_S(t - t_0)\right] \tag{1.10}$$

Inserting (1.9) or (1.10) into (1.6), the state vector of the system at time t can be obtained. Thus, at time t the probability of the system in the eigenket $|u_n\rangle$ of the operator A gives

$$|\langle u_n|\Psi_S(t)\rangle|^2 = |\langle u_n|U(t, t_0)|\Psi_S(t_0)\rangle|^2 \tag{1.11}$$

and the expectation value of A_S at time t is

$$\langle A\rangle_S = \langle\Psi_S(t)|A_S|\Psi_S(t)\rangle \tag{1.12}$$

It must be mentioned that when the two Hermitian operators A and B do not commutate (shown in equation (1.4b)), which means that they do not have the same eigenket set, then the physical quantities represented by A and B can not be measured simultaneously. In this case, the mean-square deviations or fluctuations $(\Delta A)^2 = \langle A^2\rangle - \langle A\rangle^2$ and $(\Delta B)^2 = \langle B^2\rangle - \langle B\rangle^2$ satisfy the inequality

$$(\Delta A)^2(\Delta B)^2 \geq \frac{1}{4}|\langle C\rangle|^2 \tag{1.13}$$

where

$$\langle C\rangle = \langle\Psi(t)|C|\Psi(t)\rangle \tag{1.14}$$

$$(\Delta A)^2 = \langle\Psi(t)|A^2|\Psi(t)\rangle - \langle\Psi(t)|A|\Psi(t)\rangle^2 \tag{1.15}$$

The inequality (1.13) is called the Heisenberg uncertainty relation. It expresses a fundamental relation between quantities corresponding to noncommutative operators. If $\langle C\rangle = 0$, namely $C=0$, then A and B commute. In this case, the physical observables A and B can be measured simultaneously and both have precise values. If operator A is the coordinate operator q and B is the momentum operator p, then the commutation relation between q and p is

$$[q, p] = i\hbar \tag{1.16}$$

correspondingly, (1.13) may be written as

$$(\Delta p)^2(\Delta q)^2 \geq \frac{1}{4}\hbar^2 \tag{1.17}$$

This is the well-known Heisenberg momentum-position uncertainty relation.

When A is the x-component L_x of angular momentum and B is the y-component L_y , the commutation relation between L_x and L_y gives

$$[L_x, L_y] = i\hbar L_z \tag{1.18}$$

Evidently

$$(\Delta L_x)^2(\Delta L_y)^2 \geq \frac{1}{4}\hbar^2|\langle L_z\rangle|^2 \tag{1.19}$$

Equation (1.19) is the uncertainty relation of $x - y$ components of the angular momentum, so that the product $(\Delta L_x)^2(\Delta L_y)^2$ is determined by the expectation value $\langle L_z\rangle$.

1.2 The Heisenberg picture

In the Schrödinger picture the basis vectors are visualized as a fixed set of vectors and the state vector $|\Psi_S(t)\rangle$ as changing in time. Vice versa, the same system can be described equally well by time-dependent observable operators and by a fixed state vector. This picture is physically equivalent to the Schrödinger picture and is called the Heisenberg picture. Thus, in the Heisenberg picture the state vector is

$$|\Psi_H(t)\rangle = |\Psi_S(t_0)\rangle \tag{1.20}$$

here the subscript H designates the Heisenberg picture. It is clear that the state vector $|\Psi_H(t)\rangle$ coincides with the state vector at time $t = t_0$ in the Schrödinger picture. Therefore, the state vectors in the two pictures are connected by equation (1.6)

$$|\Psi_S(t)\rangle = U(t,t_0)|\Psi_H\rangle \tag{1.21}$$

Since the expectation values of the physical observables correspond to the experimental results and they are actually independent of the pictures. That is to say,

$$\begin{aligned}\langle A\rangle &= \langle\Psi_S(t)|A_S|\Psi_S(t)\rangle = \langle\Psi(t_0)|U^\dagger(t,t_0)A_S U(t,t_0)|\Psi(t_0)\rangle \\ &= \langle\Psi_H|U^\dagger(t,t_0)A_S U(t,t_0)|\Psi_H\rangle = \langle\Psi_H|A_H|\Psi_H\rangle\end{aligned} \tag{1.22}$$

It is easy to see from the above equation that the operators in the two pictures have the following transformation law

$$A_H(t) = U^\dagger(t,t_0)A_S U(t,t_0) \tag{1.23}$$

From (1.23) we see that the operators, which are stationary in the Schrödinger picture, become time-dependent ones in the Heisenberg picture according to this transformation law.

Now we examine how the eigenvectors of the operator $A_H(t)$ evolve in the Heisenberg picture. In the Schrödinger picture, the eigenvalue equation of the operator A_S is

$$A_S|u_n^S\rangle = \lambda_n|u_n^S\rangle \tag{1.24}$$

By means of (1.23) we obtain

$$U(t,t_0)A_H(t)U^\dagger(t,t_0)|u_n^S\rangle = \lambda_n|u_n^S\rangle$$

Noticing that $U^\dagger U = 1$, the above equation becomes

$$A_H(t)|u_n^H(t)\rangle = \lambda_n|u_n^H(t)\rangle \tag{1.25}$$

where $|u_n^H(t)\rangle$ is defined as

$$|u_n^H(t)\rangle = U^\dagger(t,t_0)|u_n^S\rangle \tag{1.26}$$

This is just the transformation law of the eigenvectors of the operator A in the two pictures. It shows that the time-independent eigenvectors $|u_n^S\rangle$ in the Schrödinger picture become time-dependent eigenvectors $|u_n^H(t)\rangle$ in the Heisenberg picture. Comparing (1.21) with (1.26) we see that in the Schrödinger picture the state vector $|\Psi_S(t)\rangle$ varies along a certain direction with the time development, but in the Heisenberg picture, the eigenvectors of the operator A_H vary along the opposite direction.

In the Heisenberg picture the time evolution of the system can be obtained by solving the Heisenberg equation of the operator $A_H(t)$. Differentiating both sides of (1.25) with respect to t, we obtain

$$\begin{aligned} i\hbar\frac{d}{dt}A_H =& U^\dagger A_S H_S U - U^\dagger H_S A_S U + i\hbar U^\dagger\frac{\partial}{\partial t}A_S U \\ =& U^\dagger A_S U U^\dagger H_S U - U^\dagger H_S U U^\dagger A_S U + i\hbar U^\dagger\frac{\partial}{\partial t}A_S U \\ =& [A_H, H_H] + i\hbar U^\dagger\frac{\partial}{\partial t}A_S U \end{aligned} \tag{1.27}$$

where we have used eq.(1.25) and we have defined

$$H_H(t) = U^\dagger(t,t_0)H_S U(t,t_0) \tag{1.28}$$

which is the Hamiltonian in the Heisenberg picture. Eq.(1.27) is interpreted as the Heisenberg equation of motion for the operator A_H. Like the Schrödinger equation in the Schrödinger picture, (1.27) is the fundamental equation describing the time evolution of the system in the Heisenberg picture. When the expression of $A_H(t)$ is explicitly obtained, we can obtain the expectation value of the operator A_H.

If $\frac{d}{dt}A_H = 0$, then A_H is a constant of the motion. When A_S has no explicit time dependence, (1.27) reduces to

$$\frac{d}{dt}A_H = \frac{1}{i\hbar}[A_H, H_H] \tag{1.29}$$

In fact, for a conservative system, the following relation in the Schrödinger picture is valid: $\frac{d}{dt}H_S$ =0. As a special case, we let $A_S = H_S$ and note $U(t, t_0)$ is defined by (1.10), then $[H_S, U] = 0$. In this case, eq.(1.23) gives that $H_S = H_H$. This means that for a conservative system, the Hamiltonians both in the Schrödinger picture and in the Heisenberg picture are identical. According to (1.29), we then have

$$\frac{d}{dt}H_H = 0 \tag{1.30}$$

which shows that H is a constant of the motion in the two pictures. In other words, H has the same form in the two pictures.

Another important property we have to point out is that the commutation relations in the two pictures should be the same form, because the commutation relations represent the correlation of physical quantities which must be independent of the different description (or picture) . As an example, we take A_S, B_S, and C_S to be three operators in the Schrödinger picture and satisfy the commutation relation

$$[A_S, B_S] = iC_S \tag{1.31}$$

If we multiply both sides from the left by $U^\dagger$ and from the right by U, we have

$$U^\dagger A_S B_S U - U^\dagger B_S A_S U = iU^\dagger C_S U$$

Moreover we insert $UU^\dagger = 1$ between A_S and B_S , which yields

$$U^\dagger A_S U U^\dagger B_S U - U^\dagger B_S U U^\dagger A_S U = iU^\dagger C_S U$$

Considering eq.(1.23), we obtain the commutation relation in the Heisenberg picture as

$$[A_H(t), B_H(t)] = iC_H(t) \tag{1.32}$$

Equation (1.32) has the same form as (1.31). Thus, the simultaneous commutation relations of the operators both in the Heisenberg picture and in the Schrödinger picture must have the same form. Since these commutation relations are independent of the pictures, then the Heisenberg uncertainty relations of the operators in the Heisenberg picture, taken at the same time, are also the same as those in the Schrödinger picture.

So far we have discussed two pictures for describing the system in quantum mechanics. It is worth to point out that for the two kinds of descriptions, the Schrödinger picture is more appropriate to describe a conservation system, because the state vector may be solved in a conservation system. But for an open system (such as the atom-field coupling system in a bad cavity), because the Hamiltonian of the system has more complicated form due to the effect of the surroundings, therefore, it is not easy to solve the state vector from equation (1.5). However, it is possible to solve the Heisenberg equations (1.29) , furthermore, the time evolution of the physical observables and their expectation values can be obtained explicitly. Thus, to deal with the problems of a quantum system, one must choose an appropriate picture in which the physical properties of the system can be easily revealed and the calculating processes are relatively simple in mathematics.

1.3 The Interaction Picture

1.3.1 Equation of motion in the interaction picture

Another picture besides the two pictures discussed in the previous sections, is the interaction picture. This picture is frequently used in quantum optics, when the Hamiltonian of the system can be written as a sum form of two terms

$$H_S = H_0^S + V_S \tag{1.33}$$

where H_0^S is independent of the time and its eigenvectors are easy to solve by the following equation

$$i\hbar\frac{\partial}{\partial t}|\psi_n\rangle = H_0^S|\psi_n\rangle \tag{1.34}$$

The term V_S can be regarded as the interaction energy operator of the system, and may depend explicitly on time although it may need not. In fact, V_S usually induces a strong influence on the time behavior of the system. The aim of introducing the interaction picture is to investigate the effects of the interaction Hamiltonian of the system on the time behavior of the system.

The method of transforming the Schrödinger picture into the interaction picture is performed by a unitary operator $U_0(t, t_0)$ such that

$$|\Psi_S(t)\rangle = U_0(t,t_0)|\Psi_I(t)\rangle \tag{1.35}$$

here the subscript I refers to the interaction picture and $U_0(t,t_0)$ satisfies

$$U_0(t,t_0) = \exp\left[-\frac{i}{\hbar}H_0^S(t-t_0)\right] \tag{1.36}$$

Evidently

$$U_0^\dagger = U_0^{-1}$$
$$U_0(t_0, t_0) = 1 \tag{1.37}$$

Differentiating (1.36) with respect to time t, we find that $U_0(t, t_0)$ obeys

$$i\hbar\frac{\partial}{\partial t}U_0 = H_0^S U_0 \tag{1.38}$$

Next we deduce the equation of evolution for the state vector $|\Psi_I(t)\rangle$ in the interaction picture. For the Hamiltonian (1.33), the equation of evolution for the state vector $|\Psi_S(t)\rangle$ in the Schrödinger picture is

$$i\hbar\frac{\partial}{\partial t}|\Psi_S(t)\rangle = (H_0^S + V_S)|\Psi_S(t)\rangle \tag{1.39}$$

By using (1.35), this gives

$$i\hbar\frac{\partial U_0}{\partial t}|\Psi_I(t)\rangle + i\hbar U_0\frac{\partial}{\partial t}|\Psi_I(t)\rangle = [H_0^S + V_S]|\Psi_I(t)\rangle$$

If we use (1.38) and multiply both sides from the left by $U_0^\dagger$, then we obtain the Schrödinger equation for the state vector $|\Psi_I(t)\rangle$ in the interaction picture to be

$$i\hbar\frac{\partial}{\partial t}|\Psi_I(t)\rangle = V_I(t)|\Psi_I(t)\rangle \tag{1.40}$$

where

$$V_I(t) = U_0^\dagger V_S U_0 \tag{1.41}$$

Eq.(1.40) shows that the time evolution of the state vector $|\Psi_I(t)\rangle$ is determined by the interaction Hamiltonian, which emphasizes the effects of the interaction energy.

Since the expectation values of operators are independent of the pictures we adopt, it must have

$$\begin{aligned}\langle A\rangle &= \langle\Psi_S(t)|A_S|\Psi_S(t)\rangle = \langle\Psi_I(t)|U_0^\dagger A_S U_0|\Psi_I(t)\rangle \\ &= \langle\Psi_I(t)|A_I(t)|\Psi_I(t)\rangle\end{aligned} \tag{1.42}$$

From the above equation, we find the transformation law for the operators between the Schrödinger picture and the interaction picture as

$$A_I(t) = U_0^\dagger(t, t_0)A_S U_0(t, t_0) \tag{1.43}$$

Since not only the state vector $|\Psi_I(t)\rangle$ but also the operators depend on time in the interaction picture, we also have to discuss the equation of motion for operators in the interaction picture. Differentiating both sides of (1.43) with respect to t, and using (1.38) and its adjoint, we obtain

$$\begin{aligned} i\hbar\frac{d}{dt}A_I &= i\hbar U_0^\dagger A_S\frac{\partial}{\partial t}U_0 + i\hbar\frac{\partial}{\partial t}U_0^\dagger A_S U_0 + U_0^\dagger i\hbar\frac{\partial}{\partial t}A_S U_0 \\ &= U_0^\dagger A_S H_0^S U_0 - U_0^\dagger H_0^S A_S U_0 + U_0^\dagger i\hbar\frac{\partial}{\partial t}A_S U_0 \end{aligned} \tag{1.44}$$

As H_0^S is time-independent, and

$$[H_0^S, U_0] = 0 \tag{1.45}$$

then

$$H_0^S = H_0^I \tag{1.46}$$

Thus eq.(1.44) becomes

$$i\hbar\frac{d}{dt}A_I = [A_I, H_0^I] + i\hbar U_0^\dagger\frac{\partial}{\partial t}A_S U_0 = [A_I, H_0^S] + i\hbar U_0^\dagger\frac{\partial}{\partial t}A_S U_0 \tag{1.47}$$

This is the equation of motion for the operator $A_I(t)$ in the interaction picture. In accordance with (1.47) and the initial condition, we can in principle obtain the time evolution of the operator $A_I(t)$.

1.3.2 A formal solution of the state vector $|\Psi_I(t)\rangle$ by the perturbation theory

We return to discuss equation (1.40) and look for an expression of the state vector $|\Psi_I(t)\rangle$. From (1.35) we know $|\Psi_I(t_0)\rangle = |\Psi_S(t_0)\rangle$ at time $t = t_0$. Introducing a unitary transformation operator $U(t, t_0)$, such that

$$|\Psi_I(t)\rangle = U(t, t_0)|\Psi_I(t_0)\rangle \tag{1.48}$$

This means that the state vector evolves into $|\Psi_I(t)\rangle$ from $|\Psi_I(t_0)\rangle$ with the time development. Substituting (1.48) into (1.40) and noticing that $|\Psi_I(t_0)\rangle$ may be chosen arbitrarily, then U satisfies

$$i\hbar\frac{d}{dt}U(t, t_0) = V_I(t)U(t, t_0) \tag{1.49}$$

Solving equation (1.49) and considering $V_I(t)$ and the initial condition

$$U(t_0, t_0) = 1 \tag{1.50}$$

we can obtain the exact or perturbation solution of $U(t,t_0)$ in principle. Thus, we can get the state vector $|\Psi_I(t)\rangle$, and furthermore the state vector $|\Psi_S(t)\rangle$ in the Schrödinger picture

$$|\Psi_S(t)\rangle = U_0^\dagger(t,t_0)|\Psi_I(t)\rangle = U_0^\dagger(t,t_0)U(t,t_0)|\Psi_S(t_0)\rangle \tag{1.51}$$

Now we may obtain an approximate solution of the operator $U(t,t_0)$ from (1.49) by the perturbation theory. Integrating both sides of (1.49) and using the initial condition (1.50) we have

$$U(t,t_0) = 1 + \frac{1}{i\hbar}\int_{t_0}^{t} V_I(t_1)U(t_1,t_0)dt_1 \tag{1.52}$$

If we let $t_1 = t_2$ and the dummy integration variable $t_1 \longrightarrow t_2$, we may rewrite $U(t,t_0)$ as

$$U(t_1,t_0) = 1 + \frac{1}{i\hbar}\int_{t_0}^{t_1} V_I(t_1)U(t_2,t_0)dt_2 \tag{1.53}$$

Substituting (1.53) into the integrand in (1.52), then obtain

$$\begin{aligned} U(t,t_0) = {} & 1 + \frac{1}{i\hbar}\int_{t_0}^{t} V_I(t_1)U(t_1,t_0)dt_1 \\ & + \left(\frac{1}{i\hbar}\right)^2 \int_{t_0}^{t} dt_1 V_I(t_1)\int_{t_0}^{t_1} dt_2 V_I(t_2)U(t_2,t_0) \end{aligned} \tag{1.54}$$

We may iterate this procedure indefinitely and obtain the series expansion of $U(t,t_0)$ as

$$U(t,t_0) = 1+\sum_{n=1}^{\infty}\left(\frac{1}{i\hbar}\right)^n \int_{t_0}^{t} dt_1 \int_{t_0}^{t_1} dt_2 \cdots \int_{t_0}^{t_{n-1}} dt_n V_I(t_1)V_I(t_2)\cdots V_I(t_n) \tag{1.55}$$

If the interaction energy $V_I(t)$ is much smaller than H_0, this series is likely to converge rapidly. Inserting (1.55) into (1.48) we may obtain the perturbation expansion of the state vector at time t in the interaction picture.

As an example, we assume that the system is initially in one eigenket $|i\rangle$ of the unperturbed Hamiltonian H_0, so that

$$H_0|i\rangle = E_i|i\rangle \tag{1.56}$$

Due to the influence of the interaction energy $V_I(t)$, the system evolves into the state vector $|\Psi_I(t)\rangle$ with the time development. Now the question arises about the probability of finding the system in another eigenket $|k\rangle$ of H_0 different from $|i\rangle$ at time t. Clearly, the probability amplitude gives

$$\langle k|\Psi_I(t)\rangle = \langle k|U(t,t_0)|i\rangle \tag{1.57}$$

Inserting (1.55) into the above equation, we obtain

$$\begin{aligned}\langle k|\Psi_I(t)\rangle = &\frac{1}{i\hbar}\int_{t_0}^{t}\langle k|V_I(t_1)|i\rangle dt_1\\ &+\left(\frac{1}{i\hbar}\right)^2\int_{t_0}^{t}dt_1\int_{t_0}^{t_1}dt_2\langle k|V_I(t_1)V_I(t_2)|i\rangle\\ &+\cdots\end{aligned} \tag{1.58}$$

Substituting equation (1.41) into (1.58) and retaining the term in the first-order terms in $V_I(t)$, we have

$$\langle k|\Psi_I(t)\rangle \approx \frac{1}{i\hbar}\int_{t_0}^{t}\langle k|V_S|i\rangle \exp(i\omega_{ki}t)dt_1 \tag{1.59}$$

where

$$\omega_{ki} = (E_k - E_i)/\hbar$$

which is the transition frequency. Similarly, we can take into account the probability amplitude expression with high order of V_I, then obtain the probability amplitude of the system from initial state $|i\rangle$ to final state $|k\rangle$ at time t by means of equation (1.58).

1.4 The density operator

From the previous discussion we have learned that if the state vector of the system at time t is known, the time behavior of the system can be determined and the measurement values of the physical observables can be also predicted. The state at given time may be determined by measuring the corresponding eigenvectors. For example the polarization state of the photon can be exactly determined by the experiments of the light beams through different polarization analyzer. But generally the state vector of the system can not always be known explicitly. As an example for the atom beam emitted by a hot filament furnace, the state with certain atomic momentum is difficult to be determined. In this case, we only know the statistical distribution of atomic momentums. That is to say, we only know the incomplete informations about the state of the system. How to predict theoretically the measurement results of the system as accuracy as possible in accordance with the incomplete informations of the system ? In order to reach this aim, we first introduce a very useful tool, namely the density operator.

1.4.1 Density operator and its general properties

There are two types of state vectors for different systems. One is the pure state; the other by a statistical-mixture or mixture pure state. If the state of the system can be determined completely by a state vector, we use the pure state description. The pure state $|\Psi(t)\rangle$ can be expanded in a set of eigenkets $\{|u_n\rangle\}$ of an arbitrary physical observable of the system as

$$|\Psi(t)\rangle = \sum_n C_n(t)|u_n\rangle \tag{1.60}$$

The statistical-mixture state of the system is a generalization of the pure state, where the state of the system is not precisely specified: only the probabilities that the system is in a range of possible states are known. For example, for a system in thermal equilibrium, we only know that the probability of finding the system in states with energies E_n is $\exp(-E_n/kT)$. If this is the case, for such a system we only know that the probability of finding the system in the state $|\Psi_1\rangle$ is P_1, that of finding it in the state $|\Psi_2\rangle$ is P_2 , $\cdots$ and

$$P_1 + P_2 + \cdots = \sum_n P_n = 1 \tag{1.61}$$

In this case, we usually say that the system is in a mixture-statistical state of the states $|\Psi_1\rangle$, $|\Psi_2\rangle,\cdots$ with corresponding probabilities $P_1, P_2, \cdots$

It is necessary to point out that the P_n in a mixture state is different from the probability $|C_n(t)|^2$ in a pure state. The later one is the reflection of the intrinsic quantum properties of the system, but P_n is not related to the intrinsic quantum properties, rather than depends on the classical statistical properties of the system such as $\exp(-E_n/kT)$ for a thermal equilibrium system or depends on the measurement results of every experiments for corresponding states. Another difference is that the state functions $|\Psi_n\rangle$ in a mixture pure state may be orthogonal or not. A system which needs to be described by a statistical-mixture state, it is more appropriate to adopt the density operator method to describe. Here we first discuss the density operator for the pure state , then generalize it to the case of the mixture pure states.

As shown in (1.60) for a statistical system described by a pure state, the state vector at time t can be expanded by a set of eigenkets $\{|u_n\rangle\}$, the probability amplitudes $C_n(t)$ satisfy the normalization condition

$$\sum_n |C_n(t)|^2 = 1 \tag{1.62}$$

Explicitly the expectation value of the physical quantity A depends on the state vector $|\Psi_S(t)\rangle$ (1.60)

$$\langle A\rangle_S = \langle\Psi_S(t)|A_S|\Psi_S(t)\rangle = \sum_{n,p} C_n^*(t)C_p(t)A_{np} \tag{1.63}$$

here we have used the completeness relation of the eigenvectors $\{|u_n^S\rangle\}$, i.e., $\sum_n |u_n^S\rangle\langle u_n^S| = 1$ and we have defined the matrix elements of an observable quantity A as

$$A_{np} = \langle u_n^S|A_S|u_p^S\rangle \tag{1.64}$$

Equation (1.63) shows that the function $C_n^*(t)C_p(t)$ related to the expectation value of A can be regarded as the matrix element of the operator $|\Psi_S(t)\rangle\langle\Psi_S(t)|$, because

$$\langle u_p^S|\Psi_S(t)\rangle\langle\Psi_S(t)|u_n^S\rangle = C_n^*(t)C_p(t) \tag{1.65}$$

Thus if we introduce the density operator for the pure state as

$$\rho_S(t) = |\Psi_S(t)\rangle\langle\Psi_S(t)| \tag{1.66}$$

then the matrix element of the density operator $\rho_S(t)$ in basis vectors $\{|u_n^S\rangle\}$ can be written as

$$\rho_{pn}^S(t) = \langle u_p^S|\rho_S(t)|u_n^S\rangle = C_n^*(t)C_p(t) \tag{1.67}$$

From (1.64) and (1.67), we may obtain the expectation value of A_S by means of the density operator as follows

$$\begin{aligned}\langle A\rangle_S &= \sum_{n,p}\langle u_p^S|\rho_S(t)|u_n^S\rangle\langle u_n^S|A_S|u_p^S\rangle \\ &= \sum_p \langle u_p^S|\rho_S(t)A_S|u_p^S(t)\rangle = \mathrm{Tr}\rho_S(t)A_S \end{aligned}\tag{1.68}$$

The normalization condition (1.62) can also be expressed in terms of the density operator as

$$\sum_n |C_n(t)|^2 = \sum_n \rho_{nn}^S(t) = \mathrm{Tr}\rho_S(t) = 1 \tag{1.69}$$

We may easily obtain the equation of motion for $\rho_S(t)$ in the Schrödinger picture from its definition and the Schrödinger equation

$$\begin{aligned} i\hbar\frac{\partial}{\partial t}|\Psi_S(t)\rangle &= H_S|\Psi_S(t)\rangle \\ -i\hbar\frac{\partial}{\partial t}\langle\Psi_S(t)| &= \langle\Psi_S(t)|H_S \end{aligned}\tag{1.70}$$

From (1.66) and (1.70) we have

$$\begin{aligned}\frac{d}{dt}\rho_S(t) &= \left(\frac{d}{dt}|\Psi_S(t)\rangle)\langle\Psi_S(t)| + |\Psi_S(t)\rangle(\frac{d}{dt}\langle\Psi_S(t)|\right) \\ &= \frac{1}{i\hbar}H_S|\Psi_S(t)\rangle\langle\Psi_S(t)| - \frac{1}{i\hbar}|\Psi_S(t)\rangle\langle\Psi_S(t)|H_S \\ &= \frac{1}{i\hbar}[H_S,\rho_S(t)] \end{aligned}\tag{1.71}$$

Analogously, the density operators in the Heisenberg picture and in the interaction picture can be respectively defined as follows

$$\rho_H(t) = |\Psi_H(t)\rangle\langle\Psi_H(t)| \tag{1.72}$$

$$\rho_I(t) = |\Psi_I(t)\rangle\langle\Psi_I(t)| \tag{1.73}$$

From the transformation relations (1.21) and (1.35) for the state vectors we can easily obtain the transformation relations for $\rho_H(t)$, $\rho_I(t)$ and $\rho_S(t)$ as

$$\rho_H(t) = U^\dagger(t,t_0)\rho_S(t)U(t,t_0) \tag{1.74}$$

$$\rho_I(t) = U_0^\dagger(t,t_0)\rho_S(t)U_0(t,t_0) \tag{1.75}$$

here $U(t,t_0)$ and $U_0(t,t_0)$ are defined by (1.9) and (1.36), respectively. Evidently, the density matrix $\rho_H(t)$ in the Heisenberg picture is independent of time, but the time evolution of the $\rho_I(t)$ in the interaction picture obeys

$$\begin{aligned}\frac{d}{dt}\rho_I(t) &= \left(\frac{d}{dt}|\Psi_I(t)\rangle\langle\Psi_I(t)| + |\Psi_I(t)\rangle\left(\frac{d}{dt}\langle\Psi_I(t)|\right)\right) \\ &= \frac{1}{i\hbar}V_I(t-t_0)|\Psi_I(t)\rangle\langle\Psi_I(t)| - \frac{1}{i\hbar}|\Psi_I(t)\rangle\langle\Psi_I(t)|V_I(t-t_0) \\ &= \frac{1}{i\hbar}[V_I(t-t_0), \rho_I(t)]\end{aligned} \tag{1.76}$$

We next discuss the properties of the density operator. From (1.66), (1.72) and (1.73) we see that the density operator is Hermitian operator, so that

$$\rho^\dagger = \rho \tag{1.77}$$

In addition, we have that

$$\rho^2 = \rho \tag{1.78}$$

$$\mathrm{Tr}\rho^2 = 1 \tag{1.79}$$

Evidently, since the pure states $|\Psi(t)\rangle$ and $\exp(i\theta)|\Psi(t)\rangle$ possess the same ρ in the three pictures, the effects of phase factors can be cancelled when we adopt ρ to describe a system.

For the system described by a statistical-mixture state, the density operator in the Schrödinger picture can be defined by

$$\rho_S(t) = \sum_k P_k|\Psi_k^S(t)\rangle\langle\Psi_k^S(t)| = \sum_k P_k\rho_k^S(t) \tag{1.80}$$

with

$$\rho_k^S(t) = |\Psi_k^S(t)\rangle\langle\Psi_k^S(t)| \tag{1.81}$$

Analogously, the definitions of the density operators in the Heisenberg picture and in the interaction picture are, respectively,

$$\rho_H(t) = \sum_k P_k|\Psi_k^H(t)\rangle\langle\Psi_k^H(t)| = \sum_k P_k\rho_k^H(t) \tag{1.82}$$

$$\rho_I(t) = \sum_k P_k|\Psi_k^I(t)\rangle\langle\Psi_k^I(t)| = \sum_k P_k\rho_K^I(t) \tag{1.83}$$

so that the expectation value of A becomes

$$\langle A\rangle = \mathrm{Tr}(\rho_S A_S) = \mathrm{Tr}(\rho_H A_H) = \mathrm{Tr}(\rho_I A_I) \tag{1.84}$$

It is clear that the operators $\rho_S(t)$ (1.80) and $\rho_I(t)$ (1.83) obey the same evolution law as that for the pure states.

From eqs.(1.80)-(1.83) we see that the operator ρ is also a Hermitian operator for a statistical-mixture state. In this case, however,

$$\rho^2 = \sum_k P_k|\Psi_k\rangle\langle\Psi_k|\sum_{k'} P_{k'}|\Psi_{k'}\rangle\langle\Psi_{k'}| = \sum_k P_k^2|\Psi_k\rangle\langle\Psi_k| \neq \rho \tag{1.85}$$

and it is easy to check that

$$\mathrm{Tr}\rho^2 \leq 1 \tag{1.86}$$

here the identity relation hold only in the pure state case. Therefore, equation (1.86) can be used to judge whether a system is in a pure state or in a statistical-mixture state.

1.4.2 Solution of the equation of motion for the density operator

In general, if we know the Hamiltonian H and the initial condition, then we can obtain the explicit expressions of the density operators $\rho_S(t)$ and $\rho_I(t)$ by solving the equations (1.71) and (1.76). Furthermore we can deduce the information about the system from the expectation values, the probability distributions of measurements and the like. Since the density operator $\rho_S(t)$ in the Schrödinger picture can be transformed into $\rho_I(t)$ in the interaction picture by use of the unitary transformation relation (1.75), we need only solve one of these operators in order to derive the others. As an example we consider the interaction picture to discuss some methods of solving the equation of the density operator.

A. Solution of $\rho_I(t)$ by use of the unitary operator U(t)

For a system described by the Hamiltonian

$$H_S = H_0^S + V_S \tag{1.87}$$

if H_0^S is explicitly independent of time , then we can define a state function $|\Psi_k^I(t)\rangle$ which transforms as

$$|\Psi_k^I(t)\rangle = U(t)|\Psi_k(0)\rangle \tag{1.88}$$

Therefore, we have

$$\rho_I(t) = \sum_k P_k U(t)|\Psi_k(0)\rangle\langle\Psi_k(0)|U^\dagger(t) = U\rho(0)U^\dagger \tag{1.89}$$

Evidently, knowledge of $U(t)$ is essential to obtain $\rho_I(t)$. Substituting (1.89) into (1.76), we have

$$i\hbar\frac{\partial}{\partial t}U(t) = V_I U(t) \tag{1.90}$$

Solving the above equation we obtain $U(t)$. If V_I is explicitly independent of time, then we have

$$U(t) = \exp\left(-\frac{i}{\hbar}V_I t\right) \tag{1.91}$$

and

$$\rho_I(t) = \exp\left(-\frac{i}{\hbar}V_I t\right)\rho(0)\exp\left(\frac{i}{\hbar}V_I t\right) \tag{1.92}$$

Using equation (1.75), the above equation can be transformed into the Schrödinger picture. If H_0^S and V_S commute, we find

$$\rho_S(t) = \exp\left(-\frac{i}{\hbar}H_S t\right)\rho(0)\exp\left(\frac{i}{\hbar}H_S t\right) \tag{1.93}$$

Expanding the above equation in the eigenvectors $\{|u_n\rangle\}$ of the Hamiltonian H_S , $\rho_S(t)$ may be rewritten as

$$\begin{aligned}\rho_S(t) &= \sum_{n,\ell}\exp\left(-\frac{i}{\hbar}H_S t\right)|u_n\rangle\langle u_n|\rho(0)|u_\ell\rangle\langle u_\ell|\exp\left(\frac{i}{\hbar}H_S t\right)\\ &= \sum_{n,\ell}\langle u_n|\rho(0)|u_\ell\rangle\exp\left[-\frac{i}{\hbar}(E_n - E_\ell)t\right]|u_n\rangle\langle u_\ell|\end{aligned} \tag{1.94}$$

The matrix elements of the density operator are

$$\rho_{n\ell}^S(t) = \langle u_n|\rho_S(t)|u_\ell\rangle = \rho_{nl}^S(0)\exp\left[-\frac{i}{\hbar}(E_n - E_\ell)t\right] \tag{1.95}$$

$$\rho_{n\ell}^S(0) = \langle u_n|\rho_S(0)|u_\ell\rangle \tag{1.96}$$

Thus if the Hamiltonian of the system is explicitly time-independent, the density operator and the matrix elements $\rho_{nl}^S(t)$ are obtained in terms of the eigenkets $\{|u_n\rangle\}$ and the initial condition $\rho(0)$.

B. Solution of the density operator by means of the perturbation theory

If the interaction Hamiltonian V_S of the system is time dependence, an exact solution of equation (1.76) is not easy to obtain. However, when V_S is sufficiently smaller than H_0^S , we may obtain an approximate solution by means of the perturbation theory. Integrating (1.76) and iterating similarly to what we have done in (1.52)-(1.55) we then obtain

$$\rho_I(t) = \rho(0) + \frac{1}{i\hbar}\int_0^t [V_I(t), \rho(0)]dt_1 + \frac{1}{2!}\left(\frac{1}{i\hbar}\right)^2 \int_0^t dt_2 \int_0^t dt_1 [V_I(t_1), [V_I(t_2), \rho(0)]] + \cdots \tag{1.97}$$

It is appropriate to point out that (1.97) is more general than the perturbation formula for $|\Psi_I(t)\rangle$ (1.55). In fact here

$$\rho_I(t) = \sum_k P_k |\Psi_I(t)\rangle\langle\Psi_I(t)|$$

can be used in the case of an initial pure state as well as in the case of an initial statistical-mixture state.

C. The reduced density operator

We have dealt with the definition and solution of the density operator at some length, because the density operator is more useful than the state vector for a system described by a statistical-mixture state. We now turn to treat two interacting systems A and B by means of the density operator. Let us assume that the two systems are isolated before time t=0, and that their Hamiltonian are H_A and H_B , respectively, and satisfy

$$[H_A, H_B] = 0 \tag{1.98}$$

Assume also that their eigenenergies obey

$$H_A|A'\rangle = E_{A'}|A'\rangle, \quad H_B|B'\rangle = E_{B'}|B'\rangle \tag{1.99}$$

If this two system are put into contact at time t=0, then at $t > 0$ the total Hamiltonian becomes

$$H = H_A + H_B + V_{AB} = H_0 + V_{AB} \tag{1.100}$$

The density operator for this two coupled system in the Schrödinger picture satisfies the equation of motion

$$i\hbar\frac{d}{dt}\rho^S_{AB} = [H, \rho^S_{AB}] \tag{1.101}$$

Since the systems are uncoupled at t=0, so that the initial density operator can be taken as

$$\rho_{AB}(0) = \rho_A(0)\rho_B(0) \tag{1.102}$$

$$\mathrm{Tr}_A\rho_A(0) = \sum_{A'}\langle A'|\rho_A(0)|A'\rangle = 1 \tag{1.103}$$

$$\mathrm{Tr}_B\rho_B(0) = \sum_{B'}\langle B'|\rho_B(0)|B'\rangle = 1 \tag{1.104}$$

$$\mathrm{Tr}_{AB}\rho_{AB}(0) = \sum_{A'B'}\langle A', B'|\rho_A(0)\rho_B(0)|A', B'\rangle \tag{1.105}$$

$$= \mathrm{Tr}_A\rho_A(0)Tr_B\rho_B(0) = 1 \tag{1.106}$$

Transforming into the interaction picture, we have

$$\rho^S_{AB}(t) = U_0(t)\rho^I_{AB}(t)U_0(t) \tag{1.107}$$

with

$$U_0(t) = \exp\left[-\frac{i}{\hbar}(H_A + H_B)t\right] \tag{1.108}$$

and the equation of motion for $\rho^I_{AB}(t)$ is

$$i\hbar\frac{d}{dt}\rho^I_{AB}(t) = [V^I_{AB}(t), \rho^I_{AB}(t)] \tag{1.109}$$

Suppose that we are only interested in making measurements on the physical quantity M in subsystem A. For example, we only measure the z-component of atomic spin angular in the atom-field coupled system. Then the expectation value of the operator M gives

$$\begin{aligned}\langle M\rangle = \mathrm{Tr}_{AB}(M_S\rho^S_{AB}(t)) &= \sum_{A',B'}\langle A', B'|M_S\rho^S_{AB}(t)|A', B'\rangle \\ = \sum_{A'}\langle A'|M_S[\sum_{B'}\langle B'|\rho^S_{AB}(t)|B'\rangle]|A'\rangle \\ = \mathrm{Tr}_A(M_S\rho^S_A(t))\end{aligned} \tag{1.110}$$

where $\rho_A^S(t)$ is called the reduced density operator and is defined by

$$\rho_A^S(t) = \mathrm{Tr}_B \rho_{AB}^S(t) \tag{1.111}$$

Evidently if we know the reduced density operator $\rho_A^S(t)$ (1.110), we can easily obtain the time evolution of the expectation value of an arbitrary operator in the subsystem A. It is also easy to obtain the approximation solution of the reduced density operator in the interaction picture as

$$\begin{aligned} \rho_A^I(t) = & \rho_A(0) + \frac{1}{i\hbar}\int_0^t dt_1 \mathrm{Tr}_B[V_I(t_1), \rho_A(0)\rho_B(0)] \\ & + \frac{1}{2!}\left(\frac{1}{i\hbar}\right)^2 \int_0^t dt_2 \int_0^t dt_1 \mathrm{Tr}_B[V_I(t_1), [V_I(t_2), \rho_A(0)\rho_B(0)]] \\ & + \cdots \end{aligned} \tag{1.112}$$

References

1. P.A.M.Dirac, *The principle of quantum meçhanics*, 4th ed., (Clarendon, Oxford, 1958)
2. L.L.Schiff, *Quantum mechanics*, 3rd ed., (McGraw-Hill, New York, 1968)
3. E.Merzbacher, *Quantum mechanics*, 2nd ed., (Wiley, New York, 1970).
4. W.H.Louisell, *Quantum statistical properties of radiation*, (Wiley, New York, 1970).
5. C.Cohen-Tannoudji, J.Dupont-Roc, G.Grynberg, *Quantum mechanics*, (Wiley and Hermann, Paris, 1977).
6. R.Shankar, *Principle of Quantum mechanics*, (Plenum Press, New York, 1981).
7. Jin-sheng Peng, *Resonance fluorescence and superfluorescence*, (Science press, Beijing, 1993).

CHAPTER 2

TWO-LEVEL ATOM AND THE OPTICAL BLOCH EQUATION

2.1 Two-level atom

The spectrum structure of the atom is very complex. For example, the Hydrogen atom which is the simplest one, has an infinite number of bound levels and a set of continuum levels. So it is impossible to find an accurate solution of energy levels for a system of the atoms interacting with a radiation field. Even if we only discuss a system of a single-atom interacting with a radiation field. In order to obtain more accurate solutions, we have to adopt some approximations. Although a monochromatic electromagnetic field may induce a number of transitions among the different levels of an atom, the dominant one is that the field frequency almost matches a transition frequency between two particular levels of the atom. So the most natural assumption is that the atom has only two nondegenerate levels, which is usually interpret as two-level atom. Evidently, when a two-level atom with transition frequency ω_0 interacts with a single-mode radiation field with frequency $\omega \approx \omega_0$, the resonant transition occurs. It is clear that the two-level atom is an ideal model of the true atom. As the assumption of a classical material point plays a very important role in classical mechanics, the assumption of the two-level atom also plays an fundamental role in studying the interaction between atom and field. In the sense of concept, a two-level atom and a particle with spin 1/2 belong to the same kind of particles, so we sometimes call the two-level atom as pseudo-spin particle with spin 1/2.

2.2 Hamiltonian of a two-level atom interacting with an electromagnetic field

The Hamiltonian describing the system of a two-level atom with dipole moment $\mathbf{D}$ interacting with a radiation field which is usually represented by the electromagnetic field $\mathbf{E}(\mathbf{r}_0)$, may be written as

$$H = H_A - \mathbf{D} \cdot \mathbf{E}(\mathbf{r}_0) \tag{2.1}$$

where H_A is the energy operator of the bare two-level atom, $\mathbf{D}$ represents the dipole moment operator of the atom, $\mathbf{E}(\mathbf{r}_0)$ is the electric field operator at the position $\mathbf{r}_0$. The two eigenkets of the atom are represented by $|+\rangle$ and $|-\rangle$, respectively, and they obey the eigenvalue equation

$$H_A|\pm\rangle = E_\pm|\pm\rangle \tag{2.2}$$

The matrix representation of H_A in the Hilbert space spanned by the set of eigenkets of the atom is

$$H_A = \begin{pmatrix} \langle +|H_A|+\rangle & \langle +|H_A|-\rangle \\ \langle -|H_A|+\rangle & \langle -|H_A|-\rangle \end{pmatrix} \tag{2.3}$$

In similarity, the matrix expressions of the interaction energy $\mathbf{D} \cdot \mathbf{E}(\mathbf{r}_0)$ can be obtained as follows. Because

$$\mathbf{D} = e\mathbf{r} \tag{2.4}$$

evidently, $\mathbf{D}$ posses odd parity. If we assume that the parities of $|+\rangle$ and $|-\rangle$ are given explicitly, then obtain

$$\begin{aligned} &\langle +|\mathbf{D}|+\rangle = 0, \quad \langle -|\mathbf{D}|-\rangle = 0 \\ &\langle +|\mathbf{D}|-\rangle = \mathbf{D}_{+-} \neq 0, \quad \langle -|\mathbf{D}|+\rangle = (\mathbf{D}_{+-})^* \neq 0 \end{aligned} \tag{2.5}$$

As an example, when we choose that $|+\rangle$ and $|-\rangle$ are the eigenstates $\Psi_{211}(\mathbf{r})$ and $\Psi_{100}(\mathbf{r})$ of a Hydrogen atom, respectively, then the dipole moment element gives

$$\mathbf{D}_{+-} = \int \Psi_{211}(\mathbf{r}) e\mathbf{r} \Psi_{100}(\mathbf{r}) d^3\mathbf{r} \tag{2.6}$$

here $\mathbf{D}_{+-}$ can be separated as the product of the radial component D_1 and the angular component D_2 , where

$$\begin{aligned} D_1 &= \int (2a_0^2)^{-3/2} \exp\left(-\frac{r}{2a_0}\right) er \left(\frac{2r}{\sqrt{3}a_0}\right) \exp\left(-\frac{r}{a_0}\right) r^2 dr \\ D_2 &= iY_{11}^*(\theta, \phi)(\mathbf{e}_x \sin\theta\cos\phi + \mathbf{e}_y \sin\theta\sin\phi + \mathbf{e}_z \cos\theta) Y_{00}(\theta, \phi) d\tau \end{aligned} \tag{2.7}$$

here $\mathbf{e}_x$, $\mathbf{e}_y$, and $\mathbf{e}_z$ are the unit vectors of Cartesian coordinates. In a broad sense, the dipole moment is a complex vector, so it can be described by

$$\begin{aligned} \mathbf{D}_{+-} &= \mathbf{d}_r + i\mathbf{d}_i \\ \mathbf{D}_{-+} &= \mathbf{d}_r - i\mathbf{d}_i \end{aligned} \tag{2.8}$$

$$\begin{aligned} \mathbf{d}_r &= \frac{2^7 e a_0 \mathbf{e}_x}{3^5} \\ \mathbf{d}_i &= \frac{2^7 e a_0 \mathbf{e}_y}{3^5} \end{aligned} \tag{2.9}$$

here $\mathbf{d}_r$ and $\mathbf{d}_i$ are real vectors. Thus, in the two-dimensional Hilbert space spanned by the atomic eigenkets $\{|\pm\rangle\}$, the Hermitian operator $\mathbf{D}$ can be rewritten as

$$\mathbf{D} = \begin{pmatrix} 0 & \mathbf{d}_r + i\mathbf{d}_i \\ \mathbf{d}_r - i\mathbf{d}_i & 0 \end{pmatrix} \tag{2.10}$$

Therefore, the matrix expression of the Hamiltonian of the whole system is given as

$$\begin{aligned} H &= \begin{pmatrix} E_+ & 0 \\ 0 & E_- \end{pmatrix} - \begin{pmatrix} 0 & \mathbf{d}_r + i\mathbf{d}_i \\ \mathbf{d}_r - i\mathbf{d}_i & 0 \end{pmatrix} \cdot \mathbf{E}(\mathbf{r}_0) \\ &= \frac{E_+ - E_-}{2} \begin{pmatrix} 1 & 0 \\ 0 & -1 \end{pmatrix} + \frac{E_+ + E_-}{2} \begin{pmatrix} 1 & 0 \\ 0 & 1 \end{pmatrix} \\ &\quad -(\mathbf{d}_r \cdot \mathbf{E}) \begin{pmatrix} 0 & 1 \\ 1 & 0 \end{pmatrix} + (\mathbf{d}_i \cdot \mathbf{E}) \begin{pmatrix} 0 & -i \\ i & 0 \end{pmatrix} \end{aligned} \tag{2.11}$$

If we choose $E_+ + E_- = 0$ and notice

$$\omega = \frac{E_+ - E_-}{\hbar} \tag{2.12}$$

then (2.11) becomes

$$H = \frac{\hbar}{2}\omega_0 \sigma_3 - (\mathbf{d}_r \cdot \mathbf{E})\sigma_1 + (\mathbf{d}_i \cdot \mathbf{E})\sigma_2 \tag{2.13}$$

here the operators σ_i are exactly the Pauli operators and are defined respectively by

$$\sigma_1 = \begin{pmatrix} 0 & 1 \\ 1 & 0 \end{pmatrix}, \quad \sigma_2 = \begin{pmatrix} 0 & -i \\ i & 0 \end{pmatrix}, \quad \sigma_3 = \begin{pmatrix} 1 & 0 \\ 0 & -1 \end{pmatrix}$$

They obey the commutation relation

$$[\sigma_i, \sigma_j] = 2i\sigma_k \tag{2.14}$$

2.3 The optical Bloch equation

As yet we have obtained the Hamiltonian (2.13) of the system of an two-level atom interacting with a electromagnetic field. By use of (2.13) we can discuss the time evolution of the atomic operators. As we know, the motion of an arbitrary operator A which has no explicit time dependence obeys

$$\frac{d}{dt}A = \frac{i}{\hbar}[H, A] \tag{2.15}$$

So substitute (2.13) and (2.14) into (2.15), we can obtain the equations of motion of the atomic operators as

$$\begin{aligned}
\frac{d}{dt}\sigma_1(t) &= -\omega_0\sigma_2(t) + \frac{2}{\hbar}(\mathbf{d}_r \cdot \mathbf{E}(t))\sigma_3(t) \\
\frac{d}{dt}\sigma_2(t) &= \omega_0\sigma_3(t) + \frac{2}{\hbar}(\mathbf{d}_r \cdot \mathbf{E}(t))\sigma_3(t) \\
\frac{d}{dt}\sigma_3(t) &= -\frac{2}{\hbar}(\mathbf{d}_r \cdot \mathbf{E}(t))\sigma_2(t) - \frac{2}{\hbar}(\mathbf{d}_i \cdot \mathbf{E}(t))\sigma_1(t)
\end{aligned} \tag{2.16}$$

It is difficult to solve the above equations exactly. Specially when the electromagnetic field is represented by the operator $\mathbf{E}(t)$, the solving processes are more complex. One simplification step is to assume that the quantum correlation between the field operator and the atomic operators may be neglected. That is to say, when we calculate the expectation values of the product $\mathbf{E}(t) \cdot \sigma_j(t)$, we may replace it by $\langle \mathbf{E}(t) \rangle \cdot \langle \sigma_j(t) \rangle$, namely,

$$\langle \mathbf{E}(t) \cdot \sigma_j(t) \rangle = \langle \mathbf{E}(t) \rangle \cdot \langle \sigma_j(t) \rangle \tag{2.17}$$

this approximation is usually called the decoupling approximation. The other approximation is to assume that the expectation value $\langle \mathbf{E}(t) \rangle$ of the operator $\mathbf{E}(t)$ may be replaced by a classical variable $\mathbf{E}(\mathbf{t})$ in the quasi-classical theory. According to the above assumptions and assuming

$$R_i(t) = \langle \sigma_i(t) \rangle \quad (i = 1, 2, 3) \tag{2.18}$$

we then obtain the dynamical equations of an atom in the atom-field coupling system in the quasiclassical theory to be

$$\begin{aligned}
\frac{d}{dt}R_1(t) &= -\omega_0 R_2(t) + \frac{2}{\hbar}\mathbf{d}_i \cdot E(t,\mathbf{r}_0)R_3(t) \\
\frac{d}{dt}R_2(t) &= \omega_0 R_1(t) + \frac{2}{\hbar}\mathbf{d}_r \cdot \mathbf{E}(t,\mathbf{r}_0)R_3(t) \\
\frac{d}{dt}R_3(t) &= -\frac{2}{\hbar}\mathbf{d}_i \cdot \mathbf{E}(t,\mathbf{r}_0)R_1(t) - \frac{2}{\hbar}\mathbf{d}_r \cdot \mathbf{E}(t,\mathbf{r}_0)R_2(t)
\end{aligned} \tag{2.19}$$

Before discussing the solutions of the above equations, we analyze the physical meaning of the expectation values $R_i(t)(i = 1,2,3)$. From equation (2.13) we know that $\hbar\omega_0 R_3(t)/2$ represents the atomic energy, so $R_3(t)$ corresponds to the expectation value of the atomic energy operator. Eqs. (2.10) and (2.13) show that $R_1(t)$ and $R_2(t)$ are the expectation values of the atomic dipole moment operators.

If multiplying both sides of equation (2.19) by $R_1(t)$, $R_2(t)$ and $R_3(t)$, respectively, then we obtain

$$R_1(t)\frac{d}{dt}R_1(t) + R_2(t)\frac{d}{dt}R_2(t) + R_3(t)\frac{d}{dt}R_3(t) = 0 \tag{2.20}$$

This means

$$R_1^2(t) + R_2^2(t) + R_3^2(t) = \text{constant}. \tag{2.21}$$

It is easy to prove that this constant in (2.21) is equal to 1 as follows. If the atom is initially in a general state

$$|\Psi\rangle = a|+\rangle + b|-\rangle \tag{2.22}$$

here $|a|^2 + |b|^2 = 1$, then $R_i(0)(i = 1,2,3)$ obey

$$\begin{aligned}
R_1(0) &= \langle\Psi|\sigma_1|\Psi\rangle = a^*b + ab^* \\
R_2(0) &= \langle\Psi|\sigma_2|\Psi\rangle = -i(a^*b - ab^*) \\
R_3(0) &= \langle\Psi|\sigma_3|\Psi\rangle = a^*a - b^*b
\end{aligned}$$

namely

$$R_1^2(0) + R_2^2(0) + R_3^2(0) = 1 \tag{2.23}$$

So (2.21) becomes

$$R_1^2(t) + R_2^2(t) + R_3^2(t) = 1 \tag{2.24}$$

This means that the atom retains the probability conservation with the time development. We call $\mathbf{R}(t)$ $(R_1(t),\ R_2(t),\ R_3(t))$ the atomic pseudo-spin vector. Eq.(2.24) displays that with the time development, the pseudo-spin vector remains on the unit spherical surface as illustrated in Fig.(2.1).

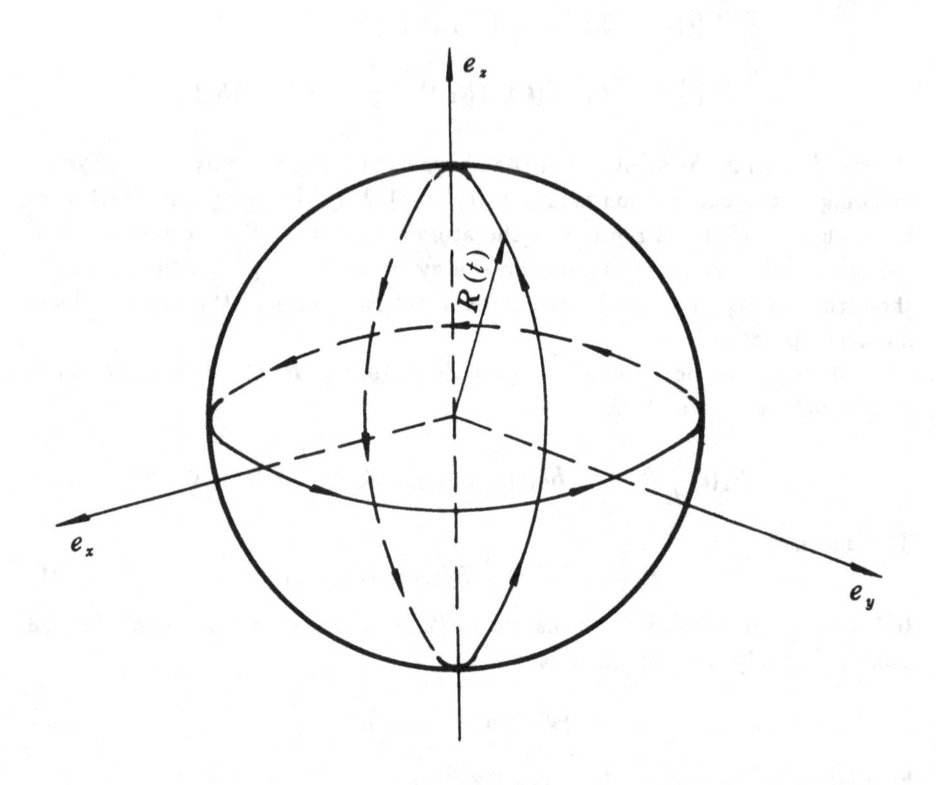

Figure 2.1: Unit sphere of the time development of pseudo-spin vector $\mathbf{R}(t)$.

As usual, the states $|+\rangle$ and $|-\rangle$ may be chosen as the transition states with $\Delta m = 0$. Thus we can choose appropriate phase factors so that the dipole moment $\mathbf{d}_i = 0$. Under this condition, we let

$$\frac{2}{\hbar}\mathbf{d}_r \cdot \mathbf{E} = \frac{2d}{\hbar}\mathbf{U}_d \cdot \mathbf{E} = \kappa E \tag{2.25}$$

here the scalar quantity E is the projection of the electric vector $\mathbf{E}$ along the unit vector $\mathbf{U}_d$, and κ is defined as

$$\frac{\hbar\kappa}{2} = d \tag{2.26}$$

$\mathbf{U}_d$ and d represent the direction and length of the dipole moment, respectively. So (2.19) reduces to

$$\begin{aligned}
\frac{d}{dt}R_1(t) &= -\omega_0 R_2(t) \\
\frac{d}{dt}R_2(t) &= \omega_0 R_1(t) + \kappa E(t, \mathbf{r}_0) R_3(t) \\
\frac{d}{dt}R_3(t) &= -\kappa E(t, \mathbf{r}_0) R_2(t)
\end{aligned} \tag{2.27}$$

These equations are very similar to those for a spin 1/2 system appearing in nuclear magnetic resonance. If we let the components of vector $\Omega(t)$ to be

$$\Omega_1(t) = -\kappa E, \quad \Omega_2(t) = 0,; \quad \Omega_3(t) = \omega_0 \tag{2.28}$$

then (2.27) can be written in the form

$$\frac{d}{dt}R_1(t) = \mathbf{\Omega}(t) \times \mathbf{R}(t) \tag{2.29}$$

We interpret equation (2.27) or (2.29) as the optical Bloch equation. Here the influence of the field on the atom induces the pseudo-spin vector $\mathbf{R}(t)$ undergoing precession on a unit spherical surface with the torque force $\mathbf{\Omega}(t)$ with the time development. The optical Bloch equation is a basic equation for discussing the properties of the atom-field coupling system in the quasi-classical theory.

2.4 Description of the dynamical behavior of a two-level atom interacting with the radiation field by the density matrix

We have discussed the dynamical behavior of a two-level atom interacting with a classical electromagnetic field without decay by means of the Heisenberg equation. In this section, we use the density operator theory to discuss the dynamic behavior of a damping two-level atom interacting with a classical field $\mathbf{E}(t)$. We first discuss the properties of the two-level atom without decay under the interaction of the classical field in the Schrödinger picture , then study the case in which the atom is a damped one.

2.4.1 Density matrix equation describing a two-level atom without decay

In the Schrödinger picture the state vector at time t for a single two-level atom system can be expanded by its eigenkets as

$$|\Psi(t)\rangle = C_+(t)|+\rangle + C_-(t)|-\rangle \tag{2.30}$$

here $C_+(t)$ and $C_-(t)$ are the probability amplitudes of the atom in the eigenkets $|+\rangle$ and $|-\rangle$ at time t, respectively. If we write the eigenkets $|+\rangle$ and $|-\rangle$ in the form

$$|+\rangle = \begin{pmatrix} 1 \\ 0 \end{pmatrix}, \qquad |-\rangle = \begin{pmatrix} 0 \\ 1 \end{pmatrix} \tag{2.31}$$

then eq.(2.30) becomes

$$|\Psi(t)\rangle = \begin{pmatrix} C_+(t) \\ C_-(t) \end{pmatrix} \tag{2.32}$$

Thus the density operator of the two-level atom can be written by

$$\rho(t) = |\Psi(t)\rangle\langle\Psi(t)| = \begin{pmatrix} C_+C_+^* & C_+C_-^* \\ C_-C_+^* & C_-C_-^* \end{pmatrix} = \begin{pmatrix} \rho_{++} & \rho_{+-} \\ \rho_{-+} & \rho_{--} \end{pmatrix} \tag{2.33}$$

It is clear that the matrix elements ρ_{++} and ρ_{--} represent the probabilities of the atom in the states $|+\rangle$ and $|-\rangle$ at time t, respectively.

Inasmuch as the atomic dipole moment in the y-axis may be equal to zero through choosing appropriate phase factors, the Hamiltonian (2.13) describing a two-level atom coupled to a classical electromagnetic field without decay reduces to

$$H = \frac{\hbar\omega_0\sigma_3}{2} - (\mathbf{d}_r \cdot \mathbf{E})\sigma_1 \tag{2.34}$$

Considering that the operators σ_1 and σ_3 and the eigenkets $|+\rangle$ and $|-\rangle$ satisfy the relations

$$\begin{aligned} \sigma_3|+\rangle &= |+\rangle, \quad \sigma_3|-\rangle = -|-\rangle, \\ \sigma_1|+\rangle &= |-\rangle, \quad \sigma_1|-\rangle = |+\rangle \end{aligned} \tag{2.35}$$

and substituting (2.32) and (2.34) into the Schrödinger equation

$$i\hbar\frac{d}{dt}|\Psi(t)\rangle = H|\Psi(t)\rangle \tag{2.36}$$

then the time-dependent equations of the probability amplitudes give

$$i\frac{d}{dt}C_+(t)=\frac{1}{2}\omega_0 C_+ - \frac{1}{2}\kappa E(C_+ + C_-) \tag{2.37}$$

$$i\frac{d}{dt}C_-(t)=-\frac{1}{2}\omega_0 C_- - \frac{1}{2}\kappa E(C_+ + C_-) \tag{2.38}$$

here we have used the orthogonal relation between $|+\rangle$ and $|-\rangle$, and κ is defined by (2.26). From eqs.(2.37), (2.38) and their complex adjoint, we can obtain the time-dependent equations of the matrix elements as follows

$$\begin{aligned}
\frac{d}{dt}\rho_{++} &= \left(\frac{d}{dt}C_+\right)C_+^* + C_+\left(\frac{d}{dt}C_+^*\right) = -i\frac{\kappa}{2}(\rho_{+-} - \rho_{-+}) \\
\frac{d}{dt}\rho_{--} &= \frac{i\kappa}{2}(\rho_{+-} - \rho_{-+}) \\
\frac{d}{dt}\rho_{+-} &= -i\omega_0\rho_{+-} - \frac{i\kappa}{2}(\rho_{+-} - \rho_{-+}) \\
\frac{d}{dt}\rho_{-+} &= i\omega_0\rho_{-+} + \frac{i\kappa}{2}(\rho_{+-} - \rho_{-+})
\end{aligned} \tag{2.39}$$

These are just the equations of motion for the density matrix of the two-level atom without decay in a classical electromagnetic field.

Now we discuss the physical meaning of (2.39). From the first two equations of (2.39) we know

$$\frac{d}{dt}(\rho_{++} + \rho_{--}) = 0 \tag{2.40}$$

i.e.,

$$\rho_{++}(t) + \rho_{--}(t) = \rho_{++}(0) + \rho_{--}(0) = 1 \tag{2.41}$$

This means that the total probability of the atom in its two eigenkets $|+\rangle$ and $|-\rangle$ is independent of time and is equal to 1. Comparing (2.41) with (2.24), we have

$$R_1^2(t) + R_2^2(t) + R_3^2(t) = \rho_{++}(t) + \rho_{--}(t) \tag{2.42}$$

The above equation shows that in the Heisenberg picture, the scalar product of the vector $\mathbf{R}(t)(R_1(t),\ R_2(t),\ R_3(t))$ represents the total probability of the atom in its two eigenkets $|+\rangle$ and $|-\rangle$.

If we introduce the real quantities

$$\rho_1 = \rho_{+-} + \rho_{-+}, \quad \rho_2 = i(\rho_{+-} - \rho_{-+}), \quad \rho_3 = \rho_{++} - \rho_{--} \tag{2.43}$$

then (2.39) becomes

$$\begin{aligned}\frac{d}{dt}\rho_1 &= -\omega_0\rho_2\\ \frac{d}{dt}\rho_2 &= \omega_0\rho_1 + \kappa E\rho_3\\ \frac{d}{dt}\rho_3 &= -\kappa E\rho_2\end{aligned} \tag{2.44}$$

Evidently (2.44) is the same as (2.27), and $\rho_i(t)(i = 1,2,3)$ correspond to the expectation values $R_i(t)(i = 1,2,3)$ of the operators $\sigma_i(t)$. So $\rho_1(t)$ and $\rho_2(t)$ have the same physical meaning as that of $R_1(t)$ and $R_2(t)$, they represent the time evolution of the atomic dipole moments. In this sense, the matrix elements $\rho_{+-}(t)$ and $\rho_{-+}(t)$ are the quantities describing the complex dipole moments of the two-level atom. The difference $\rho_3 = \rho_{++} - \rho_{--}$ of the atomic probabilities between in $|+\rangle$ and in $|-\rangle$ represent the expectation value $R_3(t)$ of the atomic energy operator. Therefore, the description of the density matrix for the system of a two-level atom interacting with the radiation field can give an explicit physical explanation.

2.4.2 Density matrix equation describing a two-level atom with decay

As we know, an atom is steady only when the case of being in its ground state. The atom, however, being in its excited state will decay to its lower states due to spontaneous emission, inelastic collisions and the like. Here we discuss the effect of decay on the atomic dynamic behavior in the system of a damped two-level atom interacting with a classical field by means of the density matrix theory.

The Hamiltonian describing a system of a two-level atom interacting with a radiation field is still represented by (2.34) even though the effect of decay is included. The influence of decay can be described well by adding phenomenological decay terms into (2.37) and (2.38). Thus we have

$$\frac{d}{dt}C_+ = -\frac{1}{2}\gamma_+C_+ + \frac{i}{2}\kappa E(C_+ + C_-) - \frac{i}{\omega_0}C_+ \tag{2.45}$$

$$\frac{d}{dt}C_- = -\frac{1}{2}\gamma_-C_- + \frac{i}{2}\kappa E(C_+ + C_-) + \frac{i}{2}\omega C_- \tag{2.46}$$

here γ_+ and γ_- are the decay constants of states $|+\rangle$ and $|-\rangle$, respectively, and correspond to the probabilities $|C_+|^2$ and $|C_-|^2$. From eqs.(2.45) and (2.46) we can see that even in the absence of the classical field (E=0), the atom is not steady, and its probability $|C_+|^2$ will decay on the form of $\exp(-\gamma_+ t)$ with the time development . The atomic lifetime is usually defined as the time which the probability has decayed to be 1/e of its original values. So the lifetime which the atom in the state $|+\rangle$ (or $|-\rangle$) is $1/\gamma_+$ (or $1/\gamma_-$).

For simplicity to calculate, we only discuss the case of equal decay constants, $\gamma_+ = \gamma_- = \gamma$. Using the similar method as deducing equation (2.39) and by means of (2.45) and (2.46), we obtain

$$\begin{aligned}
\frac{d}{dt}\rho_{++} &= -\gamma\rho_{++} - \frac{i\kappa}{2}E(\rho_{+-} - \rho_{-+}) \\
\frac{d}{dt}\rho_{--} &= -\gamma\rho_{--} + \frac{i\kappa}{2}E(\rho_{+-} - \rho_{-+}) \\
\frac{d}{dt}\rho_{+-} &= -(\gamma + i\omega_0)\rho_{+-} - \frac{i\kappa}{2}E(\rho_{++} - \rho_{--}) \\
\frac{d}{dt}\rho_{-+} &= -(\gamma - i\omega_0)\rho_{-+} + \frac{i\kappa}{2}E(\rho_{++} - \rho_{--})
\end{aligned} \tag{2.47}$$

From the first two equations of (2.47), we have

$$\frac{d}{dt}(\rho_{++} + \rho_{--}) = -\gamma(\rho_{++} + \rho_{--}) \tag{2.48}$$

Evidently

$$\rho_{++}(t) + \rho_{--}(t) = (\rho_{++}(0) + \rho_{--}(0))\exp(-\gamma t) = \exp(-\gamma t) \tag{2.49}$$

The above equation shows that the lifetime of the two-level atom is $1/\gamma$. When $t \to \infty$, $\rho_{++}(\infty) + \rho_{--}(\infty) = 0$, which means that the two-level atom may be damped completely with the time development. In fact , the real atom has a number of levels, because of the influence of some decay processes such as spontaneous emission and inelastic collision, the atom will decay from the states $|+\rangle$ and $|-\rangle$ to the other states whose energy is lower (such as ground state or sub-steady states). So the total probability of the atom in the states $|+\rangle$ and $|-\rangle$ will be equal to zero in the final time.

Substituting (2.43) into (2.47) we obtain

$$\begin{aligned}
\frac{d}{dt}\rho_1 &= -\gamma\rho_1 - \omega_0\rho_2 \\
\frac{d}{dt}\rho_2 &= -\gamma\rho_2 + \kappa E\rho_3 + \omega_0\rho_2 \\
\frac{d}{dt}\rho_3 &= -\gamma\rho_3 - \kappa E\rho_2
\end{aligned} \tag{2.50}$$

Since $\rho_i(t)$ have the same physical meaning as the expectation values $R_i(t)$, (2.50) can also be written in a form of vector like (2.29) as

$$\frac{d}{dt}\rho = -\gamma\rho + \mathbf{\Omega}(t) \times \rho(t) \tag{2.51}$$

here $\rho(t) = (\rho_1(t), \rho_2(t), \rho_3(t))$. The torque force $\mathbf{\Omega}(t)$ resulted from the interaction between the atom and the classical field is identical to (2.28). Different from the case neglecting the effect of decay, the length of the vector $\rho(t)$ (shown in (2.49)) does not remain 1 but decreases gradually to zero with the time development.

References

1. L.Allen, J.H.Eberly, *Optical Resonance and two-level atom* (Wiley, New York, 1975).

2. M.Sargent III, M.O.Scully, W.E.Lamb Jr., *Laser Physics* (Addison-Wesley, London, 1974).

3. H.Haken, *Light*, Vol.1, (North-Holland, Armsterdam, 1981).

4. P.Meystre and M.Sargent III, *Elements of quantum optics* (Springer-Verlag, New York, 1990).

5. A.C.Newell and J.V.Moloney, *Nonlinear Optics* (Addison-Wesley, New York, 1990).

6. R.W.Boyd, *Nonlinear Optics*, (Academic, New York, 1992)

7. Jin-sheng Peng, *Resonance fluorescence and superfluorescence* (Science Press, Beijing, 1993)

CHAPTER 3

QUANTIZED DESCRIPTION OF RADIATION FIELD

In the previous chapter, we have adopted operators to represent the atomic observables, however, the radiation field was still described by a classical electromagnetic field quantity. As known, the classical quantity can not exhibit the wave-particle duality of the radiation field. In order to display the wave-particle duality of light, and reveal the quantum characteristics of the interaction between light and matter, it is necessary to quantize the radiation field. In this chapter, we first give a brief statement of classical description to the radiation field, then discuss the quantization of radiation field. Finally, we introduce some state functions which may characterize the different quantum properties of light field.

3.1 Classical description of the electromagnetic field in vacuum

In Classical electrodynamics, the electromagnetic field in vaccum may be described by means of Lagrangian density

$$L = \frac{\varepsilon_0}{2}\left[\left(\frac{\partial}{\partial t}\mathbf{A}\right)^2 - c^2(\nabla \times \mathbf{A})^2\right] \tag{3.1}$$

where ε_0 is the electric permittivity in vacuum, $\mathbf{A}$ stands for the vector potential which obeys the Coulomb gauge $\nabla \cdot \mathbf{A} = 0$. The electric field strength $\mathbf{E}$ and magnetic field strength $\mathbf{B}$ may be expressed in term of $\mathbf{A}$, respectively

$$\mathbf{E} = -\frac{\partial}{\partial t}\mathbf{A}, \quad \mathbf{B} = \nabla \times \mathbf{A} \tag{3.2}$$

Noticing the relations

$$\frac{\partial}{\partial \mathbf{A}}L=-\varepsilon_0 c^2\nabla\times\nabla\times\mathbf{A}=-\varepsilon_0 c^2\nabla\times\mathbf{B}$$
$$\frac{\partial}{\partial \dot{\mathbf{A}}}L=\varepsilon_0\dot{\mathbf{A}} \tag{3.3}$$

and solving the Lagrangian equation

$$\frac{d}{dt}\left[\frac{\partial}{\partial \dot{\mathbf{A}}}L\right]-\frac{\partial}{\partial \mathbf{A}}L=0 \tag{3.4}$$

we obtain the first equation of the Maxwell's equations

$$\frac{\partial}{\partial t}\mathbf{E}=\frac{1}{c^2}\nabla\times\mathbf{B}$$
$$\frac{\partial}{\partial t}\mathbf{B}=-\nabla\times\mathbf{E}$$
$$\nabla\cdot\mathbf{B}=0 \tag{3.5}$$
$$\nabla\cdot\mathbf{E}=0$$

Here the last three equations are directly obtained from (3.2). By use of (3.2) and (3.5) we also obtain the wave motion equation

$$\frac{\partial^2}{\partial t^2}\mathbf{A}-c^2\nabla^2\mathbf{A}=0 \tag{3.6}$$

The generalized momentum $\mathbf{\Pi}$ which conjugates the vector potential $\mathbf{A}$ is defined as

$$\mathbf{\Pi}=\frac{\partial}{\partial \dot{\mathbf{A}}}L=\varepsilon_0\frac{\partial}{\partial t}\mathbf{A}=-\varepsilon_0\mathbf{E} \tag{3.7}$$

Substituting (3.7) into (3.1), the Lagrangian density may be expressed as

$$L=\frac{1}{2}\mathbf{\Pi}\cdot\frac{\partial}{\partial t}\mathbf{A}-\frac{\varepsilon_0 c^2}{2}(\nabla\times\mathbf{A})^2 \tag{3.8}$$

Therefore, the Hamiltonian function of the electromagnetic field in the vacuum becomes

$$H=\int\left(\mathbf{\Pi}\cdot\frac{\partial}{\partial t}\mathbf{A}-L\right)d^3\mathbf{r}=\frac{1}{2}\int\left[\frac{1}{\varepsilon_0}\mathbf{\Pi}^2+\varepsilon_0 c^2(\nabla\times\mathbf{A})^2\right]d^3\mathbf{r} \tag{3.9}$$

The momentum $\mathbf{P}_F$ and the angular momentum $\mathbf{J}_F$ of the field are respectively as follows

$$\mathbf{P}_F = -\int d^3\mathbf{r}\frac{\varepsilon_0}{c}(\mathbf{E}\times\mathbf{B}) = -\frac{1}{c}\int d^3\mathbf{r}[\mathbf{\Pi}\times(\nabla\times\mathbf{A})] \tag{3.10}$$

$$\mathbf{J}_F = -\int d^3\mathbf{r}\frac{1}{c}\{\mathbf{r}\times[\mathbf{\Pi}\times(\nabla\times\mathbf{A})]\} \tag{3.11}$$

We assume that the electromagnetic field is restricted in a cube with volume $V = L^3$ and the potential $\mathbf{A}(\mathbf{r},t)$ is required to obey periodic boundary conditions, namely that it takes the same value on opposite faces of the cube. In this way the solution of the wave motion equation (3.6) can be expressed in terms of the Fourier series as

$$\mathbf{A}(\mathbf{r},t) = \frac{1}{\sqrt{V}}\sum_k[\mathbf{A}_k(t)\exp(i\mathbf{k}\cdot\mathbf{r}) + \mathbf{A}_k^*(t)\exp(-i\mathbf{k}\cdot\mathbf{r})] \tag{3.12}$$

where $\mathbf{A}_k^*(t)$ and $\mathbf{A}_k(t)$ are the coefficients of the Fourier series, which represent the amplitudes of the electromagnetic field. $\mathbf{k}$ is the wave vector which describes the propagating direction of the field. The periodic condition restricts the wave vector $\mathbf{k}$ to obey

$$\mathbf{k} = \frac{2\pi}{L}(n_x\mathbf{e}_x + n_y\mathbf{e}_y + n_z\mathbf{e}_z) \qquad (n_x, n_y, n_z = 0, \pm1, \pm2, \cdots) \tag{3.13}$$

By substituting (3.12) into (3.6), it is found that the coefficient $\mathbf{A}_k(t)$ must satisfy the equation

$$\frac{d^2}{dt^2}\mathbf{A}_k(t) + \omega_k^2\mathbf{A}_k(t) = 0 \tag{3.14}$$

The solution of (3.14) gives

$$\mathbf{A}_k(t) = \mathbf{A}_k(0)\exp(-i\omega_k t) \tag{3.15}$$

here $\omega_k = ck$ is the frequency of the electromagnetic field with the wave vector $\mathbf{k}$.

Since the vector potential $\mathbf{A}(\mathbf{r},t)$ obeys the Coulomb guage, equation (3.12) shows that

$$0 = \nabla\cdot\mathbf{A}(\mathbf{r},t) = i\sum_k\{\mathbf{k}\cdot\mathbf{A}_k(t)\exp(i\mathbf{k}\cdot\mathbf{r}) \\ -[\mathbf{k}\cdot\mathbf{A}_k^*(t)]\exp(-i\mathbf{k}\cdot\mathbf{r})\} \tag{3.16}$$

so that

$$\mathbf{k} \cdot \mathbf{A}_k(t) = 0 \tag{3.17}$$

This transversal condition shows that the amplitude vectors of the field are orthogonal to the propagation direction. The coefficients $\mathbf{A}_k(t)$ being orthogonal to $\mathbf{k}$, can be specified in terms of components along two mutually orthogonal directions transverse to $\mathbf{k}$. Unit vectors along these orthogonal directions denoted by $\mathbf{e}_{kj}(j=1,2)$, obey the relations

$$\mathbf{e}_{kj} \cdot \mathbf{e}_{kj'} = \delta_{jj'}, \quad \mathbf{e}_{kj} \cdot \mathbf{k} = 0, \quad \mathbf{e}_{k1} \times \mathbf{e}_{k2} = \mathbf{k}/k \tag{3.18}$$

here $\mathbf{e}_{kj}(j=1,2)$ are usually called the unit vectors of the field polarization, they specify the polarization directions of the field. Therefore, the vector potential $\mathbf{A}(\mathbf{r},t)$ may be written as

$$\mathbf{A}(\mathbf{r},t) = \frac{1}{\sqrt{V}} \sum_{kj} [\mathbf{e}_{kj} A_{kj}(t) \exp(i\mathbf{k}\cdot\mathbf{r}) + \mathbf{e}_{kj} A^*_{kj}(t) \exp(-i\mathbf{k}\cdot\mathbf{r})] \tag{3.19}$$

The generalized momentum $\mathbf{\Pi}(\mathbf{r},t)$, the electric field strength $\mathbf{E}(\mathbf{r},t)$ and the magnetic field strength $\mathbf{B}(\mathbf{r},t)$ corresponding to $\mathbf{A}(\mathbf{r},t)$ are respectively written as

$$\begin{aligned} \mathbf{\Pi}(\mathbf{r},t) &= \frac{1}{\sqrt{V}} \sum_{kj} [\mathbf{e}_{kj} \Pi_{kj}(t) \exp(-i\mathbf{k}\cdot\mathbf{r}) + \mathbf{e}_{kj} \Pi^*_{kj}(t) \exp(i\mathbf{k}\cdot\mathbf{r})] \\ &= i\varepsilon_0 \frac{1}{\sqrt{V}} \sum_{kj} \omega_k [\mathbf{e}_{kj} A^*_{kj}(t) \exp(-i\mathbf{k}\cdot\mathbf{r}) \\ &\quad - \mathbf{e}_{kj} A_{kj}(t) \exp(i\mathbf{k}\cdot\mathbf{r})] \end{aligned} \tag{3.20}$$

$$\begin{aligned} \mathbf{E}(\mathbf{r},t) &= i \frac{1}{\sqrt{V}} \sum_{kj} \omega_k [\mathbf{e}_{kj} A_{kj}(t) \exp(i\mathbf{k}\cdot\mathbf{r}) \\ &\quad - \mathbf{e}_{kj} A^*_{kj}(t) \exp(-i\mathbf{k}\cdot\mathbf{r})] \end{aligned} \tag{3.21}$$

$$\begin{aligned} \mathbf{B}(\mathbf{r},t) &= i \frac{1}{\sqrt{V}} \sum_{kj} \omega_k [\mathbf{k} \times \mathbf{e}_{kj} A_{kj}(t) \exp(i\mathbf{k}\cdot\mathbf{r}) \\ &\quad - \mathbf{k} \times \mathbf{e}_{kj} A^*_{kj}(t) \exp(-i\mathbf{k}\cdot\mathbf{r})] \end{aligned} \tag{3.22}$$

By substitution of eqs.(3.20)-(3.22) into (3.9)-(3.11), we can obtain the expressions of the energy, momentum and angular momentum of the electromagnetic field. It is easily seen that the electromagnetic field is composed of an infinite number of transverse plane traveling wave modes, each mode is characterized by the wave vector $\mathbf{k}$ and the polarization vector $\mathbf{e}_{kj} (j = 1, 2)$.

The polarization of the electromagnetic field is a concept which must be made clear. Eq.(3.18) shows that the electromagnetic wave can be composed of two mutually orthogonal polarization components which are transverse to the wave vector $\mathbf{k}$. It is conventional to refer the polarization of a wave to the directional properties of its electric field component. For a electromagnetic field with the wave vector $\mathbf{k}$, the positive frequency component is able to express in terms of the superposition of the unit polarization vectors $\mathbf{e}_{kj} (j = 1, 2)$ as

$$\mathbf{E}_{+}(\mathbf{r}, t) = \mathbf{e}_{k1} E_1 \exp[i(\mathbf{k} \cdot \mathbf{r} - \omega t + \delta_1)] + \mathbf{e}_{k2} E_2 \exp[i(\mathbf{k} \cdot \mathbf{r} - \omega t + \delta_2)] \quad (3.23)$$

here E_1 and E_2 are the amplitudes of the electric field, and δ_1 and δ_2 are the phase angles.

If the phase difference $\delta_1 - \delta_2 = 0$ or π, and $E_1 = E_2 = E$, then (3.23) reduces to

$$\mathbf{E}_{+}(\mathbf{r}, t) = \frac{1}{\sqrt{2}} (\mathbf{e}_{k1} \pm \mathbf{e}_{k2}) \sqrt{2} E \exp[i(\mathbf{k} \cdot \mathbf{r} - \omega t + \delta_1)] \quad (3.24)$$

It is easily found that the electric field vector remains parallel to a fixed direction $\frac{1}{\sqrt{2}} (\mathbf{e}_{k1} \pm \mathbf{e}_{k2})$ in space. This electric field is called the linearly polarized light.

If the phase difference $\delta_1 - \delta_2 = \pm\frac{\pi}{2}$, and $E_1 = E_2 = E$, then (3.23) is modified as

$$E_{+}(\mathbf{r}, t) = \frac{1}{\sqrt{2}} (\mathbf{e}_{k1} \pm i\mathbf{e}_{k2}) \sqrt{2} E \exp[i(\mathbf{k} \cdot \mathbf{r} - \omega t + \delta_1)] \quad (3.25)$$

where $\frac{1}{\sqrt{2}} (\mathbf{e}_{k1} \pm i\mathbf{e}_{k2})$ are complex polarization vectors of unit length. With the time development, the electric field vector varies in the complex plane decided by the unit vectors $\mathbf{e}_{kj} (j = 1, 2)$. This electric field is called the circularly polarized light. Evidently, the real part of (3.25) is

$$E[\mathbf{e}_{k1} \cos(\mathbf{k} \cdot \mathbf{r} - \omega t + \delta_1) \mp \mathbf{e}_{k2} \sin(\mathbf{k} \cdot \mathbf{r} - \omega t + \delta_1)] \quad (3.26)$$

We see that for the electromagnetic field of circular polarization, the magnitude of the electric field vector remains constant in time, and the vector directions rotating about the direction of propagation $\mathbf{k}$ at frequency $\omega_k = ck$. For an observer facing an oncoming wave with polarization vector $\mathbf{e}_L = \frac{1}{\sqrt{2}} (\mathbf{e}_{k1} + i\mathbf{e}_{k2})$, the rotation is counter-clockwise and conventionally the wave is said to be the left-circularly polarized. A wave with polarization $\mathbf{e}_R = \frac{1}{\sqrt{2}}(\mathbf{e}_{k1} - i\mathbf{e}_{k2})$ is called the right-circular polarization light field.

3.2 Quantization of the radiation field

Now we discuss the quantization of the radiation field. As we know, the light is electromagnetic field. So to quantize the electromagnetic field is equivalent to quantize the light field.

3.2.1 Quantization of the electromagnetic field

The quantization treatment of the electromagnetic field is that the classical quantities $A_{kj}(t)$ and $\Pi_{kj}(t)$ in (3.19) and (3.20) are replaced by the operators $A_{kj}(t)$ and $\Pi_{kj}(t)$. That is to say, the vector potential $\mathbf{A}(\mathbf{r},t)$ and the generalized momentum $\mathbf{\Pi}(\mathbf{r},t)$ of the electromagnetic field are expressed by the operators $\mathbf{A}(\mathbf{r},t)$ and $\mathbf{\Pi}(\mathbf{r},t)$, respectively, and these operators are required to obey the commutation relation

$$[\mathbf{A}(\mathbf{r},t), \mathbf{\Pi}(\mathbf{r},t)] = i\hbar \tag{3.27}$$

However, it is very useful to introduce two new operators $a_{kj}(t)$ and $a_{kj}^\dagger(t)$ to replace the operators $A_{kj}(t)$ and $\Pi_{kj}(t)$, these two operators are defined as

$$A_{kj}(t) = \left(\frac{\hbar}{2\varepsilon_0\omega_k}\right)^{1/2} a_{kj}(t) \tag{3.28}$$

$$\Pi_{kj}(t) = i\left(\frac{\hbar\omega_k\varepsilon_0}{2}\right)^{1/2} a_{kj}^\dagger(t) \tag{3.29}$$

It is easy to obtain from (3.27)-(3.29) the commutation relations

$$[a_{kj}, a_{k'j'}^\dagger] = \delta_{kk'}\delta_{jj'}, \qquad [a_{kj}, a_{k'j'}] = [a_{kj}^\dagger, a_{k'j'}^\dagger] = 0 \tag{3.30}$$

Substituting eqs.(3.28) and (3.29) into (3.19) and (3.20), we obtain the vector potential operator $\mathbf{A}(\mathbf{r},t)$ and the generalized momentum operator $\mathbf{\Pi}(\mathbf{r},t)$ as

$$\mathbf{A}(\mathbf{r},t) = \sum_{kj}\left(\frac{\hbar}{2\varepsilon_0\omega_k V}\right)^{1/2} \mathbf{e}_{kj}[a_{kj}\exp(i\mathbf{k}\cdot\mathbf{r}) + a_{kj}^\dagger \exp(-i\mathbf{k}\cdot\mathbf{r})] \tag{3.31}$$

$$\mathbf{\Pi}(\mathbf{r},t) = i\sum_{kj}\left(\frac{\hbar\omega_k\varepsilon_0}{2V}\right)^{1/2}\mathbf{e}_{kj}[-a_{kj}\exp(i\mathbf{k}\cdot\mathbf{r}) + a_{kj}^{\dagger}\exp(-i\mathbf{k}\cdot\mathbf{r})] \tag{3.32}$$

By inserting the above equations into (3.9) and taking integration over the volume V, it gives

$$H_F = -\frac{\hbar}{4V}\sum_{kj}\sum_{k'j'}(\omega_k\omega_{k'})^{1/2}\left[\mathbf{e}_{kj}\cdot\mathbf{e}_{k'j'} + (\mathbf{e}_{kj}\times\mathbf{k})\cdot\frac{(\mathbf{e}_{k'j'}\times\mathbf{k}')}{kk'}\right] \times\int dV[a_{kj}e^{i\mathbf{k}\cdot\mathbf{r}} - a_{kj}^{\dagger}e^{-i\mathbf{k}\cdot\mathbf{r}}][a_{k'j'}e^{i\mathbf{k}'\cdot\mathbf{r}} - a_{k'j'}^{\dagger}e^{-i\mathbf{k}'\cdot\mathbf{r}}] \tag{3.33}$$

Here we have used the relation

$$\nabla\times[\mathbf{e}_{kj}\exp(\pm i\mathbf{k}\cdot\mathbf{r})] = \pm i(\mathbf{e}_{kj}\times\mathbf{k})\exp(\pm i\mathbf{k}\cdot\mathbf{r}) \tag{3.34}$$

and (3.18). Noticing the relation

$$\int\exp[i(\mathbf{k}-\mathbf{k}')\cdot\mathbf{r}]d^3\mathbf{r} = V\delta_{kk'} \tag{3.35}$$

then obtain

$$H_F = -\sum_{kj}\sum_{kj'}\frac{\hbar\omega_k}{4}\{(a_{kj}a_{kj}^{\dagger} + a_{kj'}^{\dagger}a_{kj'})[(\mathbf{e}_{kj}\cdot\mathbf{e}_{kj'}) + (\mathbf{e}_{kj}\times\mathbf{k})\cdot(\mathbf{e}_{kj'}\times\mathbf{k})/k^2] - (a_{kj'}a_{kj'}^{\dagger} + a_{kj}^{\dagger}a_{kj})[(\mathbf{e}_{kj}\cdot\mathbf{e}_{kj'}) + (\mathbf{e}_{kj}\times\mathbf{k})\cdot(\mathbf{e}_{kj'}\times\mathbf{k})/k^2]\}$$

By use of the following relation

$$(\mathbf{e}_{kj}\times\mathbf{k})\times(\mathbf{e}_{kj'}\times\mathbf{k}) = k^2\delta_{jj'} \tag{3.36}$$

the Hamiltonian of the free field can be rewritten as

$$H_F = \sum_{kj}\hbar\omega_k(a_{kj}^{\dagger}a_{kj} + 1/2) \tag{3.37}$$

The commutation relations eqs.(3.27) and (3.30) are the typical ones which bosons obey. This means that the electromagnetic field is consisted of a set of bosons, in other words, the electromagnetic field is composed of a set of harmonic oscillators with discrete spectra, these oscillators are described by the operators a_{kj} and $a^{\dagger}_{kj}$. By means of the Heisenberg equation and (3.37), we have

$$i\hbar\frac{d}{dt}a_{kj} = [a_{kj}, H_F] = \hbar\omega_k a_{kj} \tag{3.38}$$

the time-dependence of the operator a_{kj} gives

$$a_{kj}(t) = a_{kj}(0)\exp(-i\omega_k t) \tag{3.39}$$

Employing the above equation and its adjoint equation, and inserting the expression of $\mathbf{A}(\mathbf{r},t)$ into (3.2), we obtain the electric field operator and magnetic field operator to be

$$\begin{aligned}
\mathbf{E}(\mathbf{r},t) &= i\sum_{kj}\left(\frac{\hbar\omega_k}{2\varepsilon_0 V}\right)^{1/2}\mathbf{e}_{kj}[a_{kj}(t)\exp(i\mathbf{k}\cdot\mathbf{r}) - a^{\dagger}_{kj}(t)\exp(-i\mathbf{k}\cdot\mathbf{r})] \\
\mathbf{B}(\mathbf{r},t) &= i\sum_{kj}\left(\frac{\hbar}{2\omega_k\varepsilon_0 V}\right)^{1/2}(\mathbf{k}\times\mathbf{e}_{kj})[a_{kj}(t)\exp(i\mathbf{k}\cdot\mathbf{r}) \\
&\quad - a^{\dagger}_{kj}(t)\exp(-i\mathbf{k}\cdot\mathbf{r})]
\end{aligned} \tag{3.40}$$

As we know, the eigenvector of the Hamiltonian H_F in equation (3.37) is $|n_{k_1j_1}, n_{k_2j_2}, \ldots\rangle$ or $|\{n_{kj}\}\rangle$, here n_{kj} must be positive number. This means that the eigenvector of H_F makes up of the $n_{k_1j_1}$ bosons with the wave vector $\mathbf{k}_1$ and the polarization vector $\mathbf{e}_{k_1j_1}$, the $n_{k_2j_2}$ bosons with $\mathbf{k}_2$ and $\mathbf{e}_{k_2j_2}$ and so on. These bosons are exactly the photons.

From the commutation relation we may discuss the properties of the operators a_{kj} and $a^{\dagger}_{kj}$ of the harmonic oscillators. They obey

$$\begin{aligned}
a_{k_\iota j_\iota}|\{n_{kj}\}\rangle &= \sqrt{n_{k_\iota j_\iota}}|n_{k_1j_1}, n_{k_2j_2}, \cdots, n_{k_\iota j_\iota} - 1, \cdots\rangle \\
a^{\dagger}_{k_\iota j_\iota}|\{n_{kj}\}\rangle &= \sqrt{n_{k_\iota j_\iota} + 1}|n_{k_1j_1}, n_{k_2j_2}, \cdots, n_{k_\iota, j_\iota} + 1, \cdots\rangle \\
a^{\dagger}_{k_\iota j_\iota}a_{k_\iota j_\iota}|\{n_{kj}\}\rangle &= n_{k_\iota j_\iota}|\{n_{kj}\}\rangle \quad (j = 1, 2)
\end{aligned} \tag{3.41}$$

The above equations show that the function of the operator $a_{k_\iota j_\iota}$ on the eigenstate of H_F is to annihilate a photon with the mode $k_\ell j_\ell$, and the function of $a^{\dagger}_{k_\iota j_\iota}$ is to create a photon with the same mode. So we interpret a_{kj} and

$a_{kj}^\dagger$ (a and $a^\dagger$) as annihilation and creation operators respectively, and call the multiply operator $a_{kj}^\dagger a_{kj}$ ($a^\dagger a$) as photon number operator.

So far the radiation field is quantized, therefore we can use the annihilation operator a_{kj} and creation operator $a_{kj}^\dagger$ to describe the light field. In the following we will discuss the properties of the momentum and the spin of photons.

3.2.2 Momentum and spin of the photon

By substituting eqs.(3.31) and (3.32) into (3.10), it is found that the quantized formulation of the field momentum gives

$$\mathbf{P} = \sum_{kj} \frac{\hbar \mathbf{k}}{2}(a_{kj}^\dagger a_{kj} + a_{kj} a_{kj}^\dagger) = \sum_k \hbar \mathbf{k} \left(a_{kj}^\dagger a_{kj} + \frac{1}{2} \right)$$

The varying range of k is $(-\infty, \infty)$, so that $\sum\limits_{k=-\infty}^{\infty} \hbar \mathbf{k}/2 = 0$. Then the above equation reduces to

$$\mathbf{P} = \sum_{kj} \hbar \mathbf{k} a_{kj}^\dagger a_{kj} \tag{3.42}$$

If the field is in the mode $\mathbf{k}j$ with n_{kj} photons, then from (3.42) the momentum of the field is $n_{kj}\hbar\mathbf{k}$. When $n_{kj} = 1$, which means that there is only one photon in this mode, the momentum of the field is $\hbar\mathbf{k}$. So we see that each photon with wave vector $\mathbf{k}$ has momentum $\hbar\mathbf{k}$.

Similarly, inserting eqs.(3.31) and (3.32) into (3.11), we can obtain the quantized expression of the angular momentum. Since there exists

$$[\mathbf{\Pi} \times (\nabla \times \mathbf{A})]_k = \sum_{\ell,m} \varepsilon_{k\ell m} \Pi_\ell (\nabla \times \mathbf{A})_m = \sum_{\ell,m,i,j} \varepsilon_{k\ell m} \varepsilon_{mij} \Pi_\ell \nabla_i A_j$$

by means of the relation $\sum\limits_m \varepsilon_{k\ell m} \varepsilon_{mij} = \delta_{ki}\delta_{\ell j} - \delta_{kj}\delta_{\ell i}$, we have

$$[\mathbf{\Pi} \times (\nabla \times \mathbf{A})]_k = \sum_\ell [\Pi_\ell \nabla_k A_\ell - \nabla_\ell (\Pi_\ell A_k)]$$

So that the i-th component of the angular momentum density reduces to

$$\{\mathbf{r} \times [\mathbf{\Pi} \times (\nabla \times \mathbf{A})]\}_i = \sum_{jk} \varepsilon_{ijk} r_j [\mathbf{\Pi} \times (\nabla \times \mathbf{A})]_k = \sum_{\ell jk} \varepsilon_{ijk} r_j [\Pi_\ell \nabla_k A_\ell$$

$$-\nabla_\ell(\Pi_\ell A_k)] = \sum_\ell \Pi_\ell (\mathbf{r} \times \mathbf{A})_i A_\ell - \sum_{\ell jk} \nabla_\ell (\Pi_\ell \varepsilon_{ijk} r_j A_k) + \sum_{\ell k} \varepsilon_{i\ell k} \Pi_\ell A_k$$

Replacing the above equation into (3.11), the second term is modified as an surface integration whose integrating value is equal to zero. So the angular momentum of the electromagnetic field may be rewritten as

$$\mathbf{J}_F = -\frac{1}{c}\sum_{\ell=1}^{3}\int d^3\mathbf{r}\Pi_\ell(\mathbf{r}\times\nabla)A_\ell - \frac{1}{c}\int d^3\mathbf{r}\mathbf{\Pi}\times\mathbf{A} \tag{3.43}$$

It is evident that $\mathbf{J}_F$ can be decomposed into two terms. The first term in (3.43) depends on the point of origin, which is called the orbit angular momentum. The second term is referred to the intrinsic part of the angular momentum, namely the spin angular momentum. Now we discuss the properties of the spin angular momentum of the field with frequency ω_0 and wave vector $\mathbf{k}_0$.

Considering the two modes with the polarization directions $\mathbf{e}_1$ and $\mathbf{e}_2$ belonging to the wave vector $\mathbf{k}_0$ may be chosen to satisfy

$$\mathbf{e}_1\times\mathbf{e}_2 = \mathbf{k}_0/k_0, \quad \mathbf{e}_1\cdot\mathbf{e}_2 = 0, \quad \mathbf{e}_1\cdot\mathbf{k}_0 = \mathbf{e}_2\cdot\mathbf{k}_0 = 0 \tag{3.44}$$

Therefore, from (3.31) and (3.32) we find that the vector potential $\mathbf{A}(\mathbf{r})$ and $\mathbf{\Pi}(\mathbf{r})$ in the Schrödinger picture can be taken in the form

$$\mathbf{A}(\mathbf{r}) = \left(\frac{\hbar}{2\varepsilon_0 V\omega_0}\right)^{1/2}\{\mathbf{e}_1 a_1\exp(i\mathbf{k}_0\cdot\mathbf{r}) + \mathbf{e}_1 a_1^\dagger\exp(-i\mathbf{k}_0\cdot\mathbf{r}) + \mathbf{e}_2 a_2\exp(i\mathbf{k}_0\cdot\mathbf{r}) + \mathbf{e}_2 a_2^\dagger\exp(-i\mathbf{k}_0\cdot\mathbf{r})\} \tag{3.45}$$

$$\mathbf{\Pi}(\mathbf{r}) = -i\left(\frac{\hbar\omega_0\varepsilon_0}{2V}\right)^{1/2}\{\mathbf{e}_1 a_1\exp(i\mathbf{k}_0\cdot\mathbf{r}) - \mathbf{e}_1 a_1^\dagger\exp(-i\mathbf{k}_0\cdot\mathbf{r}) + \mathbf{e}_2 a_2\exp(i\mathbf{k}_0\cdot\mathbf{r}) - \mathbf{e}_2 a_2^\dagger\exp(-i\mathbf{k}_0\cdot\mathbf{r})\} \tag{3.46}$$

Inserting the above equation into the second term in (3.43), we have

$$\mathbf{J}_S = -\frac{1}{c}\int d^3\mathbf{r}\mathbf{\Pi}\times\mathbf{A} = i\hbar\frac{\mathbf{k}_0}{k_0}(a_2^\dagger a_1 - a_1^\dagger a_2) \tag{3.47}$$

This is just the spin angular momentum operator of the light. It shows that the spin angular momentum is parallel or antiparallel to the direction of the wave vector $\mathbf{k}_0$ of the field. Suppose we have only one photon with wave vector $\mathbf{k}_0$. A general one-photon state at a given $\mathbf{k}_0$ can be represented as a linear combination of the kets $|1,0\rangle$ and $|0,1\rangle$. Here the ket $|1,0\rangle$ means that the state has one photon of the polarization direction $\mathbf{e}_1$, and the ket $|0,1\rangle$ means that of one photon with the polarization direction $\mathbf{e}_2$. Evidently, the spin angular momentum operator is

$$S = i\hbar(a_2^\dagger a_1 - a_1^\dagger a_2) \tag{3.48}$$

it satisfies the following relations

$$\begin{aligned} \langle 1,0|S|1,0\rangle &= 0, \qquad \langle 0,1|S|0,1\rangle = 0, \\ \langle 1,0|S|0,1\rangle &= -i\hbar, \quad \langle 0,1|S|1,0\rangle = i\hbar \end{aligned} \tag{3.49}$$

Hence the matrix representation of S in the Hilbert space spanned by the kets $|1,0\rangle$ and $|0,1\rangle$ is

$$S = \begin{pmatrix} 0 & -i\hbar \\ i\hbar & 0 \end{pmatrix} \tag{3.50}$$

Solving its eigenvalue equation

$$\begin{pmatrix} 0 & -i\hbar \\ i\hbar & 0 \end{pmatrix} \begin{pmatrix} \alpha \\ \beta \end{pmatrix} = \lambda \begin{pmatrix} \alpha \\ \beta \end{pmatrix} \tag{3.51}$$

we obtain the eigenvalues and the corresponding eigenkets as

$$\begin{aligned} \lambda_+ &= i\hbar, \quad |1_+\rangle = \frac{1}{\sqrt{2}}(|1,0\rangle + i|0,1\rangle) \\ \lambda_- &= -i\hbar, \quad |1_-\rangle = \frac{1}{\sqrt{2}}(|1,0\rangle - i|0,1\rangle) \end{aligned} \tag{3.52}$$

The above equations show that each photon in one of the eigenkets has a fixed value of the spin angular momentum along its direction of propagation. If the value of the spin angular momentum is $\hbar$, which means the photon is in state $|1_+\rangle$, we speak of it as a left-circular polarization photon; if this value is $-\hbar$, which means that the photon is in state $|1_-\rangle$, we speak of it as a right-circular polarization photon. Eq.(3.52) also show that the quantum number λ of spin angular momentum equals to 1, here the photon has boson's characteristic. However, according to the theory of angular momentum, the value of the z-component of spin angular momentum may be $\hbar$, 0, $-\hbar$, but (3.52) shows that there does not exist the photon whose eigenvalue of the z-component of spin angular momentum is zero. The reason is that the electromagnetic field we discuss here obeys the Coulomb gauge, which leads to the field has only two independent polarization directions $\mathbf{e}_1$ and $\mathbf{e}_2$ perpendicular to the wave vector

$\mathbf{k}_0$, therefore, there is no photon with eigenvalue zero of S_z. Nevertheless, the photon may be in the superposition of the eigenstates $|1_+\rangle$ and $|1_-\rangle$

$$\frac{1+i}{2}|1_+\rangle + \frac{1-i}{2}|1_-\rangle = \frac{1}{\sqrt{2}}(|1,0\rangle + |0,1\rangle)$$

or

$$\frac{1-i}{2}|1_+\rangle + \frac{1+i}{2}|1_-\rangle = \frac{1}{\sqrt{2}}(|1,0\rangle - |0,1\rangle)$$

the expectation value of the spin angular momentum in this superposition states gives $\langle S\rangle$=0. We call the photon in $\frac{1+i}{2}|1_+\rangle + \frac{1-i}{2}|1_-\rangle$ or $\frac{1-i}{2}|1_+\rangle - \frac{1+i}{2}|1_-\rangle$ as linearly polarized photon.

3.3 State functions describing the light field

In order to exhibit the quantum characteristics of light, we may use different state functions. There are a number of state functions describing the light field. For example, the eigenstate of the photon number operator $N = a^\dagger a$, namely, the photon-number state $|n\rangle$; the eigenstate of the annihilation operator a, i.e., the coherent state $|\alpha\rangle$; the eigenstate of the phase operator Ψ, i.e., the phase state $|\theta\rangle$, and the squeezed state related to the squeezed operator $S(\xi) = \exp[\frac{1}{2}(\xi a^{\dagger 2} - \xi^* a^2)]$, and so on. However, the basic state is the number state $|n\rangle$, because the coherent state , the phase state and the squeezed state can be expressed by some linear combinations of the number states $\{|n\rangle\}$. We call these states as pure states. A more radical generalization of the pure states is required to treat statistical-mixture state of the radiation field. For the statistical-mixture state there is no precise specification of the state of the field, but only of the classical probabilities that the field is in some states $\{|n\rangle\}$, therefore, statistical-mixture state can not be expressed definitely by states $\{|n\rangle\}$. This section is devoted to discuss the number state, the coherent state and the phase state, and a typical statistical-mixture state, namely chaotic state. As for the squeezed state, it will be discussed in detail in Chapter 7.

3.3.1 Photon-number states

First we consider a single-mode field with frequency ω. According to eq.(3.37) the Hamiltonian of the single-mode field gives

$$H = \hbar\omega\left(a^\dagger a + \frac{1}{2}\right) \tag{3.53}$$

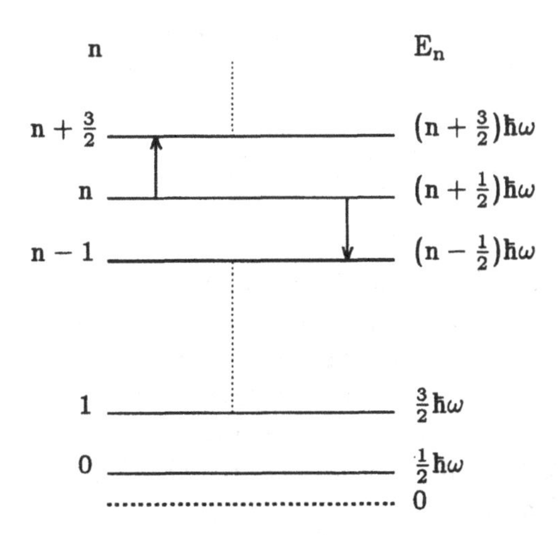

Figure 3.1: Eigenenergy spectrum a of single-mode field and the transition processes for creating and annihilating photons

and the eigenvalue equation of the photon-number operator $N = a^{\dagger}a$ is

$$N|n\rangle = n|n\rangle \tag{3.54}$$

From the commutation relation $[a, a^{\dagger}] = 1$, we know that the eigenvalue n is positive number, i.e., n=0,1,2,$\cdots$. According to the eigenvalue equation

$$H|n\rangle = E_n|n\rangle \tag{3.55}$$

we obtain the energy eigenvalue of the single-mode field

$$E_n = \hbar\omega\left(n + \frac{1}{2}\right) \tag{3.56}$$

The diagram of the energy-level is illustrated in Fig.(3.1). When the energy of the field is fixed by (3.56), the system is in the n-th excited state. If n=0, which means the system is in the vacuum state (ground state), but there exists the zero-point energy $\hbar\omega/2$. In the view of experiments, the observation results are the relatively excited degree above the ground state, so the zero-point energy has no practical meaning. For a state $|n\rangle$ with energy E_n, it has n photons and attachs the zero-point energy $\hbar\omega/2$, here each photon has energy $\hbar\omega$. In this sense, the radiation field is said to be in the excited state $|n\rangle$ with n photons. The operators a and $a^\dagger$ satisfy the equations

$$\begin{aligned} a|n\rangle &= \sqrt{n}|n-1\rangle \\ a^\dagger|n\rangle &= \sqrt{n+1}|n+1\rangle \end{aligned} \tag{3.57}$$

These show that the function of the operator a is to transfer the system from the excited state $|n\rangle$ with n photons into the state $|n-1\rangle$ with n-1 photons, in this sense the operator a is called annihilation operator for one photon; in verse, the function of the operator $a^\dagger$ transfers the system from the state $|n\rangle$ into $|n+1\rangle$, so $a^\dagger$ is called the creation operator. Thus there is a possibility to reveal the particle property of field clearly when we adopt the number states to describe a light field.

However, can the number state $|n\rangle$ exhibit the wave property of the field ? Now we discuss the expectation value $\langle n|E(\mathbf{r},t)|n\rangle$ of the electric field operator, which corresponds to the electric strength representing the wave property in the classical electrodynamics. It is found from (3.40) that the electric field operator for a single-mode radiation field is

$$E(\mathbf{r},t) = i\left(\frac{\hbar\omega}{2\varepsilon_0 V}\right)^{1/2}\{a\exp[i(\mathbf{k}\cdot\mathbf{r}-\omega t)] - a^\dagger\exp[-i(\mathbf{k}\cdot\mathbf{r}-\omega t)]\} \tag{3.58}$$

By use of (3.57) we obtain the expectation value of $\mathbf{E}(\mathbf{r},t)$ in the state $|n\rangle$ as

$$\langle n|E(\mathbf{r},t)|n\rangle = 0 \tag{3.59}$$

Since the mean value of the field strength vanishes for all $\mathbf{r}$ and t, a state with fixed photon number is not appropriate for describing wavelike phenomenon. For the field described by the number state $|n\rangle$, it is easy to verify that the uncertainty of photon number is zero. Because the fluctuations of photon number is

$$(\Delta n)^2 = \langle n|(N-n)^2|n\rangle = 0 \tag{3.60}$$

so that

$$\Delta n = 0$$

It means that the fluctuations of photon number in the number state $|n\rangle$ are equal to zero, and the system has certain photon nùmber which corresponds to the intensity of light.

Since the set of eigenstates $\{|n\rangle\}$ of the Hermitian operator N satisfies the closure relation

$$\sum_n |n\rangle\langle n| = I$$

an arbitrary state of the field can be expressed in the number state representation as

$$|\Psi\rangle = \sum_n C_n |n\rangle \tag{3.61}$$

where $|C_n|^2 = |\langle n|\Psi\rangle|^2$ represents the probability of the field in state $|n\rangle$.

In the above we have introduced the number state $|n\rangle$ which is the eigenstate of the Hamiltonian for a single-mode field. Now we may generalize it to the case of multimode field. For the multimode field described by (3.37), the eigenstate of the system may have n_1 photons in the first mode, n_2 photons in the second mode, n_ℓ photons in the ℓ-th mode and so on. So the eigenstate may be written as the direct product

$$|n_1\rangle|n_2\rangle\cdots|n_\ell\rangle\cdots$$

or

$$|n_1, n_2, \cdots, n_\ell, \cdots\rangle = |\{n_\ell\}\rangle \tag{3.62}$$

Clearly the state $|\{n_\ell\}\rangle$ is the number state of the multimode field. It must be mentioned that there is no coupling among the modes in (3.37), which means that the operator a_ℓ $(a_\ell^\dagger)$ in the Hamiltonian only affects the number state $|n_\ell\rangle$ belonging to the ℓ-th mode. For example,

$$a_\ell|\{n_\ell\}\rangle = \sqrt{n_\ell}|n_1, n_2, \cdots, n_\ell - 1, \cdots\rangle$$

In general, an arbitrary state of the multimode field can be expanded by the complete set of eigenstates $|\{n_\ell\}\rangle$ as

$$\begin{aligned} |\Psi\rangle &= \sum_{n_1}\sum_{n_2}\cdots\sum_{n_\ell}\cdots C_{n_1,n_2,\cdots,n_\ell,\cdots}|n_1, n_2, \cdots, n_\ell, \cdots\rangle \\ &= \sum_{\{n_\ell\}} C_{\{n_\ell\}}|\{n_\ell\}\rangle \end{aligned} \tag{3.63}$$

3.3.2 The coherent states of light

As we know, we usually adopt the electromagnetic wave to describe light field in classical electrodynamics. A single-mode classical standing wave may be written as

$$E(x,t) = q(t)\sin kx \tag{3.64}$$

with the wave amplitude

$$q(t) = A\sin(\omega t + \phi)$$

In the classical theory, the field has definite phase $\omega t + \phi$ and intensity $I(\propto A^2)$ simultaneously. But in quantum optics, the photon number n corresponding to the intensity I belongs to the particle picture and the phase belongs to the wave concepts, both of them can not be determined simultaneously. For example in a single-mode field described by the number state $|n\rangle$, the photon number fixed value, but the phase is not fixed. It is of interest to examine another state describing the quantized radiation field, this state has approximately fixed phase without fixed value of photon number. Such state will be denoted by $|\alpha\rangle$, and called as the coherent state of the radiation field. Coherent state is the eigenstate of the annihilation operator a, i.e.,

$$a|\alpha\rangle = \alpha|\alpha\rangle \tag{3.65}$$

Since the operator a is non-Hermitian, $\alpha = |\alpha|\exp(i\theta)$ is a complex.

A. Expression of the coherent state $|\alpha\rangle$

In order to obtain the expression of the coherent state $|\alpha\rangle$, we must solve the eigenvalue equation (3.65). By means of the complete set of the eigenstates $\{|n\rangle\}$, $|\alpha\rangle$ may be written as

$$|\alpha\rangle = \sum_n |n\rangle\langle n|\alpha\rangle = \sum_n C_n(\alpha)|n\rangle \tag{3.66}$$

where

$$C_n(\alpha) = \langle n|\alpha\rangle \tag{3.67}$$

is the transformation between the number state representation and the coherent state representation. $|C_n(\alpha)|^2$ corresponds to the probability finding n photons with frequency ω in the state $|\alpha\rangle$. Inserting (3.66) into (3.65), we have

$$a|\alpha\rangle = \sum_{n=1}^{\infty} C_n(\alpha)\sqrt{n}|n-1\rangle = \sum_{n=0}^{\infty} \alpha C_n(\alpha)|n\rangle \tag{3.68}$$

The first sum runs from 1 to ∞ since n=0 term gives zero, we may therefore let n be replaced by n+1 in the first summation, so (3.68) becomes

$$\sum_{n=0}^{\infty} C_{n+1}(\alpha)\sqrt{n+1}|n\rangle = \sum_{n=0}^{\infty} \alpha C_n(\alpha)|n\rangle \tag{3.69}$$

If we multiply both sides from the left by $\langle m|$ and notice $\langle m|n\rangle = \delta_{mn}$, then obtain the recurrence relation

$$C_{n+1}(\alpha)\sqrt{n+1} = \alpha C_n(\alpha) \tag{3.70}$$

or

$$\begin{aligned} &C_1 = \alpha C_0/\sqrt{1} \\ &C_2 = \alpha C_1/\sqrt{2} = \alpha^2 C_0/\sqrt{2!} \\ &C_3 = \alpha^3 C_0/\sqrt{3!} \\ &\quad\vdots \end{aligned}$$

so that

$$C_n(\alpha) = \frac{\alpha^n}{\sqrt{n!}} C_0 \tag{3.71}$$

Therefore, we have from (3.71) and (3.66)

$$|\alpha\rangle = C_0 \sum_{n=0}^{\infty} \frac{\alpha^n}{\sqrt{n!}}|n\rangle \tag{3.72}$$

Considering the normalization condition

$$\begin{aligned} \langle\alpha|\alpha\rangle = 1 = |C_0|^2 \sum_{n=0}^{\infty}\sum_{m=0}^{\infty} \frac{\alpha^{*m}\alpha^n}{\sqrt{m!n!}}\langle m|n\rangle \\ = |C_0|^2 \sum_{n=0}^{\infty} \frac{|\alpha|^{2n}}{n!} = |C_0|^2 \exp(|\alpha|^2) \end{aligned}$$

we obtain

$$C_0 = \exp(-|\alpha|^2/2) \tag{3.73}$$

Consequently, the coherent state is given by

$$|\alpha\rangle = \exp(-|\alpha|^2/2) \sum_{n=0}^{\infty} \frac{\alpha^n}{\sqrt{n!}} |n\rangle \tag{3.74}$$

Noticing that

$$|n\rangle = \frac{a^{\dagger n}}{\sqrt{n!}} |0\rangle \tag{3.75}$$

then the coherent state $|\alpha\rangle$ can be rewritten as

$$\begin{aligned} |\alpha\rangle &= \exp(-|\alpha|^2/2) \sum_{n=0}^{\infty} \frac{(\alpha a^\dagger)^n}{n!} |0\rangle \\ &= \exp(-|\alpha|^2/2) \exp(\alpha a^\dagger) |0\rangle \end{aligned} \tag{3.76}$$

$|\alpha\rangle$ may be generated from $|0\rangle$ by the displacement operator $D(\alpha)$

$$\begin{aligned} |\alpha\rangle &= D(\alpha)|0\rangle \\ D(\alpha) &= \exp(\alpha a^\dagger - \alpha^* a) \end{aligned} \tag{3.77}$$

The reason is as follows. By means of the relation

$$\exp(A + B) = e^A e^B \exp\left(-\frac{1}{2}[A, B]\right) \tag{3.78}$$

we have

$$\exp(\alpha a^\dagger - \alpha^* a) = \exp(-|\alpha|^2/2) \exp(\alpha a^\dagger) \exp(-\alpha^* a) \tag{3.79}$$

so that

$$\exp(\alpha a^\dagger - \alpha^* a)|0\rangle = \exp(-|\alpha|^2/2) \exp(\alpha a^\dagger) \exp(-\alpha^* a)|0\rangle \tag{3.80}$$

Considering the identity

$$\exp(-\alpha^* a)|0\rangle = |0\rangle - \alpha^* a|0\rangle + \cdots = |0\rangle \tag{3.81}$$

Eq.(3.80) is reduced to

$$\exp(\alpha a^+ - \alpha^* a)|0\rangle = \exp(-|\alpha|^2/2) \exp(\alpha a^\dagger)|0\rangle = |\alpha\rangle \tag{3.82}$$

So that (3.76) and (3.77) are equivalent.

B. Properties of the coherent states

First we demonstrate that the coherent states are not orthogonal. By use of (3.74) and its adjoint, we have

$$\begin{aligned}\langle\beta|\alpha\rangle &= \exp\left[-\frac{1}{2}(|\alpha|^2+|\beta|^2)\right]\sum_{n=0}^{\infty}\sum_{m=0}^{\infty}\frac{\beta^{*n}\alpha^m}{\sqrt{n!m!}}\langle n|m\rangle \\ &= \exp\left[-\frac{1}{2}(|\alpha|^2+|\beta|^2)\right]\sum_{n=0}^{\infty}\frac{(\alpha\beta^*)^n}{n!} \\ &= \exp\left[-\frac{1}{2}(|\alpha|^2+|\beta|^2)+\alpha\beta^*\right] \end{aligned} \tag{3.83}$$

Evidently, when $\alpha = \beta$, there exists

$$\langle\alpha|\alpha\rangle = 1$$

But for $\alpha \neq \beta$, we obtain from (3.83)

$$|\langle\beta|\alpha\rangle|^2 = \exp[-(|\alpha|^2+|\beta|^2)+\alpha\beta^*+\alpha^*\beta] = \exp[-|\alpha-\beta|^2] \neq 0 \tag{3.84}$$

This means that the coherent states with different eigenvalues are not orthogonal. However, for large $|\alpha-\beta| \gg 1$, $|\langle\beta|\alpha\rangle|^2 \to 0$. So they become approximately orthogonal as $|\alpha-\beta|^2$ increases.

Although the coherent states are not orthogonal, they do form a complete set of states. Because α is complex, the closure relation may be written as

$$\frac{1}{\pi}\int|\alpha\rangle\langle\alpha|d^2\alpha = I \tag{3.85}$$

where I is the identify operator. The integration is over the entire complex plane. If we let $\alpha = x+iy = re^{i\theta}$, then $d^2\alpha = dxdy = rdrd\theta$. In order to verify (3.85), we may use (3.74) and its adjoint in the left side in (3.85). Thus (3.85) gives

$$\frac{1}{\pi}\int|\alpha\rangle\langle\alpha|d^2\alpha = \frac{1}{\pi}\sum_{n=0}^{\infty}\sum_{m=0}^{\infty}|n\rangle\langle m|\frac{1}{\sqrt{n!m!}}\int\exp(-|\alpha|^2)\alpha^{*m}\alpha^n d^2\alpha$$

Transferring to polar coordinations, it becomes

$$\frac{1}{\pi}\int |\alpha\rangle\langle\alpha| d^2\alpha = \sum_{n=0}^{\infty}\sum_{m=0}^{\infty} |n\rangle\langle m| \frac{1}{\pi\sqrt{n!m!}}\int e^{-r^2} r^{m+n+1} dr$$

$$\times \int d\theta \exp[i(n-m)\theta] = \sum_{n=0}^{\infty} |n\rangle\langle n| \frac{1}{n!}\int d\xi \exp(-\xi)\xi^n$$

where we let $\xi = r^2$. Because the integral in the above equation equals to $n!$, so we obtain

$$\frac{1}{\pi}\int |\alpha\rangle\langle\alpha| d^2\alpha = \sum_{n=0}^{\infty} |n\rangle\langle n| = I$$

We next show that the coherent state is a minimum uncertainty state for the amplitude of light field, so that it is a perfect certainty state for describing the wavelike property of the field. In quantum mechanics, we know that there exists the following relations among the coordinate operator q, the momentum operator p and the annihilation operator a, the creation operator $a^\dagger$ of the harmonic oscillator

$$q = \sqrt{\frac{\hbar}{2\omega}}(a + a^\dagger) \tag{3.86}$$

$$p = -i\sqrt{\frac{\hbar\omega}{2}}(a - a^\dagger) \tag{3.87}$$

The operators q and p satisfy the commutation relation

$$[q, p] = i \tag{3.88}$$

So that the Hamiltonian (3.53) of a single-mode field is identified as that of a harmonic oscillator with mass 1 and frequency ω

$$H = \frac{1}{2}(p^2 + \omega^2 q^2) \tag{3.89}$$

Thus the operators q and p are the canonical momentum and coordinate operators describing the field in the coordinate representation.

From the relations among a, $a^\dagger$ and q, p we see that the expectation values of p, q, p^2 and q^2 in state $|\alpha\rangle$ are

$$\langle q\rangle = \sqrt{\frac{\hbar}{2\omega}}\langle\alpha|a + a^\dagger|\alpha\rangle = \sqrt{\frac{\hbar}{2\omega}}(\alpha + \alpha^*)$$

$$\begin{aligned}
\langle p\rangle &= i\sqrt{\frac{\hbar\omega}{2}}\langle\alpha|a^\dagger - a)|\alpha\rangle = i\sqrt{\frac{\hbar\omega}{2}}(\alpha^* - \alpha) \\
\langle q^2\rangle &= \frac{\hbar}{2\omega}\langle\alpha|a^2 + a^{\dagger 2} + aa^\dagger + a^\dagger a|\alpha\rangle \\
&= \frac{\hbar}{2\omega}(\alpha^2 + \alpha^{*2} + 2\alpha\alpha^* + 1) \\
\langle p^2\rangle &= -\frac{\hbar\omega}{2}\langle\alpha|a^2 + a^{\dagger 2} - aa^\dagger - a^\dagger a|\alpha\rangle \\
&= -\frac{\hbar\omega}{2}(\alpha^2 + \alpha^{*2} - 2\alpha\alpha^* - 1)
\end{aligned} \tag{3.90}$$

where we have used $a|\alpha\rangle = \alpha|\alpha\rangle$ and its adjoint. The variances are therefore

$$(\Delta q)^2 = \langle q^2\rangle - \langle q\rangle^2 = \frac{\hbar}{2\omega} \tag{3.91}$$

$$(\Delta p)^2 = \langle p^2\rangle - \langle p\rangle^2 = \frac{\hbar\omega}{2} \tag{3.92}$$

so that

$$(\Delta p)(\Delta q) = \hbar/2 \tag{3.93}$$

which is just the minimum uncertainty value allowed by the Heisenberg uncertainty principle for the momentum and coordinate operators.

In order to realize clearly the physical meaning of the canonical momentum operator p and the canonical coordinate operator q of the field, we introduce two new Hermitian operators X_1 and X_2 as

$$X_1 = \sqrt{\frac{2\omega}{\hbar}}q, \qquad X_2 = \sqrt{\frac{1}{2\hbar\omega}}p \tag{3.94}$$

From (3.86) and (3.87) we have

$$X_1 = \frac{1}{2}(a + a^\dagger), \qquad X_2 = \frac{1}{2i}(a - a^\dagger) \tag{3.95}$$

Substituting eqs.(3.94) and (3.95) into (3.58), we obtain

$$E(\mathbf{r}, t) = -\sqrt{\frac{\hbar\omega}{2\varepsilon_0 V}}[X_1 \sin(\mathbf{k}\cdot\mathbf{r} - \omega t) + X_2 \cos(\mathbf{k}\cdot\mathbf{r} - \omega t)] \tag{3.96}$$

This means that X_1 and X_2 are the amplitude operators of the electric field operator $E(\mathbf{r}, t)$, whose phases are orthogonal. So the operators q and p are proportional to the two amplitude operators of the field, respectively.

From the commutation relation (3.88) between q and p, we know that X_1 and X_2 obey

$$[X_1, X_2] = \frac{i}{2} \tag{3.97}$$

The corresponding uncertainty relation gives

$$(\Delta X_1)^2(\Delta X_2)^2 \geq \frac{1}{16} \tag{3.98}$$

If the field is in the coherent state $|\alpha\rangle$, it is easy to find that

$$(\Delta X_1)^2 = \frac{1}{4}, \quad (\Delta X_2)^2 = \frac{1}{4} \tag{3.99}$$

thus

$$(\Delta X_1)^2(\Delta X_2)^2 = \frac{1}{16} \tag{3.100}$$

The above equations show that the coherent state is the minimum uncertainty state for the Hermitian operators X_1 and X_2, and the variances of X_1 and X_2 equal to each other. That is to say, the fluctuations of the amplitude operators of the field in any coherent states are identical. It is found from (3.76) that the vacuum state ($\alpha = 0$) is a special case of coherent states, so the quantum fluctuations of the amplitude operators of the field result from the fluctuations of the vacuum field. In the process of transformation the vacuum state into coherent state through the displacement operator $D(\alpha)$, the quantum fluctuations of the field keep from changing. The variances of the amplitude operator of the field in α-plane are illustrated in Fig.(3.2).

It must be pointed out that for the field described by a coherent state has large uncertainty in photon number. Because

$$\langle\alpha|N|\alpha\rangle = \exp(-|\alpha|^2)\sum_{n=0}^{\infty}\frac{|\alpha|^{2n}}{n!}n = |\alpha|^2 \tag{3.101}$$

$$\langle\alpha|N^2|\alpha\rangle = \exp(-|\alpha|^2)\sum_{n=0}^{\infty}\frac{n^2|\alpha|^{2n}}{n!} = |\alpha|^4 + |\alpha|^2 \tag{3.102}$$

so the fluctuations of the photon number gives

$$\Delta n = |\alpha| \tag{3.103}$$

We can see that for large $|\alpha|$, the variance of photon number is large, which means that the photon number has large uncertainty for the field in coherent

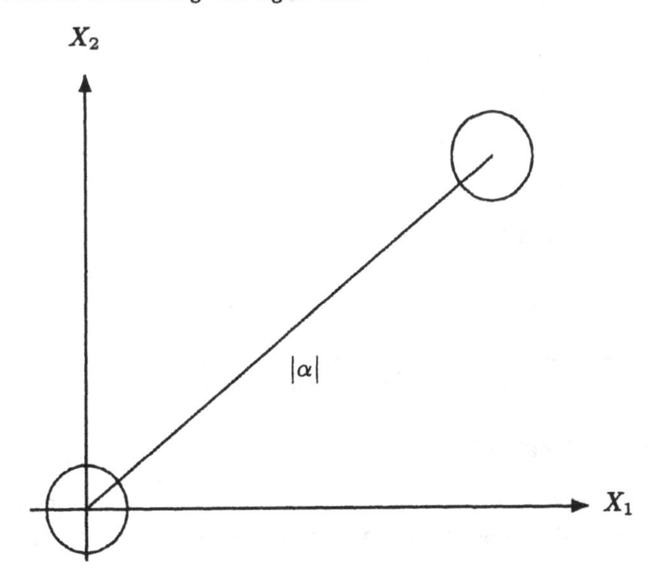

Figure 3.2: Amplitude fluctuations of coherent field in α-plane

state. In addition, (3.101) also shows that $|\alpha|^2$ is the mean photon number of the coherent field, which corresponds to the intensity of light field.

(3.71) and (3.73) show that in a coherent state, the probability of finding the field in number state $|n\rangle$ is

$$P_n = |\langle n|\alpha\rangle|^2 = \exp(-|\alpha|^2)\frac{|\alpha|^{2n}}{n!} \tag{3.104}$$

This is just a Poisson probability distribution.

In analogy to the procedure to form the multimode photon-number state, the multimode coherent state can also be constructed. Since the respective operator a_ℓ of the different modes commute with each other, the multimode coherent state can be defined by the direct product

$$|\{\alpha_\ell\}\rangle = |\alpha_1\rangle|\alpha_2\rangle\cdots|\alpha_\ell\rangle\cdots = |\alpha_1, \alpha_2, \cdots, \alpha_\ell, \cdots\rangle \tag{3.105}$$

which means that the first mode of the field is in the coherent state $|\alpha_1\rangle$, the second in $|\alpha_2\rangle, \cdots$, the ℓ-th in $|\alpha_\ell\rangle, \cdots$. By the aid of (3.74), equation (3.105) can be expanded in the number state representation as

$$|\{\alpha_\ell\}\rangle = \sum_{n_1}\sum_{n_2}\cdots\sum_{n_\ell}\ldots \exp\left[-\frac{1}{2}(|\alpha_1|^2 + |\alpha_2|^2 + \cdots + |\alpha_\ell|^2 + \cdots)\right]$$
$$\times \frac{\alpha_1^{n_1}}{\sqrt{n_1!}}\frac{\alpha_2^{n_2}}{\sqrt{n_2!}}\cdots\frac{\alpha_\ell^{n_\ell}}{\sqrt{n_\ell!}}\cdots|\{n_\ell\}\rangle \tag{3.106}$$

3.3.3 The phase operators and the phase states

In the classical theory of radiation field, it is convenient to write the electric field as a product of a real amplitude and a phase factor

$$E = E_0\{\exp[-i(\omega t - \mathbf{k}\cdot\mathbf{r} - \phi)] + \exp[i(\omega t - \mathbf{k}\cdot\mathbf{r} - \phi)]\} \tag{3.107}$$

where $\mathbf{k}$ is the wave vector, $\mathbf{r}$ the coordinate and ϕ the phase angle. In quantum theory the operator describing a single-mode electric field is written as

$$E = i\left(\frac{\hbar\omega}{2\varepsilon_0 V}\right)^{1/2}\{a\exp[-i(\omega t - \mathbf{k}\cdot\mathbf{r})] + a^\dagger\exp[i(\omega t - \mathbf{k}\cdot\mathbf{r})]\} \tag{3.108}$$

If we separate the field operator into a product of amplitude and phase operators, then (3.108) may be similar to (3.107). Considering the phase operator should have the same significance as the classical phase in an appropriate limit, and it is also an observable quantity, so that it should be associated with Hermitian operators. If we define the phase operator ϕ by the relations

$$\begin{aligned} a &= (N+1)^{1/2}\exp(i\phi) \\ a^\dagger &= \exp(-i\phi)(N+1)^{1/2} \end{aligned} \tag{3.109}$$

where $N = a^\dagger a$. The basic properties of this phase operator can be calculated from the known properties of the creation, annihilation, and number operators. According to (3.109) we know

$$\begin{aligned} \exp(i\phi) &= (N+1)^{-1/2}a \\ \exp(-i\phi) &= a^\dagger(N+1)^{-1/2} \end{aligned} \tag{3.110}$$

It is evident that

$$\begin{aligned} \exp(i\phi)\exp(-i\phi) &= 1 \\ \exp(-i\phi)\exp(i\phi) &= a^\dagger(N+1)^{-1}a \neq 1 \end{aligned} \tag{3.111}$$

Applying (3.111) to the states $|n\rangle$ we obtain

$$\exp(i\phi)|n\rangle = (N+1)^{-1/2}\sqrt{n}|n-1\rangle = \begin{cases} 0, & n=0 \\ |n-1\rangle, & n \neq 0 \end{cases}$$
$$\exp(i\phi)|n\rangle = a^{\dagger}(n+1)^{-1/2}|n\rangle = |n+1\rangle \tag{3.112}$$

The two exponential phase operators thus have non-vanishing matrix elements

$$\langle n-1|\exp(i\phi)|n\rangle = 1$$
$$\langle n+1|\exp(-i\phi)|n\rangle = 1 \tag{3.113}$$

but they do not satisfy the relation

$$\langle i|A|j\rangle = \langle j|A|i\rangle^{*} \tag{3.114}$$

So the exponential phase operators are non-Hermitian ones and they can not be used to represent observable phase properties of the radiation field. However, they can be combined to produce another pair of operators as

$$\cos\phi = \frac{1}{2}[\exp(i\phi) + \exp(-i\phi)]$$
$$\sin\phi = \frac{1}{2i}[\exp(i\phi) - \exp(-i\phi)] \tag{3.115}$$

whose non-vanishing matrix elements are

$$\langle n|\cos\phi|n-1\rangle = \langle n-1|\cos\phi|n\rangle = \frac{1}{2}$$
$$\langle n|\sin\phi|n-1\rangle = -\langle n-1|\sin\phi|n\rangle = \frac{1}{2i} \tag{3.116}$$

Thus the operators $\cos\phi$ and $\sin\phi$ are Hermitian and we can adopt them as the quantum-mechanical operators which represent the observable phase of the radiation field.

By means of (3.109) and (3.115), it is easily verified that

$$[N, \cos\phi] = -i\sin\phi \tag{3.117}$$
$$[N, \sin\phi] = i\cos\phi \tag{3.118}$$

The above commutation relations show that the number and phase operators do not commutate and it is therefore impossible, in principle, to measure them exactly at the same moment. The uncertainties satisfy the Heisenberg uncertainty relations

$$\Delta n \Delta \cos\phi \geq \frac{1}{2}|\langle \sin\phi \rangle| \tag{3.119}$$

$$\Delta n \Delta \sin\phi \geq \frac{1}{2}|\langle \cos\phi \rangle| \tag{3.120}$$

here the fluctuations of $\cos\phi$ and $\sin\phi$ are defined respectively by

$$(\Delta \cos\phi)^2 = \langle \cos^2\phi \rangle - \langle \cos\phi \rangle^2$$

$$(\Delta \sin\phi)^2 = \langle \sin^2\phi \rangle - \langle \sin\phi \rangle^2$$

Since $\cos\phi$ and $\sin\phi$ do not commute with each other

$$[\cos\phi, \sin\phi] = \{a^\dagger(N+1)a - 1\}/(2i) \neq 0 \tag{3.121}$$

it is impossible to set up the states of the radiation field which are the common eigenstates for both operators. However, it is possible to form states which are the common eigenstates of $\cos\phi$ and $\sin\phi$ in a certain limiting sense, these states are called the phase states. They are defined as follows

$$|\phi\rangle = \lim_{s\to\infty}(s+1)^{-1/2}\sum_{n=0}^{s}\exp(in\phi)|n\rangle \tag{3.122}$$

where $s+1$ is the dimension of the Hilbert space spanned by the photon number states $\{|n\rangle\}$, and the limit $s \to \infty$ means that the dimension of the Hilbert space will go to infinite. The normalization and orthogonality of the $|n\rangle$ ensure that $|\phi\rangle$ is normalized

$$\langle \phi|\phi \rangle = 1$$

It is not difficult to obtain

$$\begin{aligned}\cos\phi|\phi\rangle = \cos\phi|\phi\rangle + \frac{1}{2}\lim_{s\to\infty}(s+1)^{-1/2}\{&\exp(is\phi)|s+1\rangle \\ &- \exp[i(s+1)\phi]|s\rangle - \exp(-i\phi)|0\rangle\}\end{aligned} \tag{3.123}$$

In the limit of $s \to \infty$, eq.(3.123) gives

$$\cos\phi|\phi\rangle = \cos\phi|\phi\rangle \tag{3.124}$$

Similarly we have

$$\sin\phi|\phi\rangle = \sin\phi|\phi\rangle \tag{3.125}$$

The diagonal matrix elements of $\cos\phi$ and $\sin\phi$ are easily obtained from (3.122) and (3.124)

$$\begin{aligned}\langle\phi|\cos\phi|\phi\rangle &= \cos\phi[1 - \lim_{s\leftarrow\infty}(s+1)^{-1}] = \cos\phi \\ \langle\phi|\sin\phi|\phi\rangle &= \sin\phi[1 - \lim_{s\leftarrow\infty}(s+1)^{-1}] = \sin\phi\end{aligned} \tag{3.126}$$

These results show that the state $|\phi\rangle$ defined by (3.122) behaves as a common eigenstate of the operators $\cos\phi$ and $\sin\phi$, and ϕ is the observable phase angle.

As an example, we next discuss the phase properties of the coherent field. By use of (3.110) and (3.115), we obtain

$$\begin{aligned}\langle\alpha|\cos\phi|\alpha\rangle &= \langle\alpha|\exp(i\phi) + \exp(-i\phi)|\alpha\rangle \\ &= \frac{1}{2}\exp(-|\alpha|^2/2)\sum_n \frac{\alpha^{*n}\alpha^{m+1} + \alpha^{*n+1}\alpha^n}{[n!(n+1)!]^{1/2}} \\ &= |\alpha|\cos\xi\exp(-|\alpha|^2)\sum_n \frac{|\alpha|^{2n}}{(n+1)!}\sqrt{n+1}\end{aligned} \tag{3.127}$$

with

$$\alpha = |\alpha|\exp(i\xi) \tag{3.128}$$

where ξ is the phase angle of eigenvalue α of the coherent state. So the expectation value of the phase operator $\cos\phi$ is proportional to $cos\xi$.

If we only consider the case of $|\alpha| \gg 1$, then the photon number uncertainty in coherent state satisfies from (3.101) and (3.103)

$$\frac{\Delta n}{\overline{n}} = |\alpha|^{-1} \ll 1 \tag{3.129}$$

So $\sqrt{n+1}$ can be approximated as

$$\begin{aligned}\sqrt{n+1} &= \sqrt{|\alpha|^2 + n + 1 - |\alpha|^2} \approx |\alpha|\sqrt{1 + \frac{n - |\alpha|^2}{|\alpha|^2}} \\ &\approx \left[1 + \frac{n-|\alpha|^2}{2|\alpha|^2} - \frac{1}{8}\frac{(n-|\alpha|^2)^2}{|\alpha|^4}\right]\end{aligned} \tag{3.130}$$

Substituting the above into (3.127), we have

$$\langle\alpha|\cos\phi|\alpha\rangle = \cos\xi e^{-|\alpha|^2}\sum_n\left[1+\frac{n-|\alpha|^2}{2|\alpha|^2}-\frac{1}{8}\frac{(n-|\alpha|^2)^2}{|\alpha|^4}\right]\frac{|\alpha|^{2(n+1)}}{(n+1)!}$$
$$= \cos\xi[1-(8|\alpha|^2)^{-1}] \tag{3.131}$$

similarly,

$$\langle\alpha|\sin\phi|\alpha\rangle = \sin\xi \tag{3.132}$$

$$\langle\alpha|\cos^2\phi|\alpha\rangle = \cos^2\xi - \frac{\cos(2\xi)}{4|\alpha|^2} \tag{3.133}$$

Subsequently the phase uncertainty is therefore

$$(\Delta\cos\phi)^2 = \langle\alpha|\cos^2\phi|\alpha\rangle - (\langle\alpha|\cos\phi|\alpha\rangle)^2 = \frac{1}{4|\alpha|^2}\sin^2\xi \tag{3.134}$$

From eqs.(3.129), (3.132) and (3.134), the product of uncertainties of the photon number operator and the phase operator gives

$$\Delta n\Delta\cos\phi = \frac{1}{2}|\langle\alpha|\sin\phi|\alpha\rangle| \tag{3.135}$$

Thus for a large mean photon number, the coherent state is the minimum uncertainty state for the photon number operator and the phase operator.

It is worthy to mention that the phase operators $\cos\phi$ and $\sin\phi$ do not commutate with each other, so the observable phase must be described by these two Hermitian operators indirectly and the phase operator corresponding to the classical phase ϕ is not presented. In the following we introduce the phase operator and phase states proposed by Pegg and Barnett, which can directly describe the observable phase.

Considering that a phase state can be written as a superposition of number states, a complete orthogonal basis of the phase states may be defined as

$$|\theta_m\rangle = (s+1)^{-1/2}\sum_{m=0}^{s}\exp(in\theta_m)|n\rangle \tag{3.136}$$

here

$$\theta_m = \theta_0 + \frac{2\pi m}{s+1} \qquad (m = 0, 1, 2, \cdots, s) \tag{3.137}$$

where θ_0 is a reference phase, whose value is arbitrary. The new phase state $|\theta_m\rangle$ is defined by the combination of the $(s+1)$ number states. That is to say, $|\theta_m\rangle$ is represented by the $(s+1)$-dimensional Hilbert space spanned by the complete number states $\{|n\rangle\}$, every number state has phase weight factor $\exp(in\theta_m)$. The phase operator is therefore defined in terms of the phase state basis as

$$\Phi_\theta = \sum_{m=0}^{s} \theta_m |\theta_m\rangle\langle\theta_m| \tag{3.138}$$

it is surely Hermitian. From (3.136) and (3.138) we know

$$\Phi_\theta|\theta_m\rangle = \theta_m|\theta_m\rangle \tag{3.139}$$

which means that Φ_θ is a Hermitian phase operator with eigenstates $|\theta_m\rangle$ and the corresponding eigenvalues θ_m. Although the phase eigenstates $|\theta_m\rangle$ are defined in the finite dimensional Hilbert space, s would be tend to infinite in the end. So the set of phase eigenstates $\{|\theta_m\rangle\}$ can also form a complete basis like the set of number states $\{|n\rangle\}$ or $\{|\alpha\rangle\}$. In this sense, an arbitrary state of the field such as coherent state $|\alpha\rangle$ can be expanded in terms of the set of phase eigenstates $\{|\theta_m\rangle\}$.

Now we discuss the commutation relation between the number operator N and the phase operator Φ_θ. According to (3.136) the projection operator $|\theta_m\rangle\langle\theta_m|$ may be expressed as

$$|\theta_m\rangle\langle\theta_m| = (s+1)^{-1}\sum_{n,n'} \exp[i(n'-n)\theta_m]|n'\rangle\langle n| \tag{3.140}$$

Inserting the above into (3.138), the phase operator Φ_θ may be rewritten as

$$\Phi_\theta = \theta_0 + \frac{2\pi}{s+1} + \frac{2\pi}{s+1}\sum_{n\neq n'} \frac{\exp[i(n'-n)\theta_0]|n'\rangle\langle n|}{\exp[i(n'-n)\frac{2\pi}{s+1}]-1} \tag{3.141}$$

Thus we see that the phase operator Φ_θ can be represented by the $(s+1)$ number states $\{|n\rangle\}$. Similarly, the number operator N can also be written as

$$N = \sum_{n=0} n|n\rangle\langle n| \tag{3.142}$$

By means of (3.141) and (3.142), it is found that the commutation relation between Φ_θ and N gives

$$[\Phi_\theta, N] = \frac{2\pi}{s+1}\sum_{n\neq n'} \frac{2\pi}{s+1}\frac{(n-n')\exp[i(n'-n)\theta_0]}{\exp[i(n'-n)\frac{2\pi}{s+1}]-1}|n'\rangle\langle n| \tag{3.143}$$

Clearly

$$\langle n|[\Phi_\theta, N]|n\rangle = 0$$
$$\langle n'|[\Phi_\theta, N]|n\rangle = \frac{2\pi}{s+1}\frac{(n-n')\exp[i(n'-n)\theta_0]}{\exp[i(n'-n)\frac{2\pi}{s+1}]-1} \quad (3.144)$$

The last equation is exact but so complex, and for many purposes we need to simplify the expression (3.144). With this aim in mind we define a physical state as one which can be excited from the vacuum state by a finite interactions of a limited energy source for a finite time . Such a physical state has an upper bound limit $n_{max}\hbar\omega$ of energy when it is expanded in terms of $\{|n\rangle\}$, the corresponding eigenstate is denoted as $|n_{max}\rangle$. In this sense, the probability of the physical system in $|n_{max}\rangle$ tends to zero. So that the state describing a physical system can be written in terms of the number state representation as

$$|\Psi_p\rangle = \sum_n C_n|n\rangle \quad (n = 0, 1, 2, \cdots, n_{max}) \quad (3.145)$$

where the subscript p labels the state of the physical system with finite excitation. For a state of the field we can make s much larger than the number n associated with any significant number state component $|n\rangle$ of the state. In this case ($s \gg n, n'$) the last one of (3.144) reduces to

$$\langle n'|[\Phi_\theta, N]|n\rangle \approx \frac{2\pi}{s+1}\frac{(n-n')\exp[i(n'-n)\theta_0]}{\{\frac{2\pi}{s+1}i(n'-n)+1-1\}}$$

Combining (3.144) and the above equation, then we have

$$\langle n'|[N, \Phi_\theta]|n\rangle \approx i(1-\delta_{n,n'})\exp[i(n'-n)\theta_0] \quad (3.146)$$

This approximation equality becomes exact for all finite n and n' in the limit as s tend to infinity. Therefore, for a physical system the commutation relation between the phase operator Φ_θ and the number operator N satisfies

$$[\Phi_\theta, N]_p = i\sum_{n',n}|n'\rangle\langle n|(1-\delta_{n,n'})\exp[i(n'-n)\theta_0]$$
$$= -i + i\sum_{n'}\exp(in'\theta_0)|n'\rangle\sum_n\exp(-in\theta)\langle n| \quad (3.147)$$

By means of (3.136), the above equation reduces to

$$[\Phi_\theta, N]_p = -i[1-(s+1)|\theta_0\rangle\langle\theta_0|] \quad (3.148)$$

The expectation value of the phase-number commutator in any state $|\Psi\rangle_p$ gives

$$_p\langle\Psi|[N,\Phi_\theta]|\Psi\rangle_p = -i[1-(s+1)|_p\langle\Psi|\theta_0\rangle|^2] \tag{3.149}$$

where $|_p\langle\Psi|\theta_0\rangle|^2$ is the probability that the phase of the state is θ_0. In the continuum limit (as $s\to\infty$) this may be expressed as

$$|_p\langle\Psi|\theta_0\rangle|^2 = 2\pi\frac{P(\theta_0)}{s+1} \tag{3.150}$$

where $P(\theta_0)$ is the probability density and $\frac{s+1}{2\pi}$ is the density of states. With this substitution, the expectation value (3.149) becomes

$$_p\langle\Psi|[N,\Phi_\theta]|\Psi\rangle_p = -i[1-2\pi P(\theta_0)] \tag{3.151}$$

In this case, when we measure the phase and the photon number of a physical system, their uncertainties obey

$$\Delta N\Delta\Phi \geq \frac{1}{2}|1-2\pi P(\theta_0)| \tag{3.152}$$

This uncertainty relation depends on the reference phase θ_0. In the view of physical significance, we can choose an appropriate range so that $P(\theta_0)\ll 1$, then (3.152) reduces to

$$\Delta N\Delta\Phi \geq \frac{1}{2} \tag{3.153}$$

This is exactly the Heisenberg number-phase uncertainty relation.

Now we generalize the phase state of a single-mode field to the case of a multimode field. For a multimode field described by the Hamiltonian (3.37), its phase eigenstates may be written as

$$\begin{aligned} |\{\theta_{m_\ell}\}\rangle &= |\theta_{m_1}\rangle|\theta_{m_2}\rangle\cdots|\theta_{m_\ell}\rangle\cdots \\ &= (s_1+1)^{-1/2}(s_2+1)^{-1/2}\cdots(s_\ell+1)^{-1/2}\cdots\sum_{m_1=0}^{s_1}\sum_{m_2}^{s_2}\cdots\sum_{m_\ell}^{s_\ell}\cdots \\ &\times\exp\{i[n_1\theta_{m_1}+n_2\theta_{m_2}+\cdots+n_\ell\theta_{m_\ell}+\cdots]\}|\{n_\ell\}\rangle \end{aligned} \tag{3.154}$$

where

$$\theta_{m_\ell} = \theta_0 + \frac{2\pi m_\ell}{s_\ell+1} \quad (m_\ell = 0,1,2,\cdots,s_\ell) \tag{3.155}$$

Correspondingly, the phase operator of the ℓ-th mode is defined as

$$\Phi_\theta = \sum_{m_1=0}^{s_1} \sum_{m_2=0}^{s_2} \cdots \sum_{m_\ell}^{s_\ell} \cdots \theta_{m_\ell} |\{\theta_{m_\ell}\}\rangle\langle\{\theta_{m_\ell}\}| \tag{3.156}$$

Next we give an example of a single-mode coherent field propagating in nonlinear medium to discuss how to describe the phase properties of field by means of the phase operator and phase state. The Hamiltonian describing the propagation of a single-mode field through nonlinear medium such as the Kerr medium is

$$H = \hbar\omega a^\dagger a + \frac{1}{2}\hbar\chi a^{\dagger 2} a^2 \tag{3.157}$$

where χ is a parameter related to the nonlinear medium property, which describes the interaction strength between field and medium.

Supposing the field initially in a coherent state $|\alpha\rangle$, the state of the field at time t may be expanded in terms of the number representation as

$$|\Psi(t)\rangle = \sum_n C_n(t)|n\rangle \tag{3.158}$$

Substituting (3.158) into the Schrödinger equation

$$i\hbar\frac{\partial}{\partial t}|\Psi(t)\rangle = H|\Psi(t)\rangle \tag{3.159}$$

we obtain

$$\begin{aligned} i\hbar\frac{\partial}{\partial t}\sum_n C_n(t)|n\rangle &= \hbar(\omega a^\dagger a + \chi a^{\dagger 2}a^2/2)\sum_n C_n(t)|n\rangle \\ &= \hbar\sum_n C_n(t)[n\omega + \chi n(n-1)/2]|n\rangle \end{aligned} \tag{3.160}$$

Multiplying both side from the left by $\langle m|$ and noticing $\langle m|n\rangle = \delta_{m,n}$, then have

$$i\hbar\frac{d}{dt}C_n(t) = [n\omega + \chi n(n-1)/2]C_n(t) \tag{3.161}$$

The solution of the above equation gives

$$C_n(t) = C_n(0)\exp\{-i[n\omega + \chi n(n-1)/2]t\} \tag{3.162}$$

where $C_n(0)$ is undetermined constant. Substituting (3.162) into (3.158), we obtain

$$|\Psi(t)\rangle = \sum_n C_n(0)\exp\{-i[n\omega + \chi n(n-1)/2]t\}|n\rangle \tag{3.163}$$

Noticing the initial condition

$$|\Psi(0)\rangle = |\alpha\rangle = \sum_n \exp(-|\alpha|^2/2)\frac{\alpha^n}{\sqrt{n!}}|n\rangle \tag{3.164}$$

yield directly

$$C_n(0) = \exp(-|\alpha|^2/2)\frac{\alpha^n}{\sqrt{n!}} \tag{3.165}$$

Therefore, the state of the field at time t gives

$$|\Psi(t)\rangle = \exp(-|\alpha|^2/2)\sum_n \frac{\alpha^n}{\sqrt{n!}}\exp\{-i[n\omega + \chi n(n-1)/2]t\}|n\rangle \tag{3.166}$$

Since the phase states can form a complete basis, the state $|\Psi(t)\rangle$ may be rewritten in terms of the complete set of phase states as

$$|\Psi(t)\rangle = \sum_m \langle\theta_m|\Psi(t)\rangle|\theta_m\rangle \tag{3.167}$$

with

$$\begin{aligned}\langle\theta_m|\Psi(t)\rangle &= (s+1)^{-1}\sum_{n,n'}\exp(-in'\theta_m)\exp(-|\alpha|^2/2)\frac{\alpha^n}{\sqrt{n!}}\\ &\quad\times\exp\{-i[n\omega + \chi n(n-1)/2]t\}\langle n'|n\rangle\\ &= (s+1)^{-1}\exp(-\overline{n}/2)\sum_n \frac{\overline{n}^{\frac{n}{2}}}{n!}\\ &\quad\times\exp\{i[n\xi - n\theta_m x - n\omega t - \chi n(n-1)t/2]\}\end{aligned} \tag{3.168}$$

here $\alpha = \sqrt{\overline{n}}\exp(i\xi)$.

According to (3.104), the distribution of the photon number exhibits the Poissonian distribution, so the distribution function P_n of the field reaches to its maximum at $n = \overline{n}$, and for the case $\overline{n} \gg 1$, the photon number variance obeys

$$\Delta n/\overline{n} = \overline{n}^{-1/2} \ll 1 \tag{3.169}$$

That is to say, for $\overline{n} \gg 1$, the photons of the field is mainly located at the sharp range $\overline{n} - \overline{n}^{1/2} \sim \overline{n} + \overline{n}^{1/2}$. So we can expand $\ln P_n$ as a Taylor series in the neighborhood of $n = \overline{n}$ and only retain to the second-order terms of $n - \overline{n}$

$$\ln P_n \approx \ln P_n|_{n=\overline{n}} + \frac{1}{2}\left[\frac{\partial^2}{\partial n^2}\ln P_n\right]_{n=\overline{n}}(n-\overline{n})^2 \tag{3.170}$$

Substituting (3.104) into the above and using the Sterlin formula

$$\ln N! = N - N\ln N \tag{3.171}$$

we have

$$P_n = \exp(-\overline{n})\overline{n}^n/n! \approx (2\pi\overline{n})^{-1/2}\exp[-(n-\overline{n})^2/2\overline{n}] \tag{3.172}$$

The right hand in (3.172) is the typical Gaussian distribution function. Therefore

$$\exp(-\overline{n}/2)\overline{n}^{n/2}/\sqrt{n!} \approx (2\pi\overline{n})^{-1/4}\exp[-(n-\overline{n})^2/2\overline{n}] \tag{3.173}$$

By means of (3.173), (3.168) reduces to

$$\begin{aligned}\langle\theta_m|\Psi(t)\rangle \approx{}& \frac{1}{s+1}\sqrt{\frac{\pi}{\frac{1}{4\overline{n}} + i\chi t/2}}\exp\left\{-\frac{(\chi-\theta_m+\chi t/2-\overline{n}t-\omega t)^2}{4(\frac{1}{4\overline{n}}+i\chi t/2)}\right\}\\ &\times\exp[i\overline{n}(\xi-\omega t-\theta_m+\chi t/2-\overline{n}\chi t/2)](2\pi\overline{n})^{-1/4}\end{aligned} \tag{3.174}$$

In the continuum limit, i.e., $s \to \infty$, the discrete variation θ_m becomes to the continuum variation θ, then the phase probability distribution can be approximated as

$$P(\theta,t) = |\langle\theta|\Psi(t)\rangle|^2 \approx \frac{2\pi}{s+1}(2\pi\sigma^2)^{-1/2}\exp[-(\theta-\overline{\theta})^2/(2\sigma^2)] \tag{3.175}$$

with

$$\overline{\theta} = \xi - \omega t - \chi t(\overline{n} - 1/2)$$
$$\sigma^2 = \frac{1}{4\overline{n}} + \overline{n}\chi^2 t^2$$

Evidently, $P(\theta,t)$ is also a Gaussian distribution function. For $\overline{n} \gg 1$, although the values of θ belong to the range $\theta_0 \to \theta_0 + 2\pi$ when we calculate the integration with respect to $P(\theta,t)$, it is reasonable to expand the integration range to $-\infty \to \infty$, namely

$$\int_{\theta_0}^{\theta_0+2\pi} P(\theta,t)\frac{s+1}{2\pi}d\theta \approx \int_{-\infty}^{\infty} P(\theta,t)\frac{s+1}{2\pi}d\theta = 1$$

where $\frac{s+1}{2\pi}$ is the density of the states.

Using (3.175), we can easily obtain the time evolution of the average value of the phase operator as

$$\langle\Phi\rangle = \int \theta P(\theta, t)\frac{s+1}{2pi}d\theta = \xi - \omega t - \chi t(\overline{n} - 1/2) \tag{3.176}$$

$$\langle\Phi^2\rangle = \int \theta^2 P(\theta, t)\frac{s+1}{2\pi}d\theta = \langle\Phi\rangle^2 + \chi^2 t^2 \overline{n} + \frac{1}{4\overline{n}} \tag{3.177}$$

The phase fluctuations are therefore to be

$$(\Delta\Phi)^2 = \langle\Phi^2\rangle - \langle\Phi\rangle^2 = \frac{1}{4\overline{n}} + \overline{n}\chi^2 t^2 \tag{3.178}$$

If we let $\chi = 0$ in the Hamiltonian (3.157), namely neglecting the interaction between the field and the medium, then (3.176) and (3.178) reduce to

$$\langle\Phi\rangle = \xi - \omega t \tag{3.179}$$

$$(\Delta\Phi)^2 = \frac{1}{4\overline{n}} \tag{3.180}$$

Eq.(3.179) shows that the time evolution of the phase develops with rate ω. For $\overline{n} \gg 1$, the phase variance $(\Delta\Phi)^2 \to 0$, which means that for an intensive coherent field, its phase can be measured exactly. Combining (3.180) with (3.103), the number-phase uncertainty relation gives

$$\Delta N \Delta\Phi = \frac{1}{2} \tag{3.181}$$

This is the minimum uncertainty relation determined by (3.153). So we see that for $\overline{n} \gg 1$, the coherent states are also the minimum number-phase uncertainty states.

However, if we do not neglect the interaction between the field and the medium, i.e., $\chi \neq 0$, we can find from (3.178) that with the time development, the phase variance are enhanced due to the interaction between the field and the medium. From (3.166) we also have

$$P_n(t) = |\langle n|\Psi(t)\rangle|^2 = \exp(-\overline{n})\frac{\overline{n}^n}{n!} \tag{3.182}$$

This equation is in the agreement with (3.104), which means that the Poissonian distribution of the photon-number keeps from time when the field propagates through the medium, so the number variance is also (3.103). However,the phase-number uncertainty relation becomes

$$(\Delta N)^2(\Delta\Phi)^2 = \frac{1}{4} + \bar{n}\chi^2 t^2 > \frac{1}{4} \tag{3.183}$$

which means that the field, which is initially in a coherent state, does not retain its coherence with the time development because of the effect of the field-medium interaction.

3.3.4 Chaotic states of light

By now we have discussed the number states , the coherent states and the phase states. All these states are concerned with pure states of radiation field, that is, these states may be expressed by some definite linear combination of the basic number states $\{|n\rangle\}$. Since the number states $\{|n\rangle\}$, coherent states $\{|\alpha\rangle\}$ and phase states $\{|\theta_m\rangle\}$ have complete or overcomplete property, an arbitrary state can be expressed as the linear superpositions of these states. However, for a chaotic field, there is no specification of the state of the field, but only of the probabilities that the field will be observed to be in a range of possible states, it is impossible to determine the chaotic state of the field as the pure state. To describe this kind of field, the density matrix theory is required.

According to the density matrix theory, the radiation field in a statistical-mixture state can be adequately described by the density operator

$$\rho = \sum_i P_{\psi_i} |\psi_i\rangle\langle\psi_i| \tag{3.184}$$

where the states $|\psi_i\rangle$ are normalized pure photon states and P_{ψ_i} is the probability of the radiation field state $|\psi_i\rangle$, obviously P_{ψ_i} is a classical statistical quantity.

Of course, the density operator of field in the pure state such as number state $|n\rangle$ and coherent state $|\alpha\rangle$ may be regarded as a special case of (3.184). For example, as for a field in one of the number states $|n\rangle$, where n photons are definitely present, the density operator (3.184) becomes

$$\rho = |n\rangle\langle n| \tag{3.185}$$

Similarly, if the field in one of the coherent states $|\alpha\rangle$, the density operator can be constructed as

$$\rho = |\alpha\rangle\langle\alpha| \tag{3.186}$$

For the field in a pure state, the results when we describe it by use of its density operator are the same as those by use of its state function.

The great utility of the density operator is made apparent in its application to treat the statistical-mixture states. As an example, we consider a single-mode thermal field to illustrate how to utilize the density operator to describing chaotic field.

A single-mode thermal field can be produced in a single-mode cavity at temperature T in thermal equilibrium. In this case, there is a Boltzmann distribution of photons emitted by the cavity in number states $\{|n\rangle\}$, that is, the probability of finding the field in one of the number states $|n\rangle$ is

$$\begin{aligned} P_n &= \frac{\exp(-E_n\beta)}{\sum\limits_{n=0}^{\infty}\exp(-E_n\beta)} = \frac{\exp(-n\hbar\omega\beta)}{\sum\limits_{n=0}^{\infty}\exp(-n\hbar\omega)} \\ &= [1-\exp(-\hbar\omega\beta)]\exp(-n\hbar\omega\beta) \end{aligned} \tag{3.187}$$

with

$$\beta = (k_B T)^{-1} \tag{3.188}$$

where k_B is the Boltzmann's constant, ω is the frequency of photons radiated by the single-mode cavity. Substituting (3.187) into (3.184), we obtain the density operator for the single-mode thermal field to be

$$\rho = \sum_n [1-\exp(-\hbar\omega\beta)]\exp(-n\hbar\omega\beta)|n\rangle\langle n| \tag{3.189}$$

According to (1.68) and (3.189), the mean photon number of the field gives

$$\begin{aligned} \langle N\rangle &= \mathrm{Tr}(\rho a^\dagger a) = [1-\exp(-\hbar\omega\beta)]\sum_n n\exp(-n\hbar\omega\beta) \\ &= [\exp(\hbar\omega\beta)-1]^{-1} \end{aligned} \tag{3.190}$$

The above equation shows that the mean photon number is determined by the temperature of the cavity. The probability distribution P_n may also be expressed in terms of $\langle N\rangle$ as

$$P_n = \frac{1}{1+\langle N\rangle}\left(\frac{\langle N\rangle}{1+\langle N\rangle}\right)^n \tag{3.191}$$

Using (3.68) and (3.189), it is easy to find that the variance of the photon number in the single-mode thermal field gives

$$(\Delta N)^2 = \langle N^2\rangle - \langle N\rangle^2 = 2\langle N\rangle^2 - \langle N\rangle^2 = \langle N\rangle^2 \tag{3.192}$$

i.e.,

$$\Delta N = \langle N \rangle \tag{3.193}$$

Comparing (3.193) with (3.103), we find that the variance of thermal field is disagreement with that of coherent field. This means that the quantum fluctuations of the thermal field are different from that of the coherent field.

For the thermal excitations system of all the cavity modes in the thermal equilibrium, the number states $\{|n_k\rangle\}$ of the total modes are formed by the products of number states. Since the different field modes are independent, the combined density operator must be a product of the contributions of the different modes. Thus the density operator may be expressed in general as

$$\rho = \prod_\ell \rho_\ell = \Pi_\ell \sum_{n_\ell} P_{n_\ell} |n_\ell\rangle\langle n_\ell| = \sum_{\{n_\ell\}} P_{\{n_\ell\}} |\{n_\ell\}\rangle\langle\{n_\ell\}| \tag{3.194}$$

with

$$P_{\{n_\ell\}} = \prod_\ell \frac{\langle N_\ell\rangle^{n_\ell}}{(1+\langle N_\ell\rangle)^{n_\ell+1}} \tag{3.195}$$

$$\sum_{\{n_\ell\}} = \sum_{n_1}\sum_{n_2}\cdots \tag{3.196}$$

here $\langle N_\ell\rangle$ is the mean photon number of the ℓ-th mode. And the mean photon number of the multi-mode thermal field gives

$$\langle N\rangle = \sum_\ell \langle N_\ell\rangle \tag{3.197}$$

References

1. W.H.Louisell, *Quantum statistical properties of radiation* (Wiley, New York, 1973).

2. M.Sargent III, M.O.Scully, W.E.Lamb Jr. *Laser Physics* (Addison-Wesley, London, 1974).

3. A.S.Davydov, *Quantum mechanics* (Pergaman Press, London, 1976).

4. R.Loudon, *The quantum theory of light* (Clarendon, Oxford, 1983).

5. P.L.Knight, L.Allen, *Concepts of quantum optics* (Pergaman Press, Oxford, 1983).

6. D.P.Craig, *Molecular quantum electrodynamics- an introduction to radiation-molecular interactions* (Academic, London, 1984).

7. M.Schubert, B.Wilhelmi, *Nonlinear optics and quantum electronics* (Wiley, New York, 1986).

8. C.Cohen-Tannoudji, J.Dupont-Roc, G.Grynberg, *Photons and atoms- introduction to quantum electrodynamics* (Wiley, New York, 1984).

9. Jin-sheng Peng, *Resonance fluorescence and superfluorescence* (Science oress, Beijing, 1993).

10. L.Susskind, J.Glogower, Physics **1**, 49 (1964).

11. P.Garruthers, M.M.Nieto, Rev.Mod.Phys. **40**, 441 (1968).

12. D.T.Pegg, S.M.Barnett, Europhys, Lett. **6**, 483 (1989).

13. S.M.Barnett, D.T.Pegg, J.Mod.Opt. **36**, 7 (1989).

14. D.T.Pegg, S.M.Barnett, Phys.Rev. **A39**, 1665 (1989).

15. D.T.Pegg, S.M.Barnett, Phys.Rev. **A43**, 2579 (1991).

16. C.C.Gerry, Opt.Commun. **75**, 168 (1990).

17. R.Tanas, Ts.Gantsog, Phys.Rev. **A45**, 5031 (1992).

18. Jin-sheng Peng, Gao-xiang Li, Phys.Rev. **A45**, 3289 (1992).

19. Jin-sheng Peng, Gao-xiang Li, Phys.Rev. **A46**, 1516 (1992).

20. Gao-xiang Li, Jin-sheng Peng, ACTA Phys.Sinica **41**, 766 (1992).

21. Gao-xiang Li, Jin-sheng Peng, ACTA Opt. Sinica **13**, 120 (1993).

CHAPTER 4

DICKE HAMILTONIAN AND JAYNES-CUMMINGS MODEL

Investigating the interaction between quantized field and matter and its effects are the fundamental contents in quantum optics. How to describe the interaction between field and matter theoretically ? In this chapter, we introduce the theoretical model to describe the interaction between a single atom (or many atoms) and a single-mode field (or multi-mode field). First we give the Dicke model, which is a typical model to describe an atom interacting with multi-mode radiation field. Then we discuss the spontaneous radiation of an excited atom by means of the Dicke model. Finally, we introduce the Jaynes-Cummings model to focus our attention on the interaction properties between a single-mode field and a two-level atom.

4.1 Dicke Hamiltonian for the system of an atom interacting with the radiation field

Considering the system consisting of an atom (for example, a Hydrogen atom) with a single electron of charge e and mass m in a potential $V(r)$, and a radiation field described by vector potential $\mathbf{A}(\mathbf{r},t)$, the Hamiltonian of this atom-field coupling system is

$$H = \frac{1}{2m}\left(\mathbf{p} - \frac{e\mathbf{A}}{c}\right)^2 + V(r) + H_F = H_A + H_F + H_I \tag{4.1}$$

here $H_A = \frac{\mathbf{p}_2}{2m} + V(r)$ represents the Hamiltonian of the free atom, $\mathbf{p}$ is the electron momentum, $\mathbf{r}$ is the position of the electron relative to the atomic nucleus. The Hamiltonian describing the free field is $H_F = \sum\limits_{kj} \hbar\omega_k a_{kj}^{\dagger} a_{kj}$, and the Hamiltonian characterizing the atom-field interaction is

$$H_I = -\frac{e}{2mc}(\mathbf{p}\cdot\mathbf{A} + \mathbf{A}\cdot\mathbf{p}) + \frac{1}{2m}\left(\frac{e\mathbf{A}}{c}\right)^2 \tag{4.2}$$

Since the second term in the right hand of the above equation contains e^2, it is very weak comparing to the first term. In fact, the second term represents the interaction between different radiation oscillators of the field through the coupling of the electron to the field. Since this kind of the interaction belongs to two-photon transition process, we can neglect the second term in (4.2), consequently H_I can be approximated as

$$H_I = -\frac{e}{2mc}(\mathbf{p}\cdot\mathbf{A} + \mathbf{A}\cdot\mathbf{p}) \tag{4.3}$$

Because the operator

$$\mathbf{A}(\mathbf{r},t) = \sum_{kj}\lambda_{kj}[a_{kj}\exp(i\mathbf{k}\cdot\mathbf{r}) + a_{kj}^{\dagger}\exp(-i\mathbf{k}\cdot\mathbf{r})] \tag{4.4}$$

contains the coordinate operator $\mathbf{r}$, it means $[\mathbf{A},\mathbf{p}] \neq 0$. However, in the range of atomic dimension, the amplitude of r usually restricts to a Bohr radius $5.3\times 10^{-11}m$ (metre), but the wavelength of visible light is $10^{-6}m$, i.e., $k = \frac{2\pi}{\lambda} = 2\pi\times 10^{6}m^{-1}$, so $\mathbf{k}\cdot\mathbf{r} \approx 3.5\times 10^{-4} \ll 1$. In this case, $e^{i\mathbf{k}\cdot\mathbf{r}} \approx e^{0} = 1$ and $\mathbf{A}(\mathbf{r}) \approx \mathbf{A}(0)$. This means that we can neglect the effects of the position of the electron when we discuss the interaction between atom and field. This approximation is usually called the dipole approximation. In the dipole approximation, there exists the relation

$$[\mathbf{A},\mathbf{p}] = 0 \tag{4.5}$$

So the interaction Hamiltonian of the atom-field coupling system reduces to

$$H_I = -\frac{e}{mc}\mathbf{A}(0)\cdot\mathbf{p} \tag{4.6}$$

with

$$\mathbf{A}(0) = \sum_{kj}\lambda_{kj}(a_{kj}^{\dagger} + a_{kj})$$

$$\lambda_{kj} = \left(\frac{2\pi\hbar c^2}{V\omega_k}\right)^{\frac{1}{2}}\mathbf{e}_{kj} \tag{4.7}$$

Now we discuss how to express the electron momentum operator **p** in terms of the atomic pseudo-spin operators. As we know, in the Schrödinger picture the eigenvalue equation of the free Hamiltonian H_A is

$$H_A|n\rangle = E_n|n\rangle \tag{4.8}$$

here E_n is the eigenvalue of H_A and $|n\rangle$ is its corresponding eigenvector. In the representation of the eigenvectors $\{|n\rangle\}$, the Hamiltonian of the atom may be expressed as

$$H_A = \sum_n E_n|n\rangle\langle n| \tag{4.9}$$

If we introduce a group of generalized atomic operators as

$$\sigma_{nm} = |n\rangle\langle m| \tag{4.10}$$

then (4.9) may be rewritten as

$$H_A = \sum_n E_n \sigma_{nn} \tag{4.11}$$

and every operator describing the atomic behavior such as the atomic dipole moment, the momentum **p** and the position **r** can be generally written as

$$G = \sum_{n,m} \langle n|G|m\rangle \sigma_{nm} \tag{4.12}$$

Evidently, the atomic operators σ_{nm} satisfy the commutation relation

$$[\sigma_{ij}, \sigma_{k\ell}] = \delta_{jk}\delta_{i\ell} - \delta_{i\ell}\delta_{kj} \tag{4.13}$$

For a two-level atom, there are only two energy eigenvectors $|+\rangle$ (or $|2\rangle$) and $|-\rangle$ (or $|1\rangle$). In this case, σ_{nm} can be explicitly written as follows

$$\begin{aligned} \sigma_{11} &= \begin{pmatrix} 0 & 0 \\ 0 & 1 \end{pmatrix}, \quad \sigma_{22} = \begin{pmatrix} 1 & 0 \\ 0 & 0 \end{pmatrix}, \\ \sigma_{21} &= \begin{pmatrix} 0 & 1 \\ 0 & 0 \end{pmatrix}, \quad \sigma_{12} = \begin{pmatrix} 0 & 0 \\ 1 & 0 \end{pmatrix} \end{aligned} \tag{4.14}$$

All of these operators certainly satisfy the commutation relation (4.13) and

$$\sigma_{11} + \sigma_{22} = \begin{pmatrix} 1 & 0 \\ 0 & 1 \end{pmatrix} = I \tag{4.15}$$

here I is the unit matrix. Define

$$S_z = \frac{1}{2}(\sigma_{22} - \sigma_{11}), \quad S_+ = \sigma_{21}, \quad S_- = \sigma_{12} \tag{4.16}$$

then S_z and $S_\pm$ are just the pseudo-spin operators, which can characterize the two-level atom. By means of eqs.(4.14) and (4.16) we obtain the commutation relations

$$[S_z, S_\pm] = \pm S_\pm, \quad [S_+, S_-] = 2S_z, \quad [S_+, S_-]_+ = S_+S_- + S_-S_+ = I \tag{4.17}$$

Utilizing (4.11) and (4.16), the Hamiltonian of the free two-level atom becomes

$$H_A = E_+\sigma_{22} + E_-\sigma_{11} = \frac{\hbar\omega_0}{2}(\sigma_{22} - \sigma_{11}) = \hbar\omega_0 S_z \tag{4.18}$$

here we have chosen

$$E_+ + E_- = 0, \quad \hbar\omega_0 = E_+ - E_-$$

From (4.12), the momentum operator $\mathbf{p}$ of the two-level atom may be written as

$$\mathbf{p} = \langle +|\mathbf{p}|-\rangle S_+ + \langle -|\mathbf{p}|+\rangle S_- \tag{4.19}$$

Noticing the position operator $\mathbf{r}$ obeys

$$\frac{d}{dt}\mathbf{r} = \frac{1}{i\hbar}[\mathbf{r}, H_A] = \frac{1}{i\hbar}(\mathbf{r}H_A - H_A\mathbf{r}) = \frac{\mathbf{p}}{m} \tag{4.20}$$

we have

$$\begin{aligned} \langle +|\mathbf{p}|-\rangle &= \frac{im}{\hbar}(E_+ - E_-)\langle +|\mathbf{r}|-\rangle = \frac{im\omega_0}{\hbar e}\langle +|e\mathbf{r}|-\rangle \\ \langle -|\mathbf{p}|+\rangle &= -\frac{im\omega_0}{\hbar e}\langle -|e\mathbf{r}|+\rangle \end{aligned} \tag{4.21}$$

Clearly $\langle +|e\mathbf{r}|-\rangle$ is the matrix element of the atomic dipole moment, in general it is a complex vector. If we assume

$$\langle +|e\mathbf{r}|-\rangle = i\mathbf{D} \tag{4.22}$$

where $\mathbf{D}$ is a real vector, then from (4.18) and (4.20)-(4.22), the momentum operator $\mathbf{p}$ can be expressed by the atomic pseudo-spin operators as

$$\mathbf{p} = -\frac{m\omega_0}{e}\mathbf{D}(S_+ + S_-) \tag{4.23}$$

Substituting (4.7a) and (4.23) into (4.6), the interaction Hamiltonian of a single two-level atom coupled to a radiation field becomes

$$\begin{aligned} H_I &= -\frac{e}{mc}\sum_k (a_k^\dagger + a_k)\lambda_k \cdot \left(-\frac{m\omega_0 \mathbf{D}}{e}\right)(S_+ + S_-) \\ &= \sum_k \varepsilon_k (a_k^\dagger + a_k)(S_+ + S_-) \end{aligned} \tag{4.24}$$

For simplicity, here we have used the symbol k to replace kj, and the atom-field coupling constant obeys

$$\varepsilon_k = \left(\frac{2\pi\hbar}{V\omega_k}\right)^{1/2} \omega_0 \mathbf{D} \cdot \mathbf{e}_{kj} \qquad (j = 1, 2) \tag{4.25}$$

Therefore, the total Hamiltonian of this atom-field coupling system is

$$H = \sum_k \hbar\omega_k a_k^\dagger a_k + \hbar\omega_0 S_z + \sum_k \varepsilon_k (a_k^\dagger + a_k)(S_+ + S_-) \tag{4.26}$$

The above equation shows that the energy of the radiation field is the superposition of the photons of infinite modes with wave vector $\mathbf{k}$ and frequency ω_k, the energy of the bare atom is decided by the z-component of the atomic pseudo-spin operators. The atom-field coupling Hamiltonian is

$$V = \sum_k \varepsilon_k (a_k^\dagger S_- + a_k S_+ + a_k^\dagger S_+ + a_k S_-) \tag{4.27}$$

The first term in (4.27) corresponds to the interaction process which the atom transits from the upper level $|+\rangle$ to the lower level $|-\rangle$ and radiates one photon simultaneously. Conversely, the second term describes the process which the atom excites from $|-\rangle$ to $|+\rangle$ and simultaneously absorbs one photon. The third term represents the atomic excitation with the emission of a photon. The last term stands for the atomic deexcitation process with the absorption of a photon.

Evidently, the energy variation of the transition processes described by the first two terms in (4.27) is

$$\Delta E_1 = \hbar(\omega_k - \omega_0) \tag{4.28}$$

In the nearly resonant case, i.e., $\omega_k \approx \omega_0$, the energy variation $\Delta E \approx 0$, which means that the energy conservation holds. According to the energy-time uncertainty relation

$$\Delta E \cdot \Delta \tau_1 \geq \hbar \tag{4.29}$$

we know when $\Delta E_1 \rightarrow 0$, $\Delta \tau_1 \rightarrow \infty$. This means that these transition processes can create long lifetime photons which are called the real photons.

However, the energy uncertainty of the transition processes characterized by the last two terms is

$$\Delta E_2 = \hbar(\omega_k + \omega_0) \tag{4.30}$$

It is sufficiently large, which means that the energy conservation is not held in such transition processes. According to the energy-time uncertainty relation

$$\Delta E_2 \cdot \Delta \tau_2 \geq \hbar \tag{4.31}$$

we know that $\Delta \tau_2$ is very small since ΔE_2 is large. This means that the lifetime of the photons producing in these processes is very short. We call these photons as virtual photons. If we drop the last two terms in (4.27) which violate temporally the energy conservation, then the total Hamiltonian of the system becomes

$$H_{RWA} = \hbar\omega_0 S_z + \hbar \sum_k \omega_k a_k^\dagger a_k + \sum_k \varepsilon_k (a_k S_+ + a_k^\dagger S_-) \tag{4.32}$$

This approximation is termed as the rotating-wave approximation. This approximation is adopted frequently in quantum optics. The Hamiltonian (4.32) is typical one and is often called as the Dicke Hamiltonian for describing the atom-field coupling system.

4.2 Spontaneous emission of an excited atom

As we know, an excited atom always spontaneously transfers to the stable ground state with emitting photon (or photons). This spontaneous emission can not be understood in the frame of classical theory. In the following, we treat the spontaneous emission of an excited atom by use of the quantum theory starting from the Dicke Hamiltonian (4.32).

We have shown that the Dicke Hamiltonian of a two-level atom interacting with the radiation field is defined by (4.32). Supposing that at time t=0 the atom is in the upper state $|+\rangle$ and there are no photons in any mode of the field, i.e., the field is in vacuum state, and there is no coupling between field and atom, then the initial state vector of the atom-field coupling system gives

$$|\Psi(0)\rangle = |+, \{0_k\}\rangle \tag{4.33}$$

With the time development, what does the atomic behavior evolve when the interaction between the atom and the field takes place ?

According to the Heisenberg equation, the atomic operator S_z obeys

$$\frac{d}{dt}S_z = -\frac{i}{\hbar}[S_z, H_{RWA}] = -i\sum_k \frac{\varepsilon_k}{\hbar}(a_k S_+ - a_k^\dagger S_-) \tag{4.34}$$

Here we have used the commutation relation (4.17). In order to solve $S_z(t)$, it is necessary to define the evolution expressions of $a_k S_+$ and $a_k^\dagger S_-$. Similarly,

$$\begin{aligned}\frac{d}{dt}a_k S_+ = &-\frac{i}{\hbar}[a_k S_+, H_{RWA}] = -i(\omega_0 - \omega_k)a_k S_+ \\ &-\frac{i}{\hbar}\varepsilon_k(S_z + 1/2) - 2\frac{i}{\hbar}\sum_{k'}\varepsilon_{k'} S_z a_{k'}^\dagger a_k\end{aligned} \tag{4.35}$$

Since at time t=0 the field is in the vacuum state, we can neglect the term $a_{k'}^\dagger a_k$ when we calculate the expectation value of the above equation. That is to say we neglect the second-order infinitesimal term, which means that the effects of the radiation field created by the atom acting back on the atom are ignored. Thus (4.35) reduces to

$$\frac{d}{dt}a_k S_+ = -i(\omega_0 - \omega_k)a_k S_+ - \frac{i}{\hbar}\varepsilon_k\left(S_z + \frac{1}{2}\right) \tag{4.36}$$

Following the same procedure, we have

$$\frac{d}{dt}a_k^\dagger S_- = i(\omega_0 - \omega_k)a_k^\dagger S_- + \frac{i}{\hbar}\varepsilon_k\left(S_z + \frac{1}{2}\right) \tag{4.37}$$

The solutions of equations (4.36) and (4.37) can be formally expressed as

$$\begin{aligned}a_k S_+(t) = &\, a_k S_+(0)\exp[-i(\omega_0 - \omega_k)t] \\ &-i\frac{\varepsilon_k}{\hbar}\int_0^t \exp[i(\omega_0 - \omega_k)(t' - t)]\left[S_z(t') + \frac{1}{2}\right]dt'\end{aligned} \tag{4.38}$$

$$\begin{aligned}a_k^\dagger S_-(t) = &\, a_k^\dagger S_-(0)\exp[i(\omega_0 - \omega_k)t] \\ &+i\frac{\varepsilon_k}{\hbar}\int_0^t \exp[-i(\omega_0 - \omega_k)(t' - t)]\left[S_z(t') + \frac{1}{2}\right]dt'\end{aligned} \tag{4.39}$$

Inserting (4.38) and (4.39) into (4.34) and calculating the expectation value with respect to the state $|\Psi(0)\rangle = |+, \{0_k\}\rangle$, we have

$$\frac{d}{dt}\langle S_z(t)\rangle = -\frac{2}{\hbar^2}\sum_k \varepsilon_k^2 \int_0^t \cos[(\omega_0 - \omega_k)(t - t')] \left[\langle S_z(t')\rangle + \frac{1}{2}\right] dt' \tag{4.40}$$

Eq.(4.40) shows that the expectation value $\langle S_z(t)\rangle$ at time t is related to the value of S_z at the time range from 0 to t. However if considering the atomic decay time is sufficiently larger than ω_0^{-1}, we can choose the approximation $\langle S_z(t')\rangle \approx \langle S_z(t)\rangle$ in the integration expression of (4.40). This approximation is usually called the Markoff approximation, which corresponds physically to that $\langle S_z(t)\rangle$ is unrelated to its history. Therefore,

$$\frac{d}{dt}\langle S_z(t)\rangle = -\left[\langle S_z(t)\rangle + \frac{1}{2}\right]\frac{2\pi}{\hbar^2}\sum_k \varepsilon_k^2 \delta(\omega_k - \omega_0) \tag{4.41}$$

here we have chosen

$$\pi\delta(x) = \lim_{x\to 0}\int_0^t \cos(x\tau)d\tau \tag{4.42}$$

Letting

$$\Gamma = \frac{2\pi}{\hbar}\sum_k \varepsilon_k^2 \delta(\omega_k - \omega_0) \tag{4.43}$$

then (4.41) becomes

$$\frac{d}{dt}\langle S_z(t)\rangle = -\Gamma\left[\langle S_z\rangle + \frac{1}{2}\right] \tag{4.44}$$

The solution of (4.44) yields

$$\langle S_z(t)\rangle = -\frac{1}{2} + \left[\langle S_z(0)\rangle + \frac{1}{2}\right]\exp(-\Gamma t) \tag{4.45}$$

It shows that the expectation value $\langle S_z(t)\rangle$ decays exponentially with the time evolution. The decay lifetime is

$$\tau = \Gamma^{-1} \tag{4.46}$$

This decay process is illustrated in Fig.(4.1). On the basis of the energy-

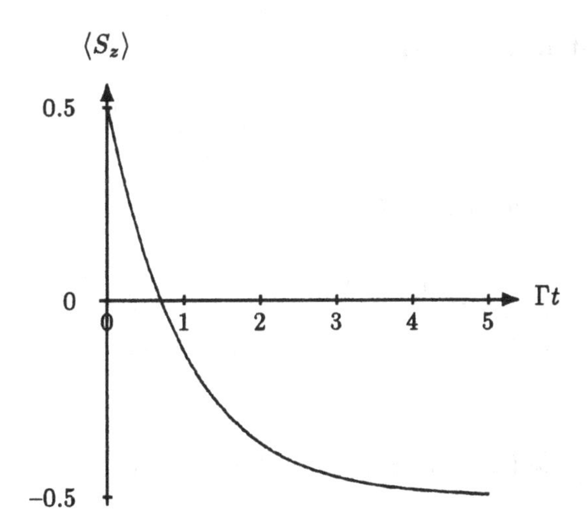

Figure 4.1: Illustration of spontaneous decay of a two-level atom initially in its excited state

time uncertainty relation, we see that the corresponding line-width of this spontaneous radiation process is Γ. Since

$$\Delta E \cdot \tau = \hbar \Delta\omega \cdot \tau \approx \hbar$$

then

$$\Delta\omega \approx \tau^{-1} = \Gamma \tag{4.47}$$

Eq.(4.43) also shows that the frequency ω_k of the photon created in the spontaneous radiation process is nearly equal to ω_0. The physical source of the spontaneous radiation is the fluctuations of vacuum field which stimulates the photon emission of the atom. For the initial condition $\langle S_z(0)\rangle = 1/2$, the time evolution of the expectation value of the atomic operator S_z becomes

$$\langle S_z(t)\rangle = e^{-\Gamma t} - \frac{1}{2} \tag{4.48}$$

which means that the atom initially in its upper state $|+\rangle$ can spontaneously decay into its lower state $|-\rangle$ when $t \to \infty$.

Next we discuss the time evolution of the atomic operators S_+ and S_- .

The equation of motion for S_+ is

$$\frac{d}{dt}S_+(t) = i\omega_0 S_+ - \frac{2i}{\hbar}\sum_k \varepsilon_k a_k^\dagger S_z \tag{4.49}$$

and the time-dependence of $a_k^\dagger S_z$ obeys

$$\frac{d}{dt}a_k^\dagger S_z = i\omega_k a^\dagger S_z + \frac{i}{2\hbar}\varepsilon_k S_+ \tag{4.50}$$

Here the relation $S_z S_+ = \frac{1}{2}S_+$ has been used. Intergrating formally the above equation, we have

$$a_k^\dagger S_+(t) = \frac{i\varepsilon_k}{2\hbar}\exp(i\omega_k t)\int_0^t \exp(-i\omega_k t')S_+(t')dt' + a_k^\dagger S_z(0)\exp(i\omega_k t) \tag{4.51}$$

In order to solve (4.51), we must have $S_+(t')$. Since the interaction Hamiltonian V is very weak comparing to H_0, as an approximation the Hamiltonian H (4.32) can be replaced by

$$H_0 = \hbar\omega_0 S_z + \sum_k \hbar\omega a_k^\dagger a_k \tag{4.52}$$

In this approximation, $S_+(t')$ obeys

$$\frac{d}{dt}S_+(t') = -\frac{i}{\hbar}[S_+(t'), H_0] = i\omega_0 S_+(t') \tag{4.53}$$

Evidently,

$$S_+(t') = S_+(t)\exp[i\omega_0(t'-t)] \tag{4.54}$$

Substituting the above equation into (4.51), we obtain

$$\begin{aligned} a_k^\dagger S_z(t) &= \frac{i}{2\hbar}\varepsilon_k S_+ \int_0^t \exp[-i(\omega_k - \omega_0)\tau]d\tau + a_k^\dagger S_z(0)\exp(i\omega_k t) \\ &= \frac{\varepsilon_k}{\hbar}S_+(t)\left[P\left(\frac{1}{\omega_k - \omega_0}\right) + i\pi\delta(\omega_k - \omega_0)\right] + a_k^\dagger S_z(0)\exp(i\omega_k t) \end{aligned} \tag{4.55}$$

where $P(\frac{1}{\omega_k - \omega_0})$ is the principle function. Inserting (4.55) into (4.49) and calculating the expectation value with respect to $|\Psi(0)\rangle$, we have

$$\frac{d}{dt}\langle S_+(t)\rangle = i\omega_0\langle S_+(t)\rangle - (i\Omega + \Gamma/2)\langle S_+(t)\rangle \tag{4.56}$$

where

$$\Omega = \frac{1}{\hbar^2} \sum_k \varepsilon_k^2 P \left(\frac{1}{\omega_k - \omega_0} \right) \tag{4.57}$$

and Γ is denoted by (4.32). The solution of (4.56) gives

$$\langle S_+(t) \rangle = \langle S_+(0) \rangle \exp[i(\omega_0 - \Omega)t] \exp(-\Gamma t/2) \tag{4.58}$$

We see that the transverse component $\langle S_+(t) \rangle$ of the atomic operators decays exponentially with speed $\Gamma/2$. The oscillating frequency of $\langle S_+(t) \rangle$ is no longer equal to ω_0, and has the shift Ω. That is to say, the eigenenergy of the two-level atom has been shifted due to the interaction of the vacuum field, this shift is usually called the Lamb shift. Since the existence of this shift, the atomic energy levels have uncertainty shift, thus the spectrum line have certain width.

4.3 The Jaynes-Cummings model

From the above section we can see that for the system of a two-level atom interacting with a mutli-mode field described by the Dicke Hamiltonian (4.32). In general, its eigenvalue equation is not easily solved exactly. Now we introduce a simple but exact solvable model named as J-C model, which proposed by Jaynes and Cummings in 1963. This model is consisted of a single two-level atom (or molecule) and a single-mode field. It is an ideal model to describe the atom-field coupling system. Because this model can be solved exactly except making the rotating-wave approximation, it plays an important role not only in quantum optics but also in laser physics, nuclear magnetic resonance and quantum field theory.

According to (4.32) we can easily obtain the Hamiltonian of the J-C model with the rotating-wave approximation as

$$H = \omega_0 S_z + \omega a^\dagger a + \varepsilon (a^\dagger S_- + a S_+) \quad (\hbar = 1) \tag{4.59}$$

Here $a^\dagger$, a are the creation and annihilation operators for the single-mode field with frequency ω, respectively, S_z and $S_\pm$ are the pseudo-spin operators for the two-level atom with eigen frequency ω_0, and ε is the atom-field coupling constant which represents the strength of the atom-field coupling. Evidently, the first term in (4.59) corresponds to the energy of the bare atom, the second one describes the energy of the field, and the third one is the atom-field interaction energy V

$$V = \varepsilon (a^\dagger S_- + a S_+) \tag{4.60}$$

These interaction terms represent the processes creating and absorbing the photons when the atom transits. In order to stress this interaction, we divide (4.59) into two parts

$$H = H_0 + V \tag{4.61}$$

here H_0 is the free energy operator of the atom and the field

$$H_0 = \omega_0 S_z + \omega a^\dagger a \tag{4.62}$$

From (4.60) and (4.62), it is easy to verify that

$$[H_0, V] = 0 \tag{4.63}$$

That is to say, H_0 and V commute with each other.

When the atom and the field is uncoupled, we have a complete set of basis vectors $\{|+,n\rangle, |-,n\rangle\}$ by H_0. By means of these vectors, we can expand an arbitrary state of the atom-field coupling system. Evidently, the eigenvalue equation of H_0 can be expressed explicitly as

$$H_0|\pm,n\rangle = (\omega_0 S_z + \omega a^\dagger a)|\pm,n\rangle = (\omega n \pm \omega_0/2)|\pm,n\rangle \tag{4.64}$$

here $|\pm,n\rangle$ represents that the field is in the n photon number state and the atom in the eigenstate $|\pm\rangle$. And for the eigenstate $|\pm,n\rangle$, the eigenvalue of H_0 is $\omega n \pm \omega_0/2$. So in this representation, H_0 has been to be diagonal. When $\omega_0 = \omega$, the eigenlevels of H are double degenerated except the ground state $|-,0\rangle$ (shown in Fig.4.2). But for $\Delta\omega = \omega - \omega_0 \neq 0$, this degeneration will be relieved. However, from (4.60) we know that $|\pm,n\rangle$ are not the eigenstates of the interaction Hamiltonian V. Since H_0 and V commute with each other, we can choose the linear superposition of the eigenvectors of H_0 to diagonize V.

The method to diagonize V is as follows. First we write the matrix expression of H in the Hilbert space $\{|\pm,n\rangle\}$, then we diagonize it. From the matrix elements $\langle\pm,n|H|\pm,n'\rangle$ $(n,n' = 0,1,2,\cdots)$ we know that in the subspace $\{|-,n+1\rangle, |+,n\rangle\}$, the matrix expression of H is

$$H = \begin{pmatrix} \omega n + \omega_0/2 & \varepsilon\sqrt{n+1} \\ \varepsilon\sqrt{n+1} & \omega(n+1) - \omega_0/2 \end{pmatrix} \tag{4.65}$$

its eigenstate can be written as

$$\begin{aligned} |u_{n+1}\rangle &= \sin\theta_{n+1}|-,n+1\rangle + \cos\theta_{n+1}|+,n\rangle \\ &= \sin\theta_{n+1}|n+1\rangle \begin{pmatrix} 0 \\ 1 \end{pmatrix} + \cos\theta_{n+1}|n\rangle \begin{pmatrix} 1 \\ 0 \end{pmatrix} \end{aligned} \tag{4.66}$$

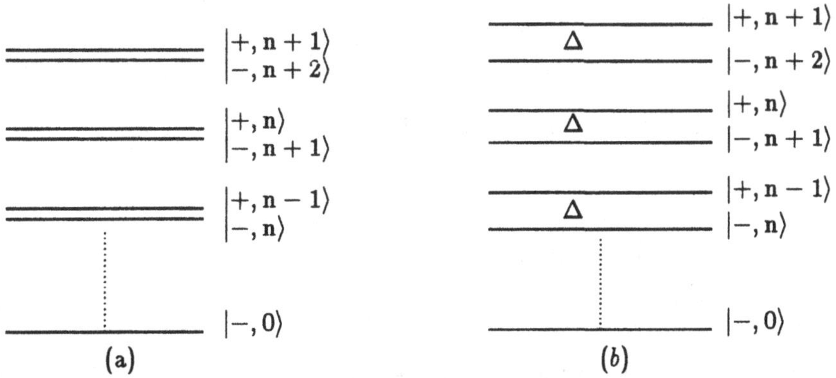

Figure 4.2: Eigenenergy spectrum of H_0: (a). the eigenenergy spectrum in resonant case $\omega = \omega_0$; (b). the eigenenergy spectrum in nonresonant case $\omega \neq \omega_0$.

The normalization condition of the eigenstates demands

$$\sin^2\theta_{n+1} + \cos^2\theta_{n+1} = 1 \tag{4.67}$$

According to the eigenvalue equation

$$H|u_{n+1}\rangle = E_{n+1}|u_{n+1}\rangle \tag{4.68}$$

we obtain

$$\begin{vmatrix} n\omega + \omega_0/2 - E_{n+1} & \varepsilon\sqrt{n+1} \\ \varepsilon\sqrt{n+1} & (n+1)\omega - \omega_0/2 - E_{n+1} \end{vmatrix} = 0 \tag{4.69}$$

Its solutions give

$$E^{\pm}_{n+1} = \omega(n+1) \pm \Delta_{n+1} \tag{4.70}$$

with

$$\Delta_{n+1} = \sqrt{\left(\frac{\omega - \omega_0}{2}\right)^2 + \varepsilon^2(n+1)} \tag{4.71}$$

The corresponding eigenstates are

$$|u^{\pm}_{n+1}\rangle = \cos\theta_{n+1}|+, n\rangle \pm \sin\theta_{n+1}|-, n+1\rangle \tag{4.72}$$

Here θ_{n+1} obeys

$$\tan\theta_{n+1} = \frac{\frac{\omega_0-\omega}{2} + \Delta_{n+1}}{\varepsilon\sqrt{n+1}} \tag{4.73}$$

For the ground state, the eigenvalue of H is

$$E_0 = -\frac{\omega_0}{2} \tag{4.74}$$

and the corresponding eigenvector is

$$|u_0\rangle = |-,0\rangle \tag{4.75}$$

Therefore, the eigenvalues and the corresponding eigenstates of H are completely determined by (4.70) and (4.72), and here H has been diagonized.

The energy diagram of the system in the resonant case ($\omega = \omega_0$) is shown in Fig.(4.3). From Fig.(4.3) we can see that the energy degeneration of H_0

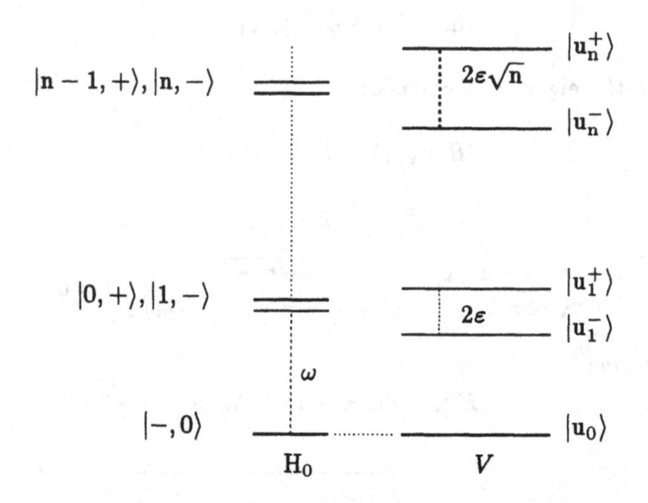

Figure 4.3: Eigenenergy of $H_0 + V$ for $\omega = \omega_0$ and larger n

has been relieved due to the atom-field interaction. We call the atomic system comprised by the bare atom and the atom-field interaction as dressed atom.

The eigenvalues of the dressed atom are decided by (4.70) and (4.74), and the eigenstates are determined by (4.72). Eq.(4.71) shows that the gap Δ_{n+1} depends on the atom-field coupling constant and the photon number n. From (4.71) we know that the first excited state of the system is

$$|u_1^-\rangle = \cos\theta_1|-,1\rangle - \sin\theta_1|+,0\rangle \tag{4.76}$$

and the corresponding eigenenergy gives

$$E_1^- = \frac{\omega}{2} - \Delta_1 \tag{4.77}$$

In general case, $E_0 < E_1^-$, namely

$$-\frac{\omega_0}{2} < \frac{\omega}{2} - \Delta_1$$

i.e.,

$$\omega + \omega_0 > 2\Delta_1 \tag{4.78}$$

or

$$(\omega + \omega_0)^2 > (\omega - \omega_0)^2 + 4\varepsilon^2 \tag{4.79}$$

That is to say, when the coupling constant ε satisfies

$$\varepsilon^2 < \omega\omega_0 \tag{4.80}$$

the ground energy of the atom-field coupling system is E_0 , and the ground state is $|-,0\rangle$. However when the atom-field coupling is so strong, i.e., ε is very large and ω is quite small, so that (4.80) may be violated. In this case, the first excited state $|u_1^-\rangle$ becomes a new ground state of the system, and the energy E_1^- of the new ground state is less than that of the old ground state $|-,0\rangle$. Moreover, for the case of very large ε, even $n > 1$ the energy E_n^- may be less than E_0. So we see that for the case of very strong field and strong coupling, the energy of the ground state have unstable property. It has been proved theoretically that this unstable property of the old ground state is the origin of the phase transition of the Dicke superradiation.

References

1. L.Allen and J.H.Eberly, *Optical resonance and two-level atom* (Wiley, New York, 1973).

2. H.Haken, *Light* Vol.1 (North-Holland, Amsterdam, 1981).

3. R.Loudon, *The quantum theory of light* (Clarendon, Oxford, 1983).

4. P.Meystre and M.Sargent III, *Elements of quantum optics* (Springer-Verlag, New York, 1990).

5. C.Cohen-Tannoudji, J.Dupont-Roc, and G.Grynberg, *Photons and atoms–Introduction to quantum electrodynamics* (Wiley, New York, 1989).

6. C.Leonardi, F.Persico, and G.Vetri, *Dicke model and theory of driven and spontaneous emission* (La Rivista de Nupvo Cimento, 1986).

7. Jin-sheng Peng, *Resonance fluorescence and superfluorescence* (Science Press, Beijing, 1993).

8. R.H.Dicke, *Phys.Rev.* **93** (1954) 99.

9. E.T.Jaynes and F.W.Cummings, *Proc.IEEE.* **51** (1963) 89.

10. P.W.Milonni, *Phys.Rep.* **25** (1976) 1.

11. P.L.Knight and P.W.Milonni, *Phys.Rep.* **66** (1980) 21.

12. M.Hillery, *Phys.Rev.* **A24** (1981) 933.

13. R.R.Puri, *J.Opt.Soc.Am.* **B2** (1985) 444.

14. Jin-sheng Peng and Gao-xiang Li, *ACTA Physica Sinica* **41** (1992) 1590.

CHAPTER 5

QUANTUM THEORY OF A SMALL SYSTEM COUPLED TO A RESERVOIR

In quantum optics, a large number of investigations need to take into account the effects of circumambient matter called reservoir (R) on a small system (S). The small system may be considered as a two-level atom or a harmonic oscillator, etc., and the reservoir may be regarded as the radiation field of black body or the wall of cavity, etc.. Since the interactions between a small system S and the reservoir R are to be stochastic, here we discuss the basic theory to describe the stochastic processes. We first introduce the Langevin equation describing the classical Brownian motion and the Fokker-Planck equation. Then through discussing the system of a harmonic as well as a two-level atom coupled a reservoir, we give a quantum theoretical treatment for the behavior of a small system coupled to a reservoir. Finally we briefly discuss the characteristic function and quasi-probability distribution function, which play important roles in theoretical treatment for the system of a small system coupled to a reservoir.

5.1 Classical Langevin equation and Fokker-Planck equation

The Brownian motion is a simple but typical stochastic motion, it can be regarded as an example for a classical small system coupled to a reservoir. The Brownian motion may be considered as a particle (small system S) which is immersed in a fluid (reservoir R). The Brownian particle undergoes to a large number of collisions with much smaller particles of the fluid. Since the interaction processes between the system S (particle) and the reservoir R have stochastic characteristic, the velocity V of the particle is a stochastic variable. Usually there are two possible treatments to describe the time evolution of the velocity $V(t)$ of the Brownian particle. The first way is to use the Langevin equation. The second treatment is by use of a Fokker-Planck equation. Although two treatments are different, the physical results of the solutions are the same. Next we introduce these two kinds of equations.

5.1.1 Langevin equation

If the Brownian particle S with mass m is immersed in a fluid R, its velocity $V(t)$ is affected by two types of forces. The first one is a friction force which describes the cumulative effect of the collisions. The simplest expression for such a friction or damping force may be given by

$$k_f = -\gamma_0 V \tag{5.1}$$

where γ_0 is a constant which corresponds to diffusion coefficient of the system. The second one is a Langevin random or stochastic force, which results from every collision of the light particle acting on the heavy one. For the one-dimension motion, this force can be described by

$$F_0(t) = \psi \sum_{j=1}^{2} \delta(t - t_j)\varepsilon_j \tag{5.2}$$

here $\varepsilon_j (j = 1, 2)$ characterize the two possible directions of the one-dimension motion, ψ stands for the collision strength. Because the stochastic force $F_0(t)$ has the space homogeneity, its average should be zero, i.e.,

$$\langle F_0(t) \rangle = 0 \tag{5.3}$$

If we discuss the correlation of the stochastic force $F_0(t)$ at different time, there exists the relation

$$\langle F_0(t) F_0(t') \rangle = c\delta(t - t') \tag{5.4}$$

which means that there is no correlation of the random forces at different time. The constant c is proportional to ψ^2, which is related to the strength of every collision and describes the strength of the fluctuations. Therefore, the equation of motion for the particle S reads

$$m\frac{d}{dt}V = -\gamma_0 V + F_0(t) \tag{5.5}$$

The corresponding Langevin equation which describes the stochastic motion of the system S is

$$\frac{d}{dt}V = -\gamma V + F(t) \tag{5.6}$$

with

$$\gamma = \gamma_0/m, \qquad F = F_0/m$$

Eq.(5.6) represents the time evolution of the particle S driven by both the friction force and the stochastic force. The stochastic force $F(t)$ obeys

$$\langle F(t)\rangle = 0 \tag{5.7}$$

$$\langle F(t)F(t')\rangle = Q\delta(t-t') \tag{5.8}$$

where Q is the noise strength of the stochastic force.

If the initial velocity of the particle S is given to be V_0, then the solution of (5.6) yields

$$V(t) = V_0 \exp(-\gamma t) + \int_0^t \exp[-\gamma(t-t')]F(t')dt' \tag{5.9}$$

The average value of $V(t)$ gives

$$\langle V(t)\rangle = V_0 \exp(-\gamma t) \tag{5.10}$$

Eq.(5.10) shows that the mean velocity of the particle S at time t decreases according to the exponential law. Furthermore, the velocity correlation function may be obtained from (5.7) and (5.8) as

$$\langle V(t_1)V(t_2)\rangle = V_0^2 \exp[-\gamma(t_1+t_2)] + \int_0^t dt_1' \int_0^t dt_2' \exp[-\gamma(t_1+t_2-t_1'-t_2')]Q\delta(t_1'-t_2') \tag{5.11}$$

To evaluate the double integral in the above equation, we integrate over t_2' first. It is clear that the integration is not equal to zero only when $t_1' = t_2'$. If $t_2 \le t_1$, then the integration over t_1' runs only from 0 to t_2 as shown in Fig.(5.1), whereas $t_1 \le t_2$, the integration over t_1' then runs only from 0 to t_1. That is to say, when we calculate the double integration and integrate over t_2'

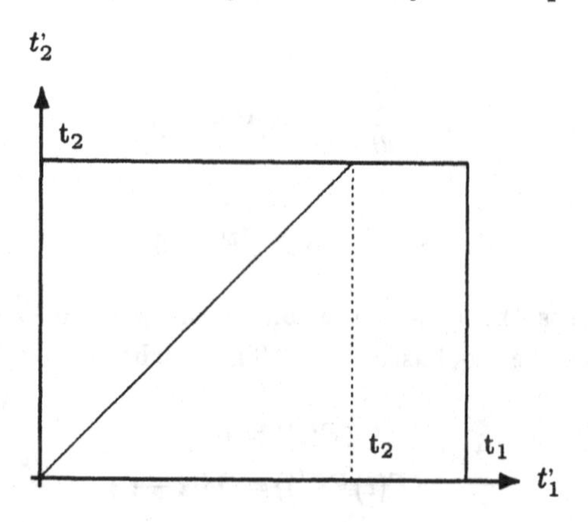

Figure 5.1: The integral range of (5.11)

first, then for integrating over t_1', the upper bound is the minimum one of t_1 and t_2, i.e., $\min(t_1, t_2)$. We therefore have

$$\begin{aligned}\langle V(t_1)V(t_2)\rangle &= V_0^2 \exp[-\gamma(t_1+t_2)] + Q \int_0^{\min(t_1,t_2)} \exp[-\gamma(t_1+t_2-2t_1')]dt_1' \\ &= V_0^2 \exp[-\gamma(t_1+t_2)] + \frac{Q}{2\gamma}\{\exp[-\gamma|t_1-t_2|] \\ &- \exp[-\gamma|t_1+t_2|]\}\end{aligned} \tag{5.12}$$

When we are only interested in the long time range which is larger than γ^{-1}, i.e., $\gamma t_1 \gg 1$, $\gamma t_2 \gg 1$, then the velocity correlation function is independent of the initial velocity V_0 and is only a function of the time difference $|t_1 - t_2|$, i.e.,

$$\langle V(t_1)V(t_2)\rangle = \frac{Q}{2\gamma} \exp[-\gamma|t_1-t_2|] \tag{5.13}$$

Eq.(5.13) presents the mean square value of the velocity, and also shows that the velocity correlation will disappear according to the exponential law (shown in Fig.(5.2)). In the stationary state the average energy of the Brownian particle is therefore given by

$$\langle E\rangle = \frac{1}{2}m\langle V^2(t)\rangle = \frac{mQ}{4\gamma} \tag{5.14}$$

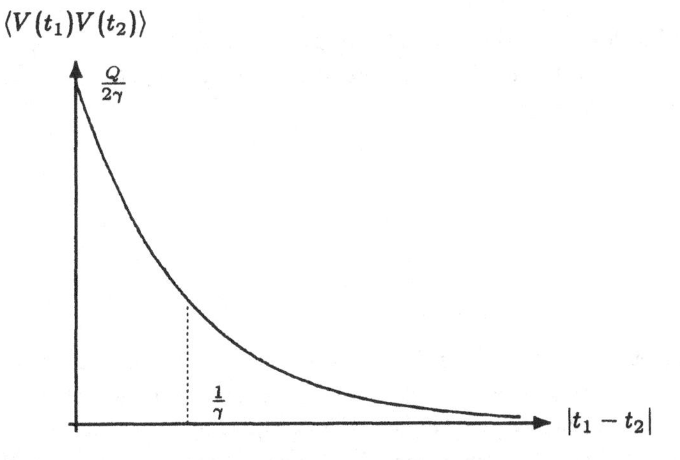

Figure 5.2: Relation between the velocity correlation function $\langle V(t_1)V(t_2)\rangle$ and the time gap $|t_1 - t_2|$

From the equipartition law of classical statistical mechanics, the mean energy of the particle reads (in one dimension)

$$\langle E\rangle = \frac{1}{2}k_B T \tag{5.15}$$

where k_B is the Boltzman's constant and T is the temperature of the reservoir R. Hence we obtain the constant Q representing the noise strength of $F(t)$ to be

$$Q = \frac{2\gamma k_B T}{m} \tag{5.16}$$

Here γ represent the diffusion strength of the reservoir to the small system. Since γ is a determinate quantity, the noise strength Q describing the strength of the stochastic force can be determined by (5.16). Eq.(5.16) is just the simple diffusion-fluctuation relation.

The Langevin equation (5.6) may be generalized to a system with a set of stochastic variables. If the system has a set of the stochastic variables $X = \{X_1(t), X_2(t), \cdots, X_n(t)\}$, the Langevin equation becomes

$$\frac{d}{dt}X_i = \sum_k M_{ik}(X)X_k + \Gamma_i(t) \tag{5.17}$$

Here M_{ik} describes the general damping or friction processes, as well as the action of driven force. $\Gamma_i(t)$ are the stochastic forces, they have the properties similar to (5.8) and (5.9)

$$\begin{aligned} \langle \Gamma_i(t) \rangle &= 0 \\ \langle \Gamma_i(t_1)\Gamma_j(t_2) \rangle &= G_{ij}\delta(t_1 - t_2) \end{aligned} \tag{5.18}$$

5.1.2 Fokker-Planck equation

Except using the Langevin equation to describe the Brownian motion, we can also use the Fokker-Planck equation to deal with the stochastic motion.

The basic idea of the Fokker-Planck equation is as follows. Suppose that there are a large number of identical systems, in which every system contains an identical Brownian particle, and at time t one can measure the velocities of all Brownian particles simultaneously. We therefore may ask for the number $N(V)dV$ of systems of the ensemble whose velocities are in the interval $(V, V + dV)$. Because V is a continuous variable we may ask for the probability $P(V,t)dV$ to find the velocity in the ineterval $(V, V + dV)$, here $P(V,t)$ is the probability distribution function of the velocity. Evidently, $P(V,t)$ is proportional to $N(V)$ and must be normalized, i.e.,

$$\int_{-\infty}^{\infty} P(V,t)dV = 1 \tag{5.19}$$

If we find the time evolution of $P(V,t)$, then we may obtain the time evolution of the velocity $V(t)$ of the Brownian particle. According to the classical statistical mechanics, one can prove strictly that the probability distribution function $P(V,t)$ of the velocity of the Brownian particle must obey the Fokker-Planck equation

$$\frac{\partial}{\partial t}P(V,t) = \left[\frac{\partial}{\partial V}\gamma V + \frac{Q}{2}\frac{\partial^2}{\partial V^2}\right]P(V,t) \tag{5.20}$$

We now derive the Fokker-Planck equation (5.20). For the particle with velocity V_0 at time t_0, because of the effect of the friction force and the stochastic force, the probability to find its velocity equal to V is $P(V,t;V_0,t_0)$ when time reaches to t. If we are only interested in the case that the probability distribution function has the invariant property as time displacement, then the value of t_0 is arbitrary. Supposing $t_0 = 0$, then t represents the time difference, the probability distribution $P(V,t;V_0,t_0)$ is only dependent on the time difference t, so it can therefore denoted as $P(V,t;V_0)$. To derive the equation of $P(V,t;V_0)$, we first discuss $P(V,t+\Delta t;V_0)$ which describes the probability of the particle velocity scattered from V_0 to V in the time interval $t+\Delta t$. The scattering process can be regarded as a two-step one in which the particle velocity first scattered to the intermediate value V_1 in the inetrval t_1, and then to V in the interval Δt. Hence the probability distribution $P(V,t+\Delta t;V_0)$ at time $t+\Delta t$ and $P(V_1,t;V_0)$ at time t are connected by

$$P(V,t+\Delta t;V_0) = \int dV_1 P(V,\Delta t;V_1)P(V_1,t;V_0) \tag{5.21}$$

Evidently, the velocity distribution function $P(V,t+\Delta t;V_0)$ is dependent on the time variable as well as on the velocity variable. What is the effects of the variations of these two variables on the velocity probability function ? First we discuss the case for small variation of time, i.e., $\Delta t \to 0$. In this case $P(V,t+\Delta t;V_0)$ may be expanded in a Taylor series and be retained only to the first order of Δt

$$P(V,t+\Delta t;V_0) \approx P(V,t;V_0) + \frac{\partial}{\partial t}P(V,t;V_0)\Delta t \tag{5.22}$$

So that we have

$$\frac{\partial}{\partial t}P(V,t;V_0) = \frac{P(V,t+\Delta t;V_0) - P(V,t;V_0)}{\Delta t} \tag{5.23}$$

Next we examine the evolution of the velocity. If we let $\Delta V = V - V_1$, then $P(V,\Delta t;V_1) = P(V,\Delta t;V-\Delta V)$ and $P(V_1,t;V_0) = P(V-\Delta V,t;V_0)$. Thus the integrand function in (5.21) may be expanded in a Taylor series according to

$$\begin{aligned} &P(V,\Delta t;V-\Delta V)P(V-\Delta V,t;V_0) \\ &\quad = P(V+\Delta V-\Delta V,\Delta t;V-\Delta V)P(V-\Delta V,t;V_0) \\ &\quad = \sum_{n=0}^{\infty}\frac{(-\Delta V)^n}{n!}\frac{\partial^n}{\partial V^n}[P(V+\Delta V,\Delta t;V)P(V,t;V_0)] \end{aligned}$$

Substituting the above equation and (5.21) into (5.23), we have

$$\frac{\partial}{\partial t}P(V,t;V_0) = \frac{1}{\Delta t}\left\{\int dV_1 P(V,\Delta t;V_1)P(V_1,t;V_0) - P(V,t;V_0)\right\}$$

$$=\frac{1}{\Delta t}\left\{\int d(\Delta V)\sum_{n=0}^{\infty}\frac{(-\Delta V)^n}{n!}\frac{\partial^n}{\partial V^n}[P(V+\Delta V,\Delta t;V)P(V,t;V_0)]-P(V,t;V_0)\right\}$$

Since the term of n=0 in the summation is equal to the last term, the above equation reduces to

$$\frac{\partial}{\partial t}P(V,t;V_0)=\sum_{n=1}^{\infty}\frac{(-1)^n}{n!}\frac{\partial^n}{\partial V^n}[M_nP(V,t;V_0)] \tag{5.24}$$

here M_n is called n-order moment, it reads

$$M_n=\frac{1}{\Delta t}\int_{-\infty}^{\infty}d(\Delta V)(\Delta V)^nP(V+\Delta V,\Delta t;V)=\frac{\langle(\Delta V)^n\rangle}{\Delta t} \tag{5.25}$$

For sufficiently small Δt, that $\langle(\Delta V)^n\rangle$ tends to zero may be faster than that $\Delta t\to 0$. So we can only retain the first two terms in right side of (5.24). Therefore, from the general formula (5.24) describing the time evolution of the velocity probability distribution function, we can obtain the Fokker-Planck equation which the velocity probability of the particle obeys

$$\frac{\partial}{\partial t}P(V,t)=-\frac{\partial}{\partial V}(M_1P)+\frac{1}{2}\frac{\partial^2}{\partial V^2}(M_2P) \tag{5.26}$$

Now we derive the first-order moment M_1 and the second-order moment M_2 corresponding to (5.6) by the aid of the moment definition (5.25). From the Langevin equation (5.6) of the Brownian motion, the velocity increment $\Delta V(t)$ in a very small time interval Δt is

$$\Delta V(t)=-\gamma\int_t^{t+\Delta t}dt'V(t')+\int_t^{t+\Delta t}dt'F(t') \tag{5.27}$$

Inasmuch as the time interval is very small, for example it is less than the characteristic time $1/\gamma$, the velocity increment $\Delta V(t)$ is also very small. We therefore have

$$\begin{aligned}M_1&=\frac{\langle(\Delta V)\rangle}{\Delta t}=-\gamma\langle(V(t)\rangle+\frac{1}{\Delta t}\int_t^{t+\Delta t}dt'\langle F(t')\rangle\\&=-\gamma\langle V(t)\rangle=-\gamma V(t)\end{aligned} \tag{5.28}$$

Similarly, the second moment M_2 is given by

$$M_2 = \frac{\langle (\Delta V)^2 \rangle}{\Delta t}$$

$$= \frac{1}{\Delta t} \left\{ \gamma^2 \langle V^2 \rangle (\Delta t)^2 - 2\gamma \int_t^{t+\Delta t} dt' \int_t^{t+\Delta t} dt'' \langle V(t') F(t'') \rangle \right.$$

$$\left. + \int_t^{t+\Delta t} dt' \int_t^{t+\Delta t} dt'' \langle F(t') F(t'') \rangle \right\} \tag{5.29}$$

The first term in braces is proportional to $(\Delta t)^2$ and yields a negligible value as $\Delta t \to 0$. The integrated function $\langle V(t')F(t'') \rangle$ in the second term is different from zero only when $t' = t''$, but the integrating result of this correlation function is equal to zero. From (5.9) we know that the third term in (5.29) is equal to Q. Hence (5.29) may be simplified as

$$M_2 = Q \tag{5.30}$$

Inserting eqs.(5.28) and (5.30) into (5.26), equation (5.26) becomes the Fokker-Planck equation (5.20) which describes the velocity probability of the Brownian particle.

If we have the solution of (5.20), then according to

$$\langle V^n \rangle = \int V^n P(V,t) dV \quad (n = 1, 2, \cdots) \tag{5.31}$$

we can obtain the n-order moments $\langle V^n \rangle$ such as the mean values of the velocity $\langle V \rangle$ and the energy $\frac{1}{2} m \langle V^2 \rangle$ at time t. Next we discuss the solution of the Fokker-Planck equation. In general, the solution of (5.20) is quite complex, which depends on different initial and boundary conditions. However, its steady-state solution is very simple, so we treat the steady-state case first. For $\frac{\partial}{\partial V} P(V,t) = 0$, the right side of (5.20) gives

$$\frac{\partial}{\partial V} \left[\gamma V P(V) + \frac{Q}{2} \frac{\partial}{\partial V} P(V) \right] = 0$$

So that we have

$$\gamma V P(V) + \frac{Q}{2} \frac{\partial}{\partial V} P(V) = C \tag{5.32}$$

here C is an undetermined constant. Considering that the probability distribution of the velocity is usually located in a limit space range, then there exists the boundary condition

$$P(V)|_{V=\pm\infty} = 0, \quad \frac{\partial}{\partial V} P(V)|_{V=\pm\infty} = 0 \tag{5.33}$$

thus C is zero. Then (5.32) becomes

$$\frac{Q}{2}\frac{\partial}{\partial V} P(V) = -\gamma V P(V)$$

The solution of the above equation is given by

$$P(V) = k \exp\left(\frac{-\gamma V^2}{Q}\right) \tag{5.34}$$

Noticing the normalization condition

$$\int_{-\infty}^{\infty} P(V) dV = 1 \tag{5.35}$$

we may obtain the constant $k = (\gamma/\pi Q)^{1/2}$. The steady-state solution of the Fokker-Planck equation therefore is

$$P(V) = \left(\frac{\gamma}{\pi Q}\right)^{1/2} \exp\left(\frac{-\gamma V^2}{Q}\right) \tag{5.36}$$

Next we derive the transient solution of the Fokker-Planck equation (5.20) with the initial condition $P(V, t_0) = \delta(V - V_0)$. To solve (5.20), we may use the technique of the Fourier transformation. Define the Fourier transformation

$$\Phi(s, t) = \int dV P(V, t) \exp(isV) \tag{5.37}$$

the corresponding inverse transformation is

$$P(V, t) = \frac{1}{2\pi} \int ds \Phi(s, t) \exp(-isV) \tag{5.38}$$

Differentiating (5.36) partially with respect to the variable t, then we have

$$\frac{\partial}{\partial t} P(V, t) = \int ds \frac{\partial \Phi(s, t)}{\partial t} \exp(-isV) \tag{5.39}$$

Differentiating (5.36) with respect to the variable V, we have

$$\frac{\partial^2}{\partial V^2}P(V,t)=\frac{1}{2\pi}\int ds\Phi(s,t)\frac{\partial^2}{\partial V^2}\exp(-isV)$$
$$=-\frac{1}{2\pi}\int ds\Phi(s,t)s^2\exp(-isV) \tag{5.40}$$

Differentiating (5.37) with respect to the variable s, we obtain

$$\frac{\partial}{\partial s}\Phi(s,t)=i\int dV\,P(V,t)V\exp(isV)$$

The Fourier inverse transformation of the above equation gives

$$VP(V,t)=-\frac{i}{2\pi}\int ds\frac{\partial\Phi(s,t)}{\partial s}\exp(-isV) \tag{5.41}$$

From (5.41) we have

$$\frac{\partial}{\partial V}(VP(V,t))=-\frac{i}{2\pi}\int ds\frac{\partial\Phi(s,t)}{\partial s}(-is)e^{-isV}$$
$$=-\frac{1}{2\pi}\int dss\frac{\partial\Phi(s,t)}{\partial s}e^{-isV} \tag{5.42}$$

Substituting eqs.(5.39), (5.40) and (5.42) into (5.20), $\Phi(s,t)$ obeys

$$\frac{\partial}{\partial t}\Phi(s,t)=-\gamma s\frac{\partial}{\partial s}\Phi(s,t)-\frac{Q}{2}s^2\Phi(s,t) \tag{5.43}$$

Calling to the theory of partial differential equation, (5.43) can be solved by use of the method of characteristics equation. Rewriting the above equation as

$$\frac{\partial}{\partial t}\Phi(s,t)+\gamma s\frac{\partial}{\partial s}\Phi(s,t)=-\frac{Q}{2}s^2\Phi(s,t)$$

The characteristic equations are therefore

$$\frac{dt}{1}=\frac{ds}{\gamma s}=\frac{d\Phi(s,t)}{-\frac{Q}{2}s^2\Phi(s,t)} \tag{5.44}$$

Integrating the first identity in (5.44), we may obtain a characteristic curve in the (s,t) plane. Namely from

$$\frac{dt}{1}=\frac{ds}{\gamma s} \tag{5.45}$$

we have

$$\gamma t = \ln s - \ln C$$

i.e.,

$$s = C \exp[\gamma(t - t_0)]$$

So the undetermined parameter C obeys

$$C = s \exp[-\gamma(t - t_0)] \tag{5.46}$$

Next we integrate the second identity in (5.44), that is, solve the equation

$$\frac{ds}{\gamma s} = \frac{d\Phi}{-\frac{Q}{2}s^2\Phi}$$

Integrating the above equation , yields

$$\ln \Phi - \ln \Omega = -\frac{Qs^2}{4\gamma}$$

i.e.,

$$\Phi = \Omega \exp\left(-\frac{Qs^2}{4\gamma}\right) \tag{5.47}$$

Here the parameter Ω has different values for different characteristic curves, so it is the function of C, that is, it is the function of the variables s and t. Therefore, (5.47) may be modified as

$$\Phi(s,t) = \Omega(s \exp[-\gamma(t - t_0)]) \exp\left(-\frac{Q}{4\gamma}s^2\right) \tag{5.48}$$

From the initial condition

$$P(V, t_0) = \delta(V - V_0) \tag{5.49}$$

we obtain from eq.(5.37)

$$\Phi(s, t_0) = \exp(isV_0) \tag{5.50}$$

Combining (5.48) with (5.40), it gives

$$\exp(isV_0) = \Omega(s) \exp\left(-\frac{Q}{4\gamma}s^2\right)$$

so that

$$\Omega(s) = \exp\left(isV_0 + \frac{Q}{4\gamma}s^2\right) \tag{5.51}$$

Substituting (5.51) into (5.48) and replacing s in (5.51) by $s\exp[-\gamma(t-t_0)]$, we obtain the expression of $\Phi(s,t)$ as

$$\begin{aligned}\Phi(s,t) &= \exp\left\{isV_0\exp[-\gamma(t-t_0)] + \frac{Q}{4\gamma}s^2\exp[-2\gamma(t-t_0)]\right\}\exp\left(-\frac{Q}{4\gamma}s^2\right) \\ &= \exp\left\{-\frac{Q}{4\gamma}s^2[1-\exp[-2\gamma(t-t_0)]] + isV_0\exp[-\gamma(t-t_0)]\right\}\end{aligned} \tag{5.52}$$

Inverting the above equation, then the transient solution of the Fokker-Planck equation (5.20) gives

$$\begin{aligned}P(V,t) &= \frac{1}{2\pi}\int_{-\infty}^{\infty} ds\exp\left\{-\frac{Q}{4\gamma}s^2[1-\exp[-2\gamma(t-t_0)]]\right\} \\ &\quad\times\exp\{isV_0\exp[-\gamma(t-t_0)] - isV\} \\ &= \frac{1}{2\pi}\int_{-\infty}^{\infty}\exp\left\{-\frac{Q}{4\gamma}[1-\exp[-2\gamma(t-t_0)]]\left[s+\frac{i2\gamma(V-V_0e^{-\gamma(t-t_0)})}{Q(1-e^{-2\gamma(t-t_0)})}\right]^2\right\} \\ &\quad\times\exp\left\{-\frac{\gamma[V-V_0e^{-\gamma(t-t_0)}]^2}{Q(1-e^{-2\gamma(t-t_0)})}\right\}\end{aligned} \tag{5.53}$$

Using the identity

$$\int_{-\infty}^{\infty} e^{-\alpha x^2}dx = \sqrt{\frac{\pi}{\alpha}}$$

(5.53) reduces to

$$P(V,t) = \sqrt{\frac{\gamma}{\pi Q[1-\exp[-2\gamma(t-t_0)]]}}\exp\left\{-\frac{\gamma[V-V_0\exp[-\gamma(t-t_0)]]^2}{Q[1-\exp[-2\gamma(t-t_0)]]}\right\} \tag{5.54}$$

This is the transient solution of the Fokker-Planck equation for the initial condition (5.49). Eq.(5.54) gives the time evolution of the probability distribution function $P(V,t)$. Evidentily, the probability distribution function $P(V,t)$ is normalized

$$\int_{-\infty}^{\infty} P(V,t)dV$$

$$= \sqrt{\frac{\gamma}{\pi Q[1-\exp[-2\gamma(t-t_0)]]}} \int_{-\infty}^{\infty} dV \exp\left\{-\frac{\gamma[V-V_0\exp[-\gamma(t-t_0)]]^2}{Q[1-\exp[-2\gamma(t-t_0)]]}\right\}$$

$$= \sqrt{\frac{\gamma}{\pi Q[1-\exp[-2\gamma(t-t_0)]]}} \left\{\frac{\gamma}{\pi Q[1-\exp[-2\gamma(t-t_0)]]}\right\}^{\frac{-1}{2}} = 1$$

When $t \to \infty$, (5.54) reduces to

$$P(V,t) = \sqrt{\frac{\gamma}{\pi Q}} \exp\left[-\frac{\gamma V^2}{Q}\right] \tag{5.55}$$

This is exactly (5.35). So when $t \to \infty$, the probability distribution function $P(V)$ reaches to its steady-state value, meanwhile the Brownian particle and the reservoir reach to the thermal equilibrium .

By the aid of eqs.(5.54) and (5.31), we can calculate the ℓ-th moment $\langle V^\ell \rangle$ of the velocity. However, calculating the mean value

$$C(\xi,t) = \langle e^{i\xi V} \rangle = \int dV \exp(i\xi V) P(V,t) \tag{5.56}$$

may be easier than calculating (5.31), here ξ is a real parameter. And all the moments of V can be found from $C(\xi,t)$ because

$$\langle V^\ell \rangle = \frac{\partial^\ell}{\partial(i\xi)^\ell} C(\xi,t)|_{\xi=0} \qquad (\ell = 0,1,2,\cdots) \tag{5.57}$$

That is, the ℓ-th moment is obtained by differentiating $C(\xi,t)$ ℓ times and then letting $\xi = 0$. In this sense $C(\xi,t)$ is usually called as the moment generating function or the characteristic function for V.

Like the Langevin equation, the Fokker-Planck equation with one stochastic variable can be generalized to the case for describing the system with a number of stochastic variables. The Fokker-Planck equation corresponding to the Langevin equation (5.17) or (5.18) is usually written as

$$\frac{\partial}{\partial t} P(x,t) = -\sum_i \frac{\partial}{\partial x_i} [B_i(x)P] + \frac{1}{2} \sum_{i,j} \frac{\partial^2}{\partial x_i \partial x_j} [Q_{ij}(x)P] \tag{5.58}$$

with

$$B_i(x) = \lim_{\Delta t \to 0} \frac{\langle \Delta x_i(t) \rangle}{\Delta t} \tag{5.59}$$

$$Q_{ij}(x) = \lim_{\Delta \to 0} \frac{1}{\Delta t} \langle \Delta x_i(t) \Delta x_j(t) \rangle \tag{5.60}$$

It is easy to verify that

$$Q_{ij} = G_{ij} \tag{5.61}$$

here G_{ij} is the matrix element measuring the noise strengthe of the stochastic force defined by (5.18).

5.2 Master equation for a quantum harmonic oscillator and a two-level atom

From the above section we know that for a classical small system which couples to a large reservoir, by solving the Fokker-Planck equation we can obtain the probability distribution function $P(V,t)$, furthermore from (5.31) we can know the mean value of the stochastic quantity describing the characteristic of the small system with the time evolution. However, for a quantized system coupled to a resevoir, can we find an equation whose characteristic is similar to the Fokker-Planck equation ?

As we know, in quantum mechanics for two systems A and B which couple to each other, if we are only interested in the expectation value of quantity M belonging to the system A, then we can adopt the reduced density operator ρ_A to describe the system A. ρ_A is defined by

$$\rho_A = \mathrm{Tr}_B \rho_{AB} \tag{5.62}$$

here ρ_{AB} is the density operator describing the total system A and B, Tr_B means tracing for the system B only. From (5.62), it is easy to calculate the expectation value of the quantity M in system A

$$\langle M \rangle = \mathrm{Tr}_A \rho_A M \tag{5.63}$$

Comparing (5.63) with (5.31) we find that ρ_A has the same function as $P(V,t)$ in the classical statistic mechanics, which allows us to calculate mean values for the small quantized system. Thus for a small quantized system coupled to a reservoir, we only need to solve the time evolution of the reduced density operator. We next introduce a method to treat the behavior of a small quantized system coupled to the reservoir in quantum mechanics through discussing a quantum harmonic oscillator as well as a two-level atom coupled to a radiation field, respectively.

5.2.1 Master equation for a quantum harmonic oscillator

Supposing that the small system S is a quantized harmonic oscillator, and the reservoir R is a radiation field with infinite modes. The Hamiltonian describing the oscillator coupling to the reservoir is

$$H = H_R + H_S + H_{RS} \tag{5.64}$$

where

$$H_S = \hbar\omega_0 a^\dagger a \tag{5.65}$$

$$H_R = \sum_k \hbar\omega_k a_k^\dagger a_k \tag{5.66}$$

$$H_{RS} = \hbar\sum_k (\varepsilon_k a_k^\dagger a + \text{h.c.}) \tag{5.67}$$

stand for the corresponding Hamiltonians of the oscillator, the radiation field and their interaction, respectively. According to the theory of the density operator introduced in Section 1.4, the density operator $\rho^I(t)$ of the above coupling system in the interaction picture obeys

$$i\hbar\frac{\partial}{\partial t}\rho^I(t) = [H_{RS}^I(t), \rho^I(t)] \tag{5.68}$$

here

$$H_{RS}^I(t) = \exp\left[\frac{i}{\hbar}(H_R + H_S)(t - t_0)\right] H_{RS} \exp\left[-\frac{i}{\hbar}(H_R + H_S)(t - t_0)\right] \tag{5.69}$$

Since we are only interested in the evolution of the system S, we can use the reduced density operator. The reduced density operator is defined by

$$R(t) = \text{Tr}_R\rho^I(t) \tag{5.70}$$

If we obtain the expression of $R(t)$, then we have all informations of the system S. Assuming the interaction between S and R takes place from time t=0, then the initial condition of the total system is

$$\rho^I(t) = R(0)\Gamma(0) \tag{5.71}$$

here $\Gamma(0)$ is the initial density operator of the reservoir. Inasmuch as the interaction between the oscillator and the radiation field does not happen at the initial time, we can regard the reservoir in the equilibrium state, that is, $\Gamma(0)$ may be written as

$$\Gamma(0) = \frac{\exp(-H_R/k_BT)}{\mathrm{Tr}_R[\exp(-H_R/k_BT)]} \tag{5.72}$$

Integrating (5.68) from 0 to t yields

$$\rho^I(t) = \frac{1}{i\hbar}\rho^I(0) + \frac{1}{i\hbar}\int_0^t [H_{RS}(t'), \rho^I(t')]dt' \tag{5.73}$$

Inserting (5.73) into (5.68), it gives

$$\frac{\partial}{\partial t}\rho^I(t) = \frac{1}{i\hbar}[H^I_{RS}(t), \rho^I(0)] + \left(\frac{1}{i\hbar}\right)^2 \int_0^t [H^I_{RS}(t), [H^I_{RS}(t'), \rho(t')]]dt' \tag{5.74}$$

By taking the trace with respect to R in (5.70), then obtain the above equation of the reduced density operator $R(t)$ as

$$\frac{\partial}{\partial t}R(t) = -\frac{1}{\hbar^2}\int_0^t \mathrm{Tr}_R[H^I_{RS}(t), [H^I_{RS}(t'), \rho^I(t')]]dt' \tag{5.75}$$

Here we considered that there is no interaction between the small system and the reservoir at initial time, therefore

$$\mathrm{Tr}_R[H^I_{RS}(t), \rho^I(0)] = 0 \tag{5.76}$$

Eq.(5.75) is just the master equation describing the time evolution of a small system coupled to a reservoir. If the reservoir is so large and the interaction with the small system is so weak that the variation of the reservoir may be neglgible with the time development, in this sense, it is reasonable to factorize the density operator $\rho^I(t)$ as

$$\rho^I(t) = R(t)\Gamma(0) \tag{5.77}$$

Therefore, the master equation can be rewritten as

$$\frac{\partial}{\partial t}R(t) = -\frac{1}{\hbar^2}\int_0^t \mathrm{Tr}_R[H^I_{RS}(t), [H^I_{RS}(t'), R(t')\Gamma(0)]]dt'$$

If we assume that the part which is explicitly unequal to zero in the integrating formula occurs at $t' \approx t$, then as a reasonable approximation, we may replace $R(t')$ by $R(t)$ in the integrand function, this approximation is called the Markov approximation. After making the Markov approximation, we obtain the generalized master equation to be

$$\frac{\partial}{\partial t} R(t) = -\frac{1}{\hbar^2} \int_0^t \mathrm{Tr}_R [H^I_{RS}(t), [H^I_{RS}(t'), R(t)\Gamma(0)]] dt' \tag{5.78}$$

Noticing when we derive (5.78), the expression of the interaction Hamiltonian $H_{RS}(t)$ is not specialized, so (5.78) can be used to treat the time behavior of a number of small different systems which coupled to a reservoir.

We next discuss the master equation of a quantized harmonic oscillator. According to (5.65)-(5.67), the interaction Hamiltonian between a harmonic oscillator and the radiation field in the interaction picture gives

$$H^I_{RS}(t) = \hbar \left\{ a^\dagger \sum_k \varepsilon_k a_k \exp[-i(\omega_k - \omega_0)t] + \text{h.c.} \right\} \tag{5.79}$$

here we have used

$$\exp(\chi a^\dagger a) a \exp(-\chi a^\dagger a) = a e^{-\chi} \tag{5.80}$$

$$\exp(\chi a^\dagger a) a^\dagger \exp(-\chi a^\dagger a) = a^\dagger e^{-\chi} \tag{5.81}$$

Substituting (5.79) into (5.78), the commutation relation in (5.78) yields

$$\begin{aligned}
&[H^I_{RS}(t), [H^I_{RS}(t'), R(t)\Gamma(0)]] \\
&= \{H^I_{RS}(t) H^I_{RS}(t') R(t)\Gamma(0) - H^I_{RS}(t) R(t)\Gamma(0) H^I_{RS}(t') \\
&- H^I_{RS}(t') R(t)\Gamma(0) H^I_{RS}(t) + R(t)\Gamma(0) H^I_{RS}(t') H^I_{RS}(t)\}
\end{aligned} \tag{5.82}$$

the first term in the right hand of the above equation is

$$\begin{aligned}
&H^I_{RS}(t) H^I_{RS}(t') R(t)\Gamma(0) \\
&= \hbar^2 \left\{ a^\dagger \sum_k \varepsilon_k a_k \exp[-i(\omega_k - \omega_0)t] + \text{h.c.} \right\} \\
&\times \left\{ a^\dagger \sum_k \varepsilon_k a_k \exp[-i(\omega_k - \omega_0)t'] + \text{h.c.} \right\} R(t)\Gamma(0)
\end{aligned}$$

$$
\begin{aligned}
&= \hbar^2 a a^\dagger \sum_k \varepsilon_k^2 a_k^\dagger a_k R(t)\Gamma(0) \exp[-i(\omega_k - \omega_0)(t - t')] \\
&+ \hbar^2 a^\dagger a \sum_k \varepsilon_k^2 a_k a_k^\dagger R(t)\Gamma(0) \exp[i(\omega_k - \omega_0)(t - t')] \\
&+ \hbar^2 a^{\dagger 2} \sum_k \varepsilon_k^2 a_k^2 R(t)\Gamma(0) \exp[-i(\omega_k - \omega_0)(t + t')] \\
&+ \hbar a^2 \sum_k \varepsilon_k^2 a_k^{\dagger 2} R(t)\Gamma(0) \exp[i(\omega_k - \omega_0)(t + t')]
\end{aligned} \tag{5.83}
$$

Calculating the trace of the above equation with respect to the reservoir, the third and fourth terms are zero. But calculating the trace of the first term yields

$$
\begin{aligned}
\mathrm{Tr}_R a a^\dagger a_k^\dagger a_k R(t)\Gamma(0) &= \sum_{n_k} a a^\dagger R(t) \langle n_k | a_k^\dagger a_k \Gamma(0) | n_k \rangle \\
&= \sum_{n_k} a a^\dagger R(t) \left\langle n_k \middle| a^\dagger a_k \frac{\exp(-H_R/k_B T)}{\mathrm{Tr}_R \exp(-H_R/k_B T)} \middle| n_k \right\rangle \\
&= \sum_{n_k} a a^\dagger R(t) \overline{n}_k
\end{aligned} \tag{5.84}
$$

here we have used (5.72), and the mean photon number of the radiation field in the thermal equilibrium is

$$
\overline{n}_k = [\exp(\hbar\omega/k_B T) - 1]^{-1} \tag{5.85}
$$

Using the relation $a_k a_k^\dagger = a_k^\dagger a_k + 1$ and following the similar method, we can also obtain the trace of the second term. Substituting all these results into (5.78) and using

$$
\int_0^t \sum_k a a^\dagger R(t) \overline{n}_k \varepsilon_k^2 \exp[i(\omega_0 - \omega_k)(t - t')] dt' = a a^\dagger \overline{n}_0 \gamma_h R(t) \tag{5.86}
$$

with

$$
\gamma_h = 2\pi \sum_k |\varepsilon_k|^2 \delta(\omega_k - \omega_0) \tag{5.87}
$$

$$
\overline{n}_0 = [\exp(\hbar\omega_0/k_B T) - 1]^{-1} \quad (when\ \omega_k = \omega_0,\ \overline{n}_k = \overline{n}_0) \tag{5.88}
$$

we then obtain the master equation of a quantized harmonic oscillator coupled to the radiation field to be

$$\frac{\partial}{\partial t}R(t) = \frac{\gamma_h}{2}\{[a, R(t)a^\dagger] + [aR(t), a^\dagger]\} + \gamma_h \bar{n}_0 [a, [R(t), a^\dagger]] \tag{5.89}$$

It is also called the master equation of a damped quantized harmonic oscillator.

There are several different methods to solve (5.89), here we solve it as follows. In the photon number representation, the time evolution of the matrix elements $\langle n|R(t)|m\rangle$ are written from (5.89) as

$$\begin{aligned}
\frac{d}{dt}\langle n|R(t)|m\rangle &= \frac{\gamma_h}{2}\{\langle n|[a, R(t)a^\dagger]|m\rangle + \langle n|[aR(t), a^\dagger]|m\rangle\} \\
&\quad + \gamma_h \bar{n}_0 \langle n|[a, [R(t), a^\dagger]]|m\rangle \\
&= \frac{\gamma_h}{2}\{\langle n|aR(t)a^\dagger|m\rangle - \langle n|R(t)a^\dagger a|m\rangle + \langle n|aR(t)a^\dagger|m\rangle \\
&\quad - \langle n|a^\dagger aR(t)|m\rangle\} + \gamma_h \bar{n}_0 \{\langle n|aR(t)a^\dagger|m\rangle - \langle n|aa^\dagger R(t)|m\rangle \\
&\quad + \langle n|a^\dagger R(t)a|m\rangle - \langle n|R(t)a^\dagger a|m\rangle\} \\
&= \frac{\gamma_h}{2}\{2\sqrt{(n+1)(m+1)}\langle n+1|R(t)|m+1\rangle - 2m\langle n|R(t)|m\rangle\} \\
&\quad + \gamma_h \bar{n}_0 \{\sqrt{(n+1)(m+1)}\langle n+1|R(t)|m+1\rangle - (n+1)\langle n|R(t)|m\rangle \\
&\quad - m\langle n|R(t)|m\rangle + \sqrt{nm}\langle n-1|R(t)|m-1\rangle\}
\end{aligned} \tag{5.90}$$

The diagonal elements $P_n = \langle n|R(t)|n\rangle$ represent the population or the probability of the system in $|n\rangle$. From (5.90) we know that the time-dependence of P_n obeys

$$\begin{aligned}
\frac{d}{dt}P_n(t) &= \gamma_h[(n+1)P_{n+1} - nP_n] \\
&\quad + \gamma_h \bar{n}_0[(n+1)P_{n+1} - (n+1)P_n - nP_n + nP_{n-1}] \\
&= \gamma_h(\bar{n}_0 + 1)(n+1)P_{n+1} \\
&\quad - \gamma_h[\bar{n}_0(n+1) + n(\bar{n}_0 + 1)]P_n + \gamma_h n \bar{n}_0 P_{n-1}
\end{aligned} \tag{5.91}$$

This is one-step equation of the populations for a damped oscillator, its solution depends on different physical conditions. For the thermal equilibrium case, the population is independent of time. Therefore, we may have the steady-state solution of equation (5.91).

$$\frac{d}{dt}P_n = 0 \tag{5.92}$$

On the basis of the steady-state condition (5.92), we may find the exact expression of P_n . Since for n=0, (5.91) and (5.92) give

$$0 = -\gamma_h \bar{n}_0 P_0 + \gamma_h(\bar{n}_0 + 1)P_1$$

so that we have

$$P_1 = \frac{\bar{n}_0}{\bar{n}_0 + 1} P_0 \tag{5.93}$$

For n=1, eqs.(5.91) and (5.92) yield

$$-\gamma_h(\bar{n}_0 + 1 + 2\bar{n}_0)P_1 + \gamma_h(\bar{n}_0 + 1)2P_2 + \gamma\bar{n}_0 P_0 = 0$$

then

$$P_2 = \frac{\bar{n}_0^2}{(\bar{n}_0 + 1)^2} P_0 \tag{5.94}$$

Iterating this calculating process, we obtain

$$P_n = \frac{\bar{n}_0^n}{(\bar{n}_0 + 1)^n} P_0 \tag{5.95}$$

P_0 can be determined by the normalization condition

$$P_0 = \frac{1}{\bar{n}_0 + 1} \tag{5.96}$$

So the stationary solution of (5.91) gives

$$P_n = \frac{\bar{n}_0^n}{(\bar{n}_0 + 1)^{n+1}} \tag{5.97}$$

In the thermal equilibrium case, the reduced density operator of the quantized oscillator coupled to the radiation field is therefore

$$R = \sum_n P_n |n\rangle\langle n| = \sum_n \frac{\bar{n}_0^n}{(\bar{n}_0 + 1)^{n+1}} |n\rangle\langle n| \tag{5.98}$$

This equation is the same as (3.191).

A particular interesting representation of the reduced density operator $R(t)$ (5.98) is the coherent state representation. Inasmuch as a complete set of coherent states have overcomplete property, we may expand $R(t)$ in the diagonal representation of coherent states. In this sense, R can be expanded as

$$R = \int P(\alpha)|\alpha\rangle\langle\alpha| d^2\alpha \tag{5.99}$$

Using the expression of the coherent state $|\beta\rangle$ in number state representation

$$|\beta\rangle = \exp(-|\beta|^2/2)\sum_m \frac{\beta^m}{\sqrt{m!}}|m\rangle \tag{5.100}$$

the expectation value of R in state $|\beta\rangle$ reads

$$\begin{aligned}\langle\beta|R|\beta\rangle &= \sum_n \frac{\bar{n}_0^n}{(\bar{n}_0+1)^{n+1}}\sum_{m,k}\frac{\beta^{*m}\beta^k}{\sqrt{m!k!}}\langle m|n\rangle\langle n|k\rangle\exp(-|\beta|^2)\\ &= \sum_n \frac{\bar{n}_0^n}{(\bar{n}_0+1)^{n+1}}\exp(-|\beta|^2)\frac{|\beta|^{2n}}{n!} = \frac{1}{\bar{n}_0+1}\sum_n\left(\frac{\bar{n}_0|\beta|^2}{\bar{n}_0+1}\right)^n\frac{1}{n!}\exp(-|\beta|^2)\\ &= \frac{1}{\bar{n}_0+1}\exp\left(-\frac{|\beta|^2}{\bar{n}_0+1}\right)\end{aligned} \tag{5.101}$$

Whereas the diagonal matrix elements of R in the coherent representation is

$$\begin{aligned}\langle\beta|R|\beta\rangle &= \int P(\alpha)\langle\beta|\alpha\rangle\langle\alpha|\beta\rangle d^2\alpha\\ &= \int d^2\alpha\exp(-|\beta-\alpha|^2)P(\alpha)\end{aligned} \tag{5.102}$$

here we have used (3.83). Comparing (5.102) with (5.101), $P(\alpha)$ obeys

$$\frac{1}{\bar{n}_0+1}\exp\left(-\frac{|\beta|^2}{\bar{n}_0+1}\right) = \int d^2\alpha\exp(-|\beta-\alpha|^2)P(\alpha) \tag{5.103}$$

If we let $\alpha = x+iy$, then the Fourier transformation of $P(\alpha)$ is defined by

$$C(k_x,k_y) = \int_{-\infty}^{\infty}\int_{-\infty}^{\infty} P(\alpha)\exp[-i(k_x x+k_y y)]dxdy \tag{5.104}$$

the corresponding inverse transformation gives

$$P(\alpha) = \left(\frac{1}{2\pi}\right)^2\int_{-\infty}^{\infty}\int_{-\infty}^{\infty} C(k_x,k_y)\exp[i(k_x x+k_y y)]dk_x dk_y \tag{5.105}$$

Noticing that the integration in (5.103) is a convolution of $P(\alpha)$ with a Gaussian function $e^{-|\alpha|^2}$, and the Fourier transformation of the function of $e^{-\lambda|\alpha|^2} =$

$\exp[-\lambda(x^2+y^2)]$ reads

$$C_\lambda(k_x,k_y)=\int_{-\infty}^{\infty}\int_{-\infty}^{\infty}\exp(-\lambda|\alpha|^2)\exp[-i(k_x x+k_y y)]dxdy$$
$$=\frac{\pi}{\lambda}\exp\left[-\frac{1}{4\lambda}(k_x^2+k_y^2)\right] \quad (5.106)$$

then the Fourier transformation expression of (5.103) gives

$$C(k_x,k_y)=\pi\exp\left[-\frac{\overline{n}_0}{4}(k_x^2+k_y^2)\right] \quad (5.107)$$

The inverse transformation of the above equation is obtained from (5.105) as

$$P(\alpha)=\frac{1}{\pi\overline{n}_0}\exp\left(-\frac{|\alpha|^2}{\overline{n}_0}\right) \quad (5.108)$$

Therefore, in the diagonal representation of the coherent state representation, the reduced density operator R of the quantized harmonic oscillator in thermal equilibrium limit is

$$R(\alpha)=\frac{1}{\pi\overline{n}_0}\int d^2\alpha\exp\left(-\frac{|\alpha|^2}{\overline{n}_0}\right)|\alpha\rangle\langle\alpha| \quad (5.109)$$

This shows that the population probability distribution function $P(\alpha)$ (5.109) obeys the Gaussian distribution as well as P_n (5.97). That is to say, both the density operators (5.98) and (5.109) describe the stochastic motion of the quantized oscillator with Gaussian statistical characteristic. In this case, the probability distribution (5.97) or (5.109) may be regarded as the statistical distribution superposed by a large number of single-mode electromagnetic fields with random phases, so it has the Gaussian statistical property.

If the oscillator does not have random motion and is in a fixed coherent state $|\alpha\rangle$, then we will find that the population of the oscillator has different statistical property. Supposing the oscillator in the coherent state $|\alpha_0\rangle$, the density operator is written as

$$R(\alpha_0)=|\alpha_0\rangle\langle\alpha_0| \quad (5.110)$$

Expanding it in the diagonal representation of coherent states, we have

$$|\alpha_0\rangle\langle\alpha_0|=\int P(\alpha)|\alpha\rangle\langle\alpha|d^2\alpha \quad (5.111)$$

Evidently, (5.111) requests

$$P(\alpha) = \delta(\alpha - \alpha_0) \tag{5.112}$$

Thus $P(\alpha)$ is a delta function in the coherent state representation. But in the number state representation, (5.110) is rewritten as

$$R(\alpha_0) = \sum_n \frac{|\alpha_0|^{2n}}{n!} \exp(-|\alpha_0|^2)|n\rangle\langle n| \tag{5.113}$$

and the probability distribution reads

$$P_n = \frac{|\alpha_0|^{2n}}{n!} \exp(-|\alpha_0|^2) \tag{5.114}$$

Clearly this probability distribution function is completely different from (5.98). Eq.(5.114) has the Poissonian distribution property, it represents the quantized statistical behavior of the system in a coherent state. However, the Gaussian distribution function (5.98) describes the statistical behavior of the quantum system in a chaotic state.

5.2.2 Master equation for a two-level atom coupled to a bath field

In quantum optics, another typical small system coupled to a reservoir may be considered as a single two-level atom coupling to a thermal field with infinite modes. Now we derive the master equation for this two-level atom.

In the rotating-wave approximation, the Hamiltonian describing a two-level atom interacting with a heat bath field is

$$H = H_A + H_R + V_{A-R} \tag{5.115}$$

with

$$\begin{aligned} H_R &= \sum_k \omega_k a_k^\dagger a_k \\ H_A &= \omega_0 S_z \\ V_{A-R} &= \sum_k g_k (a_k S_+ + a_k^\dagger S_-) \end{aligned} \tag{5.116}$$

In the interaction picture, the atom-field interaction Hamiltonian becomes

$$V_{A-R}^I = \sum_k g_k \{a_k S_+ \exp[i(\omega_0 - \omega_k)t] + a_k^\dagger S_- \exp[-i(\omega_0 - \omega_k)t]\} \tag{5.117}$$

Substituting (5.117) into (5.78) and iterating the process similar to eqs.(5.82)-(5.85), we obtain the master equation for the two-level atom damping in the bath field as

$$\frac{d}{dt}R = \frac{\Gamma}{2}[2\bar{n}_0 S_+ R S_- + 2(\bar{n}_0+1)S_- R S_+ - \bar{n}_0 R S_- S_+ \\ -(\bar{n}_0+1)RS_+S_- - \bar{n}_0 S_- S_+ R - (\bar{n}_0+1)S_+S_-R] \tag{5.118}$$

here

$$\Gamma = 2\pi \sum_k g_k^2 \delta(\omega_k - \omega_0) \tag{5.119}$$

is the spontaneous decay rate of the two-level atom in free vacuum field. Starting from (5.118) we can discuss the effect of the bath field on the behavior of the two-level atom. Next we discuss the time evolution of population inversion of a two-level atom under the interaction of bath field.

Noticing that in the interaction picture the atomic inversion operator S_z obeys

$$S_z^I = S_z^S = S_z \tag{5.120}$$

then from (5.118) and (4.17) it is easy to obtain

$$\frac{d}{dt}\langle S_z\rangle = -\Gamma[(2\bar{n}_0+1)\langle S_z\rangle + 1/2] \tag{5.121}$$

If the atom initially in its excited state $|+\rangle$, namely, $\langle S_z(0)\rangle = 1/2$, then the solution of (5.121) gives

$$\langle S_z(t)\rangle = \frac{1}{2\bar{n}_0+1}\{(\bar{n}_0+1)\exp[-(2\bar{n}_0+1)\Gamma t] - 1/2\} \tag{5.122}$$

Evidently, when n=0, (5.122) is just (4.48). That is to say, the two-level atom which initially in its excited state should spontaneously decay with rate Γ to its ground state due to the effect of vacuum fluctuations. However, if the mean photon number of the bath field is different from zero, then we see from (5.122) that the decay rate of the atom is $(2\bar{n}_0+1)\Gamma$. When $t \to \infty$, (5.122) becomes

$$\langle S_z(\infty)\rangle = -\frac{1}{2(2\bar{n}_0+1)} \tag{5.123}$$

So with the increasing of the mean photon number of the bath field, the atomic decay rate becomes fast, but the atomic decay to its ground state is incomplete.

5.3 Characteristic function and quasi-probability distribution for the quantum harmonic oscillator

From Section 5.1 we see that for the classical Brownian motion, if the probability distribution function $P(V,t)$ of the Brownian particle is obtained, then we can calculate an arbitrary order moment of the velocity by means of the characteristic function $C(\xi,t)$. However, for the quantized harmonic oscillator coupled to a reservoir, does there exist a similar characteristic function ? Although the density operator R describing a quantized harmonic oscillators has the similar property as the classical probability distribution, from (5.89) we see that R is a function of the operators a and $a^\dagger$. On the other hand, some quantities such as the product of m-order power of operator a and n-order power of operator $a^\dagger$ have $(n+m)!/(n!m!)$ permutation orders (for example $a^m a^{\dagger n}$, $a^{\dagger n} a^m$, $a^\dagger a^m (a^\dagger)^{n-1}$, $(a^\dagger)^\ell a^k (a^\dagger)^{n-\ell} a^{m-k}, \cdots$). Inasmuch as a and $a^\dagger$ do not commutate with each other, we must consider the different permutation order of a and $a^\dagger$ when we define the characteristic function of the quantized harmonic oscillator.

As usual, there are three types of ordering of the operators a and $a^\dagger$, which need to be considered.

(1) Normal ordering– A product of m annihilation operators and n creation operators is normally ordered if all of the annihilation operators are on the right, i.e., if it is in the form $(a^\dagger)^n a^m$.

(2) Anti-normal ordering– A product of m annihilation operators and n creation operators is anti-normally ordered if all of the annihilation operators are on the left, i.e., if it is of the form $a^m (a^\dagger)^n$.

(3) Symmetric ordering–A product of m annihilation operators and n creation operators can be ordered in $(n+m)!/(n!m!)$ ways. The symmetrical ordering product of these operators, denoted by $\{(a^\dagger)^n a^m\}$, is just to average all of these different order products. For example

$$\{a^\dagger a\} = \frac{1}{2}(aa^\dagger + a^\dagger a)$$

$$\{a^\dagger a^2\} = \frac{1}{3}(a^\dagger a^2 + aa^\dagger a + a^2 a^\dagger)$$

$$\{a^{\dagger 2} a^2\} = \frac{1}{6}(a^{\dagger 2} a^2 + a^\dagger aa^\dagger a + a^\dagger a^2 a^\dagger + aa^{\dagger 2} a + aa^\dagger aa^\dagger + a^2 a^{\dagger 2})$$

Next we introduce the corresponding characteristic functions and the expressions of the density operator for given operators in three orderings. We will find that these operators in three orderings are associated with C-number functions.

5.3.1 Normal ordering representation

Let us suppose that we can expand a given operator $A(a, a^\dagger)$ in a normal ordering power series as

$$A = \sum_{n,m=0}^{\infty} C_{nm}(a^\dagger)^n a^m \tag{5.124}$$

Inasmuch as the set of coherent states $\{|\alpha\rangle\}$ has overcomplete property, we further suppose that the density matrix can be expressed as

$$\rho = \int d^2\alpha P(\alpha)|\alpha\rangle\langle\alpha| \tag{5.125}$$

here $P(\alpha)$ is a C-number and is usually called the P-representation of the density operator. Because

$$\langle\alpha|(a^\dagger)^n a^m|\alpha\rangle = (\alpha^*)^n \alpha^m \tag{5.126}$$

the expectation value of the normal ordering operator A in the P-representation yields

$$\begin{aligned}\langle A\rangle = \mathrm{Tr}\rho A = \int d^2\alpha P(\alpha)\langle\alpha|A|\alpha\rangle = \int d^2\alpha P(\alpha)\sum_{n,m=0}^{\infty} C_{nm}(\alpha^*)^n\alpha^m \\ = \int d^2\alpha P(\alpha) A_N(\alpha, \alpha^*)\end{aligned} \tag{5.127}$$

with

$$A_N(\alpha, \alpha^*) = \langle\alpha|A|\alpha\rangle = \sum_{n,m=0}^{\infty} C_{nm}(\alpha^*)^n\alpha^m \tag{5.128}$$

Hence when we evaluate the expectation value of the normal ordering operator $A(a, a^\dagger)$ in the P-representation of the density operator, we can replace $A(a, a^\dagger)$ by the C-number function $A_N(\alpha, \alpha^*)$.

Noticing that

$$(a^\dagger)^n a^m = \frac{\partial^{n+m}}{\partial(i\eta\xi^*)^n \partial(i\eta\xi)^m} \exp(i\eta\xi^* a^\dagger)\exp(i\eta\xi a^\dagger)|_{\eta=0} \tag{5.129}$$

if we obtain the expectation value of the normal ordering operator $e^{i\eta\xi^* a^\dagger} e^{i\eta\xi a}$, we can calculate the expectation value of arbitrary operator $a^{\dagger n} a^m$. By means of (5.127) we have the characteristic function which has the similar property as (5.56)

$$\chi_N(\xi) = \mathrm{Tr}[\rho \exp(i\eta\xi^* a^\dagger) \exp(i\eta\xi a)] = \int d^2\alpha P(\alpha) \exp[i\eta(\xi^*\alpha^* + \xi\alpha)] \tag{5.130}$$

This function $\chi_N(\xi)$ is known as the normally ordering characteristic function. By the aid of the above equation, we can easily obtain the moment of the harmonic operator in normal ordering as

$$\langle (a^\dagger)^n a^m \rangle = \mathrm{Tr}\rho (a^\dagger)^n a^m = \frac{\partial^{n+m}}{\partial(i\eta\xi^*)^n \partial(i\eta\xi)^m} \chi_N(\xi)|_{\eta=0} \tag{5.131}$$

Clearly, (5.130) has the same property as (5.56). Notice that $P(\alpha)$ is just a two-dimensional Fourier transform of $\chi_N(\xi)$, so it can be written as

$$P(\alpha) = \left(\frac{\eta}{2\pi}\right)^2 \int\int d^2\xi \chi_N(\xi) \exp[-i\eta(\xi^*\alpha^* + \xi\alpha)] \tag{5.132}$$

Substituting (5.132) into (5.125), it gives the expression of the density operator in the P-representation.

We now discuss the property of the characteristic function $\chi(\xi)$. From (3.78) we know that

$$e^{i\eta\xi^* a^\dagger} e^{i\eta\xi a} = \exp[i\eta(\xi^* a^\dagger + \xi a)] \exp(\eta^2|\xi|^2/2) \tag{5.133}$$

Since $\exp[i\eta(\xi^* a^\dagger + \xi a)]$ is an unitary operator, we have

$$|\chi_N(\xi)| = \exp(\eta^2|\xi|^2/2)|\mathrm{Tr}(\rho \exp[i\eta(\xi^* a^\dagger + \xi a)]| \leq \exp(\eta^2|\xi|^2/2)$$

For $\rho = |n\rangle\langle n|$, if $|\xi| \gg 1$, then

$$|\chi_N(\xi)| \sim |\xi|^{2n}$$

This means that $\chi(\xi)$ may diverge. The P-representation is therefore not appropriate for any density matrices, in some cases $P(\alpha)$ may be taken as negative values or singularities. Even so, the P-representation is still useful in many of the interesting cases. In the following we discuss the property of the damped oscillator by the aid of the P-representation.

In fact, when the density operator $R(t)$ in (5.89) is replaced by ρ presented in (5.125), we have

$$\frac{\partial}{\partial t}\int P(\alpha,\alpha^*,t)|\alpha\rangle\langle\alpha|d^2\alpha = \frac{\gamma_h}{2}\int[2a|\alpha\rangle\langle\alpha|a^\dagger$$
$$-|\alpha\rangle\langle\alpha|a^\dagger a - a^\dagger a|\alpha\rangle\langle\alpha|]P(\alpha,\alpha^*,t)d^2\alpha + \gamma_h\overline{n}_0[a^\dagger[|\alpha\rangle\langle\alpha|a$$
$$+a|\alpha\rangle\langle\alpha|a^\dagger - a^\dagger a|\alpha\rangle\langle\alpha| - |\alpha\rangle\langle\alpha|aa^\dagger]P(\alpha,\alpha^*,t)d^2\alpha \tag{5.134}$$

Since

$$a^\dagger|\alpha\rangle\langle\alpha| = a^\dagger \exp(-\alpha^*\alpha)\exp(\alpha a^\dagger)|0\rangle\langle 0|\exp(\alpha^* a)$$

and

$$\frac{\partial}{\partial\alpha}\exp(\alpha a - \alpha^*\alpha) = (a^\dagger - \alpha^*)\exp[\alpha(a^\dagger - \alpha^*)]$$

we have

$$a^\dagger|\alpha\rangle\langle\alpha| = \left(\frac{\partial}{\partial\alpha} + \alpha^*\right)\exp[\alpha(a^\dagger - \alpha^*)]|0\rangle\langle 0|\exp(\alpha^* a)$$
$$= \left(\frac{\partial}{\partial\alpha} + \alpha^*\right)|\alpha\rangle\langle\alpha| \tag{5.135}$$

similarly,

$$|\alpha\rangle\langle\alpha|a = \left(\frac{\partial}{\partial\alpha^*} + \alpha\right)|\alpha\rangle\langle\alpha| \tag{5.136}$$

Using eqs.(5.135), (5.136) and $a|\alpha\rangle = \alpha|\alpha\rangle$ and its adjoint, (5.134) becomes

$$\frac{\partial}{\partial t}\int P(\alpha,\alpha^*,t)|\alpha\rangle\langle\alpha|d^2\alpha = \frac{\gamma_h}{2}\int[2|\alpha|^2 - \left(\frac{\partial}{\partial\alpha^*} + \alpha\right)\alpha^*$$
$$-\left(\frac{\partial}{\partial\alpha} + \alpha^*\right)\alpha]|\alpha\rangle\langle\alpha|P(\alpha,\alpha,t)d^2\alpha + \gamma_h\overline{n}_0[\left(\frac{\partial}{\partial\alpha} + \alpha^*\right)\left(\frac{\partial}{\partial\alpha^*} + \alpha\right)$$
$$+|\alpha|^2 - \left(\frac{\partial}{\partial\alpha} + \alpha^*\right)\alpha - \left(\frac{\partial}{\partial\alpha^*} + \alpha\right)\alpha^*]|\alpha\rangle\langle\alpha|P(\alpha,\alpha^*,t)d^2\alpha \tag{5.137}$$

Therefore, the quasi-probability distribution function $P(\alpha,\alpha^*,t)$ satisfies the Fokker-Planck equation

$$\frac{\partial}{\partial t}P(\alpha,\alpha^*,t) = \frac{\gamma_h}{2}\left(\frac{\partial}{\partial\alpha}\alpha + \frac{\partial}{\partial\alpha^*}\alpha^*\right)P(\alpha,\alpha^*,t)$$
$$+\gamma_h\overline{n}_0\frac{\partial^2}{\partial\alpha\partial\alpha^*}P(\alpha,\alpha^*,t) \tag{5.138}$$

The solution of the above equation can be determined with a given initial condition. If we assume that the oscillator is initially in a coherent state $|\alpha'\rangle$, then the corresponding quasi-probability is

$$P(\alpha,\alpha^*,t=0)=\delta(\alpha-\alpha')\delta(\alpha^*-\alpha^{*\prime}) \tag{5.139}$$

Calling the solution (5.54) of (5.20), the solution of (5.138) gives

$$P(\alpha,\alpha^*,t)=\frac{1}{\bar{n}_0[1-\exp(-\gamma_h t)]}\exp\left\{-\frac{[\alpha-\alpha'\exp(-\gamma_h t/2)]^2}{\bar{n}_0[1-\exp(-\gamma_h t)]}\right\} \tag{5.140}$$

The above equation shows that the property of the quantized harmonic oscillator coupled to a bath field, which is initially in a given coherent state with the Poissonian statistical property shown in (5.114), will be changed due to the effect of the bath with the time development. When $t\to\infty$, (5.140) will tend to (5.108). That is to say, with the time development, the oscillator will be transferred from a coherent state to a mixture pure state.

Inserting (5.140) into (5.130), one can obtain the characteristic function $\chi_N(\xi,t)$ of the oscillator, then may evaluate the arbitrary moments $\langle(a^\dagger)^n a^m\rangle$ in normal ordering. Thus to discuss the behavior of a quantized oscillator coupled to a reservoir, we can first derive the master equation for the reduced density of the oscillator, then deduce the classical Fokker-Planck equation which equivalently corresponds to the quantum master equation by the aid of the P-representation of the density operator. Further we may find the quasi-probability distribution function $P(\alpha,\alpha^*,t)$ by solving the Fokker-Planck equation, therefore obtain the characteristic function $\chi(\xi,t)$ and calculate the arbitrary moments of the oscillator operators in normal ordering.

5.3.2 Anti-normal ordering representation

Now we discuss another expression of quasi-probability distribution function $P(\alpha)$ in terms of a series expansion for the density matrix on anti-normal ordering. If we express the density matrix as an anti-normally ordered series

$$\rho=\sum_{n,m=0}^{\infty}\rho_{nm}a^m(a^\dagger)^n \tag{5.141}$$

then the expectation value of the operator $A(a,a^\dagger)$ can be written as

$$\mathrm{Tr}(\rho A)=\sum_{n,m=0}^{\infty}\sum_{r,s=0}^{\infty}\rho_{nm}C_{rs}\mathrm{Tr}[a^m(a^\dagger)^n(a^\dagger)^r a^s] \tag{5.142}$$

The trace in (5.142) can be expressed as

$$\mathrm{Tr}[a^m(a^\dagger)^n(a^\dagger)^r a^s] = \mathrm{Tr}[(a^\dagger)^{r+n}a^{m+s}] \tag{5.143}$$

and notice the identity relation

$$\frac{1}{\pi}\int |\alpha\rangle\langle\alpha| d^2\alpha = 1 \tag{5.144}$$

so that

$$\mathrm{Tr}\rho A = \frac{1}{\pi}\int d^2\alpha \sum_{n,m=0}^{\infty}\sum_{r,s=0}^{\infty} \rho_{nm}C_{rs}(\alpha^*)^n\alpha^m(\alpha^*)^r\alpha^s \tag{5.145}$$

Comparing this with (5.127), we see that

$$P(\alpha) = \frac{1}{\pi}\sum_{n,m=0}^{\infty} \rho_{nm}(\alpha^*)^n\alpha^m \tag{5.146}$$

Thus the density matrix expanded as an anti-normal ordering series may be written as

$$\rho = \frac{1}{\pi}\int d^2\alpha \sum_{n,m=0}^{\infty} \rho_{nm}(\alpha^*)^n\alpha^m|\alpha\rangle\langle\alpha| \tag{5.147}$$

If we expand a given operator $A(a, a^\dagger)$ as an anti-normal ordering series of power

$$A(a, a^\dagger) = \sum_{n,m=0}^{\infty} d_{nm}a^m(a^\dagger)^n \tag{5.148}$$

then by analogy with (5.147), we may express $A(a, a^\dagger)$ as

$$\begin{aligned} A(a, a^\dagger) &= \frac{1}{\pi}\int d^2\alpha \sum_{n,m=0}^{\infty} d_{nm}(\alpha^*)^n\alpha^m|\alpha\rangle\langle\alpha| \\ &= \frac{1}{\pi}\int d^2\alpha A_a(\alpha, \alpha^*)|\alpha\rangle\langle\alpha| \end{aligned} \tag{5.149}$$

where

$$A_a(\alpha, \alpha^*) = \sum_{n,m=0}^{\infty} d_{nm}(\alpha^*)^n\alpha^m \tag{5.150}$$

Similar to the normal ordering case, the operator $A(a, a^{\dagger})$ in the anti-normal ordering is also associated with a C-number function $A_a(\alpha, \alpha^*)$. We then have the expectation value of $A(a, a^{\dagger})$

$$\begin{aligned}\mathrm{Tr}\rho A &= \frac{1}{\pi}\int d^2\alpha A_a(\alpha, \alpha^*)\mathrm{Tr}(\rho|\alpha\rangle\langle\alpha|) \\ &= \int d^2\alpha A_a(\alpha, \alpha^*)Q(\alpha) \end{aligned} \tag{5.151}$$

with

$$Q(\alpha) = \frac{1}{\pi}\langle\alpha|\rho|\alpha\rangle \tag{5.152}$$

If let $A(a, a^{\dagger}) = \exp(i\eta\xi a)\exp(i\eta\xi^* a^{\dagger})$ in (5.151), then the characteristic function in anti-normal ordering can be defined as

$$\chi_A(\xi) = \mathrm{Tr}(\rho e^{i\eta x a} e^{i\eta\xi^* a^{\dagger}}) = \int d^2\alpha Q(\alpha)\exp[i\eta(\xi^*\alpha^* + \xi\alpha)] \tag{5.153}$$

The reason is that an arbitrary moment in the anti-normal ordering can be given from the above equation as

$$\langle a^m(a^{\dagger})^n\rangle = \mathrm{Tr}[\rho a^m(a^{\dagger})^n] = \frac{\partial^{n+m}}{\partial(i\eta\xi^*)^n\partial(i\eta\xi)^m}\chi_A(\xi)|_{\eta=0} \tag{5.154}$$

Taking the Fourier inverse transform of (5.153), the quasi-probability distribution function in the anti-normal ordering is obtained immediately

$$\begin{aligned} Q(\alpha) &= \frac{\eta^2}{4\pi}\int\int d^2\xi\chi_A(\xi)\exp[-i\eta(\xi^*\alpha^* + \xi\alpha)] \\ &= \frac{\eta^2}{4\pi}\int\int d^2\beta\frac{1}{\pi}\langle\beta|\rho|\beta\rangle e^{[-i\eta\xi^*(\alpha^*-\beta^*)]}e^{[-i\eta\xi(\alpha-\beta)]} \\ &= \frac{1}{\pi}\langle\alpha|\rho|\alpha\rangle \end{aligned} \tag{5.155}$$

$Q(\alpha)$ is usually called the Q-representation of the density operator. Evidently, the steady-state quasi-probability distribution function of the quantized harmonic oscillator gives

$$Q(\alpha) = \frac{1}{\pi}\langle\alpha|R|\alpha\rangle = \frac{1}{\pi(\overline{n}_0 + 1)}\exp\left\{-\frac{|\beta|^2}{\overline{n}_0 + 1}\right\} \tag{5.156}$$

Now we examine the properties of the characteristic function $\chi_A(\xi)$ for an anti-normal ordering. Since

$$\exp(i\eta\xi a)\exp(i\eta\xi^* a^\dagger)] = \exp[i\eta(\xi^* a^\dagger + \xi a)]\exp\left(-\frac{\eta|\xi|^2}{2}\right)$$

we have

$$|\chi_A(\xi)| \leq \exp\left(-\frac{\eta^2|\xi|^2}{2}\right) \tag{5.157}$$

So $\chi_A(\xi)$ is a square integrable function when $|\xi|^2 \to \infty$. Therefore, for some systems if in which P- representation do not exist, one can adopt the Q-representation to describe them.

5.3.3 Symmetric ordering representation

Before proceeding to discuss the distribution function and the characteristic function for the symmetric ordering case, we would like to consider some properties of the symmetric ordering scheme itself. We first note that $(\xi_1 a + \xi_2 a^\dagger)^n$ can be expanded in terms of the symmetric ordering as

$$(\xi_1 a + \xi_2 a^\dagger)^n = \sum_{\ell=0}^{\infty} \frac{n!}{\ell!(n-\ell)!}\xi_1^{n-\ell}\xi_2^{\ell}\{(a^\dagger)^{n-\ell}a^\ell\} \tag{5.158}$$

here the symbol { } represents that the product of $n-\ell$ creation operators and ℓ annihilation operators are in symmetric ordering. So that

$$\begin{aligned}\exp[i\eta(\xi a + \xi^* a^\dagger)] &= \sum_{n=0}^{\infty}\frac{1}{n!}(i\eta\xi a + i\eta\xi^* a^\dagger)^n \\ &= \sum_{n=0}^{\infty}\sum_{\ell=0}^{\infty}\frac{1}{\ell!(n-\ell)!}(i\eta\xi^*)^{n-\ell}(i\eta\xi)^\ell\{(a^\dagger)^{n-\ell}a^\ell\}\end{aligned}$$

If we let $m = n - \ell$, the above equation therefore induces

$$\exp[i\eta(\xi a + \xi^* a^\dagger)] = \sum_{\ell,m=0}^{\infty}\frac{1}{m!\ell!}(i\eta\xi^*)^m(i\eta\xi)^\ell\{(a^\dagger)^m a^\ell\} \tag{5.159}$$

The operator-function correspondence is now done in a way analogous to that of the precedent section. Expand an operator $A(a, a^\dagger)$ in a symmetric ordering power series as

$$A(a, a^\dagger) = \sum_{n,m} b_{nm}\{(a^\dagger)^n a^m\} \tag{5.160}$$

Its expectation values is

$$\langle A\rangle = \mathrm{Tr}\rho A = \sum_{n,m} b_{nm}\mathrm{Tr}\rho\{(a^\dagger)^n a^m\} = \sum_{n,m} b_{nm}\langle\{(a^\dagger)^n a^m\}\rangle \tag{5.161}$$

If we solve the moment $\langle\{(a^\dagger)^n a^m\}\rangle$, then we can obtain the expectation value of $A(a, a^\dagger)$ in symmetric ordering representation.

From (5.159) we know

$$\langle\{(a^\dagger)^n a^m\}\rangle = \frac{\partial^{n+m}}{\partial(i\eta\xi^*)^n\partial(i\eta\xi)^m}\langle\exp[i\eta(\xi a + \xi^* a^\dagger)]\rangle|_{\eta=0} \tag{5.162}$$

To calculate the moments $\langle\{(a^\dagger)^n a^m\}\rangle$, we only need to find $\langle\exp[i\eta(\xi a + \xi^* a^\dagger)]\rangle$. Noticing that $\langle\exp[i\eta(\xi a + \xi^* a^\dagger)]\rangle$ in the P-representation reads

$$\begin{aligned}
\chi_W(\xi) &= \mathrm{Tr}\rho\exp[i\eta(\xi a + \xi^* a^\dagger)] \\
&= \int\int d^2\beta P(\beta)\langle\beta|\exp[i\eta(\xi a + \xi^* a^\dagger)]|\beta\rangle \\
&= \int\int d^2\beta P(\beta)\exp(-\eta^2|\xi|^2)\langle\beta|\exp(i\eta\xi^* a^\dagger)\exp(i\eta\xi a)|\beta\rangle \\
&= \int\int d^2\beta P(\beta)\exp[-\eta^2|\xi|^2)\exp[i\eta(\xi^*\beta^* + \xi\beta)]
\end{aligned} \tag{5.163}$$

We can easily calculate the moment $\langle\{(a^\dagger)^n a^m\}\rangle$ by the aid of the above equation, thus we interpret $\chi_W(\xi)$ as the characteristic function in symmetric ordering or the Wigner characteristic function.

Taking the Fourier transform of $\chi_W(\xi)$, it gives

$$W(\alpha) = \frac{\eta^2}{\pi}\int\int d^2\xi\chi_W(\xi)\exp[-i\eta(\xi^*\alpha^* + \xi\alpha)] \tag{5.164}$$

This is just the Wigner distribution function for the symmetric ordering in the representation of coherent states. Inserting (5.163) into (5.164), we have

$$W(\alpha) = \frac{\eta^2}{\pi}\int\int d^2\xi\int\int d^2\beta P(\beta)e^{-\eta^2|\xi|^2}e^{i\eta\xi^*(\beta^*-\alpha^*)}e^{i\eta\xi(\beta-\alpha)}$$

Taking the integration in the ξ-plane, the above equation reduces to

$$W(\alpha) = 2\int\int d^2\beta P(\beta)\exp(-2|\beta-\alpha|^2) \tag{5.165}$$

Therefore, if we know the quasi-probability distribution function $P(\alpha)$ in the P-representation, then we can obtain the Wigner distribution function for the symmetric ordering. For example, for the damping oscillator in its steady-state, substituting its quais-probability distribution function $P(\alpha)$ (5.108) into (5.165), we may obtain the Wigner distribution function as

$$W(\alpha)=2\int\int d^2\beta \frac{1}{\pi\overline{n}_0}\exp(-2|\beta-\alpha|^2)\exp\left(-\frac{|\beta|^2}{\overline{n}_0}\right)$$
$$=\frac{1}{\pi(\overline{n}_0+1/2)}\exp\left(-\frac{|\alpha|^2}{\overline{n}_0+1/2}\right) \tag{5.166}$$

It is also of interest to examine the behavior of $W(\alpha)$. First we note that

$$|\chi_W(\xi)|\leq 1 \tag{5.167}$$

so that $\chi_W(\xi)$ is a square integrable function. Since $W(\alpha)$ is just the Fourier transform of $\chi_W(\xi)$, it is also square integrable. Therefore, $W(\alpha)$ is another good function like $Q(\alpha)$ and one can adopt it to describe the system in which $P(\alpha)$ does not exist.

We next discuss the Wigner distribution function in the coordinate representation. This form is particularly appropriate for obtaining the moments of the momentum p and coordinate q. From (3.87) we know that the coordinate operator q and momentum operator p of a harmonic oscillator can be expressed in terms of the annihilation operator a and creation operator $a^\dagger$ as

$$q=\sqrt{\frac{\hbar}{2\omega}}(a^\dagger+q), \quad a=\frac{1}{\sqrt{2\hbar\omega}}(\omega q+ip)$$
$$p=i\sqrt{\frac{\hbar\omega}{2}}(a^\dagger-a), \quad a^\dagger=\frac{1}{\sqrt{2\hbar\omega}}(\omega q-ip) \tag{5.168}$$

If we define two new real parameters λ and μ by the relations

$$\lambda=\eta(\xi+\xi^*)\sqrt{\frac{\omega}{2\hbar}}, \quad \mu=i\eta(\xi-\xi^*)\frac{1}{\sqrt{2\hbar\omega}} \tag{5.169}$$

and substitute (5.168) and (5.169) into (5.163), then $\chi_W(\xi)$ becomes

$$\chi_W(\lambda,\mu)=\mathrm{Tr}\rho(p,q)\exp[i(\lambda q+\mu p)] \tag{5.170}$$

Because q and p obey

$$[q,p]=i\hbar$$

we may write

$$\begin{aligned}\exp[i(\lambda q+\mu p)] &= \exp(i\lambda q)\exp(i\mu p)\exp(i\hbar\lambda\mu/2)\\ &= \exp(i\mu p/2)\exp(-i\mu p/2)\exp(i\lambda q)\exp(i\mu p/2)\exp(i\mu p/2)\exp(i\hbar\lambda\mu/2)\\ &= \exp(i\mu p/2)\exp[i\lambda e^{-i\mu p/2}q\exp i\mu p/2]\exp(i\mu p/2)\exp(i\hbar\lambda\mu/2)\end{aligned} \tag{5.171}$$

here we have used the relation

$$\exp(\xi A)F(B)\exp(-\xi A) = F(\exp(\xi A)B\exp(-\xi A)] \tag{5.172}$$

Because

$$[q, F(p)] = i\hbar\frac{\partial}{\partial p}F(p) \tag{5.173}$$

then

$$\left[q, \exp\left(i\frac{\mu p}{2}\right)\right] = i\hbar\frac{\partial}{\partial p}\exp\left(i\frac{\mu p}{2}\right) = -\frac{\hbar\mu}{2}\exp\left(i\frac{\mu p}{2}\right)$$

i.e.,

$$q\exp\left(i\frac{\mu p}{2}\right) - \exp\left(i\frac{\mu p}{2}\right)q = -\frac{\hbar\mu}{2}\exp\left(i\frac{\mu p}{2}\right) \tag{5.174}$$

When we multiply both sides from the left by the operator $\exp(-i\mu p/2)$, then we have

$$\exp\left(-i\frac{\mu p}{2}\right)q\exp\left(i\frac{\mu p}{2}\right) = q - \frac{\hbar\mu}{2} \tag{5.175}$$

Inserting (5.175) into (5.171), (5.171) becomes

$$\begin{aligned}\exp[i(\lambda q+\mu p)] &= \exp\left(i\frac{\mu p}{2}\right)\exp\left[i\lambda\left(q-\frac{\hbar\mu}{2}\right)\right]\exp\left(i\frac{\mu p}{2}\right)\exp\left(i\frac{\hbar\lambda\mu}{2}\right)\\ &= \exp(i\mu p/2)\exp(i\lambda q)\exp(i\mu p/2)\end{aligned} \tag{5.176}$$

Therefore, (5.170) reads

$$\chi_W(\lambda,\mu) = \mathrm{Tr}\rho(p,q)\exp(i\mu p/2)\exp(i\lambda q)\exp(i\mu p/2) \tag{5.177}$$

Let us evaluate the trace of (5.176) in the q-representation where

$$q|q'\rangle = q'|q'\rangle, \quad \langle q'|q''\rangle = \delta(q'-q'') \tag{5.178}$$

$$\int_{-\infty}^{\infty} dq'|q'\rangle\langle q'| = 1 \tag{5.179}$$

Furthermore, it is easy to verify that

$$\exp(i\mu p/2)|q'\rangle = |q' - \hbar\mu/2\rangle, \quad \langle q'|e^{i\mu p/2} = \langle q' + \hbar\mu/2| \tag{5.180}$$

Then we may rewrite (5.177) as

$$\begin{aligned}\chi_W(\lambda,\mu) &= \mathrm{Tr}[\exp(i\mu p/2)\rho(p,q)\exp(i\mu p/2)\exp(i\lambda q) \\ &= \int_{-\infty}^{\infty} dq''\langle q''|\exp(i\mu p/2)\rho(p,q)\exp(i\mu p/2)\exp(i\lambda q)|q''\rangle \\ &= \int_{-\infty}^{\infty} dq'' e^{i\lambda q''}\langle q'' + \hbar\mu/2|\rho(p,q)|q'' - \hbar\mu/2\rangle \end{aligned} \tag{5.181}$$

This is just the Wigner characteristic function of the quantized harmonic oscillator in the q-representation.

The corresponding Wigner distribution function $P(p', q')$ is the Fourier transform of $\chi_W(\lambda,\mu)$. Because

$$\int\int d^2\alpha \frac{1}{\pi} W(\alpha) = \int\int_{\infty}^{\infty} dq'dp' P(p', q') \tag{5.182}$$

and from (5.167) we know the parameters

$$\alpha = \frac{1}{\sqrt{2\hbar\omega}}(\omega q' + ip'), \quad \alpha^* = \frac{1}{\sqrt{2\hbar\omega}}(\omega q' - ip') \tag{5.183}$$

then have

$$P(p', q') = \frac{1}{2\pi\hbar} W(\alpha) \tag{5.184}$$

Inserting (5.183) and (5.169) into (5.164), then $P(p', q')$ reads

$$\begin{aligned} P(p', q') &= \frac{1}{4\pi^2}\int\int d\lambda d\mu \chi_W(\lambda,\mu)\exp[-i(\mu p' + \lambda q')] \\ &= \frac{1}{4\pi^2}\int dq'' \int\int d\lambda d\mu e^{i\lambda(q''-q')} e^{-i\mu p'}\langle q'' + \hbar\mu/2|\rho(p,q)|q'' - \hbar\mu/2\rangle \\ &= \frac{1}{2\pi}\int dq'' \int d\mu \delta(q' - q'')\langle q'' + \hbar\mu/2|\rho(p,q)|q' - \hbar\mu/2\rangle e^{-i\mu p'} \\ &= \frac{1}{2\pi}\int d\mu \langle q' + \hbar\mu/2|\rho(p,q)|q' - \hbar\mu/2\rangle \exp(-i\mu p') \end{aligned} \tag{5.185}$$

This is the Wigner distribution function in the coordinate representation. Finally, we let

$$\hbar\mu = s \tag{5.186}$$

then $P(p', q')$ reduces to

$$P(p', q') = \frac{1}{2\pi\hbar}\int ds \langle q' + s/2|\rho(p, q)|q' - s/2\rangle \exp(-ip's/\hbar) \tag{5.187}$$

Similarly, we can express $P(p', q')$ in the momentum representation. By use of the completeness relation

$$\int dp_1 |p_1\rangle\langle p_1| = 1 \tag{5.188}$$

(5.187) becomes

$$P(p', q') = \frac{1}{2\pi\hbar}\int ds \int dp_1 \int dp_2 \langle q' + s/2|p_1\rangle\langle p_1|\rho(p, q)|p_2\rangle \times\langle p_2|q' - s/2\rangle \exp(-ip's/\hbar) \tag{5.189}$$

Since

$$\langle p_1|q_1\rangle = \frac{1}{2\pi\hbar}\exp(-ip_1q_1/\hbar) = \langle q_1|p_1\rangle^* \tag{5.190}$$

(5.189) yields

$$\begin{aligned}
P(p', q') &= \left(\frac{1}{2\pi\hbar}\right)^2 \int ds \int dp_1 \int dp_2 \langle p_1|\rho(p, q)|p_2\rangle e^{-ip_1(q'+s/2)/\hbar} \\
&\quad \times \exp[-ip_2(q' - s/2)/\hbar] \exp(-ip's/\hbar) \\
&= \left(\frac{1}{2\pi\hbar}\right)^2 \int ds \int dp_1 \int dp_2 \langle p_1|\rho(p, q)|p_2\rangle \\
&\quad \times \exp\left[-i\frac{s}{\hbar}\left(\frac{p_1 + p_2}{2} - p'\right)\right] \exp\left[iq'\frac{p_1 - p_2}{\hbar}\right] \\
&= \frac{1}{2\pi\hbar}\int dp_1 \int dp_2 \langle p_1|\rho(p, q)|p_2\rangle \delta\left(\frac{p_1 + p_2}{2} - p\right) \exp\left[iq'\frac{p_1 - p_2}{\hbar}\right] \\
&= \frac{1}{2\pi\hbar}\int dp_1 \langle p_1|\rho(p, q)|2p' - p_1\rangle \exp\left[2iq'\frac{p_1 - p'}{\hbar}\right]
\end{aligned} \tag{5.191}$$

If we let

$$p_1 = p' + k/2 \tag{5.192}$$

then (5.191) reduces to

$$P(p',q') = \frac{1}{2\pi\hbar}\int dk\langle p'+k/2|\rho(p,q)|p'-k/2\rangle \exp(ikq'/\hbar) \tag{5.193}$$

This is the Wigner distribution function in the momentum representation.

References

1. W.H.Louisell, *Quantum statistical properties of radiation* (Wiley, New York, 1973).

2. H.Haken, *Light*, Vol.**1**, **2** (North-Holland, Amsterdam, 1981, 1985).

3. N.G.Van Kampan, *Stochastic Process in Physics and Chemistry*, (North-Holland, Amsterdam, 1981).

4. C.W.Gardiner, *Handbook of stochastic methods*, (2nd ed, Springer-Verlag, Berlin, 1989).

5. G.S.Agarwal, *Quantum Statistical Theories of spontaneous emission and their relation to other approaches*, Springer Tracts in Modern Physics, Vol.70, (Springer, Berlin, 1974).

6. H.Risken, *The Fokker-Planck equation-methods of solution and application* (2nd ed., springer-Verlag, berlin, 1989).

7. C.Leonardi, F.Persico, and G.Vetri, *Dicke model and theory of driven and spontaneous emission*, La Rivista del Nuovo Cimento (1986).

8. Jin-sheng Peng, *Resonance fluorescence and superfluorescence* , (Science Press, Beijing, 1993).

9. E.P.Wigner, *Phys.Rev.* **40** (1930) 709.

10. R.J.Cook, *Phys.Rev.A* **22** (1980) 1078.

11. P.Drummond and C.W.Gardiner, *J.Phys.A* **13** (1980) 2553.

12. M.Hillery, R.F.O'Connell, M.O.Scully, and E.P.Wigner, *Phys.Rep.* **106** (1984) 122.

13. K.Vogel and H.Risken, *Phys.Rev.A* **39** (1989) 4675.

14. J.Bergou, L.Davidovich, M.Orszag, C.Benkert, M.Hillery, and M. O. Scully , *Phys.Rev.A* **40** (1989) 5073.

PART II

THE QUANTUM PROPERTIES OF LIGHT

Nonclassical properties and effects of the light field, which have been confirmed by modern optical experiments, show that the light field has many different quantum properties. These quantum properties can not be explained by classical theory, but they can be understood in the framework of QED. This volume is devoted to discussing quantum characteristics and effects in a system of atoms interacting with a laser field. As for the evolution of the atomic behavior, it will be treated in the next volume.

In Chapter 6, we first give a brief discussion of classical coherence, followed by a discussion of the quantum coherence. Subsequently we introduce nonclassical coherent effects of light fields in quantum optics, such as photon antibunching and intermode correlation. These effects will be explained by quantum electrodynamics. Chapter 7 is devoted to investigating the squeezing of light. Squeezing of the light field is a nonclassical effect which has been extensively studied in modern quantum optics. Here we first introduce the squeezing state of light, then concentrate on discussing the squeezing properties of light in the Jaynes-Cummings model, and the amplitude square squeezing as well as the higher-order squeezing of the radiation field. As we know, resonance fluorescence is one of the most extensively investigated subjects in quantum optics, which refers to the characteristics of the fluorescence radiated by the atoms driven by quasi-resonant laser fields. In Chapter 8, we first outline the general characteristics of resonance fluorescence, then introduce the dressed canonical transformation to treat the resonance fluorescence of a two-level atom and of a three-level atom. Finally, we deal with resonance fluorescence by means of the density theory. Chapter 9 discusses superfluorescence which is a typical quantum property of cooperative spontaneous emission for a collective system of atoms. After introducing the semiclassical theory, we give a more complete quantum mechanics description of superfluorescence. Finally, we discuss superfluorescent beats. Another nonclassical effect of the light field, namely optical bistability is studied in Chapter 10. After outlining the characteristics and the generation mechanism of the optical bistability, we focus on a quantum mechanical description of this phenomenon. The last chapter of this volume is devoted to investigating the virtual photon effects which have been less considered so far. With increasing precision of optical measurements, the virtual photon effects are likely to play a role which can not be neglected in many cases, and it is important to explore the effects of virtual photon processes on the evolution of the atom-field coupling system. In Chapter 11, we give a specific statement of the effects of the virtual photon field. First we discuss the correction to the energy of the Hydrogen atom due to the effect of the virtual field, then we deal with the influences of virtual photon processes on phase fluctuations and on light squeezing for an atom interacting with a laser field.

CHAPTER 6

COHERENCE OF LIGHT

6.1 Classical coherence of light

As we know, in the overlapping range of two beams of light wave, if the interference fringes appear, one say that these two beams of light are mutual coherent; if the interference fringes disappear, one say that these two beams of light are incoherent. The results of interference experiments of light show that only when the two beams of light have the same frequencies and the steady phase difference between two beams, the interference between these two beams appear. For one beam of light, if its wavelets interfere each other, we say that this beam is a coherent light. For example, since laser is a good monochromatic light, it is a perfect coherent field. Because the thermal field is consisted of a number of light waves with different frequencies, it is an incoherent field. The classical coherence is exhibited by temporal coherence and spatial coherence. In the following sections, we discuss the temporal coherence and the spatial coherence of light respectively.

6.1.1 Temporal coherence of light

The Michelson's interferometer is a typical apparatus to measure the temporal coherence of light. Suppose that a light beam from a point source σ shown in Fig.(6.1) is divided into two beams in a Michelson interferometer and these two beams are unified by passing different paths with a path delay $\Delta s = c\Delta t$ (c is the velocity of light). The experimental results display that if Δs is sufficiently small, the interference fringes are formed in plane D, but if Δs is large, no interference fringe appear in the overlapping range of two beams. In fact, the appearance of the fringes is formed by the superposition of two light beams arisen from the different time. Therefore, this kind of interference is a manifestation of temporal coherence between the two light beams. In general

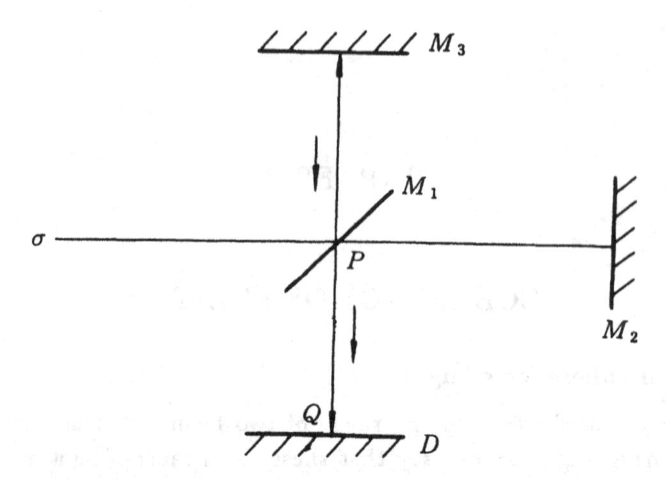

Figure 6.1: Schematic diagram of Michelson interferometer

case, each emission for a micro radiator has a delay time τ_c, so every emission corresponds to an optical path length L_c,

$$L_c = c\tau_c \tag{6.1}$$

If the path delay Δs is smaller than L_c, then the two unified beams are derived from the same original wave-train of the micro radiator, so these two beams have same frequency and constant phase difference, then induce the appearance of interference. If the path delay Δs is larger than L_c, it means that the two superposed beams originate from two different and uncorrelated wave-trains emitted by the micro radiator at different time, the interference therefore disappear. Thus L_c is called the coherence length, and the corresponding propagating delay time τ_c is called the coherence time. This coherence time τ_c also obeys

$$\tau_c = \frac{1}{\Delta\nu} \tag{6.2}$$

where $\Delta\nu$ is the effective bandwidth of light. Consequently, if the different fields are superposed at the same spatial point, only when the time difference between them is less than the coherence time τ_c, the interference fringes can be observed.

6.1.2 Spatial coherence of light

The spatial coherence of light can be exhibited by the Young's double split interference experiment. Fig.(6.2) schematically shows a double split interferometer. Here the light source is not a real point source, and has the

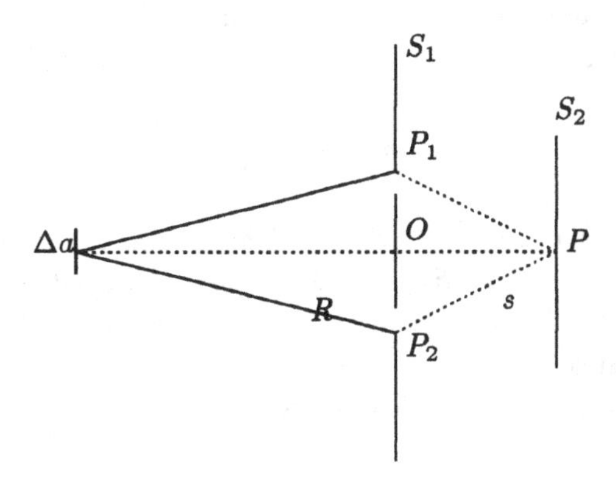

Figure 6.2: Schematic diagram of Young's double splits interference

dimension of Δa. This source can be regarded as the superposition of a number of uncorrelated point sources of light. The light waves from the source form two beams of light by passing through two pinholes P_1 and P_2 in the screen S_1. These two beams are unified at point P in the second screen S_2. The experimental results show that if the distance d between the two pinholes is small enough, there are interference fringes appearing in the neighbourhood of P. Inasmuch as the light waves from each point light source become two wavelet-trains by passing through the two pinholes in the screen S_1, the interference fringes observed at the point P are due to the superposition of this two wavelet-trains. Therefore, the interference fringes in the neighbourhood of P are actually the incoherent superposition of all the interference fringes originated from the independent point light sources. For the light source with given width Δa, if the distance d between the two pinholes in S_1 is large, the interference fringes in the neighbourhood of P will disappear. That is to say, in this case the incoherent superposition of all the interference fringes originated from the independent point light sources can not form the explicit interference fringes. The coherent property exhibiting in the neighbourhood of P indicates the spatial coherence of light. Interference fringes are generally observable if

$$\Delta\theta\Delta a < \lambda_0 \tag{6.3}$$

where $\Delta\theta$ is the angle which the line P_1P_2 subtends at the source, $\lambda_0 = c/\nu_0$ is the centre wavelength of the light. In order to observe the fringes near O, the two pinholes must be situated within an area around the point O of size

$$\Delta A \sim (R\Delta\theta)^2 \sim \frac{R^2\lambda_0^2}{(\Delta a)^2} = \frac{c^2R^2}{\nu_0^2 S} \tag{6.4}$$

where $S = (\Delta a)^2$ is the area of the source. The area ΔA is called the coherence area of the light in the plane S_1 around the point O. That is to say, only the light waves originated from the coherence area in the plane S_1 centered at the point O are mutual coherence.

The coherence area and the coherence length can be unified as the coherence volume ΔV

$$\Delta V = c\Delta A\tau_c = \frac{\lambda}{\Delta\lambda}\left(\frac{R}{\Delta a}\right)^2 \lambda_0^3 \tag{6.5}$$

where $\Delta\lambda = \Delta(c/\nu_0) = c\Delta\nu/\nu_0^3$. The coherence volume is a unified quantity describing the spatial coherence and the temporal coherence.

How to describe the degree of mutual coherence of light at two different space-time points analytically ? We now turn to discuss this item. First we introduce the first-order mutual correlation function of light from the Young's double splits interference experiment.

6.1.3 The first-order correlation function

Consider the Young's interference experiment as shown in Fig.(6.3). Evidently, the instantaneous field at a point $P(\mathbf{r},t)$ on the screen S_2 comes from the superposition of the fields at the time-space points $P_1(\mathbf{r}_1, t-t_1)$ and $P_2(\mathbf{r}_2, t-t_2)$, so the amplitude function $E(\mathbf{r},t)$ of the light at the time-space point $P(\mathbf{r},t)$ may be written as

$$E(\mathbf{r},t) = K_1E(\mathbf{r}_1, t-t_1) + K_2E(\mathbf{r}_2, t-t_2) \tag{6.6}$$

where $t_1 = \ell_1/c$ and $t_2 = \ell_2/c$ are the propagation times of the light from P_1 to P and P_2 to P, respectively. K_1 and K_2 are constant factors which depend

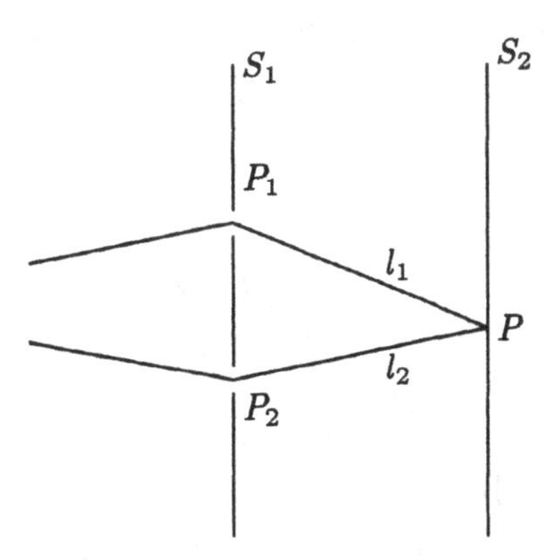

Figure 6.3: Relative position of the detecting point $P(\mathbf{r}, t)$ in screen S_2

on the space distribution and the size of the pinholes.

In the optical region, the frequency of light field is about $10^{14} \sim 10^{16} H_z$. The response time of the photo-detectors is about the order of $10^{-10}s$, no detectors can be available for determining directly the rapidly varying quantity $E(\mathbf{r}, t)$. The quantity recorded by the measurements is the square EE^* of its modulus in the neighbourhood of point P. This quantity will be referred to the intensity of light. In fact, since the response time of the detector $T > 10^{-10}s$, the measurement results of the detector are actually the mean value of the intensity of light at point P over a period of its response time T, i.e.,

$$\langle I(\mathbf{r}, t)\rangle = \langle E^*(\mathbf{r}, t)E(\mathbf{r}, t)\rangle = \frac{1}{T}\int_0^T E^*(\mathbf{r}, t)E(\mathbf{r}, t)dt \tag{6.7}$$

From (6.6) and (6.7), we have the intensity of light at point $P(\mathbf{r}, t)$ as

$$\begin{aligned}\langle I(\mathbf{r}, t)\rangle = &|K_1|^2\langle I(\mathbf{r}_1, t - t_1)\rangle + |K_2|^2\langle I(\mathbf{r}_2, t - t_2)\rangle \\ &+2|K_1K_2|Re\langle E^*(\mathbf{r}_1, t - t_1)E(\mathbf{r}_2, t - t_2)\rangle\end{aligned} \tag{6.8}$$

The first term is just the intensity of the field at P if only the aperture at P_1 is open, i.e.,

$$\langle I_1(\mathbf{r}, t)\rangle = |K_1|^2\langle E^*(\mathbf{r}_1, t - t_1)E(\mathbf{r}_1, t - t_1)\rangle = |K_1|^2\langle I(\mathbf{r}_1, t - t_1)\rangle$$

$|K_1|^2$ can be expressed in terms of the above equation

$$|K_1| = \left\{ \frac{\langle I_1(\mathbf{r},t)\rangle}{\langle I(\mathbf{r}_1, t-t_1)\rangle} \right\}^{1/2} \tag{6.9}$$

Similarly,

$$|K_2| = \left\{ \frac{\langle I_2(\mathbf{r},t)\rangle}{\langle I(\mathbf{r}_2, t-t_2)\rangle} \right\}^{1/2} \tag{6.10}$$

Consequently, (6.8) becomes

$$\begin{aligned}\langle I(\mathbf{r},t)\rangle = {} & \langle I_1(\mathbf{r},t)\rangle + \langle I_2(\mathbf{r},t)\rangle \\ & +2\left\{ \frac{\langle I_1(\mathbf{r},t)\rangle\langle I_2(\mathbf{r},t)\rangle}{\langle I(\mathbf{r}_1,t-t_1)\rangle\langle I(\mathbf{r}_2,t-t_2)\rangle} \right\}^{1/2} \mathrm{Re}\Gamma^{(1)}(\mathbf{r}_1,\mathbf{r}_2,t-t_1,t-t_2)\end{aligned} \tag{6.11}$$

here

$$\Gamma^{(1)}(\mathbf{r}_1,\mathbf{r}_2,t-t_1,t-t_2) = \langle E^*(\mathbf{r}_1,t-t_1)E(\mathbf{r}_2,t-t_2)\rangle \tag{6.12}$$

which represents the mutual correlation between the different time-space points $P_1(\mathbf{r}_1,t_1)$ and $P_2(\mathbf{r}_2,t_2)$, it is interpreted as the mutual correlation function. From (6.11) we see that the mutual correlation function $\Gamma^{(1)}(\mathbf{r}_1,\mathbf{r}_2,t-t_1,t-t_2)$ reflects the visibility of the interference fringe, or describes the interference ability of the superposed fields coming from the different time-space points.

As usual one only consider the stationary and ergodic fields, the time average in (6.7) is independent of the time origin. That is to say, the quantities $\langle I(\mathbf{r},t)\rangle$, $\langle I_1(\mathbf{r},t)\rangle$ and $\langle I_2(\mathbf{r},t)\rangle$ become time-independent, and the mutual correlation function only depends on the time difference $\tau = t_2 - t_1$. Thus (6.11) reduces to

$$\begin{aligned}\langle I(\mathbf{r})\rangle = {} & \langle I_1(\mathbf{r})\rangle + \langle I_2(\mathbf{r})\rangle \\ & +2\left\{ \frac{\langle I_1(\mathbf{r})\rangle\langle I_2(\mathbf{r})\rangle}{\langle I(\mathbf{r}_1)\rangle\langle I(\mathbf{r}_2)\rangle} \right\}^{1/2} \mathrm{Re}\Gamma^{(1)}(\mathbf{r}_1,\mathbf{r}_2,\tau)\end{aligned} \tag{6.13}$$

It is convenient to introduce the normalized function

$$g^{(1)}(\mathbf{r}_1,\mathbf{r}_2,\tau) = \frac{\Gamma^{(1)}(\mathbf{r}_1,\mathbf{r}_2,\tau)}{[\langle I(\mathbf{r}_1)\rangle\langle I(\mathbf{r}_2)\rangle]^{1/2}} \tag{6.14}$$

which is called the complex first-order coherent degree. Noticing

$$\langle |E_1(\mathbf{r}_1,t_1)+\lambda E_2(\mathbf{r}_2,t_2)|^2\rangle \geq 0$$

here λ is an arbitrary constant. For stationary and ergodic fields, the above equation becomes

$$\langle I(\mathbf{r}_1)\rangle + |\lambda|^2\langle I(\mathbf{r}_2)\rangle + \lambda\Gamma^{(1)}(\mathbf{r}_1,\mathbf{r}_2,\tau) + \lambda^*\Gamma^{(1)*}(\mathbf{r}_1,\mathbf{r}_2,\tau) \geq 0$$

If we assume

$$\lambda = -\frac{\Gamma^{(1)*}(\mathbf{r}_1,\mathbf{r}_2,\tau)}{\langle I(\mathbf{r}_2)\rangle}$$

then have

$$\langle I(\mathbf{r}_1)\rangle\langle I(\mathbf{r}_2)\rangle \geq |\Gamma^{(1)}(\mathbf{r}_1,\mathbf{r}_2,\tau)|^2 \tag{6.15}$$

This is exactly the Cauchy-Schwartz inequality. From (6.15) we obtain that the complex first-order coherent degree obeys

$$|g^{(1)}(\mathbf{r}_1,\mathbf{r}_2,\tau)| \leq 1 \tag{6.16}$$

If we let

$$g^{(1)}(\mathbf{r}_1,\mathbf{r}_2,\tau) = |g^{(1)}(\mathbf{r}_1,\mathbf{r}_2,\tau)|\exp\left[i\phi(\mathbf{r}_1,\mathbf{r}_2,\tau)\right] \tag{6.17}$$

then the field intensity (6.13) at the point P can be rewritten as

$$\begin{aligned}\langle I(\mathbf{r})\rangle = &\langle I_1(\mathbf{r})\rangle + \langle I_2(\mathbf{r})\rangle \\ &+2[\langle I_1(\mathbf{r})\rangle\langle I_2(\mathbf{r})\rangle]^{1/2}|g^{(1)}(\mathbf{r}_1,\mathbf{r}_2,\tau)|\cos\phi(\mathbf{r}_1,\mathbf{r}_2,\tau)\end{aligned} \tag{6.18}$$

The above equation shows that $|g^{(1)}(\mathbf{r}_1,\mathbf{r}_2,\tau)|$ can be obtained by measuring the maximum and minimum values of the light intensities at point P. Defining the fringe visibility V as

$$V = \frac{\langle I\rangle_{max} - \langle I\rangle_{min}}{\langle I\rangle_{max} + \langle I\rangle_{min}} \tag{6.19}$$

then the relation between $|g^{(1)}(\mathbf{r}_1,\mathbf{r}_2,\tau)|$ and V obeys

$$V = \frac{2[\langle I_1(\mathbf{r})\rangle\langle I_2(\mathbf{r})\rangle]^{1/2}}{\langle I_1(\mathbf{r}_1)\rangle\langle I_2(\mathbf{r}_2)\rangle}|g^{(1)}(\mathbf{r}_1,\mathbf{r}_2,\tau)| \tag{6.20}$$

For a special case $\langle I_1(\mathbf{r})\rangle = \langle I_2(\mathbf{r})\rangle$, the modulus of the first-order coherent degree of light field reduces to

$$|g^{(1)}(\mathbf{r}_1,\mathbf{r}_2,\tau)| = V \tag{6.21}$$

(6.20) and (6.21) show that the value of the first-order coherence of light can be determined by measuring the visibility of the interference fringes. Therefore, the first-order coherent degree indicates the capability of the light coming from two different time-space points to form interference fringes when they superposed. In general, we interpret the field satisfying $|g^{(1)}(\mathbf{r}_1,\mathbf{r}_2,\tau)| = 1$ as first-order coherent field. If $|g^{(1)}(\mathbf{r}_1,\mathbf{r}_2,\tau)| = 0$, then this field is called the incoherent one, and when $0 < |g^{(1)}(\mathbf{r}_1,\mathbf{r}_2,\tau)| < 1$, the field is said to be a partial coherent one.

From (6.14) one can find for the case of $\mathbf{r}_1 = \mathbf{r}_2$ (for example , the Michelson interferometer), there exist

$$g^{(1)}(\tau) = g^{(1)}(\mathbf{r},\mathbf{r},\tau)$$

Evidently, in this moment the first-order coherent degree of light only depends on the time difference τ. According to the definition of the coherence time, if τ is longer than the coherence time τ_c, then $g^{(1)}(\tau) = 0$, but for $\tau < \tau_c$, $g^{(1)}(\tau) \neq 0$, so $g^{(1)}(\tau)$ can characterize the temporal coherence of light. On the other hand, for $\tau = 0$ in (6.14), which means that we are concerned with the interference possibility of the field from different spatial point at the same time, then the first-order coherent degree represents the spatial coherence of light. Generally, it is impossible to express $g^{(1)}(\mathbf{r}_1,\mathbf{r}_2,\tau)$ as the product of two functions in which the spatial and temporal dependence are separated, i.e.,

$$g^{(1)}(\mathbf{r}_1,\mathbf{r}_2,\tau) \neq A(\mathbf{r}_1,\mathbf{r}_2)B(\tau)$$

This is because the light waves obey the wave equation (3.6), which connects the time varying with the spatial varying of the field.

6.1.4 The higher-order correlation function

From (6.17) and (6.18) we see that the first-order correlation function actually describes the phase correlation degree of the light waves at different time-space points, this function can not reveal the intensity correlation of the light waves at different time-space points. So the first-order correlation function can not completely display the coherent properties of light. The further study of the coherent properties of light needs to introduce the higher-order correlation function. Next we introduce the second-order correlation function by means of the Hanbury-Brown Twiss (HBT) experiment, then generalize it to the higher-order correlation function.

The HBT experiment is sketched in Fig.(6.4). Quasi-monochromatic radiation field is incident upon a 50:50 beam splitter M, so that a half of the original

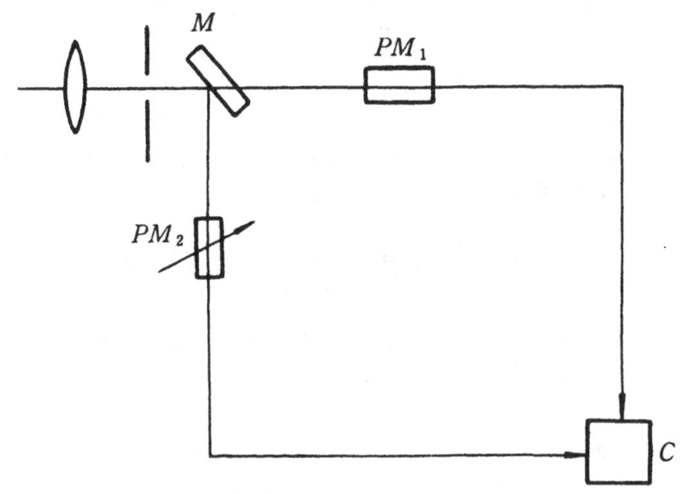

Figure 6.4: Schematic diagram of HBT experiment

intensity is directed to a photomultiplier tube PM_1, the other half to a second photomultiplier PM_2. The output signals from PM_1 and PM_2 are transported into correlator C. The quantity measured by the correlator C gives

$$\begin{aligned}&\langle (I_1(\mathbf{r}_1,t) - \langle I_1\rangle)(I_2(\mathbf{r}_2,t+\tau) - \langle I_2\rangle)\rangle \\ &= \langle I_1(\mathbf{r}_1,t)I_2(\mathbf{r}_2,t+\tau)\rangle - \langle I_1\rangle\langle I_2\rangle \end{aligned} \tag{6.22}$$

where $I_1(\mathbf{r}_1,t)$ and $I_2(\mathbf{r}_2,t+\tau)$ are the instantaneous intensities at PM_1 and PM_2, respectively. $\langle I_1\rangle$ and $\langle I_2\rangle$ are the average of $I_1(\mathbf{r}_1,t)$ and $I_2(\mathbf{r}_2,t+\tau)$ over a response time. Therefore, the output of the correlator C is actually the intensity fluctuation correlation of the fields at PM_1 and PM_2.

The first term at the right hand of (6.22) is the following correlation function

$$\begin{aligned}\Gamma^{(2)}(\mathbf{r}_1,\mathbf{r}_2,t_1,t_2) &= \langle I_1(\mathbf{r}_1,t_1)I_2(\mathbf{r}_2,t_2)\rangle \\ &= \langle E^*(\mathbf{r}_1,t_1)E^*(\mathbf{r}_2,t_2)E(\mathbf{r}_2,t_2)E(\mathbf{r}_1,t_1)\rangle \end{aligned} \tag{6.23}$$

$\Gamma^{(2)}(\mathbf{r}_1, \mathbf{r}_2, t_1, t_2)$ is the correlation of four variables, which is termed as the second-order correlation function of light. This function describes the intensity correlation of the fields at two different time-space points. As in the case of the first-order coherent degree of light, the second-order coherent degree of light is defined as

$$g^{(2)}(\mathbf{r}_1, \mathbf{r}_2, \tau) = \frac{\langle E_1^*(\mathbf{r}_1, t_1) E_2^*(\mathbf{r}_2, t_2) E_2(\mathbf{r}_2, t_2) E_1(\mathbf{r}_1, t_1)\rangle}{\langle I_1(\mathbf{r}_1, t_1)\rangle \langle I_2(\mathbf{r}_2, t_2)\rangle} \tag{6.24}$$

We usually call the field with $|g^{(1)}(\mathbf{r}_1, \mathbf{r}_2, \tau)| = g^{(2)}(\mathbf{r}_1, \mathbf{r}_2, \tau) = 1$ as the second-order coherent field.

Now we discuss the properties of the second-order coherent degree. For the stationary and ergodic field, (6.24) reduces to

$$g^{(2)}(\mathbf{r}_1, \mathbf{r}_2, \tau) = \frac{\langle E_1^*(\mathbf{r}_1, t) E_2^*(\mathbf{r}_2, t+\tau) E_2(\mathbf{r}_2, t+\tau) E_1(\mathbf{r}_1, t)\rangle}{\langle I_1(\mathbf{r}_1)\rangle \langle I_2(\mathbf{r}_2)\rangle} \tag{6.25}$$

If the two photomultiplier tubes PM_1 and PM_2 are so close that $\mathbf{r}_1 = \mathbf{r}_2 = 0$ and $\tau = 0$, then the second-order coherent degree becomes

$$g^{(2)}(\mathbf{r}_i, \mathbf{r}_i, 0) = \frac{\langle I_i^2(\mathbf{r}_i)\rangle}{\langle I_i(\mathbf{r}_i)\rangle^2} \qquad (i = 1, 2) \tag{6.26}$$

Since

$$\langle I_i^2(\mathbf{r}_i)\rangle \geq \langle I_i(\mathbf{r}_i)\rangle^2$$

then

$$g^{(2)}(\mathbf{r}_i, \mathbf{r}_i, 0) \geq 1 \tag{6.27}$$

On the other hand, for the fields at two different space-time points $(\mathbf{r}_1, t_1)$ and $(\mathbf{r}_2, t_2)$, the intensity correlation satisfies from the Cauchy-Schwartz inequality as

$$\begin{aligned} \langle I_1^2(\mathbf{r}_1)\rangle \langle I_2^2(\mathbf{r}_2)\rangle &\geq \langle I_1(\mathbf{r}_1, t) I_2(\mathbf{r}_2, t+\tau)\rangle^2 \\ &= \langle E_1^*(\mathbf{r}_1, t_1) E_2^*(\mathbf{r}_2, t_2) E_2(\mathbf{r}_2, t_2) E_1(\mathbf{r}_1, t_1)\rangle^2 \end{aligned} \tag{6.28}$$

i.e.,

$$g^{(2)}(\mathbf{r}_1, \mathbf{r}_1, 0) g^{(2)}(\mathbf{r}_2, \mathbf{r}_2, 0) \geq (g^{(2)}(\mathbf{r}_1, \mathbf{r}_2, \tau))^2 \tag{6.29}$$

This is the Cauchy-Schwartz inequality of the second-order coherent degree for the fields at different time-space points.

The second-order correlation function and the second-order coherent degree describe the intensity correlation degree of light, they reveal further the coherent properties of light. It is possible to generalize the above definitions of correlation function and the coherent degree to a more popular form. The definition of the n-th order correlation function is usually written as

$$\begin{aligned}&\Gamma^{(n)}(\mathbf{r}_1,t_1,\mathbf{r}_2,t_2,\cdots\mathbf{r}_n,t_n,\mathbf{r}_{n+1},t_{n+1}\cdots\mathbf{r}_{2n},t_{2n})\\&=\langle E_1^*(\mathbf{r}_1,t_1),\cdots E_n^*(\mathbf{r}_n,t_n)E_{n+1}(\mathbf{r}_{n+1},t_{n+1})\cdots E_{2n}(\mathbf{r}_{2n},t_{2n})\rangle\end{aligned}\tag{6.30}$$

The n-th order coherent degree of light can be defined correspondingly as

$$\begin{aligned}&g^{(n)}(\mathbf{r}_1,t_1,\mathbf{r}_2,t_2,\cdots\mathbf{r}_n,t_n,\mathbf{r}_{n+1},t_{n+1}\cdots\mathbf{r}_{2n},t_{2n})\\&=\frac{1}{D}\langle E_1^*(\mathbf{r}_1,t_1)\cdots E_n^*(\mathbf{r}_n,t_n)E_{n+1}(\mathbf{r}_{n+1},t_{n+1})\cdots E_{2n}(\mathbf{r}_{2n},t_{2n})\rangle\end{aligned}\tag{6.31}$$

where

$$\begin{aligned}D=&[\langle I_1(\mathbf{r}_1,t_1)\rangle\cdots\langle I_n(\mathbf{r}_n,t_n)\rangle\\&\times\langle I_{n+1}(\mathbf{r}_{n+1},t_{n+1})\rangle\cdots\langle I_{2n}(\mathbf{r}_{2n},t_{2n})\rangle]^{1/2}\end{aligned}$$

We call the field which obeys

$$|g^{(n)}(\mathbf{r}_1,t_1,\mathbf{r}_2,t_2,\cdots\mathbf{r}_n,t_n,\mathbf{r}_{n+1},t_{n+1}\cdots\mathbf{r}_{2n},t_{2n})|=1\tag{6.32}$$

as the n-th order coherent field. The theoretical calculation shows that the stable laser field produced by the laser above its threshold may be the n-th order coherent light.

6.2 Quantum theory of the coherence of light

Following the discussion of the classical theory on the coherence of light, we turn to discuss how to describe the coherence of light by means of the quantum theory.

6.2.1 Quantum correlation functions

Before introducing the quantum description of the coherence of light, it is necessary to analyze the photon detecting processes. The photon detectors are on the basis of the photoelectric effect. Here we consider an ideal detector which is quite small in size and its sensitivity is independent of the photon frequency. For example, an atom may be considered as such a detector. Basically, the detecting atom is assumed to be stably in its ground state, when it absorbs a photon to jump into its excited state, it must be relaxed to its ground state quickly. Therefore, the stimulation emission of the detecting atom can be ignored. This means that the detecting atom is always in its ground state, it will only absorb a photon from incident field, and has a very low probability to emit a photon with the same frequency and the same propagating vector of the incident field.

From (2.1) we know that the Hamiltonian describing the interaction between the detecting atom and the field at space-time point $(\mathbf{r}, t)$ is written in the electric-dipole approximation as

$$H_I = -\mathbf{D} \cdot \mathbf{E}(\mathbf{r}, t) \tag{6.33}$$

where $\mathbf{D}$ is the atomic dipole operator, $\mathbf{E}(\mathbf{r}, t)$ stands for the electric field operator, it may be expressed as

$$\mathbf{E}(\mathbf{r}, t) = \mathbf{E}^{(+)}(\mathbf{r}, t) + \mathbf{E}^{(-)}(\mathbf{r}, t) \tag{6.34}$$

According to (3.40), $\mathbf{E}^{(+)}(\mathbf{r}, t)$ and $\mathbf{E}^{(-)}(\mathbf{r}, t)$ are

$$\begin{aligned} \mathbf{E}^{(+)}(\mathbf{r}, t) &= i \sum_k \left(\frac{\hbar\omega_k}{2\varepsilon_0 V} \right)^{\frac{1}{2}} \varepsilon_{\mathbf{k}} \exp(i\mathbf{k} \cdot \mathbf{r}) a_k \exp(-i\omega_k t) \\ \mathbf{E}^{(-)}(\mathbf{r}, t) &= \mathbf{E}^{(+)}(\mathbf{r}, t))^{+} \end{aligned} \tag{6.35}$$

Evidently, $\mathbf{E}^{(+)}$ is the positive frequency component part of the field, which contains the annihilation operators only. And $\mathbf{E}^{(-)}$ is the negative frequency component part, which is associated with the creation operators only. So only the positive frequency component part of the field affects on the absorption processes.

Suppose that the incident field is not intense, then the detecting atom can only absorb one photon per unit time from the incident field. When we consider the atomic transition between the ground state $|g\rangle$ and an excited state $|e\rangle$, which associates with a transition of the radiation field from initial state $|i\rangle$ to a final state $|f\rangle$, then according to the perturbation theory of quantum mechanics, the transition probability per unit time under the first-order approximation will be proportional to

$$|\langle e|\mathbf{D}|g\rangle|^2 |\langle f|\mathbf{E}^{(+)}|i\rangle|^2$$

In practice, the final states $|f\rangle$ of the field are not easy to be detected, so we must sum over them all. By using the completeness relation

$$\sum_f |f\rangle\langle f| = 1$$

we obtain that the total transition probability per unit time to all states $|f\rangle$ is proportional to

$$\begin{aligned} &|\langle e|\mathbf{D}|g\rangle|^2 \langle i|\mathbf{E}^{(-)} \cdot \mathbf{E}^{(+)}|i\rangle \\ &= |\langle e|\mathbf{D}|g\rangle|^2 \mathrm{Tr}\mathbf{E}^{(-)} \cdot \mathbf{E}^{(+)}|i\rangle\langle i| \end{aligned} \tag{6.36}$$

The field under consideration may be initially in a statistical-mixture state, so we must use the density operator of the field to describe the initial state. Therefore, the transition probability per unit time between the atomic states $|g\rangle$ and $|e\rangle$ may be written as

$$P_{g\to e} = \text{constant} \times |\langle e|\mathbf{D}|g\rangle|^2 G^{(1)}(\mathbf{r}, \mathbf{r}, t, t) \tag{6.37}$$

where

$$G^{(1)}(\mathbf{r}, \mathbf{r}, t, t) = \mathrm{Tr}[\rho \mathbf{E}^{(-)}(\mathbf{r}, t)\mathbf{E}^{(+)}(\mathbf{r}, t)] \tag{6.38}$$

which corresponds to the classical correlation function $\Gamma^{(1)}(\mathbf{r}, \mathbf{r}, t, t)$, we call $G^{(1)}(\mathbf{r}, \mathbf{r}, t, t)$ as quantum self-correlation function of the field at time-space point $(\mathbf{r}, t)$. More generally, the first-order correlation function in quantum mechanics is defined as

$$G^{(1)}(\mathbf{r}_1, \mathbf{r}_2, t_1, t_2) = \mathrm{Tr}[\rho \mathbf{E}^{(-)}(\mathbf{r}_1, t_1) \cdot \mathbf{E}^{(+)}(\mathbf{r}_2, t_2)] \tag{6.39}$$

Analogous to the classical expression (6.14), the first-order coherent degree of the field in quantum mechanics is defined to be

$$\begin{aligned} &g^{(1)}(\mathbf{r}_1, \mathbf{r}_2, t_1, t_2) = \\ &\frac{\mathrm{Tr}(\rho \mathbf{E}^{(-)}(\mathbf{r}_1, t_1) \cdot \mathbf{E}^{(+)}(\mathbf{r}_2, t_2))}{\{\mathrm{Tr}(\rho \mathbf{E}^{(-)}(\mathbf{r}_1, t_1) \cdot \mathbf{E}^{(+)}(\mathbf{r}_1, t_1))\mathrm{Tr}(\rho \mathbf{E}^{(-)}(\mathbf{r}_2, t_2) \cdot \mathbf{E}^{(+)}(\mathbf{r}_2, t_2))\}^{1/2}} \end{aligned} \tag{6.40}$$

For a single-mode field with frequency ω, the operator $\mathbf{E}^{(-)}(\mathbf{r}_i, t_i)$ contains only the creation operator $a^\dagger$ and $\mathbf{E}^{(+)}(\mathbf{r}_i, t_i)$ contains only the annihilation operator a, then the modulus of the first-order coherent degree at the pairs time-space points $(\mathbf{r}, t)$ and $(\mathbf{r}, t + \tau)$ gives

$$|g^{(1)}(\tau)| = \left| \frac{\langle a^\dagger a\rangle \exp(-i\omega\tau)}{[\langle a^\dagger a\rangle\langle a^\dagger a\rangle]^{1/2}} \right| = 1 \tag{6.41}$$

These result shows that all the single-mode fields have first-order coherence, as calculated in classical theory.

The second-order coherent degree of light is expressed in a quantum mechanical form by a procedure similar to that used for the first-order coherence (6.40) as

$$g^{(2)}(\mathbf{r}_1,\mathbf{r}_2,t_1,t_2) = \frac{\mathrm{Tr}[\rho\mathbf{E}_1^{(-)}(\mathbf{r}_1,t_1)\mathbf{E}_2^{(-)}(\mathbf{r}_2,t_2):\mathbf{E}_2^{(-)}(\mathbf{r}_2,t_2)\mathbf{E}_1^{(+)}(\mathbf{r}_1,t_1)]}{\mathrm{Tr}[\rho\mathbf{E}_1^{(-)}(\mathbf{r}_1,t_1)\cdot\mathbf{E}_1^{(+)}(\mathbf{r}_1,t_1)]\mathrm{Tr}[\rho\mathbf{E}_2^{(-)}(\mathbf{r}_2,t_2)\cdot\mathbf{E}_2^{(+)}(\mathbf{r}_2,t_2)]} \tag{6.42}$$

The above equation is similar to the second-order coherent degree in classical theory which is given by (6.24). The only difference here is that the classical quantities $\mathbf{E}^*(\mathbf{r}_i,t_i)$ and $\mathbf{E}(\mathbf{r}_i,t_i)(i=1,2)$ are replaced by the operators $\mathbf{E}^{(-)}(\mathbf{r}_i,t_i)$ and $\mathbf{E}^{(+)}(\mathbf{r}_i,t_i)$. Then $g^{(2)}(\mathbf{r}_1,\mathbf{r}_2,t_1,t_2)$ represents the correlation degree of the intensity fluctuations of the field in quantum theory, and its modulus may be measured by the HBT experiment.

From (6.35) and (6.36) we obtain that for a single-mode field with frequency ω, its second-order coherent degree may be reduced as

$$g^{(2)}(t,t+\tau) = \frac{\langle a^\dagger(t)a^\dagger(t+\tau)a(t+\tau)a(\tau)\rangle}{\langle a^\dagger(t)a(t)\rangle\langle a^\dagger(t+\tau)a(t+\tau)\rangle} \tag{6.43}$$

Generally, it describes the correlation degree of the intensity of light field at two different space-time points $(\mathbf{r},t)$ and $(\mathbf{r},t+\tau)$. If the light field is a free one, i.e.,

$$a(t) = a(0)\exp(-i\omega t) \tag{6.44}$$

then eq.(6.43) becomes

$$g^{(2)}(t,t+\tau) = g^{(2)}(\tau) = \frac{\langle a^{\dagger 2}a^2\rangle}{\langle a^\dagger a\rangle^2} \tag{6.45}$$

which means that the second-order coherent degree of the field is time independent. For a two-mode field, (6.42) may be simplified as

$$g_{12}^{(2)}(t,t+\tau) = \frac{\langle a_1^\dagger(t)a_1(t)a_2^\dagger(t+\tau)a_2(t+\tau)\rangle}{\langle a_1^\dagger(t)a_1(t)\rangle\langle a_2^\dagger(t+\tau)a_2(t++\tau)\rangle} \tag{6.46}$$

Here $g_{12}^{(2)}(t,t+\tau)$ stands for the intensity correlation degree between the first mode (characterized by the operators $a_1^\dagger$ and a_1) at space-time point $(\mathbf{r},t)$ and the second mode (characterized by the operators $a_2^\dagger$ and a_2) at space-time point $(\mathbf{r}, t+\tau)$, so it is called the intermode correlation degree. For a two-mode free field, its Hamiltonian is

$$H_F = \omega_1 a_1^\dagger a_1 + \omega_2 a_2^\dagger a_2 \tag{6.47}$$

then eq.(6.46) becomes

$$g_{12}^{(2)}(t, t+\tau) = g_{(12)}^2(\tau) = \frac{\langle a_1^\dagger a_1 a_2^\dagger a_2 \rangle}{\langle a_1^\dagger a_1 \rangle \langle a_2^\dagger a_2 \rangle} \tag{6.48}$$

That is to say, the intermode correlation degree $g_{12}^{(2)}(\tau)$ of a two-mode free field is time-independent.

More generally, the n-th order coherent degree of fields in quantum theory is defined as

$$g^{(n)}(\mathbf{r}_1, \mathbf{r}_2, \cdots, \mathbf{r}_{2n}, t_1, t_2, \cdots, t_{2n}) = \frac{\mathrm{Tr}[\rho \mathbf{E}_1^{(-)}(\mathbf{r}_1, t_1) \cdots \mathbf{E}_n^{(-)}(\mathbf{r}_n, t_n) \mathbf{E}_{n+1}^{(+)}(\mathbf{r}_{n+1}, t_{n+1}) \cdots \mathbf{E}_{2n}^{(+)}(\mathbf{r}_{2n}, t_{2n})]}{(\mathrm{Tr}[\rho \mathbf{E}_1^{(-)}(\mathbf{r}_1, t_1) \cdot \mathbf{E}_1^{(+)}(\mathbf{r}_1, t_1)] \cdots \mathrm{Tr}[\rho \mathbf{E}_{2n}^{(-)}(\mathbf{r}_{2n}, t_{2n}) \cdot \mathbf{E}_{2n}^{(+)}(\mathbf{r}_{2n}, t_{2n})])^{1/2}} \tag{6.49}$$

It stands for the n-th order coherent degree of fields at space-time points $(\mathbf{r}_1, t_1)$, $(\mathbf{r}_2, t_2)$, $\cdots$, $(\mathbf{r}_{2n}, t_{2n})$.

6.2.2 Bunching and antibunching effects of light

As an example, we now discuss the coherent properties of the fields in a coherent state or in a number state and the thermal field, respectively. For the fields in a coherent state

$$|\alpha\rangle = \sum_n \exp(-|\alpha|^2/2) \frac{\alpha^n}{\sqrt{n!}} |n\rangle \tag{6.50}$$

or a number state $|n\rangle$, or for a single-mode thermal field described by the density operator

$$\rho = \sum_{n=0}^{\infty} \frac{\overline{n}}{(1+\overline{n})^{n+1}} |n\rangle\langle n| \tag{6.51}$$

the first-order coherent degrees of all these fields obey

$$|g^{(1)}(\tau)| = 1 \tag{6.52}$$

This means that all these fields are the first-order coherent fields. However, the photon distributions of these three fields are obviously different, which correspond to different photon fluctuations. The coherent properties of these three fields are also different. So we must take into account to distinguish the second-order coherence of these three fields.

From (6.45) we know that for the single-mode field in a coherent state $|\alpha\rangle$, the second-order coherent degree of the field is

$$g_{\alpha}^{(2)}(\tau)=\frac{\langle\alpha|(a^{\dagger})^2a^2|\alpha\rangle}{(\langle\alpha|a^{\dagger}a|\alpha\rangle)^2}=1 \tag{6.53}$$

It shows that the field in a coherent state $|\alpha\rangle$ is not only the first-order coherent field, but also the second-order coherent one. That is to say, the field in a coherent state $|\alpha\rangle$ has both the minimum phase fluctuations and the minimum intensity fluctuations, namely, the coherent state is a minimum number-phase uncertainty state. This is agreement with the result given in Chapter 3 by use of the phase theory. For the thermal field defined by the density operator ρ (6.51), the second-order coherent degree gives

$$g_{\rho}^{(2)}(\tau)=\frac{\mathrm{Tr}\rho(a^{\dagger})^2a^2}{(\mathrm{Tr}\rho a^{\dagger}a)^2}=2>g_{\alpha}^{(2)}(\tau)=1 \tag{6.54}$$

Clearly, although the first-order coherent degrees of the single-mode thermal field and the single-mode coherent field are equal, the second-order coherent degrees are different. From (6.54) we find that the intensity fluctuations in the single-mode thermal field are larger than those in the coherent field. Usually, if the second-order coherent degree of the fields satisfies

$$g^{(2)}(\tau)>1 \tag{6.55}$$

then it is said that this field exhibits bunching effect of photons. Eq.(6.55) is coincided with (6.27) given by the classical theory. Therefore, the bunching effect of photons is just a kind of classical effects.

For the field in a number state $|n\rangle$, the second-order coherent degree is

$$g_{n}^{(2)}(\tau)=\frac{\langle n|a^{\dagger 2}a^2|n\rangle}{(\langle n|a^{\dagger}a|n\rangle)^2}=\begin{cases}1-\frac{1}{n} & (n\geq 2),\\ 0 & (n=0,1).\end{cases} \tag{6.56}$$

This equation is clearly different from the second-order coherent degrees of both the coherent field and the thermal field. Here the second-order coherent degree obeys

$$g^{(2)}(\tau) < 1 \tag{6.57}$$

Evidently (6.57) is violated with the prediction of the classical theory, it is the reflection of the quantum characteristic of the field. Usually, we call the phenomenon which the second-order coherent degree obeys (6.57) as the antibunching effect of photons. Different from the bunching effect of photons, antibunching effect of photons is a nonclassical one. The reason for the disagreement between classical and quantum theories can be seen more clearly from an examination of the HBT experiment in the quantum theory. degree.

Before introducing the quantum theory of the HBT experiment, we first examine the classical description of the HBT experiment to distinguish the difference between the quantum theory and the classical theory. For the HBT experiment shown in Fig.(6.4), if the distances between the photomultipliers PM_1 and PM_2 and the 50:50 beam splitter M are adjusted to be equal, and the intensity of incident field is $I(z,t)$, then the instantaneous intensities received by PM_1 and PM_2 are

$$I_1(z,t) = I_2(z,t) = \frac{I(z,t)}{2} \tag{6.58}$$

their average values during the response time T read

$$\overline{I}_1 = \overline{I}_2 = \frac{\overline{I}}{2} \tag{6.59}$$

Then the output result of the correlator C gives

$$C = \langle (I_1(z,t) - \overline{I}_1)(I_2(z,t) - \overline{I}_2) \rangle = \frac{1}{4}[\langle I^2(z,t) \rangle - \overline{I}^2] \geq 0 \tag{6.60}$$

Therefore, the normalized correlation measured by the correlator C is

$$\frac{\langle I^2(z,t) \rangle - \overline{I}^2}{\overline{I}^2} = g^{(2)}(0) - 1 \geq 1 \tag{6.61}$$

This means that the measuring results of the HBT experiment are always positive or zero. Because in classical theory, no matter what the intensity of the incident field is, the instantaneous intensities $I_1(z,t)$ and $I_2(z,t)$ received by PM_1 and PM_2 are equal, which leading to $C \geq 0$ or $g^{(2)} \geq 1$. However, the quantum description of the HBT experiment will give different conclusions.

In quantum mechanical theory of the HBT experiment, a and $a^\dagger$ represent the annihilation and creation operators of the incident single-mode field, $b(c)$ and $b^\dagger$ ($c^\dagger$) stand for the annihilation and creation operator of the field received by PM_1 (PM_2). Since the incident photon is reflected or transmitted by M according to the same probability, the single-mode field from the source S may be expressed as

$$\mathbf{E}^{(+)}(\mathbf{r},t) = i\left(\frac{\hbar\omega}{2\varepsilon_0 V}\right)^{\frac{1}{2}} \varepsilon_k e^{i\mathbf{k}\cdot\mathbf{r}}(\cos\theta b + \sin\theta c)e^{-i\omega t} \tag{6.62}$$

namely, the operators a and $a^\dagger$ are expressed as

$$\begin{aligned} a &= \cos\theta b + \sin\theta c \\ a^\dagger &= \cos\theta b^\dagger + \sin\theta c^\dagger \end{aligned} \tag{6.63}$$

where the operators b and c commutate with each other, i.e., $[b,c]=0$. That is to say, the photons described by the operator b is only transmitted by the mirror M, and the photons characterised by the operator c is only reflected by the mirror M. The introduction of the factors $\cos\theta$ and $\sin\theta$ guarantees the commutation relation $[a, a^\dagger] = 1$.

Supposing the incident field in a number state $|n\rangle$, $|n\rangle$ may be expressed by means of (6.63) as

$$\begin{aligned} |n\rangle &= \frac{1}{\sqrt{n!}}(a^\dagger)^n|0\rangle = \frac{1}{\sqrt{n!}}(\cos\theta b^\dagger + \sin\theta c^\dagger)^n|0,0\rangle \\ &= \frac{1}{\sqrt{n!}}\sum_{n_b=0}^{n}\frac{n!}{n_b!(n-n_b)!}(\cos\theta)^{n_b}(\sin\theta)^{n-n_b}(b^\dagger)^{n_b}(c^\dagger)^{n-n_b}|0,0\rangle \\ &= \sum_{n_b=0}^{n}\sqrt{\frac{n!}{n_b!(n-n_b)!}}(\cos\theta)^{n_b}(\sin\theta)^{n-n_b}|n_b, n-n_b\rangle \end{aligned} \tag{6.64}$$

Therefore, the mean photon numbers measured by the detectors PM_1 and PM_2 are, respectively

$$\begin{aligned} \langle b^\dagger b\rangle &= \langle n|b^\dagger b|n\rangle = \sum_{n_b=0}^{\infty}\frac{n! n_b}{n_b!(n-n_b)!}(\cos^2\theta)^{n_b}(\sin^2\theta)^{n-n_b} \\ &= n\cos^2\theta \end{aligned} \tag{6.65}$$

$$\langle c^\dagger c\rangle = n\sin^2\theta \tag{6.66}$$

The mirror M is a splitter of light beam for half transmission and half reflection, this implies $\langle b^\dagger b\rangle = \langle c^\dagger c\rangle$, so that

$$\sin^2\theta = \cos^2\theta = \frac{1}{2}$$

Without loss of generality, we choose $\theta = \pi/4$, thus (6.63) reduces to

$$\begin{aligned} a &= \frac{1}{\sqrt{2}}(b+c) \\ a^\dagger &= \frac{1}{\sqrt{2}}(b^\dagger + c^\dagger) \end{aligned} \tag{6.67}$$

the mean photon numbers measured by PM_1 and PM_2 become

$$\langle b^\dagger b\rangle = \langle c^\dagger c\rangle = \frac{n}{2} \tag{6.68}$$

and the second-order correlation function measured by the correlator C gives

$$\langle b^\dagger c^\dagger c b\rangle = \langle n_b(n - n_b)\rangle = \frac{1}{4}n(n-1) \tag{6.69}$$

Hence the second-order coherent degree measured by the correlator C obeys

$$g^{(2)}(\tau) = \frac{\langle b^\dagger c^\dagger c b\rangle}{\langle b^\dagger b\rangle\langle c^\dagger c\rangle} = 1 - \frac{1}{n} \tag{6.70}$$

which is agreement with (6.56). So $g^{(2)}(\tau) < 1$ reflects one of the quantum coherent properties of the light field.

Table 6.1 shows respectively the number of incident photons in a HBT experiment, the photon numbers detected by two phototubes, the average values, their correlation, and the second-order coherent degree. From the table we see that for the case of $n=1$, i.e., the incident field is in the number state $|1\rangle$, at this moment the incident field has only one photon, this photon should be transmitted or reflected by the mirror M. Therefore, the possible value of n_b is 0 or 1, correspondingly, $n - n_b$ is 1 or 0. That is to say, the incident field is split as two unequal beams when it passes through the mirror, this result is different from (6.58) in the classical theory. In this sense, the correlation function $\langle b^\dagger c^\dagger c b\rangle = 0$, and $g^{(2)}(\tau) = 0$. Thus the intensity correlation value measured by the correlator gives negative. In other words, after the incident photon passes through the mirror M, it will arrive either at $PM_1(\mathbf{r}_1, t_1)$ or

Table 6.1: The number of incident photons in a HBT experiment, the photon numbers detected by the two phototubes, the average values, their correlation, and the second-order degree

n	n_b	$n-n_b$	$\langle b^\dagger b\rangle=\langle c^\dagger c\rangle$	$\langle b^\dagger c^\dagger cb\rangle$	$g^{(2)}(\tau)$
1	1	0	1/2	0	0
	0	1			
	2	0			
2	1	1	1	1/2	1/2
	1	1			
	0	2			
3	–	–	3/2	3/2	2/3
4	–	–	2	3	3/4

at $PM_2(\mathbf{r}_2, t_2)$, but it can never arrive at these two points simultaneously. Therefore, for the field which satisfies $g^{(2)}(\tau) < 1$, its photons has well-spaced characteristic. That is to say, the photon-antibunching effect results from the well-spaced distribution of the photons. So photon-antibunching effect is not allowed by the classical theory, it is a quantum property of field. In fact, the photon antibunching effect is a nonclassical effect observed first in atomic resonance fluorescence experiment in 1976 by Kimble *et al*[9].

Next we analyze the second-order coherence property of the coherent field from the HBT experiment. If the incident field is in a coherent state $|\alpha\rangle$, from (6.50) we know that the probability of the photon-number distribution gives

$$P_n = \frac{1}{n!}\exp(-\overline{n})\overline{n}^n \tag{6.71}$$

here $\overline{n} = |\alpha|^2$. Applying (6.50) and (6.64), it is easy to find

$$\langle b^\dagger b\rangle = \langle c^\dagger c\rangle = \frac{\overline{n}}{2} \tag{6.72}$$

$$\langle b^\dagger c^\dagger cb\rangle = \frac{\overline{n}^2}{4} \tag{6.73}$$

$$g^{(2)}(\tau) = \frac{\langle b^\dagger c^\dagger cb\rangle}{\langle b^\dagger b\rangle\langle c^\dagger c\rangle} = 1 \tag{6.74}$$

Eq.(6.74) is coincident with (6.53). From (6.72) and (6.73) we find that the correlation function obeys $\langle b^\dagger c^\dagger cb \rangle = \langle b^\dagger b \rangle \langle c^\dagger c \rangle$. That is to say, the photons passing through the mirror M arrive at the detector PM_1 or PM_2 randomly, and there is no correlation among them. This leads to the output result of the correlator $\langle b^\dagger c^\dagger cb \rangle - \langle b^\dagger b \rangle \langle c^\dagger c \rangle$ to be zero, thus $g^{(2)}(\tau) = 1$.

However, for the thermal field described by (6.51), the mean photon numbers measured by the detectors PM_1 and PM_2 and the correlation function are respectively as follows

$$\langle b^\dagger b \rangle = \langle c^\dagger c \rangle = \frac{\overline{n}}{2} \tag{6.75}$$

$$\langle b^\dagger c^\dagger cb \rangle = \frac{\overline{n}^2}{2} \tag{6.76}$$

$$g^{(2)}(\tau) = \frac{\langle b^\dagger c^\dagger cb \rangle}{\langle b^\dagger b \rangle \langle c^\dagger c \rangle} = 2 \tag{6.77}$$

Evidently, $\langle b^\dagger c^\dagger cb \rangle = 2\langle b^\dagger b \rangle \langle c^\dagger c \rangle$. The output result measured by the correlator is

$$C = \langle b^\dagger c^\dagger cb \rangle - \langle b^\dagger b \rangle \langle c^\dagger c \rangle = \frac{\overline{n}^2}{4} > 0 \tag{6.78}$$

This result shows that a thermal field exhibits excess intensity fluctuations. It indicates that the photons of the incident field may not only arrive at the detector PM_1 at time t but also at the detector PM_2 at time $t + \tau$. That is to say, if the two phototubes are close together or coincident, i.e., $\tau = 0$, then we may say that the photons of thermal field will arrive in pairs to the detectors PM_1 and PM_2 simultaneously, and exhibit bunching effect. Obviously, different from the photon-antibunching effect, the photon-bunching effect is a classical effect of light.

6.2.3 Intermode correlation property for the two-mode field

In the previous section we have discussed the property of the second-order coherence for a single-mode field, now we turn to discuss the intermode correlation property for a two-mode field by considering free two-mode number field $|n_1, n_2\rangle$ and two-mode thermal field, respectively.

For a free two-mode thermal field, its density operator is

$$\rho = \sum_{n_1} \sum_{n_2} \frac{\overline{n}_1{}^{n_1}}{(1+\overline{n}_1)^{n_1+1}} \frac{\overline{n}_2{}^{n_2}}{(1+\overline{n}_2)^{n_2}} |n_1, n_2\rangle\langle n_1, n_2| \tag{6.79}$$

Here the second-order coherent degree for the i-th mode is defined by $g_{ii}(\tau)$ (6.54). The intermode correlation degree may be obtained from (6.48) as

$$g_{12}^{(2)}(\tau) = \frac{\mathrm{Tr}(\rho a_1^\dagger a_2^\dagger a_2 a_1)}{Tr(\rho a_1^\dagger a_1)\mathrm{Tr}(\rho a_2^\dagger a_2)} = 1 \tag{6.80}$$

Clearly, $g_{ij}^{(2)}(\tau)$ obeys

$$g_{11}^{(2)}(0) g_{22}^{(2)}(0) > (g_{12}^{(2)}(\tau))^2 \tag{6.81}$$

That is to say, for a two-mode thermal field, the intermode correlation satisfies the Cauchy-Schwartz inequality (6.29) shown in classical theory.

But for a two-mode number field, the second-order coherent degree for i-th mode field gives

$$g_{ii}^{(2)}(\tau) = 1 - \frac{1}{n_i} \qquad (i = 1, 2) \tag{6.82}$$

and the intermode correlation degree is

$$g_{12}^{(2)}(\tau) = \frac{\langle n_1, n_2 | a_1^\dagger a_2^\dagger a_2 a_1 | n_1, n_2 \rangle}{\langle n_1 | a_1^\dagger a_1 | n_1 \rangle \langle n_2 | a_2^\dagger a_2 | n_2 \rangle} = 1 \tag{6.83}$$

From (6.82) and (6.83) we find that

$$\begin{aligned} g_{11}^{(2)}(0) g_{22}^{(2)}(0) &= \frac{(n_1 - 1)(n_2 - 1)}{n_1 n_2} \\ &= 1 - \frac{(n_1 + n_2 + 1)}{n_1 n_2} < 1 = (g_{12}^{(2)}(\tau))^2 \end{aligned} \tag{6.84}$$

Obviously, (6.84) does not satisfy the Cauchy-Schwartz inequality (6.29). This means that the intermode correlation of the two-mode number field is also nonclassical. This is a kind of nonclassical effects of the multi-mode field. This nonclassical effect can also be explained by using the similar quantum theory as shown in explaining the HBT experiment. The reader interested in this method may find in the references [7,10].

References

1. J.Perina, *Coherence of Light*, (2nd Ed.) (D.Reidel, Holland, 1985).
2. P.L.Knight and L.Allen, *Concept of Quantum Optics*, (Pergamon, Oxford, 1983).
3. R.Loudon, *The quantum theory of light*, (2nd Ed.), (Clarendon, Oxford, 1983).
4. M.Born and E.Wolf, *Princples of Optics*, (2nd Ed.), (Pergamon, Oxford, 1964).
5. L.Mandel and E.Wolf, *Rev.Mod.Phys.* **37** (1965) 231.
6. G.J.Troup and R.G.Turner, *Rep.Pro.Phys.* **37** (1974) 771.
7. J.F.Clauser, *Phys.Rev.* **D9** (1974) 853.
8. D.F.Walls, *Am.J.Phys.* **45** (1977) 952.
9. H.J.Kimble, M.Dagenais, and L.Mandel, *Phys.Rev.Lett.* **39** (1977) 691.
10. R.Loudon, *Rep.Prog.Phys.* **43** (1980) 913.
11. P.Zhou and J.S.Peng, *ACTA Opt.Sin.* **10** (1990) 837.
12. G.X.Li and J.S.Peng, *ACTA Phys.Sin.* **42** (1993) 1443.
13. Gao-xiang Li and Jin-sheng Peng, *Phys.Lett.* **A219** (1996) 41.

CHAPTER 7

SQUEEZED STATES OF LIGHT

Another characteristic of the quantized field is that the fluctuations of some operators describing the field may be squeezed. In this chapter, we focus our attention on this subject. From Chapter 3 we know for a single-mode field

$$E(t) = \lambda[a \exp(-i\omega t) + a^\dagger \exp(i\omega t)] \tag{7.1}$$

if we define two Hermitian operators X_1 and X_2 to replace the operators a and $a^\dagger$,

$$X_1 = \frac{1}{2}(a + a^\dagger) \qquad X_2 = \frac{1}{2i}(a - a^\dagger) \tag{7.2}$$

then (7.1) becomes

$$E(t) = \lambda(X_1 \cos\omega t + X_2 \sin\omega t) \tag{7.3}$$

Clearly, X_1 and X_2 are two mutual orthogonal operators describing the field amplitude, they obey the commutation relation

$$[X_1, X_2] = \frac{i}{2} \tag{7.4}$$

According to the Heisenberg uncertainty relation, the product of their quantum uncertainties must satisfy

$$(\Delta X_1)^2(\Delta X_2)^2 \geq \frac{1}{16} \tag{7.5}$$

here

$$(\Delta X_i)^2 = \langle X_i^2\rangle - \langle X_i\rangle^2 \qquad (i = 1, 2) \tag{7.6}$$

For a field in a coherent state $|\alpha\rangle$, the uncertainties of the operators X_1 and X_2 reach to their minimum value $1/4$, i.e.,

$$(\Delta X_1)^2 = (\Delta X_2)^2 = \frac{1}{4} \tag{7.7}$$

This means that the coherent state is a minimum uncertainty state whose fluctuations in X_i are equal to the minimum value 1/4, this value is usually called the vacuum fluctuations of the field amplitude. Eq.(7.7) is regarded as a minimum limitation of quantum fluctuations of the field for a long time. However, from the practical point of view this minimum limitation of quantum fluctuation of the radiation field is a bad restriction for practical applications. For example, in optical communication system, since the signal noises result from not only the equipment but also the thermal noise and the quantum noise, how to improve the noise ratio of the signal is very important. Even using a component X_i of the field in coherent state to carry communication signals, the quantum fluctuations shown in (7.7) is at least $\hbar\omega/4$. For a visible light, the optical frequency is about $10^{14} \sim 10^{15} Hz$, then the quantum noise in one component X_i of the light field is about 1ev, meanwhile the energy $k_B T$ of the thermal noise for room-temperature (T=300K) gives 2.6×10^{-2}ev. So we can see that the effect of the quantum noise is a main source to restrict the improving of the signal noise ratio in optical communication system. However, can one find a quantum state of light field which may be below the minimum quantum fluctuations for one component of the radiation field ? The answer is positive, and this kind of state is called the squeezed state of radiation field. By applying this kind of radiation field to the optical communication system, the quantum noise can be less than the minimum limitation value in coherent state of the field. This chapter is devoted to discussing the squeezed state of light.

7.1 Squeezed states of a single-mode field

7.1.1 Squeezed coherent states

A. Definition of squeezed coherent states

In 1976, Yuen defined two operators b and $b^\dagger$ as

$$b = \mu a + \nu a^\dagger; \quad b^\dagger = \mu^* a + \nu^* a^\dagger \tag{7.8}$$

here the pair of C numbers μ and ν obey

$$|\mu|^2 - |\nu|^2 = 1 \tag{7.9}$$

It follows from (7.8) and (7.9) that

$$[b, b^\dagger] = 1 \tag{7.10}$$

The commutator (7.10) provides that the characteristics of the operators b and $b^\dagger$ are exactly similar to those of a and $a^\dagger$. The transformation of variables from the operators a and $a^\dagger$ to b and $b^\dagger$ according to (7.8) and (7.9) is just a linear canonical transformation, this canonical transformation can be represented as a unitary transformation, i.e.,

$$b = SaS^\dagger = \mu a + \nu a^\dagger \tag{7.11}$$

Inasmuch as b and $b^\dagger$ obey the commutation relation (7.10), the operator $N_g = b^\dagger b$ is also the number operator, here we call N_g as the quasi-photon number operator. The eigenvalue equation of N_g is

$$N_g|m_g\rangle_g = m_g|m_g\rangle_g; \qquad N_g|0\rangle_g = 0 \tag{7.12}$$

where $|0\rangle_g$ is termed the quasi-vacuum state.

Now we examine the eigenstates of b with eigenvalues β

$$b|\beta\rangle_g = \beta|\beta\rangle_g \tag{7.13}$$

here $|\beta\rangle_g$ is called the quasi-coherent state. Following the same procedure as (3.76), $|b\rangle_g$ may be expressed as

$$|\beta\rangle_g = D_g(\beta)|0\rangle_g = \exp(\beta b^\dagger - \beta^* b)|0\rangle_g \tag{7.14}$$

Next we calculate the fluctuations of the operators X_1 and X_2 of field in the quasi-coherent state. From (7.8) and (7.9) we know

$$a = \mu^* b - \nu b^\dagger; \qquad a^\dagger = \mu b^\dagger - \nu b \tag{7.15}$$

so that

$$\langle a\rangle_g =_g \langle\beta|a|\beta\rangle_g =_g \langle\beta|\mu^* b - \nu b^\dagger|\beta\rangle_g = \mu^*\beta - \nu\beta^* \tag{7.16}$$

$$\langle a^\dagger a\rangle =_g \langle\beta|(\mu b^\dagger - \nu^* b)(\mu^* b - \nu b^\dagger)|\beta\rangle_g = |\mu^*\beta - \nu\beta^*|^2 + |\nu|^2 \tag{7.17}$$

$$\langle a^2\rangle =_g \langle\beta|(\mu^* b - \nu b^\dagger)^2|\beta\rangle_g = (\mu^*\beta - \nu\beta^*)^2 - \mu^*\nu \tag{7.18}$$

Therefore, the fluctuations of the components X_1 and X_2 are respectively expressed as

$$(\Delta X_1)^2 = \frac{1}{4}\langle(a^\dagger + a)^2\rangle - \frac{1}{4}\langle a^\dagger + a\rangle^2 = \frac{1}{4}|\mu - \nu|^2 \tag{7.19}$$

$$(\Delta X_2)^2 = \frac{1}{4}|\mu + \nu|^2 \tag{7.20}$$

Eqs.(7.19) and (7.20) show that there exist the parameters μ and ν which can ensure the fluctuations in one of the components less than 1/4, i.e.,

$$(\Delta X_i)^2 < \frac{1}{4} \qquad (i = 1 \quad or \quad 2) \tag{7.21}$$

and correspondingly the other ones larger than 1/4. However, the product of $(\Delta X_1)^2$ and $(\Delta X_2)^2$ obeys the Heisenberg uncertainty principle. That is to say, for the field in a quasi-coherent state $|\beta\rangle_g$, the variation in one of its two quadrature components can be less than that of the corresponding one of the coherent field, this reduction is at the cost of the fluctuations in another component to be larger than those in the coherent field. Usually, we call the quasi-coherent state whose fluctuations in one of two quadrature components are less than the minimum limitation in the coherent state as the squeezed coherent state or two-photon coherent state. Hence, when we make the linear canonical transformation with respect to the operators a and $a^\dagger$ as (7.11), we may find a new state in which the fluctuations in one of two quadrature components of the field can be reduced sufficiently. Now, the question arisen here is that what is the expression of this canonical transformation ? we discuss this in the following section.

B. Single-mode squeezing operator $S(\xi)$

Define the unitary transformation operator as

$$S(\xi) = \exp\left[\frac{1}{2}\xi^* a^2 - \frac{1}{2}\xi(a^{\dagger 2}\right] \tag{7.22}$$

here $\xi = re^{i\theta}$ is an arbitrary complex. Evidently, it follows that

$$S^\dagger(\xi) = S^{-1}(\xi) = S(-\xi) \tag{7.23}$$

By use of the identity relation

$$\exp(\lambda A)B\exp(-\lambda A) = B + \lambda[A, B] + \frac{\lambda^2}{2!}[A, [A, B]] + ...$$

we make the unitary transformation with respect to the operators a and $a^\dagger$ as

$$\begin{aligned} b = S(\xi)aS^\dagger(\xi) &= \exp\left[\frac{1}{2}\xi^* a^2 - \frac{1}{2}\xi(a^\dagger)^2\right] a \exp\left[\frac{1}{2}\xi(a^\dagger)^2 - \frac{1}{2}\xi^* a^2\right] \\ &= a + \left[\frac{1}{2}\xi^* a^2 - \frac{1}{2}\xi(a^\dagger)^2, a\right] + \frac{1}{2!}\left[\frac{1}{2}\xi^* a^2 - \frac{1}{2}\xi(a^\dagger)^2, \left[\frac{1}{2}\xi^* a^2 - \frac{1}{2}\xi(a^\dagger)^2, a\right]\right] \\ &\quad + \cdots \end{aligned} \tag{7.24}$$

Considering the commutators

$$\left[\frac{1}{2}\xi^* a^2 - \frac{1}{2}\xi(a^\dagger)^2, a\right] = \xi a^\dagger = r\exp(i\theta)a^\dagger$$
$$\left[\frac{1}{2}\xi^* a^2 - \frac{1}{2}\xi^*(a^\dagger)^2, a^\dagger\right] = \xi^* a = r\exp(-i\theta)a$$

with $r = |\xi|$, then (7.24) takes in the form

$$\begin{aligned} b =& a\left(1 + \frac{r^2}{2!} + \frac{r^4}{4!} + \cdots\right) + \exp(i\theta)a^\dagger\left(r + \frac{r^3}{3!} + \frac{r^5}{5!} + \cdots\right) \\ =& a\cosh r + a^\dagger \sinh r\exp(i\theta) \end{aligned} \quad (7.25)$$

Comparing the above with (7.11), yields

$$\mu = \cosh r, \qquad \nu = \exp(i\theta)\sinh r \quad (7.26)$$

and

$$|\mu|^2 - |\nu|^2 = \cosh^2 r - \sinh^2 r = 1$$

So we see that the unitary transformation operator $S(\xi)$ (7.22) is exactly a linear canonical transformation operator obeying (7.11).

Employing the unitary operator $S(\xi)$, we may express the squeezed coherent state explicitly. Multiplying $S^\dagger(\xi)$ from the left in (7.13), we have

$$S^\dagger(\xi)bS(\xi)S^\dagger(\xi)|\beta\rangle_g = S^\dagger(\xi)\beta|\beta\rangle_g$$

Noticing $a = S^\dagger(\xi)bS(\xi)$, the above equation may be rewritten as

$$aS^\dagger(\xi)|\beta\rangle_g = \beta S^\dagger(\xi)|\beta\rangle_g \quad (7.27)$$

Eq.(7.27) shows that $S(\xi)|\beta\rangle_g$ is the eigenstate of operator a with eigenvalue β, i.e., the coherent state $|\beta\rangle$. So $|\beta\rangle_g$ obeys

$$|\beta\rangle_g = S(\xi)|\beta\rangle = S(\xi)D(\beta)|0\rangle \quad (7.28)$$

here $D(\beta)$ is the displacement operator of the coherent state, its function is to transfer the vacuum state $|0\rangle$ into the coherent state $|\beta\rangle$. According to (3.77), $D(\beta)$ is

$$D(\beta) = \exp(\beta a^\dagger - \beta^* a) \quad (7.29)$$

Eq.(7.28) shows that the squeezed coherent state $|\beta\rangle_g$ results from the effect of the unitary operator $S(\xi)$ on the coherent state $|\beta\rangle$. For a squeezed coherent field, its fluctuations in one of two quadrature components are less than those for a coherent field, the reason for the noise reduction is the influence of $S(\xi)$ on $|\beta\rangle$, in this sense we call $S(\xi)$ as the squeeze operator for a single-mode field. On the other hand, inasmuch as the field operator in $S(\xi)$ (7.22) appears in the form of a^2 and $(a^\dagger)^2$, which means that the squeezed coherent state results from the processes of annihilating or creating two photons in the coherent state, thus we also call the squeezed coherent state $|\beta\rangle_g$ as the two-photon coherent state.

Next we discuss the physical meaning of the factors r and θ in the squeezed operator $S(\xi)$. Substituting (7.26) into (7.19) and (7.20), we obtain the fluctuations in X_1 and X_2 components and the expectation values of X_1 and X_2 as

$$(\Delta X_1)^2 = \frac{1}{4}[\exp(-2r)\cos^2(\theta/2) + \exp(2r)\sin^2(\theta/2)] \tag{7.30}$$

$$(\Delta X_2)^2 = \frac{1}{4}[\exp(-2r)\sin^2(\theta/2) + \exp(2r)\cos^2(\theta/2)] \tag{7.31}$$

$$\langle X_1\rangle =_g \left\langle \beta \left| \frac{1}{2}(a + a^\dagger) \right| \beta \right\rangle_g = |\beta|[\cos\phi\cosh r - \cos(\theta-\phi)\sinh r] \tag{7.32}$$

$$\langle X_2\rangle =_g \left\langle \beta \left| \frac{1}{2i}(a - a^\dagger) \right| \beta \right\rangle_g = |\beta|(\sin\phi\cosh r + \sin(\theta-\phi)\sinh r) \tag{7.33}$$

here we have used (7.16) and let $\beta = |\beta|\exp(i\phi)$. Eqs.(7.30) and (7.31) show that the fluctuations in the quadrature components only depend on r and θ, but they are not related to β. In other words, for the squeezed coherent field, although the expectation values $\langle X_1\rangle$ and $\langle X_2\rangle$ are decided by β, r and θ, the fluctuations $\langle\Delta X_1\rangle^2$ and $\langle\Delta X_2\rangle^2$ only associated with the parameter ξ. When $\theta = 0$, (7.30) and (7.31) become

$$(\Delta X_1)^2 = \frac{1}{4}\exp(-2r) \tag{7.34}$$

$$(\Delta X_2)^2 = \frac{1}{4}\exp(2r) \tag{7.35}$$

Evidently, if $r > 0$, then $(\Delta X_1)^2 < 1/4$ and $(\Delta X_2)^2 > 1/4$. This means that for a field in a squeezed coherent state, its fluctuations in the component X_1 are squeezed and correspondingly the fluctuations in the component X_2 are amplified. The reduction degree for $(\Delta X_1)^2$ depends on r. When r=0, the squeezed coherent state $|\beta\rangle_g$ reduce to the standard coherent state $|\beta\rangle$. With the increasing of r, the reduction degree for the fluctuations in the component X_1 becomes deeper, so we call r as the squeezing factor.

If $\theta = \pi$, $(\Delta X_1)^2$ and $(\Delta X_2)^2$ reduce to

$$(\Delta X_1)^2 = \frac{1}{4}\exp(2r) \tag{7.36}$$

$$(\Delta X_2)^2 = \frac{1}{4}\exp(-2r) \tag{7.37}$$

In this case, the fluctuations in the X_2 component are squeezed and those in X_1 are increased. So we see that the fluctuations in one of two quadrature components are squeezed or amplified is determined by the angle θ. Therefore, the choice of θ determines the squeezed direction of the field. In addition, we see from (7.34) and (7.37) that for $\theta = 0$ or π,

$$(\Delta X_1)^2(\Delta X_2)^2 = \frac{1}{16} \tag{7.38}$$

So when $\theta = 0$ or π, the squeezed coherent field is also a minimum-uncertainty state, but its property is disagreement with that of $|\beta\rangle$. Because for the state $|\beta\rangle_g$, its fluctuations in two quadrature components are not equal to 1/4.

In general, the condition for the fluctuations in X_1 to be squeezed is

$$\frac{1}{4}[\exp(-2r)\cos^2(\theta/2) + \exp(2r)\sin^2(\theta/2)] < \frac{1}{4}$$

i.e.,

$$\cos\theta > \tanh r \tag{7.39}$$

That is to say, for the squeezed coherent field, that the uncertainty $(\Delta X_1)^2$ can be squeezed occurs only when the factors r and θ satisfy (7.39). Similarly, the condition for $(\Delta X_2)^2 < 1/4$ is

$$\frac{1}{4}[\exp(-2r)\sin^2(\theta/2) + \exp(2r)\cos^2(\theta/2)] < \frac{1}{4}$$

i.e.,

$$\cos\theta < -\tanh r \tag{7.40}$$

Figs.(7.1)–(7.3) are illustrated the two quadrature components and their fluctuations of the fields in a coherent state $|\beta\rangle$ and a squeezed coherent state

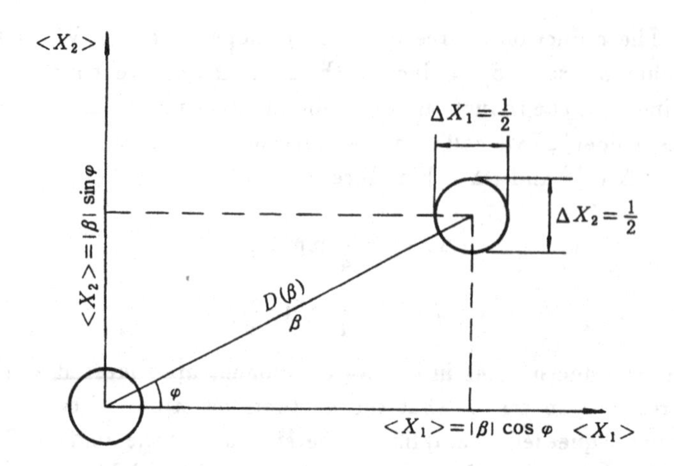

Figure 7.1: The avearge values and their fluctuations of two orthogonal components of coherent field in $X_1 - X_2$ phase space

$|\beta\rangle_g$ in the $X_1 - -X_2$ phase space. Fig.(7.1) shows that the operator $D(\beta)$ translates the vacuum error contour centered at the origin (0,0) to the position $(|\beta|\cos\phi, |\beta|\sin\phi)$, whilst leaving the shape and especially the widths $\langle(\Delta X_1)^2\rangle^{1/2}$ and $\langle(\Delta X_2)^2\rangle^{1/2}$ unchanged from the vacuum values. Fig.(7.2a) is the phase-space description of the mean values and the uncertainties of the quadrature operators X_1 and X_2 for a squeezed coherent state $|\beta\rangle_g$ when $\theta = 0$. From (7.34) and (7.35), its error contour is an ellipse whose major axis is $(\Delta X_2)^2 = \frac{1}{4}\exp(2r)$ and minor axis $(\Delta X_1)^2 = \frac{1}{4}\exp(-2r)$. The centre of this ellipse is at the position $(\langle X_1\rangle, \langle X_2\rangle)$. From eq.(7.32) and (7.33), $\langle X_1\rangle$ and $\langle X_2\rangle$ are respectively as

$$\begin{aligned}\langle X_1\rangle &= |\beta|(\cosh r - \sinh r)\cos\phi \\ \langle X_2\rangle &= |\beta|(\cosh r + \sinh r)\sin\phi\end{aligned} \tag{7.41}$$

So we see that the effect of the squeeze operator $S(\xi)$ not only converts the error circle to the ellipse, but also produces a translation of the centre $(\langle X_1\rangle, \langle X_2\rangle)$. Fig.(7.2b) illustrates the general case for $\theta \neq 0$. In this case the error contour is also an ellipse but whose major and minor axes are not parallel to the X_1 and X_2 axes, respectively, and the centre of the ellipse depends on β, ϕ, r and θ.

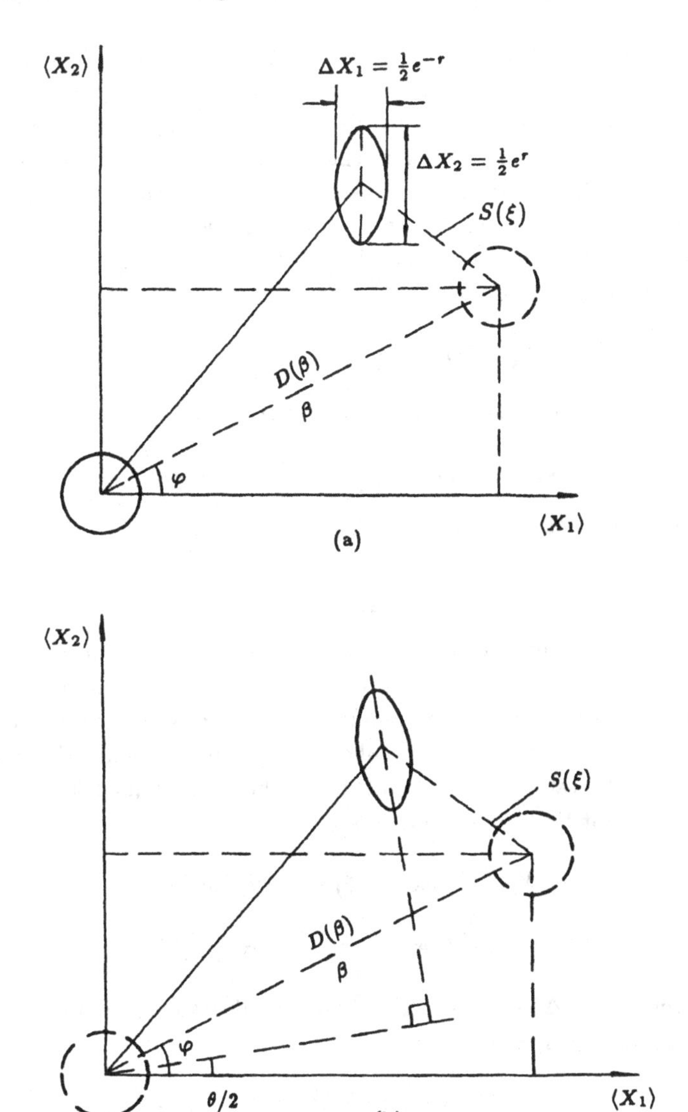

Figure 7.2: The average values and their fluctuations of two orthogonal components of (a) squeezed coherent field $|\beta\rangle_g$ for $\theta = 0$ and (b) squeezed coherent field $|\beta\rangle_g$ for $\theta \neq 0$ in $X_1 - X_2$ phase space

Clearly, Figs.(7.1) and (7.2) are the visual description for the squeezing of quantum noise of the field in the squeezed state $|\beta\rangle_g$. Fig.(7.1) shows that for a coherent state $|\beta\rangle$, the displacement operator $D(\beta)$ just translates the centre of the error contour, but not changes the quantum noise of the field. However, Figs.(7.2a-b) display that the squeezing operator $S(\xi)$ translates the error circle to a ellipse, therefore, the quantum noise of the field may decrease on one direction and increase on another vertical direction.

The uncertainty product obtained from (7.30) and (7.31) gives

$$(\Delta X_1)^2(\Delta X_2)^2 = \frac{1}{16}(\sin^2\theta\cosh^2 2r + \cos^2\theta) \tag{7.42}$$

This takes the minimum-uncertainty-state value

$$(\Delta X_1)^2(\Delta X_2)^2 = \frac{1}{16} \qquad for \quad \theta = 0, \quad \pi$$

but it is larger for other angles. When $\theta = \pi/2$ or $3\pi/2$, it gives a maximum value

$$(\Delta X_1)^2(\Delta X_2)^2 = \frac{1}{16}\cosh^2 2r \tag{7.43}$$

That is to say, a squeezed coherent state may not be a minimum uncertainty state. In general, we call the state in which the fluctuations in one component of the field is less than the corresponding fluctuations in a coherent state as the squeezed state or the squeezed coherent state of the field. However, for a squeezed coherent state, we can introduce another pair of quadrature operators Y_1 and Y_2 different from (7.2), they obey

$$\begin{aligned} Y_1 &= \frac{1}{2}[a\exp(-i\xi) + a^\dagger\exp(i\xi)] \\ Y_2 &= \frac{1}{2i}[a\exp(-i\xi) - a^\dagger\exp(i\xi)] \end{aligned} \tag{7.44}$$

then the product of $(\Delta Y_1)^2$ and $(\Delta Y_2)^2$ also satisfies the minimum uncertainty relation even for $\theta \neq 0$, π, i.e.,

$$(\Delta Y_1)^2(\Delta Y_2)^2 = \frac{1}{16} \tag{7.45}$$

From eqs.(7.44), (7.16)-(7.18) and (7.26) we easily obtain

$$\begin{aligned} (\Delta Y_1)^2 &= \frac{1}{8}[\exp(2r) + \exp(-2r) - (e^{2r} - e^{-2r})\cos(\theta - 2\xi)] \\ (\Delta Y_2)^2 &= \frac{1}{8}[\exp(2r) + \exp(-2r) + (e^{2r} - e^{-2r})\sin(\theta - 2\xi)] \end{aligned} \tag{7.46}$$

Evidently, when $\xi = \theta/2$, (7.46) become

$$(\Delta Y_1)^2 = \frac{1}{4}\exp(-2r)$$
$$(\Delta Y_2)^2 = \frac{1}{4}\exp(2r) \tag{7.47}$$

If $\xi = \theta/2 + \pi/2$, then $(\Delta Y_1)^2$ and $(\Delta Y_2)^2$ reduce to

$$(\Delta Y_1)^2 = \frac{1}{4}\exp(2r)$$
$$(\Delta Y_2)^2 = \frac{1}{4}\exp(-2r) \tag{7.48}$$

So we see that the product of $(\Delta Y_1)^2$ and $(\Delta Y_2)^2$ is exactly (7.45) for arbitrary θ. This is because

$$Y_1 = X_1 \cos\xi + X_2 \sin\xi$$
$$Y_2 = -X_2 \sin\xi + X_2 \cos\xi \tag{7.49}$$

This means that the new phase space $Y_1 - -Y_2$ is formed from the counter-clockwise rotation of the old phase space $X_1 - -X_2$ at the angle ξ. Correspondingly, the phase angle in the squeeze operator $S(\xi)$ becomes $\theta' = \theta - 2\xi$. Consequently, the squeezed coherent state defined for an arbitrary value of θ is a minimum-uncertainty state for a new pair of quadrature operators Y_1 and Y_2 for $\xi = \theta/2$ or $\xi = \theta/2 + \pi/2$. Fig.(7.2b) is also the pictorial representation of the squeezed coherent field $|\beta\rangle_g$ in the $Y_1 - -Y_2$ phase space for $\xi = \theta/2$.

C. State functions of a squeezed coherent field in the number state representation and coherent state representation

First we deduce the formula of the squeezed coherent state $|\beta\rangle$ in the number state representation. Noticing the completeness property of the set of number states $\{|n\rangle\}$, $|\beta\rangle_g$ can be expanded as

$$|\beta\rangle_g = \sum_n C_n |n\rangle \tag{7.50}$$

Here we also let $\mu = \cosh r$ and $\nu = e^{i\phi}\sinh r$ according to (7.26). Substituting (7.11) and (7.50) into (7.13) we have

$$\mu \sum_n C_n \sqrt{n}|n-1\rangle + \nu \sum_n C_n \sqrt{n+1}|n+1\rangle = \beta \sum_n C_n |n\rangle$$

Applying the orthogonal property of the set of number states, it is easy to obtain the recurrence formula of the probability amplitude C_n

$$\mu C_1 = \beta C_0 \tag{7.51}$$

$$\mu C_2\sqrt{2} + \nu\sqrt{1}C_0 = \beta C_1 \tag{7.52}$$

$$\cdots\cdots$$

$$\mu C_n\sqrt{n} + \nu\sqrt{n-1}C_{n-2} = \beta C_{n-1} \tag{7.53}$$

Taking a variable transformation in (7.53) as

$$C_n = A_n(n!)^{-1/2}\left(\frac{\nu}{2\mu}\right)^{n/2} \tag{7.54}$$

then (7.53) becomes

$$A_n - 2\beta(2\mu\nu)^{-1/2}A_{n-1} + 2(n-1)A_{n-2} = 0 \tag{7.55}$$

Evidently, (7.55) obeys the recurrence formula of the Hermitian polynomial

$$H_n(z) - 2zH_{n-1}(z) + 2(n-1)H_{n-2}(z) = 0 \tag{7.56}$$

where the Hermitian polynomial $H_n(z)$ is defined by

$$H_n(z) = \sum_{m=0}^{[n/2]} \frac{(-1)^m n!(2z)^{n-2m}}{m!(n-2m)!} = (-1)^n \exp(z^2)\frac{d^n}{dz^n}\exp(-z^2)$$

Consequently, the solution of (7.55) is

$$A_n = CH_n\left(\frac{\beta}{\sqrt{2\mu\nu}}\right)$$

where C is an undetermined constant. Therefore, C_n may be written as

$$C_n = CH_n\left(\frac{\beta}{\sqrt{2\mu\nu}}\right)\frac{1}{\sqrt{n!}}\left(\frac{\nu}{2\mu}\right)^{n/2} \tag{7.57}$$

Inasmuch as $H_0(\beta/\sqrt{2\mu\nu}) = 1$, we have $C = C_0$. Thus

$$|\beta\rangle_g = C_0\sum_{n=0}^{\infty}\frac{1}{\sqrt{n!}}H_n\left(\frac{\beta}{\sqrt{2\mu\nu}}\right)\left(\frac{\nu}{2\mu}\right)^{n/2} \tag{7.58}$$

Considering the normalization condition ${}_g\langle\beta|\beta\rangle_g = 1$, we have

$$|C_0|^2 \sum_{n=0}^{\infty} \frac{1}{n!} \left|\frac{\nu}{2\mu}\right|^n H_n\left(\frac{\beta}{\sqrt{2\mu\nu}}\right) H_n\left(\frac{\beta^*}{\sqrt{2\mu\nu}}\right) = 1 \tag{7.59}$$

Since the Hermitian polynomials satisfy the relation

$$\sum_{n=0}^{\infty} \frac{1}{n!}\left(\frac{t}{2}\right)^n H_n(x)H_n(y) = (1-t^2)^{-1/2} \exp\left\{\frac{2xyt - (x^2-y^2)t^2}{1-t^2}\right\} \tag{7.60}$$

(7.59) reduces to

$$|C_0|^2 = \frac{1}{|\mu|} \exp\left\{-|\beta|^2 + \frac{\nu^*\beta^2}{2\mu} + \frac{\nu\beta^{*2}}{2\mu^*}\right\} \tag{7.61}$$

If we choose an appropriate phase in C_0 so that

$$\frac{\nu^*\beta^2}{2\mu} = \frac{\nu\beta^{*2}}{2\mu^*}$$

then the constant C_0 becomes

$$C_0 = \frac{1}{\sqrt{\mu}} \exp\left\{-(|\beta|^2 - \frac{\nu^*}{\mu}\beta^2)/2\right\}$$

Consequently, the squeezed coherent state in the number state representation gives

$$|\beta\rangle_g = \sum_{n=0}^{\infty} (n!\mu)^{-\frac{1}{2}} \left(\frac{\nu}{2\mu}\right)^{\frac{n}{2}} \exp\left\{-(|\beta|^2 - \frac{\nu^*}{\mu}\beta^2)/2\right\} H_n(\beta/\sqrt{2\mu\nu})|n\rangle \tag{7.62}$$

Clearly, the photon-number distribution function is

$$|C_n|^2 = \frac{1}{n!\mu}\left|\frac{\nu}{2\mu}\right|^n \exp\left(-|\beta|^2 + \frac{\nu^*}{2\mu}\beta^2 + \frac{\nu}{2\mu}\beta^{*2}\right) H_n\left(\frac{\beta}{\sqrt{2\mu\nu}}\right) H_n\left(\frac{\beta^*}{\sqrt{2\mu^*\nu^*}}\right) \tag{7.63}$$

it is no longer the Poissonian distribution. So it says that the fields described by the squeezed coherent state and the coherent state have different statistical characteristic.

Next we discuss the formula of the squeezed coherent state $\beta\rangle_g$ in coherent state representation. Inserting the completeness relation

$$\frac{1}{\pi}\int |\alpha\rangle\langle\alpha| d^2\alpha = 1$$

into (7.62), we obtain

$$
\begin{aligned}
|\beta\rangle_g = & \frac{1}{\pi}\sum_n \left(\frac{1}{\mu n!}\right)^{1/2}\left(\frac{\nu}{2\mu}\right)^{n/2} \exp\left\{-\frac{1}{2}|\beta|^2+\frac{\nu^*}{2\mu}\beta^2\right\} \\
& \times H_n\left(\frac{\beta}{\sqrt{2\mu\nu}}\right)\int d^2\alpha\langle\alpha|n\rangle|\alpha\rangle \\
= & \frac{1}{\pi}\int d^2\alpha|\alpha\rangle\sum_n \frac{1}{\sqrt{\mu}}\frac{\alpha^{*n}}{n!}\left(\frac{\nu}{2\mu}\right)^{n/2} \\
& \times \exp\left\{-\frac{1}{2}|\beta|^2+\frac{\nu^*}{2\mu}\beta^2-\frac{1}{2}|\alpha|^2\right\}H_n\left(\frac{\beta}{\sqrt{2\mu\nu}}\right) \\
= & \frac{1}{\pi\sqrt{\mu}}\exp\left\{-\frac{1}{2}|\beta|^2+\frac{\nu^*}{2\mu}\beta^2\right\} \\
& \times\int d^2\alpha \exp\left\{-\frac{\nu}{2\mu}\alpha^*+\frac{\nu^*\beta}{\mu}-\frac{1}{2}|\alpha|^2\right\}|\alpha\rangle
\end{aligned}
\tag{7.64}
$$

here we have used the relation

$$
\exp(-t^2+2tz)=\sum_n H_n(z)\frac{t^n}{n!}
$$

Now we turn to discuss the orthogonal and completeness properties of the set of the squeezed coherent states $\{|\beta\rangle_g\}$. Like coherent states, the squeezed coherent states are not orthogonal to each other. The reason reads as follows. From (7.62) we obtain

$$
\begin{aligned}
{}_g\langle\beta|\beta_0\rangle_g = & \langle\beta,\mu,\nu|\beta_0,\mu_0,\nu_0\rangle \\
= & \sum_n \frac{1}{n!\sqrt{\mu^*\mu_0}}\left(\frac{1}{2}\sqrt{\frac{\nu^*\nu_0}{\mu^*\mu_0}}\right)^n H_n\left(\frac{\beta^*}{\sqrt{2\mu^*\nu^*}}\right)H_n\left(\frac{\beta_0}{\sqrt{2\mu_0\nu_0}}\right) \\
& \times\exp\left\{-\frac{1}{2}(|\beta|^2+|\beta_0|^2)+\frac{\nu_0^*}{2\mu_0}\beta_0^2+\frac{\nu}{2\mu^*}\beta^{*2}\right\} \\
= & \exp\left\{-\frac{1}{2}(|\beta|^2+|\beta_0|^2)+\frac{\nu_0^*}{2\mu_0}\beta_0^2+\frac{\nu}{2\mu^*}\beta^{*2}\right\}\frac{1}{\sqrt{\mu^*\mu_0-\nu^*\nu_0}} \\
& \times\exp\left\{\frac{\beta^*\beta_0-\frac{\nu^*}{2\mu_0}\beta_0^2-\frac{\nu_0}{2\mu^*}\beta^{*2}}{\mu^*\mu_0-\nu_0\nu^*}\right\} \\
= & \frac{1}{\sqrt{\mu^*\mu_0-\nu^*\nu_0}}\exp\left[-\frac{1}{2}|\beta_0|^2-\frac{1}{2}|\beta|^2+\frac{\beta^*\beta_0}{\mu_0\mu^*-\nu_0\nu^*}\right.
\end{aligned}
$$

$$+\frac{\beta_0^2(\mu^*\nu_0^* - \nu^*\mu_0^*) + \beta^{*2}(\mu_0\nu - \nu_0\mu)}{2(\mu^*\mu_0 - \nu^*\nu_0)}\Bigg] \neq 0 \tag{7.65}$$

here we have used (7.9) and (7.60). Eq.(7.65) shows that two arbitrary squeezed coherent states with different β, μ, and ν are not orthogonal each other. Especially, when $\mu = \mu_0$ and $\nu = \nu_0$, which means that the different coherent states are squeezed identically, the inner product of two squeezed coherent states gives

$$\langle\beta,\mu,\nu|\beta_0,\mu_0,\nu_0\rangle = \exp\left\{-\frac{1}{2}|\beta-\beta_0|^2\right\} = \langle\beta|\beta_0\rangle \tag{7.66}$$

This means that the set of squeezed coherent states $\{|\beta,\mu,\nu\rangle\}$ have the same orthogonal property as the set of coherent states, and are not connected with parameters μ, ν.

As for the completeness property of the set of squeezed coherent states, we can find out by following the similar procedure as deriving the completeness property of the set of coherent states. From (7.23) and (7.28) we know

$$\begin{aligned}\int d^2\beta|\beta\rangle_{gg}\langle\beta| &= \int |\beta,\mu,\nu\rangle\langle\beta,\mu,\nu|d^2\beta \\ &= S(\xi)\int |\beta\rangle\langle\beta|d^2\beta S^\dagger(\xi) = \pi\end{aligned} \tag{7.67}$$

here we have used $\int|\beta\rangle\langle\beta|d^2\beta = \pi$ and $S(\xi)S^\dagger(\xi) = 1$. Similar to the set of coherent states, the set of squeezed coherent states possess the over-completeness property. In other words, the squeezing operator does not affect the over-completeness property of the set of coherent states. Consequently, the completeness relation of the set of squeezed coherent states can be written as

$$I = \frac{1}{\pi}\int d^2\beta|\beta\rangle_{gg}\langle\beta| \tag{7.68}$$

D. Coherence of the squeezed coherent field

From Chapter 6 we know that the photon antibunching effect is a nonclassical property of the field. Now the question is whether or not the field described by squeezed coherent state which is a nonclassical one, can exhibit the antibunching effect. We next discuss this question.

For a squeezed coherent state $|\beta\rangle_g$, from eqs.(7.13), (7.15) and (7.26) we obtain

$$\begin{aligned}\langle a^\dagger a\rangle &= {}_g\langle\beta|(b^\dagger\cosh r - b\sinh r e^{-i\theta})(b\cosh r - b^\dagger\sinh r e^{i\theta})|\beta\rangle_g\\ &= |\beta|^2[\cosh 2r - \cos(\theta-2\phi)\sinh 2r + \sinh^2 r] \end{aligned}\tag{7.69}$$

$$\begin{aligned}\langle(a^\dagger)^2a^2\rangle &= {}_g\langle\beta|(b^\dagger\cosh r - b\sinh r e^{-i\theta})^2(b\cosh r - b^\dagger\sinh r e^{i\theta})^2|\beta\rangle_g\\ &= |\beta|^4[\cosh 2r - \cos(\theta-2\phi)\sinh 2r]^2\\ &+ |\beta|^2[4\sinh^2 r(\sinh^2 r + 2\cosh^2 r)\\ &- \cosh^2 r\sinh 2r\cos(\theta-2\phi) - 5\sinh^2 r\sinh 2r\cos(\theta-2\phi)]\\ &+ \sinh^2 r(2\sinh^2 r + \cosh^2 r)\end{aligned}\tag{7.70}$$

then the second-order coherence degree which represents the intensity correlation of the field reads

$$\begin{aligned}g^{(2)}(0) &= \frac{\langle a^{\dagger 2}a^2\rangle}{\langle a^\dagger a\rangle^2} = 1 + \frac{\langle a^{\dagger 2}a^2\rangle - \langle a^\dagger a\rangle^2}{\langle a^\dagger a\rangle^2}\\ &= 1 + \frac{\sinh^2 r\cosh 2r}{\{|\beta|^2[\cosh 2r - \sinh 2r\cos(\theta-2\phi)] + \sinh^2 r\}^2}\\ &+ \frac{2|\beta|^2\sinh^2 r[\cosh 2r - \sinh 2r\cos(\theta-2\phi)]}{\{|\beta|^2[\cosh 2r - \sinh 2r\cos(\theta-2\phi)] + \sinh^2 r\}^2}\\ &+ \frac{|\beta|^2\sinh 2r[\sinh 2r - \cosh 2r\cos(\theta-2\phi)]}{\{|\beta|^2[\cosh 2r - \sinh 2r\cos(\theta-2\phi)] + \sinh^2 r\}^2}\\ &= 1 + \frac{\sinh^2 r\cosh 2r + 2\sinh r\cosh 3r|\beta|^2[\tanh 3r - \cos(\theta-2\phi)]}{\{|\beta|^2[\cosh 2r - \sinh 2r\cos(\theta-2\phi)] + \sinh^2 r\}^2}\end{aligned}\tag{7.71}$$

We see from (7.71) that when $g^{(2)}(0) > 1$, the squeezed coherent field $|\beta\rangle_g$ exhibits the photon bunching effect, and when $g^{(2)}(0) = 1$, there is no correlation among the photons in the squeezed coherent field, but for $g^{(2)}(0) < 1$, the photon antibunching effect appears.

Now we discuss the influences of the squeezing parameters r and θ on the second-order coherence of the field. If we choose $\theta = 2\phi$, then equation (7.71) reduces to

$$g^{(2)}(0) = 1 + \frac{\sinh^2 r\cosh 2r - 2|\beta|^2\exp(-3r)\sinh r}{[|\beta|^2\exp(-2r) + \sinh^2 r]^2}\tag{7.72}$$

Evidently, if r satisfies $|\beta|^2 < \frac{1}{2}\exp(3r)\sinh r\cosh 2r$, then $g^{(2)}(0) > 1$, which means that the photons in the squeezed coherent field are bunching. When

$|\beta|^2 = \frac{1}{2}\exp(3r)\sinh r\cosh r$, $g^{(2)}(0) = 1$, which stands for that the photons are random. However, it is worth pointing out that the field is not the Poissonian distribution, so the photon distribution of the squeezed coherent field is different from that of the coherent field. If r is chosen as $|\beta|^2 > \frac{1}{2}\exp(3r)\sinh r\cosh 2r$, then $g^{(2)}(0) > 1$. In this moment, the field exhibits the antibunching effect of photons.

When $\theta - 2\phi \neq 0$, from (7.71) we find if

$$\cos(\theta - 2\phi) > \frac{\sinh r\cosh 2r + 2\sinh 3r|\beta|^2}{2\cosh 3r|\beta|^2} \tag{7.73}$$

then the photons in a squeezed coherent field $|\beta\rangle_g$ are antibunching. For other arbitrary value of θ, the photons in the squeezed field are bunching or random. Comparing (7.73) with (7.39) we find that even if $\phi = 0$, the conditions which lead to the squeezed field to exhibit the squeezing and the antibunching are different. That is to say, the squeezing effect and the antibunching effect of photons are two independent nonclassical phenomena of the field.

It is worthy pointing out that in 1981, C.M.Caves introduced another definition of the squeezed coherent state to be

$$|\alpha\rangle_g = D(\alpha)S(\xi)|0\rangle \tag{7.74}$$

Here $D(\alpha)$ is the usual displacement operator, $S(\xi)$ is a squeeze operator. Comparing (7.28) with (7.74) we know that these two definitions are obviously different. Because the operators $D(\alpha)$ and $S(\xi)$ do not commutate with each other. These two squeezed coherent states are generated by different physical processes. Eq.(7.28) shows that one can adopt some nonlinear optical process such as degenerate parametric amplification process to transfer the coherent state $|\beta\rangle$ to the squeezed coherent state $|\beta\rangle_g$ as shown in Fig.(7.2b). But (7.74) shows that the first step is to fix the system to be the squeezed vacuum state, then transfer the squeezed vacuum state to the squeezed coherent state by some physical forces as shown in Fig.(7.3). However, we can verify that $D(\alpha)$ and $S(\xi)$ may commutate with each other for appropriate parameters. Since

$$S(\xi)D(\alpha) = S(\xi)D(\alpha)S^{\dagger}(\xi)S(\xi) = D(\beta)S(\xi) \tag{7.75}$$

with

$$\beta = \alpha\cosh r - \alpha^* \exp(i\theta)\sinh r \tag{7.76}$$

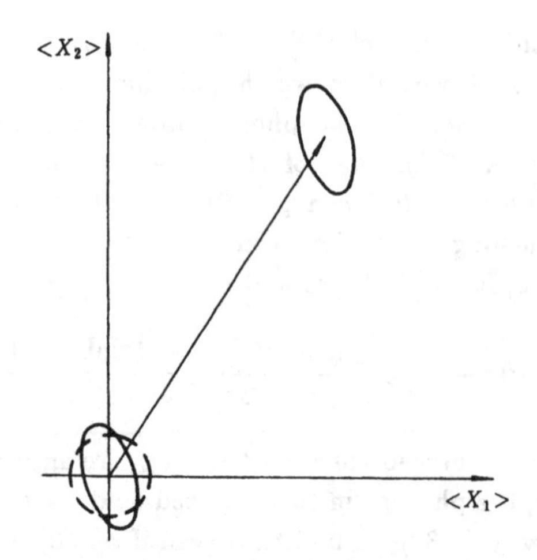

Figure 7.3: The average values and their fluctuations of two orthogonal components of a squeezed coherent field described by (7.74) in $X_1 - -X_2$ phase space

It is easy to see that these two squeezed coherent fields can display the same squeezing for one quadrature component. But the conditions of the antibunching effect and the antibunching degree for these two squeezed coherent fields are different.

7.1.2 Squeezed vacuum state

Next we discuss the squeezed vacuum state which has been produced in experiments. The definition of the squeezed vacuum state is

$$|0\rangle_g = S(\xi)|0\rangle \tag{7.77}$$

Comparing with the squeezed coherent state $|\beta\rangle_g$ (or $|\alpha\rangle_g$), here only need $\beta = 0$ (or $\alpha = 0$). That is to say, the squeezing operator $S(\xi)$ converts the vacuum state to the squeezed vacuum state, which leads to the fluctuations of one quadrature component of the field to be reduced and those of the other one to be increased. Considering the property of the Hermitian polynomial

$$H_{2n}(0) = (-1)^n \frac{(2n)!}{n!}, \quad H_{2n+1}(0) = 0 \tag{7.78}$$

then from (7.62) the state function of the squeezed vacuum state in the number state representation reads

$$|0\rangle_g = \sum_{n=0}^{\infty} \frac{(-e^{i\theta}\tanh r)^n (2n!)^{1/2}}{\sqrt{\cosh r} n! 2^n} |2n\rangle \tag{7.79}$$

The above equation shows that the squeezed vacuum state is consisted of the even-number photon states $\{|2n\rangle\}$. The inner product of different squeezed states gives

$$_g\langle 0'|0\rangle_g = \langle 0, r', \theta|0, r, \theta\rangle = sech^{1/2}(r - r') \tag{7.80}$$

Obviously, the squeezed vacuum states are not orthogonal to each other similar to the squeezed states.

The second-order coherence degree of the squeezed vacuum field can be obtained directly from (7.71) when we let $|\beta| = 0$, namely

$$g^{(2)}(0) = 1 + \frac{\cosh 2r}{\sinh^2 r} = 3 + \frac{1}{\langle n\rangle} > 1 \tag{7.81}$$

here $\langle n\rangle = \sinh^2 r$ is the mean photon number of the squeezed vacuum field. Evidently, the single-mode squeezed vacuum field exhibits the stronger bunching effect of photons than the single-mode thermal field, and the antibunching phenomenon of photons can never appear.

7.2 Squeezed states of a two-mode radiation field

As yet we have given a brief insight to the properties of the single-mode squeezed coherent field and the single-mode squeezed vacuum field, we now turn to discuss the squeezed state of a two-mode radiation field.

Similar to the definitions of the squeezed coherent state of a single-mode field, there are two definitions of the squeezed coherent state of the two-mode field with frequencies ω_+ and ω_- as

$$|\beta_+, \beta_-, \xi\rangle = S_{+-}(\xi) D_+(\beta_+) D_-(\beta_-)|0\rangle \tag{7.82}$$

$$|\alpha_+, \alpha_-, \xi\rangle = D_+(\alpha_+) D_-(\alpha_-) S_{+-}(\xi)|0\rangle \tag{7.83}$$

Here the operators D_+, D_- and S_{+-} are defined, respectively, as

$$D_\pm(\alpha_\pm) = \exp(\alpha_\pm a_\pm^\dagger - \alpha_\pm^* a_\pm) \tag{7.84}$$

$$D_\pm(\beta_\pm) = \exp(\beta_\pm a_\pm^\dagger - \beta_\pm^* a_\pm) \tag{7.85}$$

$$S_{+-}(\xi) = \exp(\xi a_+ a_- - \xi a_+^\dagger a_-^\dagger) \tag{7.86}$$

where a_+ (a_-) and $a_+^\dagger$ ($a_-^\dagger$) represent the annihilation and creation operators for the photons with frequency ω_+ (ω_-), respectively, they obey the following commutation relations

$$[a_\pm, a_\pm^\dagger] = 1, \quad [a_\pm^\dagger, a_\mp] = [a_\pm, a_\mp^\dagger] = 0$$

It is clear that D_+ and D_- represent the displacement operators for the modes whose frequencies are ω_+ and ω_-, respectively. D_+ and D_- have the same properties as the displacement operator for the single-mode field. Comparing the operator $S_{+-}(\xi)$ with the squeeze operator $S(\xi)$ for the single-mode field, we find that $S_{+-}(\xi)$ contains the two-mode field operators $a_\pm$ and $a_\pm^\dagger$, in addition, the factor 1/2 ahead of ξ is absent in $S_{+-}(\xi)$. So the operator $S_{+-}(\xi)$ is called the two-mode squeeze operator.

By means of (7.24), it is easy to prove that

$$b_\pm = S_{+-}(\xi) a_\pm S_{+-}^\dagger(\xi) = a_\pm \cosh r + a_\mp^\dagger \exp(i\theta) \sinh r \tag{7.87}$$

here $\xi = r\exp(i\theta)$. Clearly, the above equation is similar to (7.25). From eqs.(7.82), (7.83) and (7.87) we obtain

$$b_\pm|\beta_+, \beta_-, \xi\rangle = S_{+-}(\xi) a_\pm S_{+-}^\dagger(\xi) D_+(\beta_+) D_-(\beta_-)|0\rangle = \beta_\pm|\beta_+, \beta_-, \xi\rangle \tag{7.88}$$

So we see that the two-mode squeezed coherent state defined by (7.82) is the eigenstate of the operator b_+ (b_-) with eigenvalue β_+ (β_-). Consequently, for the two-mode squeezed coherent state defined by (7.82), the squeezing properties for one mode are identical to those of the single-mode squeezed coherent state defined by (7.28). We can also prove that the relation between $D_\pm(\beta_\pm)$ and $D_\pm(\alpha_\pm)$ is

$$D_\pm(\beta_\pm) = S_{+-}^\dagger(\xi) D_\pm(\alpha) S_{+-}(\xi) \tag{7.89}$$

with

$$\beta_\pm = \alpha_\pm \cosh r + \alpha_\mp^* \exp(i\theta) \sinh r \tag{7.90}$$

Thus the two definitions of the two-mode squeezed coherent state can be formally reduced to the same. In order to simplify the calculation, next we only discuss the squeezed properties of the two-mode squeezed coherent field defined by (7.82).

For a two-mode field, we are usually interested in the following two quadrature components

$$U_1 = \frac{(a_+ + a_+^\dagger + a_- + a_-^\dagger)}{2^{3/2}}$$

$$U_2 = \frac{-i(a_+ - a_+^\dagger + a_- - a_-^\dagger)}{2^{3/2}} \tag{7.91}$$

Obviously, they obey the commutation relation

$$[U_1, U_2] = \frac{i}{2} \tag{7.92}$$

and the corresponding uncertainty relation gives

$$(\Delta U_1)^2 (\Delta U_2)^2 \geq \frac{1}{16} \tag{7.93}$$

Similar to the single-mode case, for the two-mode field if the fluctuations of one quadrature component satisfy

$$(\Delta U_i)^2 < \frac{1}{4} \qquad (i = 1 \text{ or } 2) \tag{7.94}$$

then the fluctuation in the i-th component of the two-mode field is squeezed. If the field is in the two-mode squeezed coherent state $|\alpha_+, \alpha_-, \xi\rangle$, then from (7.82) and (7.87) we can easily obtain

$$\langle a_\pm \rangle = \alpha_\pm \tag{7.95}$$

$$\langle a_\pm^\dagger a_\pm \rangle = |\alpha_\pm|^2 + \sinh^2 r \tag{7.96}$$

$$\langle a_+^\dagger a_- \rangle = \alpha_+^* \alpha_-, \quad \langle a_-^\dagger a_+ \rangle = \alpha_-^* \alpha_+ \tag{7.97}$$

$$\langle a_\pm a_\pm \rangle = \alpha_\pm^2 \tag{7.98}$$

$$\langle a_+ a_- \rangle = \langle a_- a_+ \rangle = \alpha_+ \alpha_- \exp(i\theta) \sinh r \cosh r \tag{7.99}$$

From (7.95)-(7.99) we see that the squeeze operator $S_{+-}(\xi)$ just affects the mean photon numbers $\langle a_\pm^\dagger a_\pm \rangle$ of each mode and the nondiagonal elements $\langle a_+ a_- \rangle$ and $\langle a_- a_+ \rangle$ between two modes. The fluctuations in the U_1 component can be easily obtained from (7.95)-(7.99)

$$(\Delta U_1)^2 = \langle U_1^2 \rangle - \langle U_1 \rangle^2 = \frac{1}{4}[\exp(-2r)\cos^2(\theta/2) + \exp(2r)\sin^2(\theta/2)] \tag{7.100}$$

This equation is completely identical with (7.30). So the quantum fluctuation property of the two-mode squeezed coherent field is the same as that of the single-mode squeezed coherent field. It is worth pointing out that $(\Delta U_1)^2$ is not related to α_+ and α_- which represent the amplitudes of the two-mode field.

If we let $\alpha_+ = \alpha_- = 0$ in (7.83), then the two-mode squeezed coherent state is reduced to the two-mode squeezed vacuum state. The two-mode squeezed vacuum state is therefore defined as

$$|0,0,\xi\rangle = S_{+-}(\xi)|0\rangle \tag{7.101}$$

In this case, the mean photon number of each mode gives

$$\langle a_\pm^\dagger a_\pm\rangle = \sinh^2 r \tag{7.102}$$

which is also identical with that of the single-mode field case.

We now turn to verify the second-order coherent degree of the two-mode squeezed vacuum field. From (7.87) and (7.101) we see

$$\langle (a_\pm^\dagger)^2 a_\pm^2\rangle = \langle 0|S_{+-}^\dagger(\xi)(a_\pm^\dagger)^2 a_\pm^2 S_{+-}(\xi)|0\rangle = 2\sinh^2 r \tag{7.103}$$

$$\begin{aligned}\langle a_+^\dagger a_-^\dagger a_+ a_-\rangle &= \langle 0|S_{+-}^\dagger(\xi) a_+^\dagger a_-^\dagger a_+ a_- S_{+-}(\xi)|0\rangle \\ &= 2\sinh^4 r + \sinh^2 r\end{aligned} \tag{7.104}$$

Then the second-order coherent degree of the i-th mode of the two-mode squeezed vacuum field gives

$$g_1^{(2)}(0) = \frac{\langle (a_+^\dagger)^2 a_+^2\rangle}{\langle a_+^\dagger a_+\rangle^2} = 2 = g_2^{(2)}(0) \tag{7.105}$$

Eq.(7.105) means that the photons in each mode of the two-mode squeezed vacuum field are bunching, and $g_i^{(2)}(0)(i{=}1,\ 2)$ are equal to the second-order coherent degree (6.80) of the two-mode thermal field, but here $g_i^{(2)}(0)(i{=}1,\ 2)$ are less than the second-order coherence degree (7.81) of the single-mode squeezed vacuum field. In addition, the cross-correlation degree between two modes of the squeezed vacuum field can be obtained from (6.46) and (7.104) as

$$g_{12}^{(2)}(0) = \frac{\langle a_+^\dagger a_-^\dagger a_+ a_-\rangle}{\langle a_+^\dagger a_+\rangle\langle a_-^\dagger a_-\rangle} = 2 + \frac{1}{\sinh^2 r} > 1 \tag{7.106}$$

So the photons between two modes of the two-mode squeezed vacuum field are correlative and satisfy

$$(g_{12}^{(2)}(0))^2 > g_1^{(2)}(0) g_2^{(2)}(0) \tag{7.107}$$

Thus the correlation between two modes is a nonclassical one.

In order to realize the second-order coherence of the two-mode squeezed vacuum field, we have to solve the state function of the two-mode squeezed vacuum state in the number representation. First we prove an operator identity. If the operators A, B and C satisfy

$$[A,B] = C = C^{\dagger}, \quad [A,C] = 2A, \quad [B,C] = -2B \tag{7.108}$$

then there exists an identity

$$\exp[r(Ae^{i\theta} - Be^{i\theta})] = \exp(-Be^{i\theta}\tanh r)\exp[-C\ln(\cosh r)]\exp(Ae^{-i\theta}\tanh r) \tag{7.109}$$

The proof of (7.109) may be taken as follows.

Introducing two exponential operators

$$U_1(\lambda) = \exp\{\lambda[r(e^{i\theta}A - Be^{i\theta}) + i\varepsilon C]\} \tag{7.110}$$

$$U_2(\lambda) = \exp[P_1(\lambda)B]\exp[P_0(\lambda)C]\exp[P_2(\lambda)A] \tag{7.111}$$

where λ is an arbitrary real parameter, $P_j(\lambda)(j$=0,1,2$)$ are undetermined functions of the parameter λ, and ε is a real constant independent of λ, it is clear that $U_1^{\dagger}(\lambda) = U_1^{-1}(\lambda)$. If one can choose appropriate $P_j(\lambda)(j$=0,1,2$)$ so that

$$U_1(\lambda) = U_2(\lambda) \tag{7.112}$$

then (7.109) can be deduced when $\lambda = 1$ and $\varepsilon = 0$.

Differentiating (7.112) with respect to λ, we then have

$$\begin{aligned}
[r(e^{-i\theta}A - Be^{i\theta}) + i\varepsilon C]U_1 &= [r(e^{-i\theta}A - Be^{i\theta}) + i\varepsilon C]U_2 \\
&= P'B\exp(P_1B)\exp(P_0C)\exp(P_2A) \\
&\quad + P_0'\exp(P_1B)C\exp(P_0C)\exp(P_2A) \\
&\quad + P_2'\exp(P_1B)\exp(P_0C)A\exp(P_2A)
\end{aligned} \tag{7.113}$$

Here $P_i'(\lambda)$ represents $\frac{d}{d\lambda}P_i(\lambda)$. Multiplying $U_2^{-1}(\lambda)$ from the right hand in the above equation, here $U_2^{-1}(\lambda)$ defines as

$$U_2^{-1}(\lambda) = \exp(-P_2A)\exp(-P_0C)\exp(-P_1B)$$

then (7.113) becomes

$$\begin{aligned}
r(e^{-i\theta}A - Be^{i\theta}) + i\varepsilon C &= P_1'B + P_0'\exp(P_1B)C\exp(-P_1B) \\
&+ P_2'\exp(P_1B)\exp(P_0C)A\exp(-P_0C)\exp(-P_1B)
\end{aligned} \tag{7.114}$$

Noticing the operator identity

$$\exp(tF)G\exp(-tF) = G + t[F,G] + \frac{t^2}{2!}[F,[F,G]] + ...$$

and applying (7.108) we obtain

$$\begin{aligned}\exp(tB)A\exp(-tB) &= A - tC + t^2B \\ \exp(tB)C\exp(-tB) &= C - 2tB \\ \exp(tC)A\exp(-tC) &= \exp(-2t)A\end{aligned} \tag{7.115}$$

Substituting (7.115) into (7.114), we have

$$\begin{aligned}r(e^{-i\theta}A - Be^{i\theta}) + i\varepsilon C &= P_1'B + P_0'(C - 2P_1B) \\ &\quad + P_2'\exp(-2P_0)(A - P_1C + P_1^2B) \\ &= P_2'\exp(-2P_0)A + [P_0' - P_1P_2'\exp(-2P_0)]C \\ &\quad + [P_1' + P_1^2P_2'\exp(-2P_0) - 2P_1P_0']B\end{aligned} \tag{7.116}$$

The conditions for the above equation holding up may be

$$\begin{aligned}P_2'\exp(-2P_0) &= r\exp(-i\theta) \\ P_0' - P_1P_2'\exp(-2P_0) &= i\varepsilon \\ P_1' - 2P_1P_0' + P_1'^2\exp(-2P_0) &= -r\exp(i\theta)\end{aligned} \tag{7.117}$$

Since for $\lambda = 0$, $U_1(0) = U_2(0) = 1$, $P_j(0)(j = 0,1,2)$ obey

$$P_j(0) = 0 \tag{7.118}$$

Inserting the first equation of (7.117) into the second one, then

$$P_0' - P_1r\exp(-i\theta) = i\varepsilon \tag{7.119}$$

Combining (7.117) with (7.119), then P_1 satisfies the following differential equation

$$P_1' - 2i\varepsilon P_1 - r\exp(-i\theta)P_1^2 = -r\exp(i\theta) \tag{7.120}$$

Taking the variable transformation to (7.120) as

$$P_1(\lambda) = -\frac{\exp(i\theta)u'(\lambda)}{ru(\lambda)} \tag{7.121}$$

(7.120) therefore deduces to

$$u'' - 2i\varepsilon u' - r^2 u = 0 \tag{7.122}$$

This is exactly a second-order ordinary differential equation. Considering for $P_1(0) = 0$, $u'(0) = 0$, so the solution of (7.122) gives

$$u(\lambda) = E\{\exp[-(\sqrt{r^2-\varepsilon^2} - i\varepsilon)\lambda] + \frac{\sqrt{r^2-\varepsilon^2} - i\varepsilon}{\sqrt{r^2-\varepsilon^2} + i\varepsilon}\exp[(\sqrt{r^2-\varepsilon^2} + i\varepsilon)\lambda]\} \tag{7.123}$$

Inserting (7.123) into (7.121), one can yield $P_1(\lambda)$ as

$$P_1(\lambda) = -\frac{r\exp(i\theta)\sinh(\sqrt{r^2-\varepsilon^2}\lambda)}{\sqrt{r^2-\varepsilon^2}\cosh(\sqrt{r^2-\varepsilon^2}\lambda) - i\varepsilon\sinh(\sqrt{r^2-\varepsilon^2}\lambda)} \tag{7.124}$$

Combining (7.124) with (7.119) and applying (7.118) and the following integral identity

$$\int \frac{\alpha + \beta\cosh x + q\sinh x}{a + b\cosh x + c\sinh x}dx = \frac{qb - \beta c}{b^2 - c^2}\ln(a + b\cosh x + c\sinh x) + \frac{\beta b - qc}{b^2 - c^2}x$$
$$+(\alpha - a\frac{\beta b - qc}{b^2 - c^2})\int \frac{dx}{a + b\cosh x + c\sinh x} \qquad (b^2 \neq c^2)$$

we can obtain $P_0(\lambda)$ as

$$P_0(\lambda) = -\ln\left[\cosh(\sqrt{r^2-\varepsilon^2}\lambda) - \frac{i\varepsilon}{\sqrt{r^2-\varepsilon^2}}\sinh(\sqrt{r^2-\varepsilon^2}\lambda)\right] \tag{7.125}$$

Substituting (7.125) into (7.117a), and applying (7.118) and

$$\int \frac{dx}{a\cosh x + b\sinh x} = \frac{1}{a^2 - b^2}\frac{a\sinh x + b\cosh x}{a\cosh x + b\sinh x} + constant$$

then the solution of $P_2(\lambda)$ reads

$$P_2(\lambda) = \frac{re^{-i\theta}\sinh(\sqrt{r^2-\varepsilon^2}\lambda)}{\sqrt{r^2-\varepsilon^2}\cosh(\sqrt{r^2-\varepsilon^2}\lambda) - i\varepsilon\sinh(\sqrt{r^2-\varepsilon^2}\lambda)} \tag{7.126}$$

Now we have obtained the explicit formulae of $P_j(\lambda)$ which satisfy (7.112). If let $\lambda = 1$ and $\varepsilon = 0$, then we have

$$\exp[r(Ae^{-i\theta} - Be^{i\theta})] = \exp(-Be^{i\theta}\tanh r)\exp[-C\ln(\cosh r)]\exp(Ae^{-i\theta}\tanh r)$$

This is exactly (7.109).

If let $A = a_+a_-$, $B = a_+^\dagger a_-^\dagger$, $C = [A, B] = [a_+a_-, a_+^\dagger a_-^\dagger] = a_+^\dagger a_+ + a_-^\dagger a_- + 1$, then from eq.(7.109) we have

$$\begin{aligned} S_{+-}(\xi) &= \exp(-a_+^\dagger a_-^\dagger e^{i\theta} \tanh r) \exp[-(a_+^\dagger a_+ + a_-^\dagger a_- + 1) \ln(\cosh r)] \\ &\quad \times \exp(a_+ a_- e^{-i\theta} \tanh r) \\ &= \frac{1}{\cosh r} \exp(-a_+^\dagger a_-^\dagger e^{i\theta} \tanh r) \exp[-(a_+^\dagger a_+ + a_-^\dagger a_-) \ln(\cosh r)] \\ &\quad \times \exp(a_+ a_- e^{-i\theta} \tanh r) \end{aligned} \tag{7.127}$$

Substituting (7.127) into (7.101), then the state function of the two-mode squeezed vacuum state in the number representation is expressed as

$$|0, 0, \xi\rangle = \frac{1}{\cosh r} \sum_{n=0}^{\infty} (-e^{i\theta} \tanh r)^n |n, n\rangle \tag{7.128}$$

here we have used

$$\exp(a_+a_-)|0\rangle = 1, \quad \exp(a_\pm^\dagger a_\pm)|0\rangle = 1, \quad \exp(a_+^\dagger a_-^\dagger) = \sum_{n=0}^{\infty} \frac{1}{n!}(a_+^\dagger a_-^\dagger)^n$$

From (7.128) we see that the two-mode squeezed vacuum state is composed of the set of two-mode number states $\{|n, n\rangle\}$ in which the two modes have equal photon numbers. The total photon number is $2n$ for every basis vector $|n, n\rangle$, which is similar to (7.79) for the single-mode squeezed vacuum state. However, each mode has n photons in every basis vector $|n, n\rangle$, the photon-bunching degree of every mode in the two-mode squeezed vacuum state is weaker than that of the single-mode case. In addition, from (7.128) we can also see that the photon distribution of the two-mode squeezed vacuum field reads

$$P_{n_+, n_-} = \frac{\bar{n}^{n_+}}{(1+\bar{n})^{n_+ + 1}} \delta_{n_+, n_-} \tag{7.129}$$

here $\bar{n} = \langle a_+^\dagger a_+ \rangle = \langle a_-^\dagger a_- \rangle = \sinh^2 r$. But the photon distribution of the two-mode thermal field is

$$P_{n_+, n_-} = \frac{\bar{n}_+^{n_+}}{(1+\bar{n}_+)^{n_+ + 1}} \cdot \frac{\bar{n}_-^{n_-}}{(1+\bar{n}_-)^{n_- + 1}} \tag{7.130}$$

It is clear that the photon distribution of the two-mode squeezed vacuum field is equal to zero except when $n_+ = n_-$, which means that there exists strong correlation between two modes. But for the two-mode thermal field, $P_{n_+,n_-} \neq 0$ for $n_+ \neq n_-$, then the correlation degree between two modes in the two-mode squeezed vacuum field is stronger than that in the two-mode thermal field.

7.3 Higher-order squeezing of a radiation field and the amplitude square squeezing

From Sections 1 and 2 we know that a field in a squeezed state exhibits fluctuations in one quadrature component smaller than those in a coherent state (or in a vacuum state), at the cost of increasing the fluctuations in an other quadrature component. Next we introduce two kinds of squeezing, i.e., the higher-order squeezing and the amplitude square squeezing.

7.3.1 Higher-order squeezing of a radiation field

A multi-mode field can be expressed as

$$\mathbf{E}(\mathbf{r},t) = \mathbf{E}^{(+)}(\mathbf{r},t) + \mathbf{E}^{(-)}(\mathbf{r},t) \tag{7.131}$$

here $\mathbf{E}^{(+)}(\mathbf{r},t)$ and $\mathbf{E}^{(-)}(\mathbf{r},t)$ are the positive and negative frequency parts of the field. In practice, a detector can only respond the field variables within a fixed frequency band and a special polarization direction, so we assume that the field has finite modes with one polarization direction. Under this assumption, (7.131) can be rewritten in a scalar form as

$$E^{(+)}(\mathbf{r},t) = \frac{1}{\sqrt{V}} \sum_{[k]} l(k) a_k \exp[i(\mathbf{k}\cdot\mathbf{r} - \omega t)] \tag{7.132}$$

where a_k is the annihilation operator for the mode with the wave vector $\mathbf{k}$, $[k]$ represents the finite values of wave vector $\mathbf{k}$ which can be responded by the detector, ω is the centre frequency of the band, $l(\mathbf{k})$ is defined as

$$l(\mathbf{k}) = \left(\frac{\hbar\omega}{2\varepsilon}\right)^{1/2}$$

From (7.132) we can obtain the commutation relation which $E^{(+)}$ and $E^{(-)}$ obey as

$$[E^{(+)}, E^{(-)}] = \frac{1}{V} \sum_{[k]} |l(\mathbf{k})|^2 = C \tag{7.133}$$

here C is a finite positive number.

For two quadrature components E_1 and E_2 of the electric field with one polarization are defined by

$$E_1 = E^{(+)}\exp(-i\phi) + E^{(-)}\exp(i\phi)$$
$$E_2 = E^{(+)}\exp[-i(\phi+\pi/2)] + E^{(-)}\exp[i(\phi+\pi/2)] \quad (7.134)$$

they satisfy the commutation and uncertainty relations

$$[E_1, E_2] = 2iC \quad (7.135)$$
$$\langle(\Delta E_1)^2\rangle\langle(\Delta E_2)^2\rangle \geq C^2 \quad (7.136)$$

where ΔE_i stands for the deviation $E_i - \langle E_i\rangle (i = 1,2)$. If there exists some phase angle ϕ for which

$$\langle(\Delta E_i)^2\rangle < C \qquad (i = 1 \text{ or } 2) \quad (7.137)$$

then the fluctuations in the E_i component are said to be squeezed to second-order. Evidently, (7.21) and (7.94) can be regarded as special cases of (7.137). Also, with the help of the commutation relation (7.135), the squeezing condition can be rewritten in the normally ordered form

$$\langle : (\Delta E_i)^2 : \rangle = \langle(\Delta E_i)^2\rangle - C < 0 \quad (7.138)$$

here : : represents that the creation and annihilation operators are put in normal order. For example, from (7.2) we know

$$(\Delta X_1)^2 = \frac{1}{4}(\langle a^\dagger a\rangle + \langle aa^\dagger\rangle + \langle a^{\dagger 2}\rangle + \langle a^2\rangle - \langle a + a^\dagger\rangle^2)$$

then

$$: (\Delta X_1)^2 := \frac{1}{4}(2\langle a^\dagger a\rangle + \langle a^{\dagger 2}\rangle + \langle a^2\rangle - \langle a + a^\dagger\rangle^2) = (\Delta X_1)^2 - \frac{1}{4}$$

From (7.134) E_2 can be regarded as a special case of E_1 if we let $\phi \longrightarrow \phi + \pi/2$, so we may only discuss the higher-order squeezing in E_1.

We now turn to make a generalization of the foregoing squeezing concept to the $2N$-th order (N=1,2,3,$\cdots$). If there exists a phase angle ϕ, such that $\langle(\Delta E_1)^{2N}\rangle$ for the squeezed state is smaller than the corresponding value in a coherent state of the field, then the fluctuations in E_1 are squeezed to the

$2N$-th order. This higher-order squeezing may be generalized to the $(2N+1)$-th order moment of ΔE_1, and for a coherent state we have $\langle : (\Delta E_1)^{2N+1} : \rangle = 0$, but the state which obey the inequality

$$\langle : (\Delta E_1)^{2N+1} : \rangle < 0$$

is not always nonclassical. This is because for a radiation field in some state, its odd-order moments may be positive or negative. Therefore, we can determine from the even-order moments whether the radiation field is a classical or nonclassical one.

In order to exhibit the nonclassical property of the squeezed state, an appropriate way is to display the even-order moments as the normally ordered forms. Noticing (7.134), then there exists the identity

$$\exp(xE_1) = \exp(xE^{(-)}e^{i\phi}) \exp(xE^{(+)}e^{-i\phi}) \exp(x^2C/2)$$

here x is an arbitrary real parameter. By using the above equation, the expectation value of the exponential operator $\exp(x\Delta E_1)$ may be written as

$$\langle \exp(x\Delta E_1) \rangle = \langle : \exp(x\Delta E_1) : \rangle \exp(x^2C/2) \tag{7.139}$$

Expanding both sides of the above equation as a power series in x and comparing coefficient of x^{2N}, we readily obtain the relation

$$\begin{aligned} \langle (\Delta E_1)^{2N} \rangle &= \sum_{q=0}^{N} \frac{\langle : (\Delta E_1)^{2(N-q)} : \rangle}{(2N-2q)!q!} \left(\frac{C}{2}\right)^q (2N)! \\ &= \langle : (\Delta E_1)^{2N} : \rangle + \frac{(2N-1)2N}{1!} \left(\frac{C}{2}\right) \langle : (\Delta E_1)^{2(N-1)} : \rangle \\ &\quad + \frac{(2N-3)(2N-2)(2N-1)2N}{2!} \left(\frac{C}{2}\right)^2 \langle : (\Delta E_1)^{2(N-2)} : \rangle \\ &\quad + \cdots + (2N-1)!!C^N \end{aligned} \tag{7.140}$$

For a coherent state, we have $\langle : (\Delta E_1)^{2n} : \rangle = 0 (n = 0, 1, 2, \cdots)$, so that if the field satisfies the condition

$$\langle (\Delta E_1)^{2N} \rangle < (2N-1)!!C^N \tag{7.141}$$

then the quantum noise of the radiation field is to be squeezed to $2N$-order in component E_1. For example, we have the second order squeezing if

$$\langle : (\Delta E_1)^2 : \rangle < 0$$

the fourth order squeezing if

$$\langle : (\Delta E_1)^4 : \rangle + 6C\langle : (\Delta E_1)^2 : \rangle < 0 \tag{7.142}$$

the sixth order squeezing if

$$\langle : (\Delta E_1)^6 : \rangle + 15C\langle : (\Delta E_1)^4 : \rangle + 45C^2\langle : (\Delta E_1)^2 : \rangle < 0 \tag{7.143}$$

etc. If the radiation field is a single-mode one, then from (7.2) and (7.134) for $\phi = 0$ we find that E_1 and X_1 obey

$$X_1 = \frac{E_1}{2}$$

thus the condition for the fourth order squeezing is simplified to be

$$\langle X_1^4 \rangle - 4\langle X_1^3 \rangle \langle X_1 \rangle + 6\langle X_1^2 \rangle \langle X_1^2 \rangle - 2\langle X_1^4 \rangle < \frac{3}{16} \tag{7.144}$$

7.3.2 Amplitude square squeezing

Next we introduce another kind of squeezing of the radiation field. For a single-mode field, we may define two operators Y_1 and Y_2 as

$$Y_1 = \frac{a^2 + a^{\dagger 2}}{2}, \quad Y_2 = \frac{a^2 - a^{\dagger 2}}{2i} \tag{7.145}$$

Evidently, they stands for the real and imaginary parts of the amplitude square of the field, respectively, and obey the commutation relation

$$[Y_1, Y_2] = i(2N + 1) \tag{7.146}$$

here $N = a^\dagger a$. The corresponding uncertainty relation is

$$(\Delta Y_1)^2 (\Delta Y_2)^2 \geq \langle (N + 1/2) \rangle^2 \tag{7.147}$$

If the fluctuations of the field in component Y_i satisfy

$$(\Delta Y_i)^2 < \langle (N + 1/2) \rangle \qquad (i = 1,\ 2) \tag{7.148}$$

then the field is said to have the amplitude square squeezing property.

Up to now, we have introduced the different definitions of the squeezing of the field, all these squeezing properties are the reflections of the nonclassical properties of the radiation field. Are these squeezing properties of the radiation field independent of each other ? In other words, if a field can exhibit the second order squeezing under certain condition, can it exhibit the higher-order squeezing or the amplitude square squeezing simultaneously ? Next we discuss this question.

7.3.3. Independence of the different definitions of the squeezing for the radiation field

As an example, here we only discuss the second-order squeezing and the amplitude square squeezing of the radiation field to realize whether all the squeezing definitions of the field are independent or not. Suppose that a single-mode field is in a superposition state

$$|\Psi\rangle = \cos\theta|0\rangle + \sin\theta\exp(-i\phi)|1\rangle \tag{7.149}$$

We readily have

$$\begin{aligned}\langle X_1\rangle &= \left\langle\frac{a+a^\dagger}{2}\right\rangle = \cos\theta\sin\theta\exp(-i\phi)\\ \langle X_1^2\rangle &= \left\langle\left(\frac{a+a^\dagger}{2}\right)^2\right\rangle = \frac{1+2\sin^2\theta}{4}\end{aligned} \tag{7.150}$$

so that

$$\langle(\Delta X_1)^2\rangle = \langle X_1^2\rangle - \langle X_1\rangle^2 = \frac{1}{4} + \frac{1}{2}\sin^2\theta(1-2\cos^2\theta\cos^2\phi) \tag{7.151}$$

The above equation shows that if ϕ and θ satisfy

$$1-2\cos^2\theta\cos^2\phi < 0 \tag{7.152}$$

then the field described by (7.149) is squeezed in the X_1 component.

Now we examine whether the fluctuations of the amplitude square can also be squeezed. From (7.149) we obtain

$$\langle Y_1\rangle = \left\langle\frac{a^2+a^{\dagger 2}}{2}\right\rangle = 0 = \left\langle\frac{a^2-a^{\dagger 2}}{2i}\right\rangle = \langle Y_2\rangle \tag{7.153}$$

$$\begin{aligned}\langle Y_1^2\rangle &= \left\langle\left(\frac{a^2+a^{\dagger 2}}{2}\right)^2\right\rangle = 2\cos^2\theta + 6\sin^2\theta = \langle Y_2^2\rangle\\ \langle N\rangle &= \sin^2\theta\end{aligned} \tag{7.154}$$

and

$$\langle(\Delta Y_i)^2\rangle - \left\langle N + \frac{1}{2}\right\rangle = \frac{3}{2} + 3\sin^2\theta > 0 \tag{7.155}$$

It is clear that the field in the state $|\Psi\rangle$ (7.144) can not exhibit the amplitude square squeezing.

From the above discussion we see that a field can exhibit the second order squeezing under certain conditions, but it may not also exhibit the amplitude square squeezing. This shows that all kinds of squeezing definitions are independent, they reveal the different nonclassical properties of the radiation field, and give different insights to the quantum properties of the field. Thus, one can choose different states describing the radiation fields in experiments to reveal its different quantum characteristics.

7.4. Squeezing of light in the Jaynes-Cummings model

From (7.149) we see that a field in the superposition of a vacuum state and a single-photon state may exhibit the squeezing of field, this field can be produced by the atom-field coupling system via single-photon processes, which is denoted as the J-C model.

From (4.60) we know that in the rotating-wave approximation the Hamiltonian characterising the J-C model is

$$H = H_0 + V \qquad (\hbar = 1) \tag{7.156}$$

$$H_0 = \omega a^\dagger a + \omega_0 S_z \tag{7.157}$$

$$V = g(a^\dagger S_- + aS_+) \tag{7.158}$$

In order to simplify the calculation, we assume that the interaction between the field and the atom is resonant, i.e., $\omega_0 = \omega$. So in the interaction picture, the interaction Hamiltonian of the system gives

$$V^I(t) = g(a^\dagger S_- + aS_+) \tag{7.159}$$

Supposing that at time t=0, the atom is in the vacuum state and the two-level atom in the superposition of its excited state $|e\rangle$ and ground state $|g\rangle$, i.e., $\cos(\theta/2)|e\rangle + \sin(\theta/2)\exp(-i\phi)|g\rangle$, that is, the initial state of the atom-field coupling system is

$$|\Psi(0)\rangle = \cos(\theta/2)|e, 0\rangle + \sin(\theta/2)\exp(-i\phi)|g, 0\rangle \tag{7.160}$$

With the time development, the state vector of the atom-field coupling system evolves into

$$|\Psi^I(t)\rangle = C_3(t)|e,0\rangle + C_2(t)|g,1\rangle + C_1(t)|g,0\rangle \tag{7.161}$$

Substituting eqs.(7.159) and (7.161) into the Schrödinger equation in the interaction picture

$$i\hbar\frac{\partial}{\partial t}|\Psi(t)^I\rangle = V^I(t)|\Psi(t)^I\rangle \tag{7.162}$$

and utilizing the orthogonal relation among $|e,0\rangle$, $|g,1\rangle$ and $|g,0\rangle$, we can easily obtain

$$\begin{aligned} i\frac{d}{dt}C_3(t) &= gC_2(t) \\ i\frac{d}{dt}C_2(t) &= gC_3(t) \\ \frac{d}{dt}C_1(t) &= 0 \end{aligned} \tag{7.163}$$

Considering the initial condition (7.160), then the solution of (7.163) is

$$\begin{aligned} C_3(t) &= \cos(\theta/2)\cos(gt) \\ C_2(t) &= -i\cos(\theta/2)\sin(gt) \\ C_1(t) &= \sin(\theta/2)\exp(i\phi) \end{aligned} \tag{7.164}$$

The state vector at time t is obtained by the substitution of (7.164) into (7.161).
Applying (7.161) and (7.164), one can readily obtain

$$\langle a\rangle = -ie^{-i\phi}\sin(\theta/2)\cos(\theta/2)\sin(gt)e^{-i\omega t} \tag{7.165}$$

$$\langle a^2\rangle = 0 \tag{7.166}$$

$$\langle a^\dagger a\rangle = \cos^2(\theta/2)\sin^2(gt) \tag{7.167}$$

Here a^I and $a^{\dagger I}$ are the annihilation and creation operators in the interaction picture, and

$$a^I(t) = a\exp(-i\omega t) = (a^{\dagger I}(t))^\dagger \tag{7.168}$$

As we know the detector in experiment measurements can not response the fast oscillating, it can only respond the amplitude envolope with slowly varying. In order to agree with the experimental measurements, we define two Hermitian operators representing the slowly varying amplitude, i.e.,

$$X_1 = \frac{1}{2}[a\exp(i\omega t) + a^\dagger\exp(-i\omega t)]$$
$$X_2 = \frac{1}{2i}[a\exp(i\omega t) - a^\dagger\exp(-i\omega t)] \qquad (7.169)$$

From (7.165)-(7.167) we obtain that the fluctuations in X_1 and X_2 are

$$\langle(\Delta X_1)^2\rangle = \cos^2(\theta/2)\sin^2(gt)\left(\frac{1}{2} - \sin^2\phi\sin^2(\theta/2)\right) + \frac{1}{4} \qquad (7.170)$$

$$\langle(\Delta X_4)^2\rangle = \cos^2(\theta/2)\sin^2(gt)\left(\frac{1}{2} - \cos^2\phi\sin^2(\theta/2)\right) + \frac{1}{4} \qquad (7.171)$$

When $\phi = \pi/2$, $\theta = 2\pi/3$ or $4\pi/3$, $\langle(\Delta X_1)^2\rangle$ is reduced to

$$\langle(\Delta X_1)^2\rangle = \frac{1}{4} - \frac{1}{16}\sin^2(gt) \qquad (7.172)$$

This equation shows that when $gt \neq n\pi$ $(n = 0, 1, 2, \cdots)$, the fluctuations in X_1 can be squeezed, and when $t = n\pi/2$ $(n = 1, 3, 5, \cdots)$, $\langle(\Delta X_1)^2\rangle$ is minimum as shown in Fig.(7.4a). However, if $\theta = 0$, which means that the atom is initially in the excited state, then (7.170) and (7.171) are simplified as

$$\langle(\Delta X_i)^2\rangle = \frac{1}{4} + \frac{1}{16}\sin^2(gt) \geq \frac{1}{4} \qquad (7.173)$$

Therefore, for the system of an atom initially in its excited state interacting with a vacuum field, the quantum fluctuations of the field do not exhibit the squeezing. In addition, from Fig.(7.4b) we know that $\langle(\Delta X_i)^2\rangle \geq 1/4$, which means that the random spontaneous radiation will increase the quantum noise of the field.

From the above discussion we find that the atomic initial state has a strong effect on the properties of the field in the atom-field coupling system.

In addition, from (7.162) and (7.164) we see that when $t = \pi/(2g)$, the state vector of the system evolves into

$$|\Psi^I(t = \pi/(2g))\rangle = \exp(-i\phi)\sin(\theta/2)|g, 0\rangle - i\cos(\theta/2)|g, 1\rangle$$
$$= |g\rangle \otimes |\Psi_F\rangle \qquad (7.174)$$

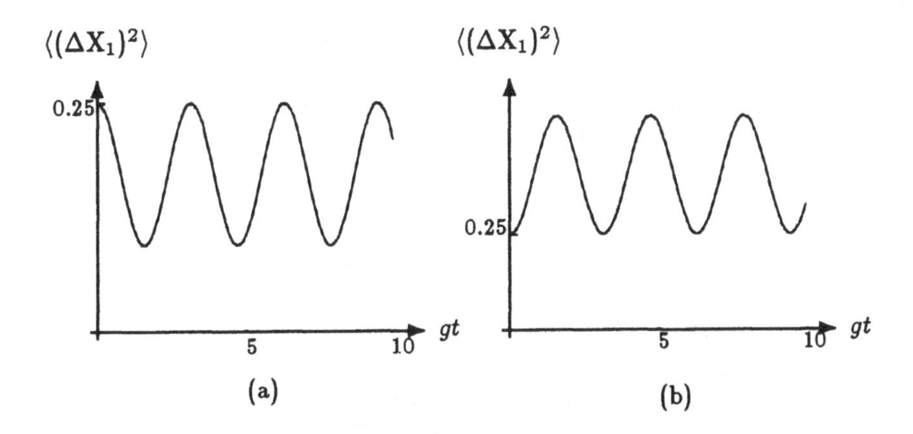

Figure 7.4: Diagram of the variance in X_1 component, (a). for $\theta = 2\pi/3$, $4\pi/3$ and $\phi = \pi/2$; (b). for $\theta = 0$

here

$$|\Psi_F\rangle = \exp(-i\phi)\sin(\theta/2)|0\rangle - i\cos(\theta/2)|1\rangle \tag{7.175}$$

$|\Psi_F\rangle$ is similar to (7.149), which is a squeezed state of the radiation field. So an atom initially in the superposition state of its excited and ground state may emit squeezed light.

The above results show that for the field interacting with the atom, its squeezing property depends strongly on the atomic initial state. A question arisen here is what the influence of the initial state of the field on the squeezing of the field in the J-C model. Now we turn to discuss this question.

Assuming that the field is initially in a coherent state

$$|\alpha\rangle = \sum_n F_n|n\rangle, \quad F_n = \exp(-|\alpha|^2/2)\alpha^n/\sqrt{n!} \tag{7.176}$$

and the atom initially in its excited state $|e\rangle$, namely, the initial state of the atom-field coupling system is expressed as

$$|\Psi(0)\rangle = \sum_n F_n|e, n\rangle \tag{7.177}$$

With the time evolution, the state vector of the atom-field coupling system can expanded as

$$|\Psi^I(t)\rangle = \sum_n a_n(t)|e,n\rangle + b_{n+1}(t)|g,n+1\rangle \tag{7.178}$$

Substituting (7.159) and (7.178) into the Schrödinger equation in the interaction picture, the probability amplitudes $a_n(t)$ and $b_{n+1}(t)$ obey

$$\begin{aligned} i\frac{d}{dt}a_n(t) &= g\sqrt{n+1}b_{n+1}(t) \\ i\frac{d}{dt}b_{n+1}(t) &= g\sqrt{n+1}a_n(t) \end{aligned} \tag{7.179}$$

Considering the initial condition (7.177), the solutions of (7.179) give

$$\begin{aligned} a_n(t) &= F_n \cos(g\sqrt{n+1}t) \\ b_{n+1}(t) &= -iF_n \sin(g\sqrt{n+1}t) \end{aligned} \tag{7.180}$$

Substituting (7.180) into (7.178), we may obtain the time-dependent state of the system in the interaction picture. Furthermore we have

$$\begin{aligned} \langle a \exp(i\omega t)\rangle = \exp(-|\alpha|^2)e^{i\eta} \sum_{n=0}^{\infty} \frac{\bar{n}^{n+1/2}}{n!}[\cos(g\sqrt{n+1}t)\cos(g\sqrt{n+2}t) \\ +\sqrt{\frac{n+2}{n+1}}\sin(g\sqrt{n+1}t)\sin(g\sqrt{n+2}t)] \end{aligned} \tag{7.181}$$

$$\begin{aligned} \langle a^2 \exp(i2\omega t)\rangle = \exp(-|\alpha|^2)e^{2i\eta} \sum_{n=0}^{\infty} \frac{\bar{n}^{n+1}}{n!}[\cos(g\sqrt{n+1}t)\cos(g\sqrt{n+3}t) \\ +\sqrt{\frac{n+3}{n+1}}\sin(g\sqrt{n+1}t)\sin(g\sqrt{n+3}t)] \end{aligned} \tag{7.182}$$

$$\langle a^\dagger a\rangle = \bar{n} + \exp(-|\alpha|^2)\sum_{n=0}^{\infty}\frac{\bar{n}^n}{n!}\sin^2(gt\sqrt{n+1}) \tag{7.183}$$

here $\alpha = \bar{n}^{1/2}\exp(i\eta)$. As previous discussion, here we are only interested in the fluctuations of the slowly varying amplitudes, and concentrate on the fluctuations in the component X_1. Evidently,

$$\begin{aligned} Q_1 &= \langle(\Delta X_1)^2\rangle - \frac{1}{4} \\ &= \frac{1}{2}\{\langle a^\dagger a\rangle + Re\langle a^2 e^{2i\omega t}\rangle - 2(Re\langle ae^{i\omega t}\rangle)^2\} \end{aligned} \tag{7.184}$$

Substituting (7.181)-(7.183) into the above equation, then we can obtain the time evolution of Q_1 . If at some special time t there exists $Q_1 < 0$, then the fluctuations in the component X_1 are squeezed, meanwhile the field is in a squeezed state.

Eq.(7.184) is so complex that the time evolution of Q_1 is not explicit. However, we can discuss by the aid of numerical method to calculate (7.184). Fig.(7.5) shows the numerical calculation results. We can see that for $\overline{n} = 10$

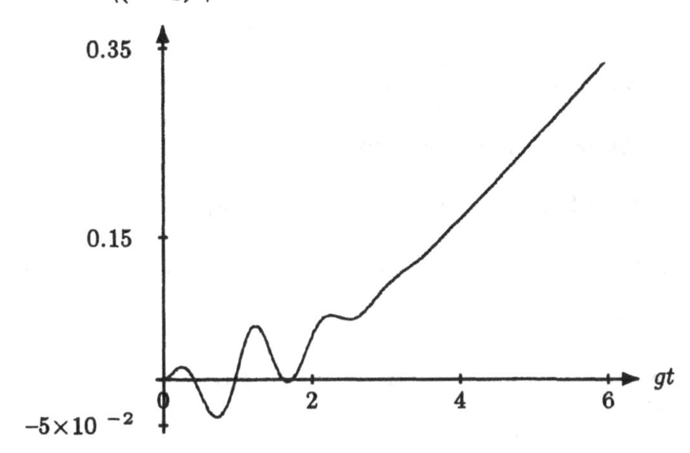

Figure 7.5: Time-dependence of Q_1 for $\overline{n} = 10$ and $\eta = 0$

and $\eta = 0$, the phenomenon of $Q_1 < 0$ appears for an initial short time range. This means that in the J-C model which consists of an excited atom and a coherent field, the fluctuations in the component X_i can be squeezed in initial period with the time development. But we even found in previous discussion that if the field is initially in a vacuum state, the excited atom can never emit the squeezed light. So we may conclude that the initial state of the field as well as the atomic initial state also have a strong effect on the squeezing of the radiation field.

References

1. D.Stoler, *Phys.Rev.* **D1** (1970) 3217.

2. H.P.Yuen, *Phys.Rev.* **A13** (1976) 2226.

3. C.M.Caves, *Phys.Rev.* **D23** (1981) 1698.

4. D.F.Walls, *Nature* **306** (1983) 141.

5. C.M.Caves and B.L.Schumaker, *Phys.Rev.* **A31** (1985) 3068.

6. B.L.Schumaker and C.M.Caves, *Phys.Rev.* **A31** (1985) 3098.

7. D.R.Turx, *Phys.Rev.* **D31** (1985) 1988.

8. C.K.Hong and L.Mandel, *Phys.Rev.Lett.* **54** (198) 323.

9. G.Compagno, J.S.Peng, and F.Persico, *Opt.Commun.* **57** (1986) 415.

10. R.Loudon and P.L.Knight, *J.Mod.Opt.* **34** (1987) 709.

11. K.Wodkiewicz, P.L.Knight, S.J.Buckle and S.M.Barnett, *Phys.Rev.* **A35** (1987) 2657.

12. L.A.Wu, H.J.Kimble, J.L.Hall and H.Wu, *Phys.Rev.Lett.* **57** (1987) 2520.

13. L.A.Wu, M.Xiao and H.J.Kimble, *J.Opt.Soc.Am.* **B4** (1987) 1465.

14. C.C.Gerry and P.J.Moyer, *Phys.Rev.* **A38** (1988) 5665.

15. M.C.Teich and B.Saleh, *Quantum Opt.* **1** (1989) 153.

16. P.Zhou and J.S.Peng, *Acta Opt.Sin.* **10** (1990) 837.

17. P.Kumar, O.Aytue and J.Huang, *Phys.Rev.Lett.* **64** (19900) 1015.

18. P.Zhou and J.S.Peng, *Sci.Sin.* **34** (1991) 585.

19. M.Rosenbluh and R.M.Shelby, *Phys.Rev.Lett.* **66** (1991) 153.

20. J.S.Peng and G.X.Li, *ACTA Phys.Sin.* **42** 568.

21. G.M.Dariano, *Int.J.Mod.Phys.* **B6** (1992) 1291.

22. C.Fabre, *Phys.Rep.* **219** (1992) 215.

23. H.J.Kimble, *Phys.Rep.* **219** (1992) 227.

24. J.S.Peng and G.X.Li, *Phys.Rev.* **A47** 91993) 3197.

25. M.J.Collett, *Phys.Rev.Lett.* **70** (1993) 3400.

26. G.X.Li and J.S.Peng, *Chin.Phys.Lett.* **12** (1995) 79.

27. G.X.Li and J.S.Peng, *ACTA Phys.Sin.* **44** (1995) 1697.

28. J.Kitching, A.Yariv and Y.Shevy, *Phys.Rev.Lett.* **74** (1995) 3372.

CHAPTER 8

RESONANCE FLUORESCENCE

From Chapter 4 we know that an excited atom can spontaneously radiate fluorescent photons due to the effect of the vacuum fluctuation, then goes back to its ground state, the correspondent fluorescence frequency is equal to the atomic transition frequency ω_0. However, if the atom is resonantly driven by an external field, what are the distribution of the fluorescence spectra ? In 1969, R.Mollow investigated theoretically the fluorescence emission of the atom under the interaction of a laser field in resonant case, namely resonance fluorescence. His theoretical prediction was confirmed by F.Schuda et.al. in experiments in 1974. Up to now the resonant fluorescence have been widely investigated both theoretically and experimentally, this give an impetus to the atomic or molecular fine structures and the high-resolution spectroscopy technique.

The schematic diagram of the resonance fluorescence experiment is illustrated in Fig.(8.1). The interaction between a beam of a strong laser field with frequency ω and a beam of two-level atoms with transition frequency $\omega_0 (\omega \approx \omega_0)$ takes place at the overlapping region. For example, the transition of Na atom between level ${}^2S_{1/2}$, F=2 and ${}^2P_{3/2}$, F=3 is resonantly driven by a laser field. The fluorescent light radiated by the Na atom in the interaction region is detected by a frequency analyser. The experimental results show that the intensity distribution of the fluorescent light yields a three-component spectrum as shown in Fig.(8.2), the frequency of the central peak is the same as the laser frequency ω, while the frequencies of the two side peaks are symmetrically displaced by a quantity $\Delta_n = 2\varepsilon\sqrt{n}$, where ε is the coupling constant between the laser field and the two-level atom, and n is the number of laser photons which is proportional to the laser power. The width of the two side lines is 3/2 that of the central component, while their peak intensity is 1/3 that of the central peak. The experimental results also show that with the increasing of

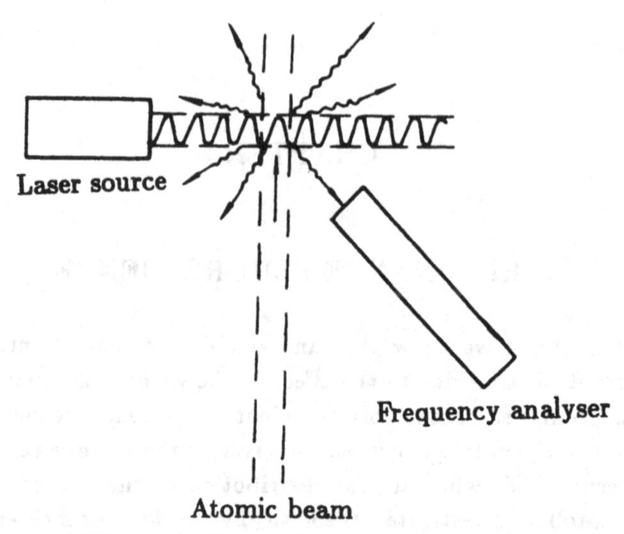

Figure 8.1: Schematic illustration of resonance fluorescence

the input laser field , namely, n is increased, the structure of the three-peak distribution is more explicit, but if n is small, i.e., Δ_n is small, then there only appears one peak of the fluorescence distribution, as in the spontaneous emission.

This chapter is devoted to discussing the theory of resonance fluorescence in detail. In order to clear the physical processes of the resonance fluorescence, we first introduce a theoretical approach dealing with the interaction between a strong laser field and a two-level atom, which is termed as the dressed canonical transformation approach. We then discuss the resonance fluorescence distribution of a two-level atom by use of the dressed canonical transformation. Subsequently, this approach is generalized to treat the resonance fluorescence spectrum of a three-level atom system. Finally, the resonance fluorescence of a two-level atom is re-examined by use of the density matrix theory.

8.1 Resonance fluorescence distribution of a two-level atom

8.1.1 Dressed canonical transformation

As we know, the Hamiltonian of a two-level atom coupling to a single-mode

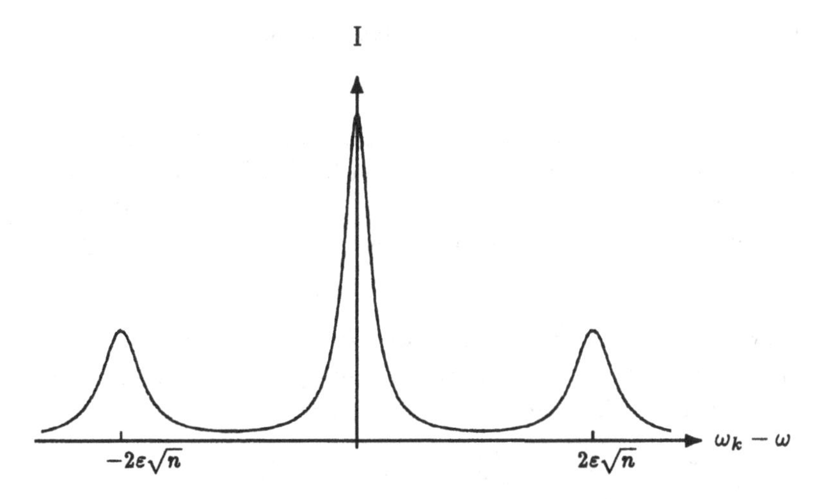

Figure 8.2: The result of resonance fluorescence experiment

field in the rotating-wave approximation is written as

$$H = H_0 + V \tag{8.1}$$

$$H_0 = \omega a^{\dagger} a + \omega_0 S_z \qquad (\hbar = 1) \tag{8.2}$$

$$V = \varepsilon(a^{\dagger} S_- + a S_+) \tag{8.3}$$

where $\omega a^+ a$ stands for the energy of the single-mode field, $\omega_0 S_z$ corresponds to the energy of the atom in the absence of the field, V represents the interaction energy between field and atom. Obviously, the eigen-energies and eigenkets of the Hamiltonian H_0 change dramatically due to the effect of the interaction energy. In Chapter 4, we have introduced a method to calculate the energy spectrum of this system. That is, combining the atom with the field and diagonalizing the Hamiltonian through solving the eigenvalue equation, then one can obtain the eigen-energies and corresponding eigenkets of the atom-field coupling system, whilst the atom interacting with the field is called the dressed atom. Here we introduce another way to diagonalizing the Hamiltonian H, which is termed the dressed canonical transformation approach. For the system described by eqs.(8.1)-(8.3), we may introduce an operator of the total number of excitations as

$$N = a^{\dagger} a + S_z + \frac{1}{2} \tag{8.4}$$

Evidently, N satisfies the commutation relation

$$[N, H_0] = [N, V] = [N, H] = 0 \tag{8.5}$$

so it is a constant of motion. In the representation of $\{|n, \pm\rangle\}$, N is diagonalized. If we extract the state $|0, -\rangle$ from the Hilbert space of the system, then the operators such as $N^{1/2}, N^{-1}, N^{-1/2}$ are definition operators, and obey the commutation relation (8.5). In this case, we can define a unitary operator as

$$T = \exp\{-\theta(4N)^{-\frac{1}{2}}(aS_+ - a^\dagger S_-)\} \tag{8.6}$$

where θ is an undetermined operator which commutates with N. Using the unitary operator we can perform a unitary transformation on an arbitrary operator A, which commutates with N, namely,

$$\begin{aligned} T^{-1}AT = A &+ \left(-\frac{\theta}{2}N^{-\frac{1}{2}}\right)[A, aS_+ - a^\dagger S_-] \\ &+ \frac{1}{2!}\left(-\frac{\theta}{2}N^{-\frac{1}{2}}\right)^2 [[A, aS_+ - a^\dagger S_-], aS_+ - a^\dagger S_-] + \cdots \end{aligned} \tag{8.7}$$

Since θ and N commutate with operator A, N and θ may be out of the Poissonian bracket in equation (8.7). If the operator A stands for S_z, a^+a or $(aS_+ + a^+S_-)$, then according to (8.7) one can obtain

$$\begin{aligned} T^{-1}S_zT = S_z &+ \left(-\frac{\theta}{2}N^{-\frac{1}{2}}\right)[S_z, aS_+ - a^\dagger S_-] \\ &+ \frac{1}{2!}\left(-\frac{\theta}{2}N^{-\frac{1}{2}}\right)^2 [[S_z, aS_+ - a^\dagger S_-], aS_+ - a^\dagger S_-] + \cdots \end{aligned} \tag{8.8}$$

Evidently, the first-order Poissonian bracket reduces to

$$[S_z, aS_+ - a^\dagger S_-] = a[S_z, S_+] - a^\dagger[S_z, S_-] = aS_+ + a^\dagger S_- \tag{8.9}$$

here the commutation relation

$$[S_z, S_\pm] = \pm S_\pm \tag{8.10}$$

has been used. And the second-order Poissonian bracket becomes

$$\begin{aligned} &[S_z, aS_+ - a^\dagger S_-]_2 = [[S_z, aS_+ - a^\dagger S_-], aS_+ - a^\dagger S_-] \\ &= [aS_+ + a^\dagger S_-, aS_+ - a^\dagger S_-] = [aS_+, -a^\dagger S_-] + [a^\dagger S_-, aS_+] \\ &= -S_-S_+ - 2aa^\dagger S_z - S_+S_- + a^\dagger a(-2S_z) \\ &= -(S_+S_- + S_-S_+) - (a^\dagger a + aa^\dagger)2S_z = -1 - (2a^\dagger a + 1)S_z \end{aligned} \tag{8.11}$$

where we have used the familiar commutation relations

$$[a, a^\dagger] = 1 \tag{8.12}$$

$$[S_+, S_-] = 2S_z, \qquad S_\pm S_\mp = \frac{1}{2} \pm S_z \tag{8.13}$$

Noticing (8.4), then (8.11) reads

$$[S_z, aS_+ - a^\dagger S_-]_2 = -4NS_z \tag{8.14}$$

Similarly, the third-order Poissonian bracket in (8.8) gives

$$[S_z, aS_+ - a^\dagger S_-]_3 = -4N[S_z, aS_+ - a^\dagger S_-] = -4N(aS_+ + a\dagger S_-] \tag{8.15}$$

here we have used the relation

$$[N, aS_+ - a^\dagger S_-] = [a^\dagger a, aS_+ - a^\dagger S_-] + [S_z, aS_+ - a^\dagger S_-] = 0 \tag{8.16}$$

and the fourth-order Poissonian bracket reduces to

$$[S_z, aS_+ - a^\dagger S_-]_4 = (-4N)^2 S_z \tag{8.17}$$

Substituting eqs.(8.9) and (8.14)-(8.17) into (8.8), we have

$$\begin{aligned}
T^{-1}S_zT = {} & S_z + \left(-\frac{\theta}{2}N^{-\frac{1}{2}}\right)(aS_+ + a^\dagger S_-) + \frac{1}{2!}\left(-\frac{\theta}{2}N^{-\frac{1}{2}}\right)^2(-4NS_z) \\
& + \frac{1}{3!}\left(-\frac{\theta}{2}N^{-\frac{1}{2}}\right)^3(-4N)(aS_+ + a^\dagger S_-) \\
& + \frac{1}{4!}\left(-\frac{\theta}{2}N^{-\frac{1}{2}}\right)^4(-4N)^2 S_z + \cdots \\
= {} & S_z\left(1 - \frac{1}{2!}\theta^2 + \frac{1}{4!}\theta^4 - \cdots\right) \\
& -(aS_+ + a^\dagger S_-)N^{-\frac{1}{2}}\left(\theta - \frac{\theta^3}{3!} + \frac{\theta^5}{5!} - \cdots\right) \\
= {} & S_z\cos\theta - \frac{1}{2}(aS_+ + a^\dagger S_-)N^{-\frac{1}{2}}\sin\theta
\end{aligned} \tag{8.18}$$

The similar calculation yields

$$T^{-1}a^\dagger aT = a^\dagger a + S_z - S_z\cos\theta + \frac{1}{2}N^{-\frac{1}{2}}(aS_+ + a^\dagger S_-)\sin\theta \tag{8.19}$$

and

$$T^{-1}(aS_+ + a^\dagger S_-)T = (aS_+ + a^\dagger S_-)\cos\theta + 2S_z N^{1/2}\sin\theta \tag{8.20}$$

Consequently, performing the unitary transformation T on H yields

$$\begin{aligned} T^{-1}HT = & \omega a^\dagger a + [\omega - (\omega-\omega_0)\cos\theta + 2\varepsilon N^{\frac{1}{2}}\sin\theta]S_z \\ & + \left[\frac{1}{2}N^{-1/2}(\omega-\omega_0)\sin\theta + \varepsilon\cos\theta\right](aS_+ + a^\dagger S_-) \end{aligned} \tag{8.21}$$

If we choose θ to obey the identity

$$\frac{1}{2}N^{-1/2}(\omega-\omega_0)\sin\theta + \varepsilon\cos\theta = 0$$

i.e.,

$$\sin\theta = -\frac{2\varepsilon N^{1/2}}{[(\omega-\omega_0)^2 + 4\varepsilon^2 N]^{1/2}} \tag{8.22}$$

$$\cos\theta = \frac{\omega-\omega_0}{[(\omega-\omega_0)^2 + 4\varepsilon^2 N]^{1/2}} \tag{8.23}$$

then (8.21) reduces to

$$T^{-1}HT = \omega a^\dagger a + (\omega - \Delta_n)S_z \tag{8.24}$$

where

$$\Delta_n = \sqrt{(\omega-\omega_0)^2 + 4\varepsilon^2 N} \tag{8.25}$$

Now we define a new operator

$$\tilde{A} = T^{-1}AT \tag{8.26}$$

which is usually called the dressed operator. From (8.24) we obtain

$$\tilde{H} = \omega a^\dagger a + (\omega - \Delta_n)S_z \tag{8.27}$$

Eq.(8.27) is identical with (8.1), this Hamiltonian has been diagonalized. The eigenkets of (8.27) can be expressed in terms of $|n,\sigma\rangle(\sigma = \pm)$ as

$$|\tilde{n},\tilde{\sigma}\rangle = T|n,\sigma\rangle \tag{8.28}$$

Since $\tilde{A}$ and $|\tilde{n}, \tilde{\sigma}\rangle$ are connected with A and $|n, \sigma\rangle$ by the unitary transformation T, the physical characteristics represented by these variables are equivalent. For simplicity, the tilde label $\tilde{}$ will be ignored in the following. From (8.27) we know the eigenvalues of H are given by

$$E_{n,+} = n\omega + \frac{\omega - \Delta_n}{2}$$
$$E_{n,-} = n\omega - \frac{\omega - \Delta_n}{2} \tag{8.29}$$

the energy spectrum are shown in Fig.(8.3).

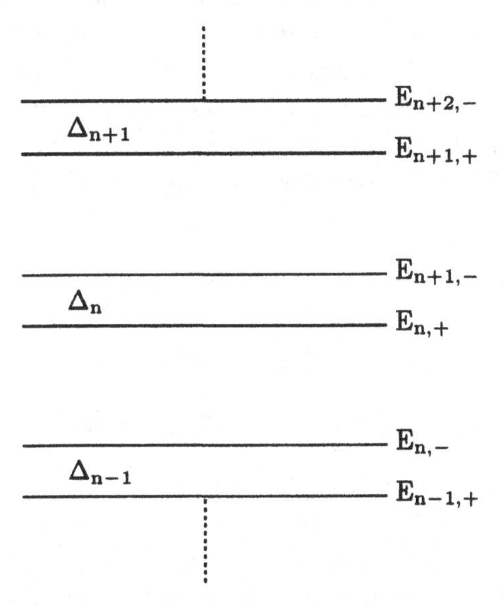

Figure 8.3: Eigenenergy spectrum of Hamiltonian H for large n

In the resonant case, it is immediately to find the relation between (8.29) and (4.70) as

$$E_{n,-} = \left(n - \frac{1}{2}\right)\omega + \varepsilon\sqrt{n} = E_n^+$$
$$E_{n-1,+} = \left(n - \frac{1}{2}\right)\omega - \varepsilon\sqrt{n} = E_n^- \tag{8.30}$$

So we see that the results obtained both by using dressing transformation (8.6) and (8.7) and by diagonalizing directly the Hamiltonian (8.1) are equivalent.

However, we will see that the dressed transformation approach may be utilized in more general.

We now turn to consider the dressed atomic operators which do not commutate with N, such as $S_{\pm}$. It is possible to verify the commutation relation

$$[N^{-\frac{1}{2}}, S_{+}] = -\frac{1}{2}(N^{\frac{3}{2}} - N^{\frac{1}{2}})^{-1}S_{+}$$

For the strong laser field case, i.e., $n \gg 1$, the nonzero matrix elements of the commutator $[N^{-1/2}, S_{+}]$ are proportional to $n^{-3/2}$, thus we can neglect this commutator in comparison with $[S_{+}, aS_{+} - a^{\dagger}S_{-}]$. It can be proved that the nonzero matrix elements of the commutator $[\theta, S_{+}]$ are proportional to $n^{-1/2}$, so it can also be neglected. This enables us to treat the operators N and θ in the operator T as C-numbers when we calculate the expanded series of $T^{-1}S_{+}T$. That is to say, the approximations $[N^{-1/2}, S_{+}] = [\theta, S_{+}] = 0$ may be held in calculating $T^{-1}S_{+}T$. Consequently, the dressed operators S_{+} and S_{-} are approximated as

$$\begin{aligned} T^{-1}S_{+}T = & \frac{1}{2}S_{+}(1+\cos\theta) + N^{-\frac{1}{2}}a^{\dagger}S_{z}\sin\theta \\ & -\frac{1}{2}N^{-1}a^{\dagger 2}S_{-}(1-\cos\theta) \end{aligned} \tag{8.31}$$

$$\begin{aligned} T^{-1}S_{-}T = & \frac{1}{2}S_{-}(1+\cos\theta) + N^{-\frac{1}{2}}aS_{z}\sin\theta \\ & -\frac{1}{2}N^{-1}a^{2}S_{+}(1-\cos\theta) \end{aligned} \tag{8.32}$$

8.1.2 Spectral distribution of the resonance fluorescence of a two-level atom

Now we adopt the dressed transformation approach to discuss the resonance fluorescence of a two-level atom driven by a strong laser field. If the density of the atomic beam is very low, the interaction among the atoms may be ignored. So as a reasonable approximation, we can consider a single atom interacting with the strong field. The Hamiltonian of this system is written as

$$H = \omega a^{\dagger}a + \omega_0 S_z + \varepsilon(aS_{+} + a^{\dagger}S_{-}) + \sum_k \omega_k a_k^{\dagger}a_k + \sum_k \varepsilon_k(a_k S_{+} + a_k^{\dagger}S_{-}) \tag{8.33}$$

The first three terms in the right-hand of (8.33) correspond to the energies of atom and field as well as their interaction, The fourth term represents the energy of the fluorescence photons. Since the spontaneous emission has infinite modes, k is chosen from 0 to infinite for the summation. The last term in (8.33) stands for the interaction between the two-level atom and the fluorescence field.

In considering the laser field is very strong, the interaction between fluorescent field and atom

$$V = \sum_k \varepsilon_k (a_k S_+ + a_k^\dagger S_-) \tag{8.34}$$

is very weak as compared to the other terms in (8.33), so we can treat it as a perturbation term. This enable us to decompose (8.33) into two parts

$$H = H_0 + V \tag{8.35}$$

Performing the unitary transformation T on H_0 and noticing that a_k commutates with T, we obtain

$$T^{-1} H_0 T = \omega a^\dagger a + (\omega - \Delta_n) S_z + \sum_k \omega_k a_k^\dagger a_k \tag{8.36}$$

We then perform the unitary transformation on V. Applying (8.31) and (8.32) yields

$$\begin{aligned} T^{-1} V T = \sum_k \varepsilon_k \Big\{ & \frac{1}{2}(1 + \cos\theta) a_k^\dagger S_- + \sin\theta a_k^\dagger N^{-\frac{1}{2}} a S_z \\ & - \frac{1}{2}(1 - \cos\theta) a_k^\dagger N^{-1} a^2 S_+ + \text{h.c.} \Big\} \end{aligned} \tag{8.37}$$

It represents the interaction energy between the dressed atom and the fluorescence field. Evidently, although the dressed transformation has eliminated the interaction terms between the atom and the laser field in Hamiltonian (8.33), the interaction terms (8.37) between the atom and the fluorescence field become more complex. However, we will see that such complex terms just reveal the processes of the resonance fluorescence transition of the atom in a strong laser field. If let

$$T^{-1} V T = V_1 + V_2 + V_3 \tag{8.38}$$

with

$$V_1 = \sum_k \varepsilon_k \left[\frac{1}{2}(1 + \cos\theta) a_k^\dagger S_- + \text{h.c.} \right] \tag{8.39}$$

$$V_2 = \sum_k \varepsilon_k [\sin\theta a_k^\dagger N^{-\frac{1}{2}} a S_z + \text{h.c.}] \tag{8.40}$$

$$V_3 = -\sum_k \varepsilon_k \left[\frac{1}{2}(1-\cos\theta) a_k^\dagger N^{-1} a^2 S_+ \text{h.c.}\right] \tag{8.41}$$

Obviously, $V_i (i = 1, 2, 3)$ correspond to three different emission and absorption processes of the fluorescence field as shown in Fig.(8.4). For the resonant

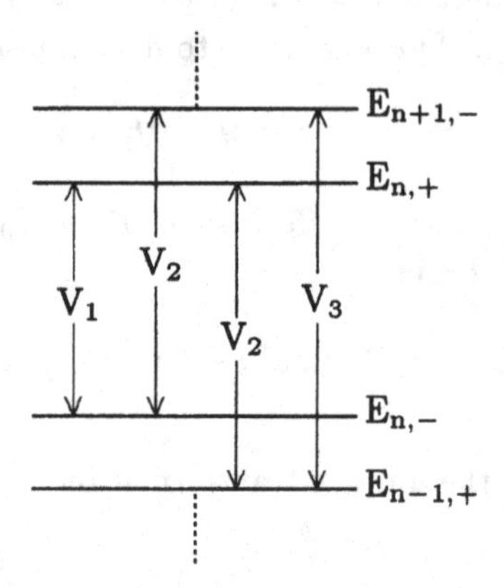

Figure 8.4: Transition processes induced by V_1, V_2 and V_3 among the dressed states

case, V_1 describes the transition process in which the atom goes from its excited state to its ground state and emit a fluorescent photon with frequency $\omega_{k_1} = \omega - \Delta_n$ simultaneously, and vice versa. V_2 corresponds to the process of the atom converting a driving laser photon into a fluorescent one with frequency $\omega_{k_2} = \omega$ and vice versa. V_3 stands for the process in which the atom is excited (de-excited) to its upper (lower) state through annihilating (creating) two laser photons and emitting (absorbing) one fluorescent photon with frequency $\omega_{k_3} = \omega + \Delta_n$. Consequently, we find from eqs.(8.38)-(8.41) that the resonance fluorescence spectrum of a two-level atom driven by a single-mode laser field contain three peaks. The qualitative analysis shows that the intensity of the central peak ($\omega_{k_2} = \omega$) is stronger than that of two side peaks ($\omega_{k_{1,3}} = \omega \pm \Delta_n$). The reason is that in the emission process described by V_2, the atom can be either in the ground state or in the excited state, but V_1 and

V_2 correspond to the induced transition processes in which the atom only in one of its two states. Because of the natural width and the Dopller shifts, the fluorescent lines are broadened. In order to observe the three-peak structure of the spectral lines, it requests that the width of the spectral lines is smaller than

$$\Delta_n = [(\omega - \omega_0)^2 + 4\varepsilon^2 n]^{\frac{1}{2}}$$

That is to say, the intensity of the driven laser must be very high ($n \gg 1$). When Δ_n is sufficiently large, the frequencies of all the fluorescent photons are located in the neighbourhoods of $\omega - \Delta_n$, ω and $\omega + \Delta_n$, respectively.

8.1.3 Linewidth of the fluorescence spectrum

Since $V_i (i = 1, 2, 3)$ represent the fluorescence transition processes, the linewidth of three peaks can be found from their corresponding transition rates. For simplicity, we let

$$S_1 = S_+, \qquad S_2 = N^{-\frac{1}{2}} a^\dagger S_z, \qquad S_3 = N^{-1} a^{\dagger 2} S_- \tag{8.42}$$

In the Heisenberg picture we have

$$\frac{d}{dt} S_i(t) = -i[S_i(t), H_0 + V] = -i[S_i(t), H_0] - i[S_i(t), V] \quad (i = 1, 2, 3) \tag{8.43}$$

After a formal integration it yields

$$S_i(t) = S_i(0) - i \int_0^t [S_i(t'), H_0] dt' - i \int_0^t [S_i(t'), V] dt' \tag{8.44}$$

Substituting (8.44) into the last terms in (8.43) then obtain

$$\begin{aligned} \frac{d}{dt} S_i(t) = &-i[S_i(t), H_0] - i[S_i(0), V] - i \int_0^t [[S_i(t'), H_0(t')], V(t)] dt' \\ &-i \int_0^t [[S_i(t'), V(t')], V(t)] dt' \end{aligned} \tag{8.45}$$

If we only remain the diagonal elements of (8.45) in the fluorescent photon representation, then the master equation of S_i can be deduced. So the nondiagonal elements in (8.45) may be neglected, that is to say, the terms containing odd-order of V_i can be neglected. Inasmuch as the different subscripts i correspond to different fluorescence transition processes, we may also neglect the terms containing $V_i V_j (i \neq j)$. It is necessary to point out that neglecting the nondiagonal elements in the fluorescent photon representation is just a kind of approximation. This is because when $t \neq 0$, these terms representing some fluorescence transitions are different from to zero. However, at the moment t=0 no transition appear, so these nondiagonal terms equal to zero. Hence for very small t, these terms are sufficiently smaller in comparison with the diagonal terms, we can therefore ignore the nondiagonal terms. On the basis of these approximations, (8,45) reduces to

$$\frac{d}{dt}S_i(t) = -i[S_i(t), H_0(t)] - \sum_j \int_0^t [[S_i(t'), V_j(t')], V_j(t)]dt' \tag{8.46}$$

Assuming $t' = t - \tau$, then

$$\begin{aligned}\frac{d}{dt}S_i(t) &= -i[S_i(t), H_0(t)] + \sum_j \int_t^0 [[S_i(t-\tau), V_j(t-\tau)], V_j(t)]d\tau \\ &= -i[S_i(t), H_0(t)] - \sum_j \int_0^t [[S_i(t-\tau), V_j(t-\tau)], V_j(t)]d\tau \end{aligned} \tag{8.47}$$

The evolution of the operator $S_i(t-\tau)V_j(t-\tau)$ in the integral may be obtained by solving the Heisenberg equation and retaining only to the zero-order. It is no difficulty to prove that $S_i(t-\tau)V_j(t-\tau)$ obeys the exponential law, namely

$$S_i(t-\tau)V_j(t-\tau) = \exp(-\tau/\tau_c)S_i(t)V_j(t) \tag{8.48}$$

here t is a parameter and τ_c is the characteristic correlation time. Substituting (8.48) into (8.47) and supposing time τ to tend to infinite, we obtain from (8.47)

$$\frac{d}{dt}S_i(t) = -i[S_i(t), H_0(t)] - \tau_c \sum_j [[S_i(t), V_j(t)], V_j(t)] \tag{8.49}$$

here we have used

$$\int_0^\infty \exp(-\tau/\tau_c)d\tau = \tau_c \tag{8.50}$$

From (8.42) and (8.36) we know that the first term in (8.49) corresponds to

$$-i[S_1,H_0]=-i[S_+,\omega a^\dagger a+(\omega-\Delta_n)S_z+\sum_k \omega_k a^\dagger a_k]$$
$$=i(\omega-\Delta_n)S_1 \tag{8.51}$$
$$-i[S_2,H_0]=-[N^{-1/2}a^\dagger S_z,H_0]=i\omega S_2 \tag{8.52}$$
$$-i[S_3,H_0]=-i[N^{-1}a^{\dagger 2}S_+,H_0]=i(\omega+\Delta_n)S_3 \tag{8.53}$$

The second terms in the right-hand of (8.49)

$$D_{ij}=[[S_i(t),V_j(t)],V_j(t)] \tag{8.54}$$

are always diagonal in the fluorescent photon number representation. These include nine terms, for example

$$\begin{aligned}D_{11}&=[[S_1,V_1],V_1]\\&=\left[\sum_{k1}\varepsilon_{k1}^2\frac{1}{2}(1+\cos\theta)2a_{k1}^\dagger S_z,\sum_{k1}\frac{1}{2}(1+\cos\theta)(a_{k1}S_++a_{k1}^\dagger S_-)\right]\\&=\sum_{k1}\varepsilon_{k1}^2\frac{1}{2}(1+\cos\theta)^2\left(a_{k1}^\dagger a_{k1}+\frac{1}{2}\right)S_1\end{aligned}$$

Similarly

$$\begin{aligned}D_{12}&=\sum_{k2}\varepsilon_{k2}^2\sin^2\theta(2a_{k2}^\dagger a_{k2}+1)S_1\\D_{13}&=\sum_{k1}^{2}\varepsilon_{k3}^2(1-\cos\theta)^2\left(a_{k3}^\dagger a_{k3}+\frac{1}{2}\right)S_1\\D_{21}&=\sum_{k1}\varepsilon_{k1}^2\frac{1}{2}(1+\cos\theta)^2\left(a_{k1}^\dagger a_{k1}+S_z+\frac{1}{2}\right)S_2\\D_{22}&=0\\D_{23}&=\sum_{k3}\varepsilon_{k3}^2(1-\cos\theta)^2\left(a_{k3}^\dagger a_{k3}-S_z+\frac{1}{2}\right)S_2\\D_{31}&=\sum_{k1}\varepsilon_{k1}^2\frac{1}{2}(1+\cos\theta)^2\left(a_{k1}^\dagger a_{k1}+\frac{1}{2}\right)S_3\\D_{32}&=\sum_{k2}\varepsilon_{k2}^2\sin^2\theta(2a_{k2}^\dagger a_{k2}+1)S_3\end{aligned} \tag{8.55}$$

$$D_{33} = \sum_{k3} \varepsilon_{k3}^2 \frac{1}{2}(1-\cos\theta)^2 \left(a_{k3}^\dagger a_{k3} + \frac{1}{2}\right) S_3$$

here the approximation $a^+a \approx N$ has been adopted and the higher-order term of $0(N^{-1/2})$ has been dropped because N is large. Now we can calculate the nonzero elements of (8.49) in the representation of the eigenkets of H_0, these elements represent the corresponding transition rates of the fluorescent photons. If we assume the fluorescence field is in vacuum state, i.e., $n_{k_i} = 0 (i = 1, 2, 3)$, then the eigenkets of H_0 may be expressed as

$$|n, \sigma\rangle = |n, \sigma, 0_{k1}, 0_{k2}, 0_{k3}\rangle \tag{8.56}$$

here n is the photon number of the laser field and $\sigma = \pm$. In this case, the nondiagonal elements corresponding to the fluorescence transition rates are

$$\left\langle n,+\left|\frac{d}{dt}S_+\right|n,-\right\rangle = \langle n,+|-i[S_1,H_0] - \tau_c(D_{11}+D_{12}+D_{13})|n,-\rangle$$
$$= i(\omega-\Delta_n)\langle n,+|S_+|n,-\rangle - \tau_c\left\{\sum_{k1}\frac{\varepsilon_{k1}^2}{4}(1+\cos\theta)^2\right.$$
$$\left. + \sum_{k2}\varepsilon_{k2}^2\sin^2\theta + \sum_{k3}^{2}\varepsilon_{k3}^2\frac{1}{4}(1-\cos\theta)^2\right\}\langle n,+|S_+|n,-\rangle \tag{8.57}$$

$$\left\langle n,+\left|\frac{d}{dt}(N^{-1/2}a^\dagger S_z)\right|n-1,+\right\rangle = i\omega\langle n,+|N^{-1/2}a^\dagger S_z|n-1,+\rangle$$
$$-\tau_c\left\{\sum_{k1}\varepsilon_{k1}^2(1+\cos\theta)^2\right\}\langle n,+|N^{-1/2}a^\dagger S_z|n-1,+\rangle \tag{8.58}$$

$$\left\langle n,-\left|\frac{d}{dt}(N^{-1/2}a^\dagger S_z)\right|n-1,-\right\rangle = i\omega\langle n,-|N^{-1/2}a^\dagger S_z|n-1,-\rangle$$
$$-\tau_c\sum_{k3}(1-\cos\theta)^2\langle n,-|N^{-1/2}a^\dagger S_z|n-1,-\rangle \tag{8.59}$$

$$\left\langle n,-\left|\frac{d}{dt}(N^{-1}a^{\dagger 2}S_-)\right|n-2,+\right\rangle = i(\omega+\Delta_n)\langle n,-|N^{-1}a^{\dagger 2}S_| n-2,+\rangle$$
$$-\tau_c\left\{\sum_{k1}\frac{1}{4}\varepsilon_{k1}^2(1+\cos\theta)^2 + \sum_{k2}\varepsilon_{k2}^2\sin^2\theta\right.$$
$$\left. + \sum_{k3}\varepsilon_{k3}^2\frac{1}{4}(1-\cos\theta)^2\right\}\langle n,-|N^{-1}a^{\dagger 2}S_-|n-2,+\rangle \tag{8.60}$$

Considering in the neighbourhood of $\omega \approx \omega_{ki}$, the coupling constant ε_{ki} is not explicitly dependent of ki, so we can let

$$\kappa = \sum_{ki} \varepsilon_{ki}^2 \tag{8.61}$$

Therefore, the solution of (8.57) is immediately obtained as

$$\langle n, +|S_1|n, -\rangle = \langle n, +|S_1|n, -\rangle_0 \exp\{[i(\omega - \Delta_n) - \Gamma_1]t\} \tag{8.62}$$

here the linewidth of the spectrum line with frequency $\omega - \Delta_n$ gives

$$\begin{aligned}\Gamma_1 =& \tau_c \kappa \left\{\frac{1}{4}(1+\cos\theta)^2 + \sin^2\theta + \frac{1}{4}(1-\cos\theta)^2\right\} \\ =& \frac{1}{2}\tau_c\kappa(3-\cos^2\theta) = \Gamma_{\omega-\Delta_n}\end{aligned} \tag{8.63}$$

Similarly, the solutions of (8.58) and (8.59) give

$$\langle n, \pm|S_2|n-1, \pm\rangle = \langle n, \pm|S_2|n-1, \pm\rangle_0 \exp[(i\omega - \Gamma_2)t] \tag{8.64}$$

and the linewidth is

$$\Gamma_2 = \tau_c\kappa(1+\cos^2\theta) = \Gamma_\omega \tag{8.65}$$

the solution of (8.60) is

$$\langle n, -|S_3|n-2, +\rangle = \langle n, -|S_3|n-2, +\rangle_0 \exp\{[i(\omega + \Delta_n) - \Gamma_3]t\} \tag{8.66}$$

with the linewidth

$$\Gamma_3 = \frac{1}{2}\tau_c\kappa(3-\cos^2\theta) = \Gamma_{\omega+\Delta_n} \tag{8.67}$$

In the resonant case, i.e., $\cos\theta=0$, we have

$$\Gamma_\omega = \tau_c\kappa, \qquad \Gamma_{\omega\pm\Delta_n} = \frac{3}{2}\tau_c\kappa \tag{8.68}$$

the ratio of the central line to the two side lines gives

$$\Gamma_\omega/\Gamma_{\omega\pm\Delta_n} = \frac{2}{3} \tag{8.69}$$

Obviously, this result is agreement with the experimental result. Now we turn to discuss the relative intensities of the spectral lines.

8.1.4 Intensity distribution of the resonance fluorescence spectrum

In order to find the intensity distribution of the resonance fluorescence spectrum, one must calculate the fluorescence emission rate, i.e.,

$$R_k = \frac{d}{dt}\langle n, \sigma, \{0_k\}|a_k^\dagger a_k|n, \sigma, \{0_k\}\rangle \tag{8.70}$$

here the fluorescence field is initially chosen in the vacuum state $\{|0_K\rangle\}$. Obviously, to calculate (8.70), it is necessary to have the expression of $a_{k_j}(t)$. The Heisenberg equation of a_k gives

$$\frac{d}{dt}a_k = -i[a_k, H] = -i\omega_k a_k - i\varepsilon_k S_- \tag{8.71}$$

here H is given by (8.33). The formal solution of the above equation yields

$$a_k(t) = a_k(0)\exp(-i\omega_k t) - i\varepsilon_k \int_0^t S_-(t')\exp[-i(\omega_k(t-t')]dt' \tag{8.72}$$

Applying (8.72) and its conjugation, we obtain

$$\begin{aligned}
&\langle a_k^\dagger(t)a_k(t)\rangle \\
&= \Bigg\langle\Bigg\{ a_k^\dagger(0)a_k(0) - i\varepsilon_k a_k^\dagger(0)e^{i\omega_k t}\int_0^t S_-(t'')\exp[-i\omega_k(t-t'')]dt'' \\
&+ i\varepsilon_k a_k(0)e^{-i\omega_k t}\int_0^t S_+(t')\exp[i\omega_k(t-t')]dt' \\
&+ \varepsilon_k^2\int_0^t S_+(t')\exp[i\omega_k(t-t')]dt'\int_0^t S_-(t'')\exp[-i\omega_k(t-t'')]dt''\Bigg\}\Bigg\rangle
\end{aligned} \tag{8.73}$$

Since the state vector of the system in the Heisenberg picture is time-independent, the fluorescent field is remained in the vacuum state at time t. Consequently, the above equation retain only the last term to be nonzero , i.e.,

$$\begin{aligned}
&\langle a_k^\dagger(t)a_k(t)\rangle \\
&= \varepsilon_k^2\Bigg\langle\Bigg\{\int_0^t S_+(t')\exp[i\omega_k(t-t')]dt'\int_0^t S_-(t'')\exp[-i\omega_k(t-t'')]dt''\Bigg\}\Bigg\rangle
\end{aligned}$$

$$= \varepsilon_k^2 \int_0^t dt' \int_0^{t'} dt'' \langle \{S_+(t')S_-(t'') + S_+(t'')S_-(t')\}\rangle \exp[-i\omega_k(t'-t'')]$$

$$= 2\mathrm{Re}\varepsilon_k^2 \int_0^t dt' \int_0^{t'} dt'' \langle S_+(t')S_-(t')\rangle \exp[-i\omega_k(t'-t'')] \tag{8.74}$$

Transforming the above equation into the dressed representation, and noticing

$$T^{-1}S_+(t')TT^{-1}S_-(t'')T = \sum_{\ell,m} R_\ell^\dagger(t')R_m(t'') \tag{8.75}$$

with

$$\begin{aligned} T^{-1}S_-(t)T &= R_1 + R_2 + R_3 \\ &= \frac{1}{2}(1+\cos\theta)S_- + \sin\theta N^{-1/2}aS_z + \frac{1}{2}(1-\cos\theta)N^{-1}a^2S_+ \end{aligned} \tag{8.76}$$

$$\begin{aligned} R_1 &= \frac{1}{2}(1+\cos\theta)S_- \\ R_2 &= \sin\theta N^{-1/2}aS_z \\ R_3 &= -\frac{1}{2}(1-\cos\theta)N^{-1}a^2S_+ \end{aligned} \tag{8.77}$$

then (8.74) becomes

$$\langle a_k^\dagger(t)a_k(t)\rangle = 2\mathrm{Re}\varepsilon_k^2 \sum_{\ell,m} \int_0^t dt' \int_0^{t'} dt'' \langle R_\ell^\dagger(t')R_m(t'')\rangle \exp[-i\omega_k(t'-t'')] \tag{8.78}$$

In order to solve the operator $R_\ell(t)$, we take into account the Hamiltonian in the dressed representation, i.e.,

$$\tilde{H} = \omega a^\dagger a + (\omega - \Delta_n)S_z + \sum_k \omega_k a_k^\dagger a_k + \sum_{k,\ell} \varepsilon_k(a_k R_\ell^\dagger + a_k^\dagger R_\ell) \tag{8.79}$$

If A is an arbitrary operator which does not contain a_k, then

$$\begin{aligned} \frac{d}{dt}A &= -i[A, H_0] - i[A, \sum_\ell V_\ell] \\ &= -i[A, H_0] - i\sum_{k,\ell} \varepsilon_k\{a_k^\dagger[A, R_\ell] + [A, R_\ell^\dagger]a_k\} \end{aligned} \tag{8.80}$$

The operators $a_k(t)$ and $a_k^\dagger(t)$ related to the fluorescence photons can be eliminated in the above equation by the aid of the expression

$$a_k(t) = a_k(0)e^{-i\omega_k t} - i\varepsilon_k \sum_\ell \int_0^t R_\ell(t-\tau)\exp(-i\omega_k\tau)d\tau \tag{8.81}$$

The time-evolution of $R_\ell(t)$ can be obtained by solving the zero-order Heisenberg equation, i.e.,

$$\frac{d}{dt}R_\ell(t) = -i[R_\ell, H_0] = -i\omega_\ell R_\ell \qquad (\ell = 1,2,3) \tag{8.82}$$

where

$$\omega_1 = \omega - \Delta_n, \qquad \omega_2 = \omega, \qquad \omega_3 = \omega + \Delta_n \tag{8.83}$$

The solution of (8.82) is readily obtained as

$$R_\ell(t) = R_\ell(0)\exp(-i\omega_\ell t) \tag{8.84}$$

Substituting (8.84) into (8.81) and taking the integration, it yields

$$a_k(t) = a_k(0)e^{-i\omega_k t} - \varepsilon_k \sum_\ell R_\ell \left\{ P\left(\frac{1}{\omega_k - \omega_\ell}\right) + i\pi\delta(\omega_k - \omega_\ell) \right\} \tag{8.85}$$

here $P(\frac{1}{\omega_\ell - \omega_k})$ is the principle function. Inserting eq.(8.85) and its conjugation into (8.80), we get

$$\begin{aligned} \frac{d}{dt}A = -i[A, H_0] + \sum_{\ell,m=1}^{3} \{&(-i\Omega_m + \gamma_m)R_m^\dagger[A, R_\ell] \\ &-(i\Omega_\ell + \gamma_\ell)[A, R_m]R_\ell^\dagger\} + F_A \end{aligned} \tag{8.86}$$

where

$$\begin{aligned} \Omega_\ell &= P\left(\sum_k \varepsilon_k^2 \frac{1}{\omega_\ell - \omega_k}\right) \\ \gamma_\ell &= \pi\sum_k \varepsilon_k^2 \delta(\omega_\ell - \omega_k) \\ F_A &= -i\sum_{\ell,k} \varepsilon_k \{a_k^\dagger(0)e^{i\omega_k t}[A, R_\ell] + [A, R_\ell^\dagger]a_k(0)e^{-i\omega_k t}\} \end{aligned} \tag{8.87}$$

Since we have assumed the fluorescence field initially in the vacuum state, $\langle 0|F_A|0\rangle=0$. Comparing (8.86) with the classical Langevin equation (5.17) we find that F_A can be regarded as a stochastic force. So (8.86) is called the quantum Langevin equation of the operator A. Because of $\langle 0|F_A|0\rangle=0$, one can neglect the term F_A in (8.86). In addition, (8.86) can be further simplified. For example, the terms containing Ω_m induce the Lamb shifts of the operator A, as an approximation these terms can be ignored. γ_ℓ correspond to decay rates of the operator A due to the influence of the fluorescence field. Considering that the difference among γ_ℓ is very small, as an approximation we can assume $\gamma_\ell = \gamma(\ell=1,2,3)$. The further approximation is that the terms such as the time-dependence products $R_\ell^+ R_m (\ell \neq m)$ may be approximated as a function $\exp[i(\omega_\ell - \omega_m)t]$, which corresponding to fast oscillation, so these terms of $\ell \neq m$ may be negligible. On the basis of all the above approximations, the quantum Langevin equation (8.86) reduces to

$$\frac{d}{dt}A = -i[A, H_0] - \gamma \sum_\ell \{[A, R_\ell^\dagger]R_\ell - R^\dagger[A, R_\ell]\} \tag{8.88}$$

When the operator A is replaced by R_ℓ, (8.88) shows that the time evolution of R_ℓ depends on two other R_ℓ, which means that these three kinds of fluorescence transitions exist interference effects. So it is not exact to regard the three decay processes as independence. From (8.88) we get

$$\frac{d}{dt}R_\ell = -[R_\ell, H_0] - \gamma \sum_{\ell'=1}^{3} \{[R_\ell, R_{\ell'}^\dagger]R_{\ell'} - R_{\ell'}^\dagger[R_\ell, R_{\ell'}]\} \tag{8.89}$$

Substituting H_0 and (8.77) into the above equation, we obtain

$$\frac{d}{dt}R_1 = -(i\omega_1 + \Gamma_1)R_1 \tag{8.90}$$

with

$$\Gamma_1 = \frac{1}{2}\gamma(3 - \cos^2\theta) \tag{8.91}$$

Similarly we have

$$\frac{d}{dt}R_2 = -(i\omega_2 + \Gamma_2)R_2 - \gamma \sin\theta \cos\theta N^{-1/2} a \tag{8.92}$$

$$\frac{d}{dt}R_3 = -(i\omega_3 + \gamma_3)R_3 \tag{8.93}$$

with

$$\Gamma_2 = \gamma(1+\cos^2\theta), \qquad \Gamma_3 = \Gamma_1 \tag{8.94}$$

Here ω_i are determined by (8.83). Eqs.(8.94) and (8.91) show that the decay rates of the three kinds of transition processes are not only related to the coupling constant γ, but also depended on the driven laser intensity and the detuning Δ. For the resonant case, $\cos\theta$=0, so the linewidths are $\Gamma_1 = \Gamma_3 = 3\gamma/2$ and $\Gamma_2 = \gamma$, which are coincided with the experimental results. In order to obtain the fluorescence intensity, it is necessary to give the explicit expressions of $R_\ell(t)$. Eqs.(8.90), (8.92) and (8.93) give

$$\begin{aligned}
R_1(t) &= R_1(0)\exp[-(i\omega_1+\Gamma_1)t] \\
R_3(t) &= R_3(0)\exp[-(i\omega_3+\Gamma_3)t] \\
R_2(t) &= R_2(0)\exp[-(i\omega_2+\Gamma_2)t] \\
&\quad -\int_0^t \gamma\sin\theta\cos\theta N^{-1/2}ae^{-i\omega t'}\exp[-(i\omega+\Gamma_2)(t-t')]dt' \\
&\quad -(R_2(0)-a_0/\Gamma_2)\exp[-(i\omega_2+\gamma_2)t]+\frac{a_0}{\Gamma_2}\exp(-i\omega t)
\end{aligned} \tag{8.95}$$

here

$$a_0 = \gamma\sin\theta\cos\theta N^{-1/2}a(0) \tag{8.96}$$

Now we turn to calculate the fluorescence intensities. Because $R_1^+R_3$ and $R_3^+R_1$ are proportional to S_+^2 and S_-^2, as well as $R_1^+R_2, R_2^+R_1, R_2^+R_3$ and $R_3^+R_2$ are proportional to aS_+ and a^+S_-, all these matrix elements are then equal to zero. Thus, (8.78) remains only the diagonal parts, namely

$$\begin{aligned}
\langle a_k^\dagger a_k(t)\rangle_m &= 2\varepsilon_k^2\mathrm{Re}\left\{\sum_\ell\int_0^t dt'\int_0^{t'} dt''\langle R_\ell^\dagger(t')R_\ell(t'')\rangle_m\right. \\
&\quad \times\exp\{-[i(\omega_k-\omega_\ell)+\Gamma_\ell](t'-t'')\} \\
&\quad \left.+\Gamma_2^{-1}\int_0^t dt'\int_0^{t'} dt''\langle a_0^\dagger(t')R_2(t'')\rangle_m\exp\{-[i(\omega_k-\omega)+\Gamma_2](t'-t'')\right\} \\
&\quad +\Gamma_2^{-1}\int_0^t dt'\int_0^{t''} dt''\langle a_0^\dagger(t')R_2(t'')\rangle_m\exp[-i(\omega_k-\omega)(t'-t'')]
\end{aligned} \tag{8.97}$$

Since

$$
\begin{aligned}
R_1^\dagger R_1 &= \frac{1}{4}(1+\cos\theta)^2 S_+ S_- = \frac{1}{4}(1+\cos\theta)^2\left(S_z+\frac{1}{2}\right) \\
R_2^\dagger R_2 &= \frac{1}{4}\sin^2\theta \\
R_3^\dagger R_3 &= -\frac{1}{4}(1-\cos\theta)^2\left(S_z-\frac{1}{2}\right) \\
a^\dagger R_2 &= \gamma\sin^2\theta\cos\theta S_z
\end{aligned}
\tag{8.98}
$$

We have

$$
\begin{aligned}
\langle a_k^\dagger(t)a_k(t)\rangle_m &= 2\varepsilon_k^2 \mathrm{Re}\left\{\int_0^t dt' \int_0^{t'} dt'' \left[(1+\cos\theta)^2\left\langle S_z(t'')+\frac{1}{2}\right\rangle_m\right.\right. \\
&\times \exp\{-[i(\omega_k-\omega_1)+\Gamma_1](t'-t'')\} \\
&+\frac{1}{4}\exp\{-[i(\omega_k-\omega_2)+\Gamma_2](t'-t'')\}\sin^2\theta \\
&\left.-\frac{1}{4}(1-\cos\theta)^2\langle S_z(t'')-1/2\rangle_m \exp\{-[i(\omega_k-\omega_3)+\Gamma_3](t'-t'')\}\right] \\
&+\Gamma_2^{-1}\gamma\sin^2\theta\cos\theta\int_0^t dt'\int_0^{t''}\langle S_z(t'')\rangle_m \exp\{-[i(\omega_k-\omega)+\Gamma_2](t'-t'')\} \\
&\left.-\Gamma_2^{-1}\gamma\sin^2\theta\cos\theta\int_0^t dt'\int_0^{t'} dt''\langle S_z(t'')\rangle_m \exp[-i(\omega_k-\omega)(t'-t'')]\right\}
\end{aligned}
\tag{8.99}
$$

In order to solve the above equation, the time-dependence of $S_z(t'')$ is required. Since

$$
\frac{d}{dt}S_z(t) = -\Gamma_2 S_z - \gamma\cos\theta = -\Gamma_2\left[S_z + \frac{\cos\theta}{1+\cos^2\theta}\right] \tag{8.100}
$$

it gives

$$
S_z(t) = \left[S_z(0) + \frac{\cos\theta}{1+\cos^2\theta}\right]\exp(-\Gamma_2 t) + C \tag{8.101}
$$

By means of the initial condition $S_z(t) = S_z(0)$, (8.101) becomes

$$
S_z(t) = \left[S_z(0) + \frac{\cos\theta}{1+\cos^2\theta}\right]\exp(-\Gamma_2 t) - \frac{\cos\theta}{1+\cos^2\theta} \tag{8.102}
$$

Noticing the observation time $t \gg \gamma^{-1}$, so the part of the integrals in (8.99) coming from the transient $t'' \ll \gamma^{-1}$ is negligible, that is to say, $S_z(t'')$ in (8.9) can be replaced by

$$S_z(t'') \approx -\frac{\cos\theta}{1+\cos^2\theta} \tag{8.103}$$

In addition, as for the following integral term in (8.99)

$$I = \int_0^t dt' \int_0^{t'} dt'' f(t'-t'') \tag{8.104}$$

one can make the change of variables

$$t' - t'' = \tau \tag{8.105}$$

i.e.,

$$t'' = t' - \tau$$

then

$$I = \int_0^t dt' \int_0^{t'} d\tau f(\tau) \tag{8.106}$$

When $\tau > \gamma^{-1}$, $f(\tau)$ is very small. And noticing $t \gg \gamma^{-1}$, the integral range with respect to τ in (8.106) may be extended to ∞, i.e.,

$$I = \int_0^t dt' \int_0^\infty d\tau f(\tau) = t \int_0^\infty d\tau f(\tau) \tag{8.107}$$

By use of (8.103) and (8.107), (8.99) becomes

$$\begin{aligned}\langle a_k^\dagger(t)a_k(t)\rangle_m = 2t\varepsilon_k^2 \mathrm{Re}\Bigg[&\frac{1}{8}\frac{\sin^4\theta}{1+\cos^2\theta}\Bigg(\int_0^\infty d\tau \exp\{-[i(\omega_k-\omega_1)+\Gamma_1]\tau\} \\ &+\int_0^\infty d\tau \exp\{-[i(\omega_k-\omega_3)+\Gamma_3]\tau\}\Bigg) \\ &+\frac{1}{4}\frac{\sin^6\theta}{(1+\cos^2\theta)^2}\int_0^\infty d\tau \exp\{-[i(\omega_k-\omega)+\Gamma_2]\tau\}\Bigg]\end{aligned}$$

$$+2\varepsilon_k^2\mathrm{Re}\frac{\sin^2\theta\cos^2\theta}{(1+\cos^2\theta)^2}\int_0^t dt'\int_0^t dt''\exp[-i(\omega_k-\omega)(t'-t'')]$$

$$=\frac{t}{4}\varepsilon_k^2\mathrm{Re}\left[\frac{\sin^4\theta}{1+\cos^2\theta}\left(\frac{1}{\Gamma_1+i(\omega_k-\omega_1)}+\frac{1}{\Gamma_3+i(\omega_k-\omega_3)}\right)\right]$$

$$+\frac{t}{2}\varepsilon_k^2\mathrm{Re}\frac{\sin^6\theta}{(1+\cos^2\theta)^2}\frac{1}{\Gamma_2+i(\omega_k-\omega_2)}+2\pi t\mathrm{Re}\varepsilon_k^2\frac{\sin^2\theta\cos^2\theta}{(1+\cos^2\theta)^2}\delta(\omega_k-\omega)$$

$$=t\gamma\varepsilon_k^2\frac{\sin^4\theta}{1+\cos^2\theta}\left[\frac{1}{4}\frac{3-\cos^2\theta}{(\omega_k-\omega_1)^2+\Gamma_1^2}+\frac{1}{4}\frac{3-\cos^2\theta}{(\omega_k-\omega_3)^2+\Gamma_3^2}\right.$$

$$\left.+\frac{\sin^2\theta}{(\omega_k-\omega_2)^2+\Gamma_2^2}\right]+2\pi\varepsilon_k^2 t\frac{\sin^2\theta\cos^2\theta}{(1+\cos^2\theta)^2}\delta(\omega_k-\omega) \tag{8.108}$$

Thus the intensity of the fluorescence gives

$$R_k=\frac{d}{dt}\langle a_k^\dagger a_k\rangle_m=\frac{\gamma\varepsilon_k^2}{2}\frac{\sin^4\theta}{1+\cos^2\theta}\left\{\frac{1}{4}\frac{3-\cos^2\theta}{(\omega_k-\omega_1)^2+\Gamma_1^2}\right.$$

$$\left.+\frac{1}{4}\frac{3-\cos^2\theta}{(\omega_k-\omega_3)^2+\Gamma_3^2}+\frac{\sin^2\theta}{(\omega_k-\omega_2)^2+\Gamma_2^2}\right\}$$

$$+2\pi\varepsilon_k^2\frac{\sin^2\theta\cos^2\theta}{(1+\cos^2\theta)^2}\delta(\omega_k-\omega) \tag{8.109}$$

In the resonant case, i.e., $\cos\theta=0$ and $\sin\theta=1$, the above equation reduces to

$$R_k=\frac{\gamma}{2}\varepsilon_k^2\left\{\frac{1}{4}\frac{3}{(\omega_k-\omega_1)^2+\Gamma_1^2}+\frac{1}{4}\frac{3}{(\omega_k-\omega_3)^2+\Gamma_3^2}+\frac{1}{(\omega_k-\omega_2)^2+\Gamma_2^2}\right\} \tag{8.110}$$

Clearly, the intensity I_2 of the central peak of the fluorescence can be obtained from (8.110) to be proportional to γ^{-2} as

$$I_2\propto\Gamma_2^{-2}=\gamma^{-2} \tag{8.111}$$

and the intensities of the two side peaks are as follows

$$I_1=I_3\propto\frac{3}{4}\Gamma_1^{-2}=\frac{\gamma^{-2}}{3} \tag{8.112}$$

Obviously, the intensity ratio of the central peak to the two side peaks is

$$I_2:I_{1,3}=3:1 \tag{8.113}$$

This is coincident with the experimental result, the last term in (8.109) represents the elastic scattering. When $t \gg \gamma^{-1}$, from (8.95) we know that the expectation values of R_1 and R_2 tend to be zero, but , $\langle R_2(\infty)\rangle$ does not tend to zero. This is because

$$\langle R_2(\infty)\rangle = \frac{1}{\Gamma_2}\langle a_0\rangle e^{-i\omega t} = \frac{\sin\theta\cos\theta}{1+\cos^2\theta}\langle N^{-1/2}a(0)\rangle e^{-i\omega t} \tag{8.114}$$

The above equation shows that when $t \gg \gamma^{-1}$, $\langle R_2\rangle$ does not tend to zero but oscillate with a special amplitude, which corresponds to the atomic stimulation absorption and emission processes. In briefly, the theoretical results of the resonance fluorescence for a two-level atom are mainly coincident with the experimental results.

8.2 Resonance fluorescence spectra of a three-level atom

8.2.1 Hamiltonian of a three-level atom under the interaction of a bimodal field

As yet all the atoms under the consideration are assumed to have two levels. But the real atoms have many levels. As a generalization, here we discuss the resonance fluorescence spectrum of a three-level atom. The level-diagram of the three-level atom is depicted in Fig.(8.5). The three-level atom has energy

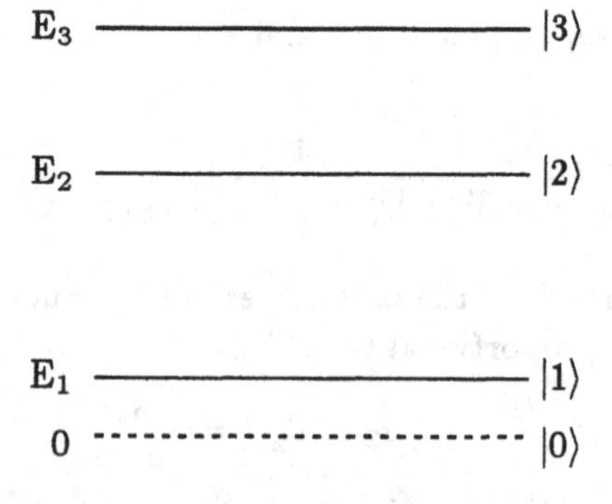

Figure 8.5: Energy levels of a three-level atom

levels $E_1 < E_2 < E_3$ with three corresponding states $|1\rangle$, $|2\rangle$ and $|3\rangle$. The operators describing the atom can be introduced by assuming that $|0\rangle$ is the atomic vacuum state, C_i^+ corresponds to an operator of creating an atomic number in the i-th level (i=1,2,3), and C_i is an operator of annihilating an atomic number in the i-th level, they obey

$$\begin{aligned} C_i^\dagger|0\rangle &= |i\rangle \\ C_i|i\rangle &= |0\rangle \end{aligned} \qquad (8.115)$$

In the atomic eigenket representation, the operators C_i can be expressed as

$$C_1 = \begin{pmatrix} 0 & 1 & 0 & 0 \\ 0 & 0 & 0 & 0 \\ 0 & 0 & 0 & 0 \\ 0 & 0 & 0 & 0 \end{pmatrix}, \qquad C_2 = \begin{pmatrix} 0 & 0 & 1 & 0 \\ 0 & 0 & 0 & 0 \\ 0 & 0 & 0 & 0 \\ 0 & 0 & 0 & 0 \end{pmatrix},$$

$$C_3 = \begin{pmatrix} 0 & 0 & 0 & 1 \\ 0 & 0 & 0 & 0 \\ 0 & 0 & 0 & 0 \\ 0 & 0 & 0 & 0 \end{pmatrix} \qquad (8.116)$$

$C_i^\dagger$ are their conjugations. According to (8.116) and its conjugation, it is directly verified that

$$C_iC_j^\dagger = 0 \qquad (i \neq j), \qquad C_iC_i^\dagger C_j = C_j \qquad (8.117)$$

The three-level atom may be interacted with a single-mode field or a bimodal field. As an example, we turn to discuss the system of a three-level atom coupled to two beams of laser fields shown in Fig.(8.6). The double arrows in Fig.(8.6) indicate transitions induced by driving laser fields, the dashed arrows indicate the dipole-allowed fluorescence transitions. The Hamiltonian of this system can be described in the rotating-wave approximation as

$$\begin{aligned} H = &\frac{1}{2}\omega_0(C_2^\dagger C_2 - C_1^\dagger C_1) + E_3C_3^\dagger C_3 + \omega_\alpha a^\dagger a + \omega_\beta b^\dagger b \\ &+\varepsilon_\alpha(aC_2^\dagger C_1 + a^\dagger C_1^\dagger C_2) + \varepsilon_\beta(bC_3^\dagger C_2 + b^\dagger C_2^\dagger C_3) + \sum_k \omega_k a_k^\dagger a_k \\ &+\sum \varepsilon_k\{a_k(C_2^\dagger C_1 + C_3^\dagger C_2) + a_k^\dagger(C_1^\dagger C_2 + C_2^\dagger C_3)\} \end{aligned} \qquad (8.118)$$

here we have chosen

$$\omega_0 = E_2 - E_1, \qquad \frac{E_2 + E_1}{2} = 0 \qquad (\hbar = 1) \qquad (8.119)$$

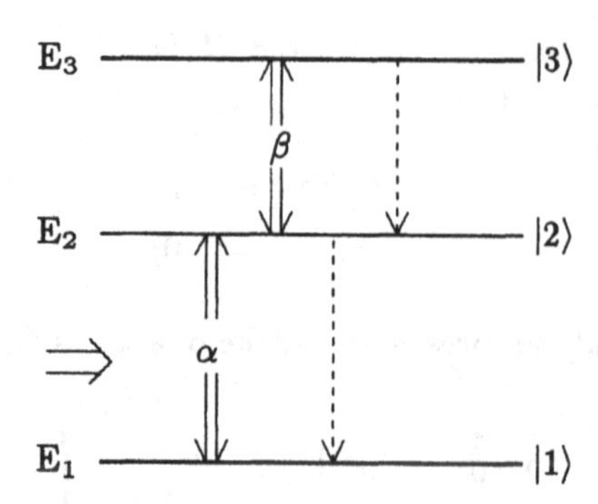

Figure 8.6: Induced transitions and fluorescence transtions of a three-level atom driven by two laser fields with frequencies ω_α and ω_β

$a^\dagger$ $(b^\dagger)$ and $a(b)$ are the creation and annihilation operators for the photons with frequency $\omega_\alpha(\omega_\beta)$.$a_K^\dagger$ and a_K stand for the creation and annihilation operators of the fluorescence field, respectively. The first and second terms in (8.118) represent the atomic energy, the third and the fourth terms stand for the energy of laser fields, seventh term represents the energy of the fluorescence field, the fifth and sixth terms describe the interactions between atom and laser fields, the last term is the Hamiltonian describing the interaction between the fluorescence field and the atom. ε_α and ε_β are the laser-atom coupling constants, ε_K is the fluorescence field-atom coupling constant.

8.2.2 Resonance fluorescence spectrum of a three-level atom interacting with a strong and a weak monochromatic laser field

Here we consider the case which the laser field with frequency ω_α is strong as well as the one with frequency ω_β is weak, that is to say,

$$\langle\omega_\alpha a^\dagger a\rangle \gg \langle\omega_\beta b^\dagger b\rangle \tag{8.120}$$

So (8.118) may be divided into two parts H_0 and H_1, here

$$H_0 = \frac{\omega_0}{2}(C_2^\dagger C_2 - C_1^\dagger C_1) + E_3 C^\dagger C_3 + \omega_\alpha a^\dagger a + \omega_\beta b^\dagger b + \sum_k \omega_k a^\dagger a_k$$
$$+\varepsilon_\alpha(a^\dagger C_2^\dagger C_1 + a^\dagger C_1^\dagger C_2) \tag{8.121}$$

The remainder H_1 is relatively small in comparison with H_0 , we may treat it as a perturbation term. Now we generalize the previous dressed transformation theory to deal with a three-level atom interacting with a strong laser field. The canonical unitary transformation operator adopted here may be

$$T = \exp\{-\theta(4N)^{-1/2}(aC_2^\dagger C_1 - a^\dagger C_1^\dagger C_2)\} \tag{8.122}$$

where

$$N = a^\dagger a + \frac{1}{2}(C_2^\dagger C_2 - C_1^\dagger C_1) + \frac{1}{2} \tag{8.123}$$

Through performing the unitary transformation T on H_0, and utilizing (8.117) and the strong field approximation, we may obtain

$$\begin{aligned}
T^{-1}H_0T = & H_0 + \left(-\frac{\theta}{(4N)^{1/2}}\right)[H_0, aC_2^\dagger C_1 - a^\dagger C_1^\dagger C_2] \\
& + \frac{1}{2!}\left(-\frac{\theta}{(4N)^{1/2}}\right)^2 [[H_0, aC_2^\dagger C_1 - a^\dagger C_1^\dagger C_2], aC_2^\dagger C_1 + a^\dagger C_1^\dagger C_2] + \cdots \\
& = \frac{1}{2}(\omega_\alpha - \Delta)(C_2^\dagger C_2 - C_1^\dagger C_1) + E_3 C_3^\dagger C_3 \\
& + \omega_\alpha a^\dagger a + \omega_\beta b^\dagger b + \sum_k \omega_k a_k^\dagger a_k
\end{aligned} \tag{8.124}$$

here we have chosen

$$\theta = atn\left(\frac{2\varepsilon_\alpha N^{1/2}}{\omega_0 - \omega_\alpha}\right) \tag{8.125}$$

$$\Delta = \sqrt{(\omega_0 - \omega_\alpha)^2 + 4\varepsilon_\alpha^2 N} \tag{8.126}$$

Equation (8.124) shows that the parts standing for the interaction between the strong laser field and the atom have been formally eliminated by the unitary dressed transformation, and the effects of the strong laser field on the atom is contained in the parameter Δ. That is to say, the interaction between the laser field and the atom has been mixed by the dressed transformation. Therefore, (8.124) may be considered as a Hamiltonian of a three-level atom dressed by the strong laser field of frequency ω_α. The next step is to treat the perturbation part H_1 in (8.121) in the dressed representation

$$H_1 = \varepsilon_\beta(bC_3^\dagger C_2 + b^\dagger C_2^\dagger C_3) + \sum_k \varepsilon_k\{a_k(C_2^\dagger C_1 + C_3^\dagger C_2) + a_k^\dagger(C_1^\dagger C_2 + C_2^\dagger C_3)\} \tag{8.127}$$

Since the operator T commutates with b, b^+, a_K and a_K^+, the T transformation on H_1 is just to perform the unitary transformation on $C_2^+C_3, C_1^+C_2$ and their conjugations. By means of the strong field approximation $\langle a^+a\rangle \gg 1$, we can neglect the order of $0(N^{-1/2})$ and $0(N^{-1})$, then obtain

$$\begin{aligned}
T^{-1}C_2^\dagger C_3 T = & C_2^\dagger C_3 - \left(\frac{\theta}{(4N)^{1/2}}\right)[C_2^\dagger C_3, aC_2^\dagger C_1 - a^\dagger C_1^\dagger C_2] \\
& + \frac{1}{2!}\left(-\frac{\theta}{(4N)^{1/2}}\right)^2 [[C_2^\dagger C_3, aC_2^\dagger C_1 - a^\dagger C_1^\dagger C_2], aC_2^\dagger C_1 - a^\dagger C_1^\dagger C_2] \\
& + \cdots \\
= & \cos(\theta/2)C_2^\dagger C_3 - N^{-1/2}\sin(\theta/2)a^\dagger C_1^\dagger C_2
\end{aligned} \tag{8.128}$$

Similarly, the expression of $T^{-1}C_3^+C_2T, T^{-1}C_1^+C_2T$ and $T^{-1}C_2^+C_1T$ can be obtained. Substituting all of these into $T^{-1}H_1T$, then we finally obtain

$$\begin{aligned}
T^{-1}HT = & T^{-1}H_0T + T^{-1}H_1T = \frac{1}{2}(\omega_\alpha - \Delta)(C_2^\dagger C_2 - C_1^\dagger C_1) + E_3C_3^\dagger C_3 \\
& + \omega_\alpha a^\dagger a + \omega_\beta b^\dagger b + \sum_k \omega_k a_k^\dagger a_k + \varepsilon_\beta \left\{ b\left[\cos\left(\frac{\theta}{2}\right) C_3^\dagger C_2 \right.\right. \\
& \left. - \sin\left(\frac{\theta}{2}\right)\frac{1}{\sqrt{N}}aC_3^\dagger C_1\right] + b^\dagger\left.\left[\cos\left(\frac{\theta}{2}\right)C_2^\dagger C_3 - \sin\left(\frac{\theta}{2}\right)\frac{1}{\sqrt{N}}a^\dagger C_1^\dagger C_3\right]\right\} \\
& + \sum_k \varepsilon_k \left\{ a_k\left[\frac{1}{2}(1+\cos\theta)C_2^\dagger C_1 + \frac{1}{2}\sin\theta\frac{1}{\sqrt{N}}a^\dagger(C_2^\dagger C_2 - C_1^\dagger C_1)\right.\right. \\
& \left. + \frac{1}{2}(\cos\theta - 1)\frac{1}{N}a^{\dagger 2}C_1^\dagger C_2 + \cos\left(\frac{\theta}{2}\right)C_3^\dagger C_2 - \sin\left(\frac{\theta}{2}\right)\frac{1}{\sqrt{N}}aC_3^\dagger C_1\right] \\
& + a_k^\dagger\left[\frac{1}{2}(\cos\theta + 1)C_1^\dagger C_2 + \frac{1}{2}\sin\theta\frac{1}{\sqrt{N}}a(C_2^\dagger C_2 - C_1^\dagger C_1)\right. \\
& + \frac{1}{2}(\cos\theta - 1)\frac{1}{N}a^2C_2^\dagger C_1 + \cos\left(\frac{\theta}{2}\right)C_2^\dagger C_3 \\
& \left.\left. - \sin\left(\frac{\theta}{2}\right)\frac{1}{\sqrt{N}}a^\dagger C_1^\dagger C_3\right]\right\}
\end{aligned} \tag{8.129}$$

From (8.129) we see that the interaction term between the atom and the strong driving field of frequency ω_α in Hamiltonian (8.118) has been eliminated, but the interaction part between the atom and the weak laser field of frequency ω_β and the one between the atom and the fluorescence field become more complex.

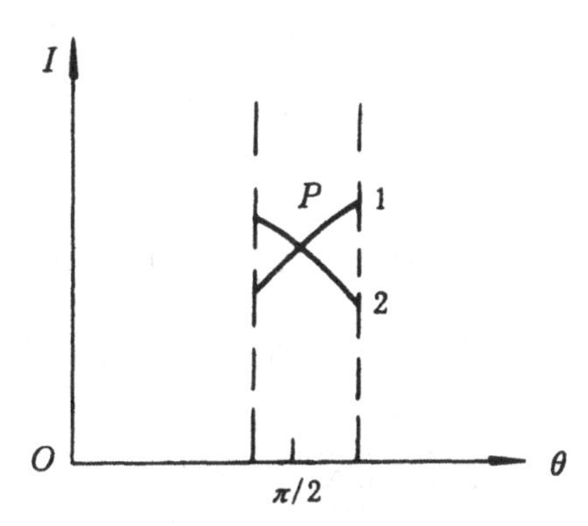

Figure 8.7: Curve 1 represents the amplitude of two-photon processes versus θ. Curve 2 represents the amplitude of single-photon processes versus θ. Point P corresponds to the resonant case

In fact, these complicated terms are the responses of the effects of the interaction between the atom and the strong laser field. However, by dissecting the properties of these complicated terms, one can reveal some physical processes of the atomic system induced by the strong driving laser.

One of the process shown in (8.129) is the two-photon process, which describes the atomic transition between the state $|1\rangle$ and the state $|3\rangle$ while a β photon and a α photon are created or annihilated simultaneously. The other is three-photon process for the atom goes from the state $|1\rangle$ to $|2\rangle$ and emits a fluorescence photon, as well as two photons of frequency ω_α are absorbed. These two kinds of multi-photon processes are strongly dependent on $\sin(\theta/2)$. Fig.(8.7) shows the amplitude of these multi-photon processes in comparison with that of one-photon processes (which is proportional to $\cos(\theta/2)$) in the nearly resonant range, i.e., $\omega_0 \approx \omega_\alpha$. In this case

$$\cos\theta = \frac{\omega_0 - \omega_\alpha}{2\varepsilon_\alpha\sqrt{n}} \approx 0, \qquad \sin\frac{\theta}{2} \approx \frac{1}{2}$$

It is easy to see that in the neighbourhood of $\omega_0 \approx \omega_\alpha$, with the increasing of the frequency of the driving field $(\omega_0 < \omega_\alpha)$, the two-photon processes dominate the one-photon processes. But with the decreasing of $\omega_\alpha (\omega_0 > \omega_\alpha)$, the intensity of the two-photon processes is weaker than that of the one-photon processes. This shows that the intensity of the two-photon processes obviously depends on the detuning $\delta = \omega_0 - \omega_\alpha$.

In order to display the characteristics of the resonance fluorescence spectrum, it is necessary to obtain the time evolution of the fluorescent operator $a_k(t)$. In the Heisenberg picture, $a_k(t)$ obeys

$$\begin{aligned}\frac{d}{dt}a_k = -i[a_k, H] = -i\omega_k a_k - i\varepsilon_k \Big[&\frac{1}{2}(\cos\theta + 1)C_1^\dagger C_2 \\ &+\frac{1}{2}\sin\theta\frac{1}{\sqrt{N}}a(C_2^\dagger C_2 - C_1^\dagger C_1) + \frac{1}{2}(\cos\theta - 1)\frac{1}{N}a^2 C_2^\dagger C_1 \\ &+ \cos\frac{\theta}{2}C_2^\dagger C_3 - \sin\frac{\theta}{2}\frac{1}{\sqrt{N}}a^\dagger C_1^\dagger C_3\Big]\end{aligned} \quad (8.130)$$

its solution gives

$$\begin{aligned}a_k(t) = a_k(0)e^{-i\omega_k t} - i\varepsilon_k \int_0^t \Big[&\frac{1}{2}(\cos\theta + 1)C_1^\dagger C_2 - \sin\frac{\theta}{2}\frac{1}{\sqrt{N}}a^\dagger C_1^\dagger C_3 \\ &+ \cos\frac{\theta}{2}C_2^\dagger C_3 + \frac{1}{2}\sin\theta\frac{1}{\sqrt{N}}a(C_2^\dagger C_2 - C_1^\dagger C_1) \\ &+\frac{1}{2}(\cos\theta - 1)\frac{1}{N}a^2 C_2^\dagger C_1\Big]_{t-\tau} e^{-i\omega_k\tau}d\tau\end{aligned} \quad (8.131)$$

To calculate the integration term in (8.131), we need to know the time-dependence of the operators $C_1^+ C_2$, $a(C_2^+ C_2 - C_1^+ C_1)$, $a^2 C_2^+ C_1$, $C_2^+ C_3$, and $a^+ C_1^+ C_3$. In the zero-order approximation (namely, the perturbation term H_1 in H is ignored), the time-dependence of these operators are obtained from the Heisenberg equations as

$$\begin{aligned}&C_1^\dagger C_2 = (C_1^\dagger C_2)_0 \exp[-i(\omega_\alpha - \Delta)t] \\ &a^2 C_2^\dagger C_1 = (a^2 C_2^\dagger C_1)_0 \exp[-i(\omega_\alpha + \Delta)t] \\ &a(C_2^\dagger C_2 - C_1^\dagger C_1) = a(C_2^\dagger C_2 - C_1^\dagger C_1)_0 \exp(-i\omega_\alpha t) \\ &C_2^\dagger C_3 = (C_2^\dagger C_3)_0 \exp\{-i[E_3 - (\omega_\alpha - \Delta)/2]t\} \\ &a^\dagger C_1^\dagger C_3 = (a^\dagger C_1^\dagger C_3)_0 \exp\{-i[E_3 - (\omega_\alpha + \Delta)/2]t\}\end{aligned} \quad (8.132)$$

Substituting (8.132) into (8.131) and subsequently integrating it we obtain

$$a_k(t) = a_k(0)e^{-i\omega_k t} + \varepsilon_k \left\{\frac{1}{2}(\cos\theta + 1)(C_1^\dagger C_2)_t \left[P\left(\frac{1}{\omega_\alpha - \Delta - \omega_k}\right)\right.\right.$$

$$-i\pi\delta(\omega_\alpha - \Delta - \omega_k)] + \frac{1}{2}(\cos\theta - 1)\frac{1}{N}(a^2 C_2^\dagger C_1)_t[-i\pi\delta(\omega_\alpha + \Delta - \omega_k)$$

$$+P\left(\frac{1}{\omega_\alpha + \Delta - \omega_k}\right)\Bigg] + \cdots\Bigg\} \tag{8.133}$$

Inasmuch as the principal functions in (8.133) represent the small Lamb shifts, we can neglect these terms. Noticing ε_K is the coupling constant between the k-th mode of fluorescence field and the atom, so that $\sum \varepsilon_k a_k$ is proportional to the amplitude of the fluorescent field, thus

$$E(t) \propto \sum_k \varepsilon_k a_k(0) e^{-i\omega_k t} - \frac{i}{2}\gamma_1(\cos\theta + 1)C_1^\dagger C_2$$

$$-i\gamma_2(\cos\theta - 1)\frac{1}{N}a^2 C_2^\dagger C_1 - \frac{i}{2}\gamma_3 \sin\theta \frac{1}{\sqrt{N}} a(C_2^\dagger C_2 - C_1^\dagger C_1)$$

$$-i\gamma_4 \cos\left(\frac{\theta}{2}\right) C_2^\dagger C_3 + i\gamma_5 \sin\left(\frac{\theta}{2}\right)\frac{1}{\sqrt{N}} a^\dagger C_1^\dagger C_3 \tag{8.134}$$

where

$$\begin{aligned}
\gamma_1 &= \pi\sum_k \varepsilon_k^2 \delta(\omega_\alpha - \Delta - \omega_k) \\
\gamma_2 &= \pi\sum_k \varepsilon_k^2 \delta(\omega_\alpha + \Delta - \omega_k) \\
\gamma_3 &= \pi\sum_k \varepsilon_k^2 \delta(\omega_\alpha - \omega_k) \\
\gamma_4 &= \pi\sum_k \varepsilon_k^2 \delta[E_3 - (\omega_\alpha - \Delta)/2 - \omega_k] \\
\gamma_5 &= \pi\sum_k \varepsilon_k^2 \delta[E_3 - (\omega_\alpha + \Delta)/2 - \omega_k]
\end{aligned} \tag{8.135}$$

These $\gamma_i (i = 1, 2, \cdots, 5)$ obviously correspond to the spontaneous relaxation rates. The corresponding fluorescence transitions among the dressed atomic levels are shown by wiggly arrows in Fig.(8.8). To summarize, the resonance fluorescence spectra of the three-level atom driven by a strong and a weak laser fields are composed by five lines with frequencies $\omega_\alpha - \Delta$, $\omega_\alpha + \Delta$, ω_α, $E_3 - (\omega_\alpha - \Delta)/2$ and $E_3 - (\omega_\alpha + \Delta)/2$. From (8.134) we can also see that the intensities given by $E^*(t)E(t)$ and the linewidthes of the spectral lines depend on the intensity of driving fields and the detuning $\delta\omega$.

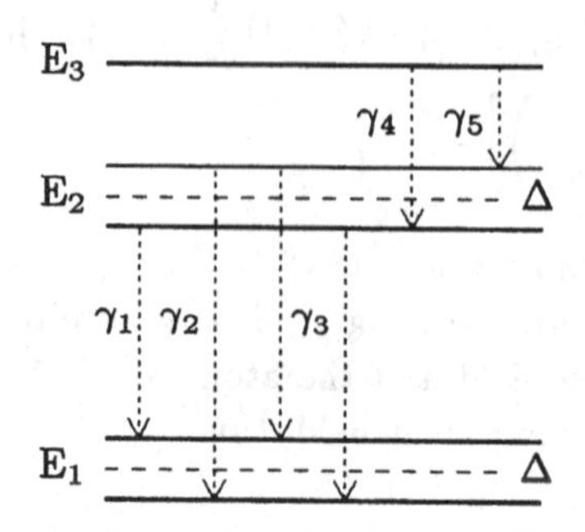

Figure 8.8: Dressed states and fluorescence transitions of a three-level atom driven by a strong and a weak laser fields

8.2.3 Resonance fluorescence spectral distribution of a three-level atom driven by two strong laser fields

In the previous section we have discussed the resonance fluorescence spectral distribution of a three-level atom driven by a strong and a weak laser fields, however, if both the driving fields are strong, what is the resonance fluorescence spectral distribution ?

The Hamiltonian describing this system is the same as (8.118), but the two driving laser fields are strong, i.e.,

$$\langle a^\dagger a\rangle \gg 1, \qquad \langle b^\dagger b\rangle \gg 1 \tag{8.136}$$

For simplifying the calculation, we may divide the Hamiltonian (8.118) into two parts H_0 and H_1, and H_0 and H_1 are defined as

$$\begin{aligned}H_0 = {}&\frac{\omega_0}{2}(C_2^\dagger C_2 - C_1^\dagger C_1) + E_3 C_3^\dagger C_3 + \omega_\alpha a^\dagger a + \omega_\beta b^\dagger b + \varepsilon_\alpha (aC_2^\dagger C_1 \\ &+ a^\dagger C_1^\dagger C_2) + \varepsilon_\beta (bC_3^\dagger C_2 + b^\dagger C_2^\dagger C_3) + \sum_k \omega_k a_k^\dagger a_k\end{aligned} \tag{8.137}$$

$$H_1 = \sum_k \varepsilon_k \{a_k (C_2^\dagger C_1 + C_3^\dagger C_2) + a_k^\dagger (C_1^\dagger C_2 + C_2^\dagger C_3)\} \tag{8.138}$$

Considering the dependence of the coupling constants on different laser frequencies is very weak, we may put $\varepsilon_\alpha = \varepsilon_\beta = \varepsilon$.

In order to diagonalize H_0, we choose the unitary transformation operator T as

$$T = \exp\left\{-\frac{\theta}{\sqrt{4N}}[(aC_2^\dagger C_1 - a^\dagger C_1^\dagger C_2) + (bC_3^\dagger C_2 - b^\dagger C_2^\dagger C_3)]\right\}$$
$$= \exp\left\{-\frac{\theta}{\sqrt{4N}}(\overline{V}_1 + \overline{V}_2)\right\} \tag{8.139}$$

where

$$N = a^\dagger a + b^\dagger b + \frac{1}{2}(C_2^\dagger C_2 - C_1^\dagger C_1) + \frac{3}{2}C_3^\dagger C_3 + \frac{1}{2} \tag{8.140}$$

is the operator associated with the total excitation number of the system, and θ is a free parameter. That the operator T performs a unitary transformation on H_0 gives

$$\tilde{H}_0 = T^{-1}H_0T = H_0 + \left(-\frac{1}{2}\theta N^{-1/2}\right)[H_0, \overline{V}_1 + \overline{V}_2]$$
$$+\frac{1}{2!}\left(-\frac{1}{2}\theta N^{-1/2}\right)^2 [[H_0, \overline{V}_1 + \overline{V}_2], \overline{V}_1 + \overline{V}_2] + \cdots \tag{8.141}$$

Because the two laser beams are high-power, we may use the strong field approximation, i.e., assume

$$N \approx a^\dagger a + b^\dagger b \approx 2a^\dagger a \approx 2b^\dagger b \tag{8.142}$$

Considering the commutation relations of operators a^+, a, b^+, b and C_i^+, C_i, and choosing the free parameter θ to satisfy the following equation

$$\sin\frac{\theta}{2} = \frac{2\varepsilon\sqrt{N}}{\Delta}, \qquad \cos\frac{\theta}{2} = \frac{\delta_\alpha + \delta_\beta}{\Delta}, \qquad \delta_\alpha = \omega_0 - \omega_\alpha$$
$$\delta_\beta = E_3 - \frac{\omega_0}{2} - \omega_\beta, \qquad \Delta = \sqrt{4\varepsilon^2 N + (\delta_\alpha + \delta_\beta)^2} \tag{8.143}$$

we can obtain after some algebra the following result

$$\tilde{H}_0 = \frac{1}{2}\omega_0(C_2^\dagger C_2 - C_1^\dagger C_1) + E_3C^\dagger C_3 + \omega_\alpha a^\dagger a + \omega_\beta b^\dagger b$$
$$+\frac{\delta_\alpha - \delta_\beta}{8}(C_3^\dagger C_3 - C_1^\dagger C_1)(\cos\theta - 1) + \frac{\delta_\alpha + \delta_\beta}{8}(C_3^\dagger C_3 - C_1^\dagger C_1)(\cos\theta - 1)$$
$$-\frac{\delta_\beta - \delta_\alpha}{4N}(abC_3^\dagger C_1 + a^\dagger b^\dagger C_1^\dagger C_3)(\cos\theta - 1)$$

$$+\frac{1}{4\sqrt{N}}(\delta_\beta-\delta_\alpha)(V_1-V_2)\sin\theta+\varepsilon\sqrt{N}(C_3^\dagger C_3$$
$$-C_1^\dagger C_1)\sin\frac{\theta}{2}+\sum_k \varepsilon_k a_k^\dagger a_k \tag{8.144}$$

In the resonant case, we have $\cos(\theta/2)=0$, so that

$$\tilde{H}_0=\frac{1}{2}\omega_0 C_2^\dagger C_2-\frac{1}{2}(\omega_0+\Delta_0)C_1^\dagger C_1+\left(E_3+\frac{\Delta_0}{2}\right)C_3^\dagger C_3$$
$$+\omega_\alpha a^\dagger a+\omega_\beta b^\dagger b+\sum_k \omega_k a_k^\dagger a_k \tag{8.145}$$

where

$$\Delta_0=2\varepsilon\sqrt{N} \tag{8.146}$$

The parts of $\varepsilon(aC_2^+C_1+a^+C_1^+C_2)$ and $\varepsilon(bC_3^+C_2+b^+C_2^+C_3)$ which indicate the interaction among the strong laser fields and the atom have been eliminated by such dressing transformation. The effects of laser fields are contained in the parameter Δ_0. Therefore, (8.145) may be considered as a Hamiltonian of the three-level atom dressed by two intensive laser fields.

The next step is to consider the effect of the fluorescence interaction term H_1 in the dressing representation. By means of the strong field approximation, we can neglect the order of $0(N^{-1/2})$ and $0(N^{-1})$, then obtain

$$\widetilde{C_1^\dagger C_2}=T^{-1}C_1^\dagger C_2 T=\left[1+\frac{1}{4}(\cos\theta-1)+\frac{1}{2}\left(\cos\frac{\theta}{2}-1\right)\right]C_1^\dagger C_2$$
$$+\frac{1}{\sqrt{N}}\left(\frac{1}{4}\sin\theta+\frac{1}{2}\sin\frac{\theta}{2}\right)a(C_2^\dagger C_2-C_1^\dagger C_1)$$
$$+\frac{1}{\sqrt{N}}\left(\frac{1}{4}\sin\theta+\frac{1}{2}\sin\frac{\theta}{2}\right)b^\dagger C_1^\dagger C_3+\frac{1}{2N}(\cos\theta-1)a^2C_2^\dagger C_1$$
$$-\frac{1}{2N}(\cos\theta-1)ab^\dagger C_2^\dagger C_3+\left[\frac{1}{N}\left(\cos\frac{\theta}{2}-1\right)-\frac{1}{2N}(\cos\theta-1)\right]abC_3^\dagger C_2$$
$$+N^{-3/2}\left[\frac{1}{2}\sin\theta-\sin\frac{\theta}{2}\right]a^2bC_3^\dagger C_1 \tag{8.147}$$

In resonant case, it reduces to

$$\widetilde{C_1^\dagger C_2}=\frac{1}{2\sqrt{N}}a(C_2^\dagger C_2-C_1^\dagger C_1)+\frac{1}{2\sqrt{N}}b^\dagger C_1^\dagger C_3-\frac{1}{N}a^2C_2^\dagger C_1$$
$$+\frac{1}{N}ab^\dagger C_2^\dagger C_3-N^{-3/2}a^2bC_3^\dagger C_1 \tag{8.148}$$

Similarly, we have

$$\tilde{C_2^\dagger C_3} = \frac{1}{2\sqrt{N}} b(C_3^\dagger C_3 - C_2^\dagger C_2) - \frac{1}{2\sqrt{N}} a^\dagger C_1^\dagger C_3 + \frac{1}{N} b^2 C_3^\dagger C_2$$
$$- \frac{1}{N} a^\dagger b C_1^\dagger C_2 + N^{-3/2} a b^2 C_3^\dagger C_1 \tag{8.149}$$

From (8.148), (8.149) and (8.145) we obtain the dressed Hamiltonian in resonant case in the form

$$\tilde{H} = \frac{1}{2}\omega_0 C_2^\dagger C_2 - \frac{1}{2}(\omega_0 + \Delta_0) C_1^\dagger C_1 + \left(E_3 + \frac{\Delta_0}{2}\right) C_3^\dagger C_3 + \omega_\alpha a^\dagger a + \omega_\beta b^\dagger b$$
$$+ \sum_k \omega_k a_k^\dagger a_k + \sum_k \varepsilon_k \left\{ a_k^\dagger \left[\frac{1}{2\sqrt{N}} a (C_2^\dagger C_2 - C_1^\dagger C_1) \right.\right.$$
$$+ \frac{1}{2\sqrt{N}} b^\dagger C_1^\dagger C_3 - \frac{1}{N} a^2 C_2^\dagger C_1 + \frac{1}{N} a b^\dagger C_2^\dagger C_3 - N^{-3/2} a^2 b C_3^\dagger C_1$$
$$+ \frac{1}{2\sqrt{N}} b (C_3^\dagger C_3 - C_2^\dagger C_2) - \frac{1}{2\sqrt{N}} a^\dagger C_1^\dagger C_3 + \frac{1}{N} b^2 C_3^\dagger C_2$$
$$\left.\left. - \frac{1}{N} a^\dagger b C_1^\dagger C_2 + N^{-3/2} a b^2 C_3^\dagger C_1 \right] + \text{h.c.} \right\} \tag{8.150}$$

We see that the interaction terms among the atom and the strong driving fields in Hamiltonian (8.137) have been eliminated by the unitary dressing transformation, but the interaction term between the atom and the fluorescent field become more complicated. However, these complex terms just reveal obviously the fluorescence processes of a three-level atom induced by the driving laser fields. In order to realize such fluorescence processes, we now discuss the time evolution of the fluorescence operator.

From the Heisenberg equation for $a_k(t)$, we obtain

$$a_k(t) = a_k(0) e^{-i\omega_k t} - i\varepsilon_k \int_0^t \left[\frac{1}{2\sqrt{N}} a (C_2^\dagger C_2 - C_1^\dagger C_1) + \frac{1}{2\sqrt{N}} b^\dagger C_1^\dagger C_3 \right.$$
$$- \frac{1}{N} a^2 C_2^\dagger C_1 + \frac{1}{N} a b^\dagger C_2^\dagger C_3 - N^{-3/2} a^2 b C_3^\dagger C_1 + \frac{1}{2\sqrt{N}} b (C_3^\dagger C_3 - C_2^\dagger C_2)$$
$$- \frac{1}{2\sqrt{N}} a^\dagger C_1^\dagger C_3 + \frac{1}{N} b^2 C_3^\dagger C_2$$
$$\left. - \frac{1}{N} a^\dagger b C_1^\dagger C_2 + N^{-3/2} a b^2 C_3^\dagger C_1 \right]_{t-\tau} e^{-i\omega_k \tau} d\tau \tag{8.151}$$

The time-dependent product term of the operators in the integral term can be obtained from the Heisenberg equations by neglecting the nondiagonal terms in (8.150) in the zero-order approximation as

$$\begin{aligned}
&[a(C_2^\dagger C_2 - C_1^\dagger C_1)]_t = [a(C_2^\dagger C_2 - C_1^\dagger C_1)]_0 \exp(-i\omega_\alpha t) \\
&[bC_1^\dagger C_3]_t = [bC_1^\dagger C_3]_0 \exp\left[-i\left(\frac{\omega_0}{2} + \Delta_0 + E_3 - \omega_\beta\right) t\right] \qquad (8.152) \\
&\vdots
\end{aligned}$$

Substitution of (8.152) into (8.151) and subsequent integration gives

$$\begin{aligned}
a_k(t) = a_k(0)e^{-i\omega_k t} &+ \varepsilon_k \left\{ \frac{1}{2\sqrt{N}} [aC_2^\dagger C_2 - C_1^\dagger C_1]_t \left[P\left(\frac{1}{\omega_\alpha - \omega_k}\right)\right.\right. \\
&+ i\pi\delta(\omega_\alpha - \omega_k)\Big] + \frac{1}{2\sqrt{N}} [bC_1^\dagger C_3]_t \left[P\left(\frac{1}{E_3 + \omega_0/2 + \Delta_0 - \omega_\beta - \omega_k}\right)\right. \\
&\left.\left.+ i\pi\delta\left(E_3 + \frac{\omega_0}{2} + \Delta_0 - \omega_\beta - \omega_k\right)\right] + \cdots \right\} \qquad (8.153)
\end{aligned}$$

We may neglect the principal parts in (8.153), because they yield only small Lamb shifts in the energy levels. ε_k is the coupling constant between the k-th mode of fluorescence field and the atom, so that $\sum_k \varepsilon_k a_k$ is proportional to the amplitude of the fluorescent field, thus

$$\begin{aligned}
E(t) \propto \sum_k \varepsilon_k a_k(t) = \sum_k a_k(0)e^{-i\omega_k t} &- i\gamma_1 \frac{1}{2\sqrt{N}} a(C_2^\dagger C_2 - C_1^\dagger C_1) \\
&- i\gamma_2 \frac{1}{2\sqrt{N}} b^\dagger C_1^\dagger C_3 + i\gamma_3 \frac{1}{N} a^2 C_2^\dagger C_1 - i\gamma_4 \frac{1}{N} ab^\dagger C_2^\dagger C_3 \\
&+ i\gamma_5 N^{-3/2} a^2 b C_3^\dagger C_1 - i\gamma_6 \frac{1}{2\sqrt{N}} b(C_3^\dagger C_3 - C_2^\dagger C_2) \\
&+ i\gamma_7 \frac{1}{2\sqrt{N}} a^\dagger C_1^\dagger C_3 - i\gamma_8 \frac{1}{N} b^2 C_3^\dagger C_2 \\
&+ i\gamma_9 \frac{1}{N} a^\dagger b C_1^\dagger C_2 - i\gamma_{10} N^{-3/2} ab^2 C_3^\dagger C_1 \qquad (8.154)
\end{aligned}$$

where

$$\gamma_1 = \pi \sum_k \varepsilon_k^2 \delta(\omega_\alpha - \omega_k)$$

$$
\begin{aligned}
\gamma_2 &= \pi \sum_k \varepsilon_k^2 \delta \left(E_3 + \frac{1}{2}\omega_0 + \Delta_0 - \omega_\beta - \omega_k \right) \\
\gamma_3 &= \pi \sum_k \varepsilon_k^2 \delta \left(2\omega_\alpha - \omega_0 - \frac{\Delta_0}{2} - \omega_k \right) \\
\gamma_4 &= \pi \sum_k \varepsilon_k^2 \delta \left(E_3 + \omega_\alpha - \omega_\beta - \frac{1}{2}\omega_0 + \frac{1}{2}\Delta_0 - \omega_k \right) \\
\gamma_5 &= \pi \sum_k \varepsilon_k^2 \delta \left(2\omega_\alpha + \omega_\beta - \frac{1}{2}\omega_0 - E_3 - \Delta_0 - \omega_k \right) \\
\gamma_6 &= \pi \sum_k \varepsilon_k^2 \delta (\omega_\beta - \omega_k) \\
\gamma_7 &= \pi \sum_k \varepsilon_k^2 \delta \left(E_3 - \omega_\alpha + \frac{\omega_0}{2} + \Delta_0 - \omega_k \right) \\
\gamma_8 &= \pi \sum_k \varepsilon_k^2 \delta \left(2\omega_\alpha + \frac{\omega_0}{2} - E_3 - \frac{\Delta_0}{2} - \omega_k \right) \\
\gamma_9 &= \pi \sum_k \varepsilon_k^2 \delta \left(\omega_0 + \omega_\beta - \omega_\alpha + \frac{\Delta_0}{2} - \omega_k \right) \\
\gamma_{10} &= \pi \sum_k \varepsilon_k^2 \delta \left(2\omega_\beta + \omega_\alpha - E_3 - \frac{\omega_0}{2} - \Delta_0 - \omega_k \right)
\end{aligned}
\tag{8.155}
$$

These $\gamma_i (i = 1, 2, \cdots, 10)$ obviously correspond to the spontaneous relaxation rates for transitions between the dressed atomic levels shown in Fig.(8.9). Here the wiggly arrows indicate the allowed fluorescence emissions.

We may see from the above results, when the detuning $\delta\omega > 2\Delta_0$, the resonance fluorescence spectrum consists of ten components symmetrically about the frequencies ω_α and ω_β, respectively. Moreover, when the frequency detuning $\delta\omega$ is less than $2\Delta_0$, the fluorescence spectral lines will overlap, if $\delta\omega = \Delta_0$ the spectral distribution will be seven peak lines. When the detuning $\delta\omega$=0, i.e., $\omega_\alpha = \omega_\beta$, we find the exact five peaks fluorescence distribution as

$$
\begin{aligned}
\gamma_1 = \gamma_6 &= \pi \sum_k \varepsilon_k^2 \delta (\omega_\alpha - \omega_k) \\
\gamma_2 = \gamma_7 &= \pi \sum_k \varepsilon_k^2 \delta \left(E_3 + \frac{\omega_0}{2} + 2\Delta_0 - \omega_\beta - \omega_k \right) \\
\gamma_3 = \gamma_6 &= \pi \sum_k \varepsilon_k^2 \delta (2\omega_\alpha - \omega_0 - \Delta_0 - \omega_k)
\end{aligned}
\tag{8.156}
$$

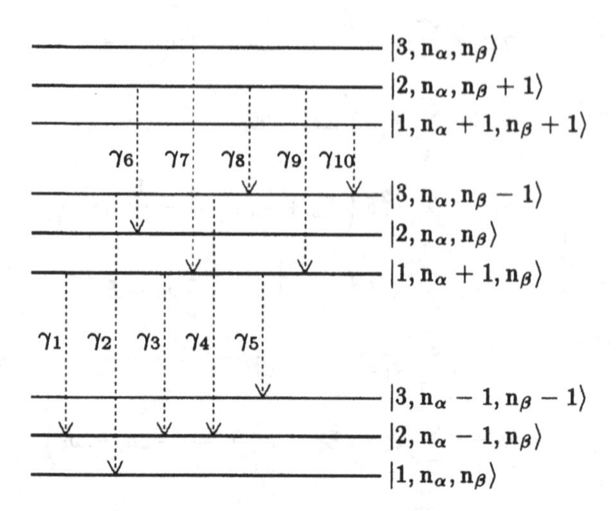

Figure 8.9: Dressed states and fluorescence transitions of a three-level atom driven by two strong laser fields

$$\gamma_4 = \gamma_9 = \pi \sum_k \varepsilon_k^2 \delta(\omega_0 + \Delta_0 - \omega_k)$$

$$\gamma_5 = \gamma_{10} = \pi \sum_k \varepsilon_k^2 \delta(3\omega_\alpha - 2\omega_0 - 2\Delta_0 - \omega_k)$$

Evidently, that the resonance fluorescence spectrum of a three-level atom driven by two strong laser fields consists of ten components is more generalized conclusion. The resonance fluorescence spectrum consisting of seven or five lines is only two special cases of the present results.

8.3 Single-atom resonance fluorescence described by the density matrix theory

According to the density matrix theory in Chapter 5, for the single-atom resonance fluorescence system, if one regard the vacuum field as a reservoir and the atom as a small system, the resonance fluorescence of a two-level or a three-level atom described by (8.1) or (8.118) can be treated by use of the density matrix theory of a small system coupled to a reservoir. As an example, we now discuss the resonance fluorescence of a two-level atom driven by a strong laser field by use of the density matrix theory.

For simplicity to calculate, here the driving laser field is still assumed to be so intensive that it can be characterized by a classical electromagnetic field. In this case, the Hamiltonian of the system may be described by

$$H = H_A + H_{AL} + H_R + H_{RS} \tag{8.157}$$

where

$$\begin{aligned} H_A &= \omega_0 S_z \\ H_{AL} &= \varepsilon(Ee^{-i\omega t}S_+ + Ee^{i\omega t}S_-) \\ H_R &= \sum_k \omega_k a_k^\dagger a_k \\ H_{RS} &= \sum_k \varepsilon_k (a_k^\dagger S_- + a_k S_+) \end{aligned} \tag{8.158}$$

here E is the amplitude of the electric field with frequency ω, which is assumed to be time-independent, H_{AL} represents the interaction Hamiltonian between the atom and the strong laser field, H_R stands for the energy of the fluorescent field, and H_{RS} describes the interaction between the atom and the fluorescent field. In the interaction picture, from (5.68) we know that the atomic density matrix ρ obeys

$$\frac{d}{dt}\rho = -i[H_{AL}^I, \rho] + \frac{\Gamma}{2}(2S_-\rho S_+ - \rho S_+ S_- - S_+ S_- \rho) \tag{8.159}$$

where Γ stands for the atomic spontaneous relaxation rate, according to (4.43) it satisfies

$$\Gamma = 2\pi \sum_k \varepsilon_k^2 \delta(\omega_k - \omega_0) \tag{8.160}$$

In the resonant case ($\omega_0 = \omega$), H_{AL}^I becomes

$$H_{AL}^I = \Omega(S_+ + S_-) \tag{8.161}$$

here $\Omega = \varepsilon E$. Eq.(8.159) may be formally rewritten as

$$\frac{d}{dt}\rho = L\rho \tag{8.162}$$

where L is called the generalized Liouvillian operator, and defined as

$$LM = -i[H_{AL}^I, M] + \frac{\Gamma}{2}(2S_- M S_+ - M S_+ S_- - S_+ S_- M) \tag{8.163}$$

where M is an arbitrary operator of the system. The evolution equation of the operator M may be written as

$$\frac{d}{dt}M = LM \tag{8.164}$$

Denoting $|+\rangle$ and $|-\rangle$ as the excited and ground states of bare atom, then in the interaction picture, the eigenstates of the laser field-atom coupling system, can be obtained from the eigenvalue equation

$$H^I_{AL}|i\rangle = E_i|i\rangle \tag{8.165}$$

The eigenvalues and corresponding eigenstates are expressed as

$$E_1 = -\Omega, \qquad |1\rangle = \frac{1}{\sqrt{2}}(|+\rangle - |-\rangle) \tag{8.166}$$

$$E_2 = \Omega, \qquad |2\rangle = \frac{1}{\sqrt{2}}(|+\rangle + |-\rangle) \tag{8.167}$$

Therefore, the atomic density matrix in the dressed representation is written as

$$\rho = \begin{pmatrix} \rho_{11} & \rho_{12} \\ \rho_{21} & \rho_{22} \end{pmatrix} \tag{8.168}$$

Substituting (8.166)-(8.168) into (8.159), the density matrix elements $\rho_{ij}(i,j = 1,2)$ obey

$$\begin{aligned}
\frac{d}{dt}\rho_{22} &= -\frac{\Gamma}{4}\rho_{22} + \frac{\Gamma}{4}\rho_{11} \\
\frac{d}{dt}\rho_{11} &= -\frac{\Gamma}{4}\rho_{11} + \frac{\Gamma}{4}\rho_{22} \\
\frac{d}{dt}\rho_{21} &= -\frac{\Gamma}{2}\rho_{22} - \frac{\Gamma}{2}\rho_{11} - \left(\frac{3}{4}\Gamma + 2i\Omega\right)\rho_{21} - \frac{\Gamma}{4}\rho_{12} \\
\frac{d}{dt}\rho_{12} &= -\frac{\Gamma}{2}\rho_{22} - \frac{\Gamma}{4}\rho_{21} - \left(\frac{3}{4}\Gamma - 2i\Omega\right)\rho_{12}
\end{aligned} \tag{8.169}$$

here we have used the relations

$$\begin{aligned}
S_+|1\rangle &= \frac{1}{\sqrt{2}}S_+(|+\rangle - |-\rangle) = -\frac{1}{\sqrt{2}}|+\rangle = -\frac{1}{2}(|2\rangle + |1\rangle) \\
S_+|2\rangle &= \frac{1}{2}(|2\rangle + |1\rangle), \qquad S_-|1\rangle = \frac{1}{2}(|2\rangle - |1\rangle),
\end{aligned}$$

$$S_-|2\rangle = \frac{1}{2}(|2\rangle - |1\rangle), \qquad S_+S_-|1\rangle = \frac{1}{2}(|2\rangle + |1\rangle),$$
$$S_+S_-|2\rangle = \frac{1}{2}(|2\rangle + |1\rangle)$$

To calculate the steady fluorescence intensity distribution (8.78), we need to calculate the two-time correlation function given by

$$\langle S_+(t)S_-(t+\tau)\rangle_S = \mathrm{Tr}_S(\rho(t)S_+(t)S_-(t+\tau))\exp(-i\omega_0\tau) \qquad (8.170)$$

So if $\rho(t)$ is solved from (8.169), then the fluorescence intensity distribution can be yielded by (8.170) and (8.78). According to (8.164), $S_-(t+\tau)$ can also be formally expressed as

$$S_-(t+\tau) = e^{L\tau}S_-(t) \qquad (8.171)$$

Substituting (8.171) into (8.170) and by use of $\mathrm{Tr}AB = \mathrm{Tr}BA$, then

$$\langle S_+(t)S_-(t+\tau)\rangle_S = \mathrm{Tr}_S S_-(t)e^{L\tau}\rho(t)S_+(t)e^{-i\omega_0\tau} \qquad (8.172)$$

In order to calculate the trace in (8.172), it is appropriate to adopt the diagonalized representation of the operator $\exp(L\tau)$. In the bare atomic representation, $\exp(L\tau)$ is not diagonalized. However, because $[H^I_{AL}, \rho]$ is more larger than $\Gamma/2(2S_-\rho S_+ - \rho S_+S_- - S_+S_-\rho)$ in the strong field limit, $\exp(L\tau)$ may be regarded as a diagonalized operator in the dressing representation. For this reason, the evaluation of (8.172) must be proceeded in a dressing representation. According to (4.10), the generalized atomic operator is defined as

$$\sigma_{nm} = |n\rangle\langle m|$$

So the atomic rasing and lowering operators in the dressed representation obey

$$D_+ = |2\rangle\langle 1| = \frac{1}{2}\begin{pmatrix} 1 & -1 \\ 1 & -1 \end{pmatrix} \qquad (8.173)$$

$$D_- = |1\rangle\langle 2| = \frac{1}{2}\begin{pmatrix} 1 & 1 \\ -1 & -1 \end{pmatrix} \qquad (8.174)$$

In fact, $D_\pm$ and $S_\pm$ can be connected by a unitary transformation U, i.e.,

$$D_+ = U^{-1}S_+U, \qquad D_- = U^{-1}S_-U \qquad (8.175)$$

Utilizing the above equation to transfer (8.172) into the dressed representation yields

$$\begin{aligned}
&\langle S_+(t)S_-(t+\tau)\rangle_S \\
&= \mathrm{Tr}_S UU^{-1}S_-(t)UU^{-1}e^{L\tau}UU^{-1}\rho(t)UU^{-1}S_+(t)UU^{-1}e^{-i\omega_0\tau} \\
&= \mathrm{Tr}_S UD_-(t)U^{-1}e^{L\tau}UU^{-1}\rho(t)UD_+(t)U^{-1}e^{-i\omega_0\tau} \\
&= \mathrm{Tr}_S D_-(t)U^{-1}e^{L\tau}UU^{-1}\rho(t)UD_+(t)e^{-i\omega_0\tau}
\end{aligned} \tag{8.176}$$

For simplicity of view, here $U^{-1}\exp(L\tau)U$ and $U^{-1}\rho(t)U$ are still labeled by $\exp(L\tau)$ and $\rho(t)$, but it must be mentioned that $\exp(L\tau)$ and $\rho(t)$ are in the dressed representation in the following. Therefore, (8.176) may be rewritten as

$$\langle S_+(t)S_-(t+\tau)\rangle_S = \mathrm{Tr}_S D_-(t)e^{L\tau}\rho(t)D_+(t)\exp(-i\omega_0\tau) \tag{8.177}$$

In order to calculate (8.177), we define a new operator $\sum(\tau)$ as

$$\Sigma(\tau) = e^{L\tau}\Sigma(0) \tag{8.178}$$

$$\Sigma(0) = \rho(t)D_+(t) \tag{8.179}$$

From (8.178) and (8.179) we see that $\Sigma(\tau)$ may be expressed as

$$\Sigma(\tau) = \begin{pmatrix} \Sigma_{22}(\tau) & \Sigma_{21}(\tau) \\ \Sigma_{12}(\tau) & \Sigma_{11}(\tau) \end{pmatrix} \tag{8.180}$$

By the use of eqs.(8.173), (8.174), (8.178) and (8.180), equation (8.177) may reduce to

$$\langle S_+(t)S_-(t+\tau)\rangle_S = \frac{1}{2}\left[(\Sigma_{22}(\tau) - \Sigma_{11}(\tau)) - (\Sigma_{21}(\tau) - \Sigma_{12}(\tau))\right]e^{-i\omega_0\tau} \tag{8.181}$$

We know from (8.181) that once the matrix elements $\sum_{ij}(\tau)(i,j = 1,2)$ are given, the two-time correlation function $\langle S_+(t)S_-(t+\tau)\rangle$ can be obtained. Differentiating (8.178) with respect to τ, we have

$$\frac{d}{d\tau}\Sigma(\tau) = L\Sigma(\tau) \tag{8.182}$$

Evidently, the equations of motion for $\rho(t)$ (8.162) and $\sum(\tau)$ (8.182) have the same form, so the solutions of $\sum(\tau)$ and $\rho(t)$ have the same form except for their different initial values. Next, we turn to solve the set of equations (8.169).

Defining the vector as

$$\tilde{\rho}(t) = \frac{1}{2}\begin{pmatrix} \rho_{22}(t)+\rho_{11}(t) \\ \rho_{22}(t)-\rho_{11}(t) \\ \rho_{21}(t)+\rho_{12}(t) \\ \rho_{21}(t)-\rho_{12}(t) \end{pmatrix} = \begin{pmatrix} \tilde{\rho}_1(t) \\ \tilde{\rho}_2(t) \\ \tilde{\rho}_3(t) \\ \tilde{\rho}_4(t) \end{pmatrix} \tag{8.183}$$

then (8.169) becomes

$$\begin{aligned} \frac{d}{dt}\tilde{\rho}_1 &= 0 \\ \frac{d}{dt}\tilde{\rho}_2 &= -\frac{\Gamma}{2}\tilde{\rho}_2 \\ \frac{d}{dt}\tilde{\rho}_3 &= -\Gamma\tilde{\rho}_1 - \Gamma\tilde{\rho}_3 - 2i\Omega\tilde{\rho}_4 \\ \frac{d}{dt}\tilde{\rho}_4 &= -\frac{\Gamma}{2}\tilde{\rho}_4 - 2i\Omega\tilde{\rho}_3 \end{aligned} \tag{8.184}$$

The solutions of the above equations can be immediately obtained as

$$\begin{aligned} \tilde{\rho}_1(t) &= \tilde{\rho}_1(0) \\ \tilde{\rho}_2(t) &= \tilde{\rho}_2(0)\exp(-\Gamma t/2) \\ \tilde{\rho}_3(t) &= \frac{i}{2\Omega}(-\Gamma/4+i\alpha)C_1\exp[(-3\Gamma/4+i\alpha)t] - \frac{\Gamma^2/2}{4\Omega^2+\Gamma^2/2}\tilde{\rho}_1(0) \\ &\quad -\frac{i}{2}(\Gamma/4+i\alpha)C_2\exp[(-3\Gamma/4-i\alpha)t] \\ \tilde{\rho}_4(t) &= C_1\exp[(-3\Gamma/4+i\alpha)t] + C_2\exp[(-3\Gamma/4-i\alpha)t] \\ &\quad +\frac{2i\Omega\Gamma}{4\Omega^2+\Gamma^2/2} \end{aligned} \tag{8.185}$$

where

$$\begin{aligned} \alpha &= \sqrt{4\Omega^2-(\gamma/4)^2} \\ C_1 &= \frac{1}{2}\left(1+\frac{i\Gamma}{4\alpha}\right)\tilde{\rho}_4(0) + \frac{\Omega}{\alpha}\tilde{\rho}_3(0) - \frac{i}{2}\frac{\Omega\Gamma}{4\Omega^2+\Gamma^2/2}\left(1+\frac{3i\Omega}{4\alpha}\right)\tilde{\rho}_1(0) \\ C_2 &= \frac{1}{2}\left(1-\frac{i\Gamma}{4\alpha}\right)\tilde{\rho}_4(0) - \frac{\Omega}{\alpha}\tilde{\rho}_3(0) - \frac{i}{2}\frac{\Omega\Gamma}{4\Omega^2+\Gamma^2/2}\left(1-\frac{3i\Omega}{4\alpha}\right)\tilde{\rho}_1(0) \end{aligned}$$

here $\tilde{\rho}_i(0)(i = 1,2,3,4)$ are the parameters related to the matrix elements $\rho_{ij}(0)$

Since $\sum_{ij}(\tau)$ obey the same equation of motion as $\rho_{ij}(t)$, the solutions of $\sum_{ij}(\tau)$ have the same form as (8.185). But it is necessary to point out that $\tilde{\rho}_i(0)$ which related to $\rho_{ij}(0)$ must be replaced by $\sum_{ij}(0)$. Utilizing (8.179), (8.169) and (8.174), $\sum(0)$ is given by

$$\Sigma(0) = \rho(t)D_+(t) = \frac{1}{2}\begin{pmatrix} \rho_{22}(t)+\rho_{21}(t) & -(\rho_{22}(t)+\rho_{21}(t)) \\ \rho_{12}(t)+\rho_{11}(t) & -(\rho_{12}(t)+\rho_{11}(t)) \end{pmatrix} \tag{8.186}$$

Similar to the definition of $\tilde{\rho}(t)$, the vector $\tilde{\Sigma}(t)$ is defined as

$$\tilde{\Sigma}(t) = \frac{1}{2}\begin{pmatrix} \Sigma_{22}(t)+\Sigma_{11}(t) \\ \Sigma_{22}(t)-\Sigma_{11}(t) \\ \Sigma_{21}(t)+\Sigma_{12}(t) \\ \Sigma_{21}(t)-\Sigma_{12}(t) \end{pmatrix} = \begin{pmatrix} \tilde{\Sigma}_1(t) \\ \tilde{\Sigma}_2(t) \\ \tilde{\Sigma}_3(t) \\ \tilde{\Sigma}_4(t) \end{pmatrix} \tag{8.187}$$

the initial value of $\tilde{\Sigma}(\tau)$ gives

$$\begin{aligned} \tilde{\Sigma}(0) &= \frac{1}{4}\begin{pmatrix} \rho_{22}(t)+\rho_{21}(t)-(\rho_{11}(t)+\rho_{12}(t)) \\ \rho_{22}(t)+\rho_{21}(t)+\rho_{11}(t)+\rho_{12}(t) \\ \rho_{11}(t)+\rho_{12}(t)-(\rho_{22}(t)+\rho_{21}(t)) \\ -(\rho_{22}(t)+\rho_{12}(t)+\rho_{21}(t)+\rho_{11}(t)) \end{pmatrix} \\ &= \frac{1}{2}\begin{pmatrix} \tilde{\rho}_2(t)+\tilde{\rho}_4(t) \\ \tilde{\rho}_1(t)+\tilde{\rho}_3(t) \\ -\tilde{\rho}_2(t)-\tilde{\rho}_4(t) \\ -\tilde{\rho}_1(t)-\tilde{\rho}_3(t) \end{pmatrix} \end{aligned} \tag{8.188}$$

According to (8.185), $\tilde{\Sigma}_2(t)$ and $\tilde{\Sigma}_4(t)$ can be expressed as

$$\tilde{\Sigma}_2(\tau) = \tilde{\Sigma}_2(0)\exp(-\Gamma\tau/2) \tag{8.189}$$

$$\begin{aligned} \tilde{\Sigma}_4(\tau) &= E_1(0)\exp[(-3\Gamma/4+i\alpha)\tau] + E_2(0)\exp[(-3\Gamma/4-i\alpha)\tau] \\ &\quad + \frac{2i\Omega\Gamma}{4\Omega^2+\Gamma^2/2}\tilde{\Sigma}_1(0) \end{aligned} \tag{8.190}$$

with

$$E_1 = \frac{1}{2}\left(1+\frac{i\Gamma}{4\alpha}\right)\tilde{\Sigma}_4(0) + \frac{\Omega}{\alpha}\tilde{\Sigma}_3(0) - \frac{i}{2}\frac{\Omega\Gamma}{4\Omega^2+\Gamma^2/2}\left(1+i\frac{3\Omega}{4\alpha}\right)\tilde{\Sigma}_1(0)$$

$$E_2 = \frac{1}{2}\left(1-\frac{i\Gamma}{4\alpha}\right)\tilde{\Sigma}_4(0) - \frac{\Omega}{\alpha}\tilde{\Sigma}_3(0) - \frac{i}{2}\frac{\Omega\Gamma}{4\Omega^2+\Gamma^2/2}\left(1-i\frac{3\Omega}{4\alpha}\right)\tilde{\Sigma}_1(0)$$

From eqs.(8.102)-(8.108) we know that the measurement result is the steady fluorescence distribution because of the observation time $t \gg \Gamma^{-1}$. So we obtain $\tilde{\Sigma}_i(0)$ as

$$\tilde{\Sigma}_1(0) = \frac{1}{2}(\tilde{\rho}_2(\infty) + \tilde{\rho}_4(\infty)) = \frac{i\Omega\Gamma}{4\Omega^2 + \Gamma^2/2}\tilde{\rho}_1(0) = -\tilde{\Sigma}_3(0) \tag{8.191}$$

$$\tilde{\Sigma}_2(0) = \frac{4\Omega^2}{8\Omega^2 + \Gamma^2}\tilde{\rho}_1(0) = -\tilde{\Sigma}_4(0) \tag{8.192}$$

where

$$\tilde{\rho}_1(0) = \frac{1}{2}(\rho_{22}(0) + \rho_{11}(0)) = \frac{1}{2} \tag{8.193}$$

In the strong laser field case, that is, $\Omega^2/\Gamma^2 \gg 1$, (8.189) and (8.190) reduce to

$$\tilde{\Sigma}_2(\tau) = \frac{1}{4}\exp(-\Gamma\tau/2) \tag{8.194}$$

$$\begin{aligned}\tilde{\Sigma}_4(\tau) = &-\frac{1}{8}\exp[(-3\Gamma/4 + 2i\Omega)\tau] - \frac{1}{8}\exp[(-3\Gamma/4 - 2i\Omega)\tau] \\ &-\frac{1}{4}\frac{\Gamma^2}{4\Omega^2 + \Gamma^2/2}\end{aligned} \tag{8.195}$$

Substituting (8.194) and (8.195) into (8.181), we have

$$\begin{aligned}\langle S_+(t)S_-(t+\tau)\rangle_S &= (\tilde{\Sigma}_2(\tau) - \tilde{\Sigma}_4(\tau))\exp(-i\omega_0\tau) \\ &= \left\{\frac{1}{4}e^{-\frac{\Gamma}{2}t} + \frac{1}{8}e^{(-\frac{3}{4}\Gamma + 2i\Omega)\tau} + \frac{1}{8}e^{(-\frac{3}{4}\Gamma - 2i\Omega)\tau} + \frac{\Gamma^2/2}{8\Omega^2 + \Gamma^2}\right\}e^{-i\omega_0\tau}\end{aligned} \tag{8.196}$$

This is exactly the steady two-time correlation function. Inserting (8.196) into (8.74) and proceeding on similar procedures of eqs.(8.102) -(8.107), then have

$$\begin{aligned}\langle a_k^\dagger a_k\rangle &= 2\varepsilon_k^2 \mathrm{Re}\int_0^t dt \int_0^\infty d\tau \langle S_+(t)S_-(t+\tau)\rangle_S e^{i\omega_k\tau} \\ &= 2\varepsilon_k^2 t \mathrm{Re}\int_0^\infty d\tau \left\{\frac{1}{4}\exp\left[-\frac{\Gamma}{2} + i(\omega_k - \omega_0)\tau\right]\right. \\ &+\frac{1}{8}\exp\left[-\frac{3}{4}\Gamma + i(\omega_k - \omega_0 + 2\Omega)\tau\right] + \frac{1}{8}\exp\left[-\frac{3}{4}\Gamma + i(\omega_k - \omega_0 - 2\Omega)\tau\right] \\ &\left.+\frac{\Gamma^2}{2}\frac{1}{8\Omega^2 + \Gamma^2}e^{i(\omega_k - \omega_0)\tau}\right\}\end{aligned}$$

$$= \frac{t\Gamma}{4}\varepsilon_k^2 \left\{ \frac{1}{4}\frac{3}{(\omega_k-\omega_0+2\Omega)^2+(3\Gamma/4)^2} + \frac{1}{(\omega_k-\omega_0)^2+(\Gamma/2)^2} \right.$$
$$\left. + \frac{1}{4}\frac{3}{(\Omega_k-\omega_0-2\Omega)^2+(3\Gamma/4)^2} + \frac{4\Gamma^2}{8\Omega^2+\Gamma^2}2\pi\delta(\omega_k-\omega) \right\} \tag{8.197}$$

Therefore, the emission intensity of the fluorescent field gives

$$R_k = \frac{d}{dt}\langle a_k^\dagger a_k \rangle = \frac{\Gamma}{4}\varepsilon_k^2 \left\{ \frac{1}{4}\frac{3}{(\omega_k-\omega_0+2\Omega)^2+(3\Gamma/4)^2} \right.$$
$$+ \frac{1}{(\omega_k-\omega_0)^2+(\Gamma/2)^2} + \frac{1}{4}\frac{3}{(\omega_k-\omega_0-2\Omega)^2+(3\Gamma/4)^2}$$
$$\left. + \frac{4\Gamma^2}{8\Omega^2+\Gamma^2}2\pi\delta(\omega_k-\omega_0) \right\} \tag{8.198}$$

Since Γ in (8.198) (defined by (8.160)) and γ in (8.110) (defined by (8.87)) obey

$$\Gamma = 2\gamma \tag{8.199}$$

the first three terms in (8.198) are coincided with (8.110). This means that the description of the resonance fluorescence by the intensity matrix theory is agreement with the experimental result.

It is worthy pointing out that the last term in (8.198) represents the processes of the laser photons scattered elastically into the fluorescent field by the atom. In this moment, the atom is not dressed by the laser field, so this term corresponds to the elastically scattering processes of the weak laser field which is absorbed and emitted by the bare atom. This term is absent in (8.110), because we have neglected the terms $\ell \neq m$ in (8.88). In the strong laser field case, i.e., $\Omega^2 \gg \Gamma^2$, the last term in (8.198) is an infinitesimal in comparison with the first three terms, so as an approximation this term can be ignored. Therefore, in the strong field limit, (8.198) is complete agreement with (8.110).

References

1. B.R.Mollow, *Phys.Rev.* **188** (1969) 1969.

2. F.Schuda, C.E.Ju Stroud, and M.Hercher, *J.Phys.* **B7** (1974) L198.

3. H.J.Carmichael and D.F.Walls, **J.Phys. B9** (1976) 1199.

4. H.J.Kimble and L.Mandel, *Phys.Rev.* **A13** (1976) 2123.

5. H.J.Kimble, M.Dagenis and L.Mandel, *Phys.Rev.Lett.* **39** (1977) 691.
6. C.Cohen-Tannoudji and S.Reynaud, *J.Phys.* **B10** (1977) 365.
7. C.Cohen-Tannoudji and S.Reynaud, *J.Phys.* **B10** (1977) 2311.
8. G.Compagno and F.Persico, *Phys.Rev.* **A22** (1980) 2108.
9. G.Compagno and F.Persico, *Phys.Rev.* **A25** (1982) 3138.
10. G.Compagno, J.S.Peng and F.Persico, *Phys.Rev.* **A26** (1982) 2065.
11. G.Compagno, J.S.Peng and F.Persico, *Phys.Lett.* **A88** (1982) 285.
12. Z.Fick, R.Tanas and S.Kielish, *Phys.Rev.* **A29** (1984) 2004.
13. C.Leonardi, F.Persico and G.Vetri, *Dicke model and theory of driven and spontaneous emission*, (La Rivista de Nuovo Cimento, 1986).
14. J.S.Peng, *Acta Phys. Sinica* **34** (1986) 788; *Chin.Phys.* **6** (1986) 941.
15. P.T.H.Fisk, H.A.Bochor and R.J.Sandermen, *Phys.Rev.* **A34** (1986) 2418.
16. H.J.Carmichel, A.S.Lane and D.F.Walls, *Phys.Rev.Lett.* **58** (1987) 2539.
17. J.S.Peng, *Science Sinica* A No.3 (1988) 331.
18. B.N.Jagatap, Q.V.Lawande and S.V.Lawande, *Phys.Rev.* **A43** (1991) 535.
19. W.Vogel and R.Blatt, *Phys.Rev.* **A45** (1992) 3319.
20. R.Vyas and S.Singh, *Phys.Rev.* **A45** (1992) 8095.
21. T.Quang and F.Helen, *Phys.Rev.* **A47** (1993) 2285.
22. H.Freedhoff and T.Quang, *Phys.Rev.Lett.* **72** (1994)474.
23. S.Swain, *Phys.Rev.Lett.* **73** (1994) 1493.
24. A.Banerjce, *Phys.Rev.* **A52** (1995) 2472.
25. J.S.Peng, *Resonance Fluorescence and Superfluorescence*, (Science Press, Beijing, 1993).

26. P.Kochan, H.J.Carmichael, P.R.Morrow, and M.G.Raizen, *Phys.Rev.Lett.* **75** (1995) 45.

CHAPTER 9

SUPERFLUORESCENCE

Superfluorescence (or named as superradiance) is one of the important subjects in quantum optics, and an elementary quantum effect in optics. Since the superfluorescence has the typicality in theoretical treatment and the application prospect in practice, it has been widely studied in quantum optics over the past twenty years. This chapter is concentrated on this subject. First we introduce the basic features of the superfluorescence, then discuss the quasi-classical description of the superfluorescence. Secondly, we give a statement of a quantum theoretical description for superfluorescence. Finally, we deal with the superfluorescent beats in Dicke model.

9.1 Elementary features of superfluorescence

Superfluorescence origins from the cooperative spontaneous emission of the multi-atom system. Generally speaking, if N atoms in their excited state, which do not couple to each other, decay independently to their ground states, the total intensity I of the radiation field is proportional to the number N of atoms. The radiation field is spatially isotropic and the emission intensity I decays with time t exponentially as shown in Fig.(9.1), the decay time is just the lifetime τ_{sp} of a single-atom. This is just the observation results in the rare gas of atoms. If the density of excited atoms is so dense that the distances between the neighbourhood atoms are very short, then the multi-atom system can form a cooperative spontaneous emission system due to the coupling of radiation field. One characteristic of the cooperative emission is that its radiation exhibits pulse-shape as shown in Fig.(9.2). The intensity of the peak is very high and proportional to the square of the atomic number N^2, this peak attenuates very fast and the peak width T_R has the magnitude of τ_{sp}/N. When the laser pumping for exciting the atoms is switched off, the peak of pulse appears

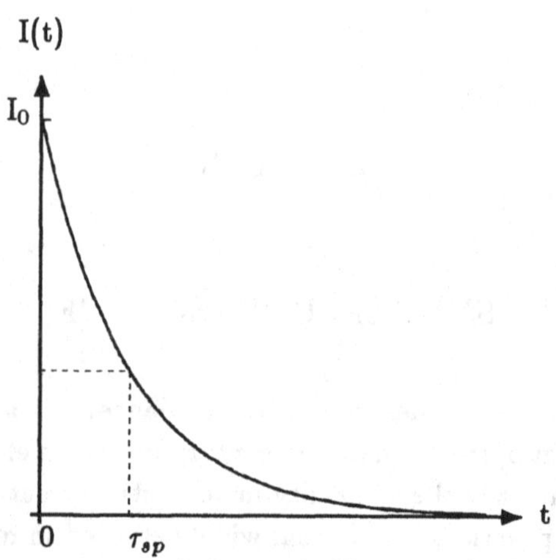

Figure 9.1: Spontaneous emission of N excited isolated atoms and time-dependence of the emission intensity I

after a delay time T_D, which is approximately equal to $\frac{1}{N}\tau \ln N$, but the value of T_D is uncertainty. The other feature of the superfluorescence is to have a directional property, for a pencil-shape sample, the superfluorescence is only around a small solid angle along the axial direction as shown in Fig.(9.2b).

Superfluorescence and its basic characteristics were first revealed by Dicke in 1954. when he discussed theoretically the cooperative spontaneous emission of a multi-atom system. Because the experimental conditions for creating cooperative emission of the multi-atom system is very difficult to achieve in early optical technology, the superfluorescence was not observed experimentally for about twenty years later. In 1973, Skribanowitz and his co-workers observed this phenomenon experimentally. A typical experimental device of the superfluorescence is shown in Fig.(9.3a), where HF gas is full of a cylinder cavity without reflection mirrors, and the HF atoms are excited from their ground states $|g\rangle$ to excited states $|e\rangle$ owing to the pumping of the laser field. Subsequently, the atoms spontaneously decay from $|e\rangle$ to the third state (final state) $|f\rangle$. Because there are no reflection mirrors, there is no feedback in the cavity. The pulse signal is detected by the detector. The experimental results as shown in Fig.(9.4) are coincided with Dicke's prediction. The superfluorescence pulse

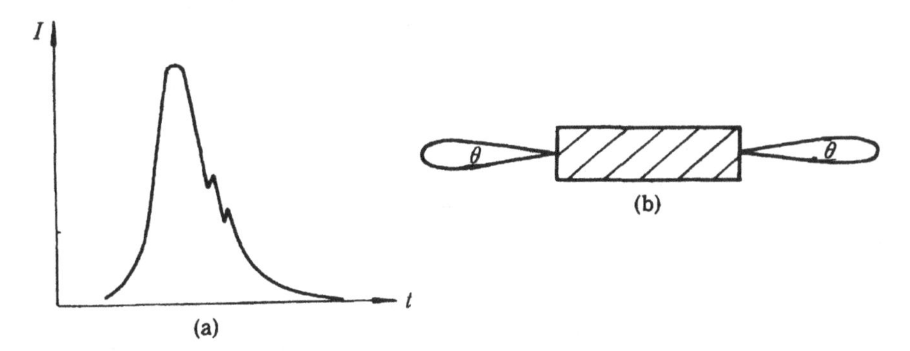

Figure 9.2: Diagrammatic illustration of superfluorescence emission, (a) Time evolution of emission intensity I; (b). Distribution of the emission along the axis of cavity

is observed by the detector, the intensity of the pulse is much stronger than that of ordinary spontaneous emission, it is proportional to the square of the atomic number N. This pulse appears after switching off the laser pumping for a certain time T_D, which is called the delay time of superfluorescence. The duration T_R of the pulse is very narrow and satisfies $T_R \ll T_D$, but both T_R and T_D are much smaller than the lifetime τ_{sp} of the ordinary spontaneous emission. In addition, the output pulse has very good orientation property. For a pencil sample, the output pulse is just along the pencil direction both front and back in a small solid angle λ^2/A, here A is the cross section of the pencil sample, and λ is the wavelength of radiation field. How to explain these properties of superfluorescence, we will deal with it by means of both quasi-classical theory and quantum electrodynamics in the following.

Before proceeding the theoretical treatments, first we analyze the physical process of generating superfluorescence. The reason why the spontaneous emission of collective atoms radiates the superfluorescence is that the coupling of collective atoms to the radiation field leads to the coherent decay of collective atoms. For a complete population inversion system (all the atoms are in their excited states), there is no macroscopic polarization at initial time t=0, the system is in sub-steady state. But a small perturbation electric field may be induced by the influences of the spontaneous emission or thermal radiation of environment, this small electric field can lead to the medium to be polarized

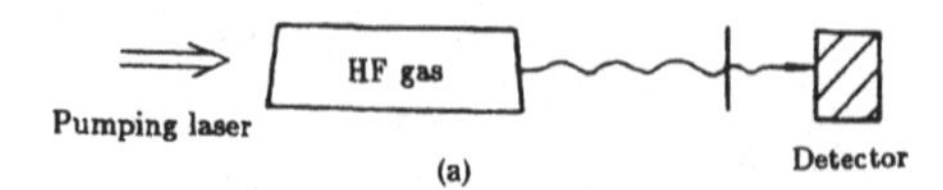

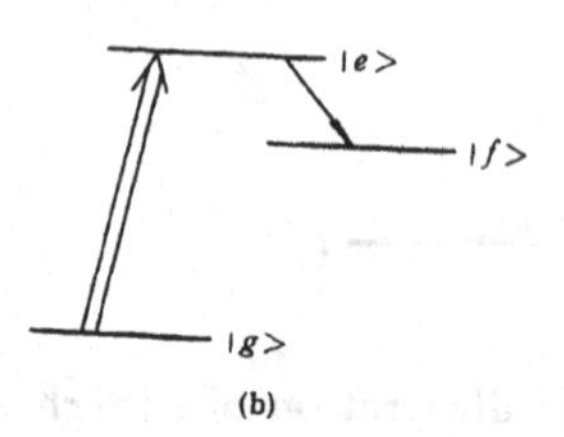

Figure 9.3: Schematic illustration of superfluorescence experimental apparatus, (a). Illustration of experimental apparatus, (b). Energy levels diagram of pumping and transition

slightly. The small polarization acts as a polarization field sources and results in the medium progressively to form a macroscopic polarization, in which the correlation among atoms become very strong, the total system is therefore macroscopic polarized. In the macroscopic system, there exists strong correlation among the atomic dipole moments, this leads to the phase of atomic dipole moments to be equal. In this case, the system radiates superfluorescence pulse with maximum intensity, the peak value of pulse is proportional to the square of the macroscopic dipole moment, so the output superfluorescence intensity I is proportional to N^2. The density of the population inversion will immediately decrease after emitting the superfluorescence pulse, and the atoms are rapidly back to their ground states, the polarization is certainly destroyed. If the pencil sample is sufficiently long, some of the destroyed polarization part may be again polarized by radiation field from the rest part, then emits other superfluorescence pulse, but the intensity of this pulse is much weaker than that of the previous one. So the output superfluorescence pulse exhibits attenuation oscillation as shown in Fig.(9.5). We call this phenomenon as "ring" effect of the superfluorescence.

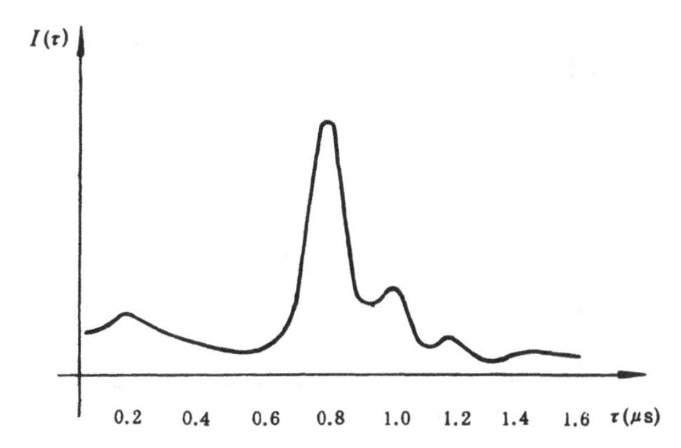

Figure 9.4: Superfluorescence pulse observed by experiment

9.2 Quasi-classical description of superfluorescence

Now we discuss the collective decay of N atoms in the absence of external field by use of the quasi-classical theory. As usual, the quasi-classical theory is referred to that the field is characterized by a classical quantity and the atom is described by a set of quantum operators when we discuss the system of atoms coupling to a radiation field. Here we assume that there are N two-level atoms with frequency $\omega_\ell(\ell = 1, 2, \cdots, N)$, the interaction between the ℓ-th atom and the radiation field is characterized by the Hamiltonian

$$H^{(\ell)} = \frac{1}{2}\hbar\omega_\ell\sigma_3^{(\ell)} - \mathbf{D}^{(\ell)} \cdot \mathbf{E}^{(\ell)} \tag{9.1}$$

The first term in the right hand of (9.1) represents the atomic energy, the second one describes the energy of the interaction between the ℓ-th atom with dipole moment $\mathbf{D}^{(\ell)}$ and the field $\mathbf{E}^{(\ell)}$ located at the position of the ℓ-th atom. The expectation value of the atomic dipole moment is

$$\mathbf{D}^{(\ell)}(t) = \langle e\mathbf{q}_\ell\rangle(t) = \mathbf{n}D^{(\ell)}(t) \tag{9.2}$$

here $\mathbf{n}$ is the unit directional vector of the dipole moment. The field $\mathbf{E}^{(\ell)}(t)$ at the position of the ℓ-th atom is the contribution of fields generated by all N atoms, i.e.,

$$\mathbf{E}^{(\ell)}(t) = \sum_{m=1}^{N} \mathbf{E}_{\ell m}(t) \tag{9.3}$$

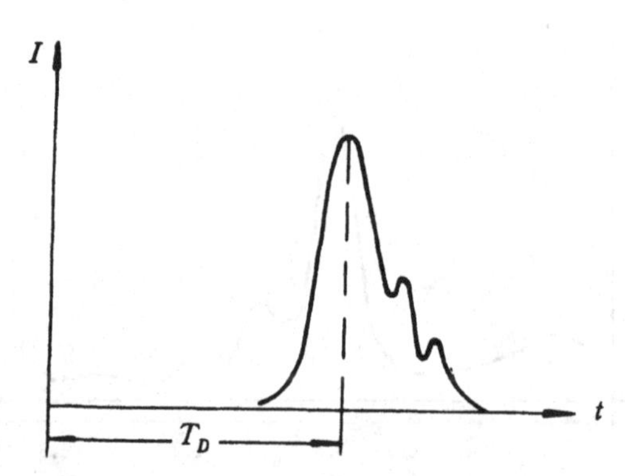

Figure 9.5: Ring effect in superfluorescence pulse

Note in the classical electrodynamics, the field at the position of the ℓ-th atom contributed by the N-1 dipoles except the ℓ-th one is expressed as

$$\mathbf{E}_{\ell m}(t) = \left\{ 3\frac{(\mathbf{n}\cdot\mathbf{R}_{\ell m})}{R_{\ell m}^5}\mathbf{R}_{\ell m} - \frac{\mathbf{n}}{\mathbf{R}_{\ell m}^3} \right\} D^{(m)} - \frac{2\omega_m^2}{3c^3}\mathbf{n}\frac{d}{dt}D^{(m)} \tag{9.4}$$

with

$$\mathbf{R}_{\ell m} = \mathbf{r}_\ell - \mathbf{r}_m, \qquad R_{\ell m} = |\mathbf{R}_{\ell m}|$$

and the field generated by the ℓ-th atom itself is written as

$$\mathbf{E}_{\ell\ell} = \frac{4\omega_\ell^2 k_{max}}{3\pi c^3}\mathbf{n}D^{(\ell)} - \frac{2\omega_\ell^2}{3c^3}\mathbf{n}\frac{d}{dt}D^{(\ell)} \tag{9.5}$$

here k_{max} is the cut-off frequency of the field. For an atom with Bohr radius a_0, it radiates the field with minimum wavelength $\lambda \approx 2\pi a_0$. Using $\lambda\omega = c$, then the cut-off frequency gives by $k_{max} \approx c/a_0$.

We now discuss the time-evolution of the ℓ-th atom. If the atomic dipole operator $\mathbf{D}^{(\ell)}$ in (9.1) is chosen as

$$\mathbf{D}^{(\ell)} = \mathbf{n}d\sigma_1^{(\ell)} \tag{9.6}$$

then (9.1) can be rewritten as

$$\begin{aligned} H_I &= -\mathbf{n}d\sigma_1^{(\ell)} \cdot \mathbf{E}^{(\ell)} \\ H_A &= \frac{1}{2}\hbar\omega_\ell\sigma_3^{(\ell)} \end{aligned} \tag{9.7}$$

In the interaction picture, the time-dependence of the ℓ-th atom obeys the Schrödinger equation

$$i\hbar\frac{\partial}{\partial t}|\Psi^I(t)\rangle = \tilde{H}_I^{(\ell)}|\Psi^I(t)\rangle \tag{9.8}$$

where

$$\begin{aligned}\tilde{H}_I^{(\ell)} &= \exp(iH_A t/\hbar)H_I\exp(-iH_A t/\hbar)\\ &= -\mathbf{n}d\cdot\mathbf{E}^{(\ell)}\exp(i\omega_1\sigma_3^{(\ell)}t/2)\sigma_1^{(\ell)}\exp(-i\omega_1\sigma_3^{(\ell)}t/2)\\ &= -d\mathbf{n}\cdot\mathbf{E}^{(\ell)}[\sigma_+^{(\ell)}\exp(i\omega_\ell t)+\sigma_-^{(\ell)}\exp(-i\omega_\ell t)]\\ &= -d\mathbf{n}\cdot\mathbf{E}^{(\ell)}[\sigma_1^{(\ell)}\cos\omega_\ell t-\sigma_2^{(\ell)}\sin\omega_\ell t]\end{aligned} \tag{9.9}$$

If we let the expectation values of atomic operators $S_i^{(\ell)}(t) = \langle\sigma_i^{(\ell)}(t)\rangle$, then from the Heisenberg equation we obtain

$$\begin{aligned}\frac{d}{dt}S_1^{(\ell)}(t) &= -\frac{2}{\hbar}d\mathbf{n}\cdot\mathbf{E}^{(\ell)}\sin\omega_\ell t S_3^{(\ell)}(t)\\ \frac{d}{dt}S_2^{(\ell)}(t) &= -\frac{2}{\hbar}d\mathbf{n}\cdot\mathbf{E}^{(\ell)}\cos\omega_\ell t S_3^{(\ell)}(t)\\ \frac{d}{dt}S_3^{(\ell)}(t) &= -\frac{2}{\hbar}d\mathbf{n}\cdot\mathbf{E}^{(\ell)}\cos\omega_\ell t S_2^{(\ell)}+\frac{2}{\hbar}d\mathbf{n}\cdot\mathbf{E}^{(\ell)}\sin\omega_\ell t S_1^{(\ell)}(t)\end{aligned} \tag{9.10}$$

Therefore, the Bloch vector $\mathbf{S}^{(\ell)}(t) = (S_1^{(\ell)}, S_2^{(\ell)}, S_3^{(\ell)})$ of the ℓ-th atom obeys the Bloch equation

$$\frac{d}{dt}\mathbf{S}^{(\ell)}(t) = \mathbf{\Omega}^{(\ell)}\times\mathbf{S}^{(\ell)}(t) \tag{9.11}$$

where the rotation vector $\mathbf{\Omega}^{(\ell)}$ is defined by

$$\mathbf{\Omega}^{(\ell)}(t) = -\frac{2}{\hbar}d\mathbf{n}\cdot\mathbf{E}^{(\ell)}(\cos\omega_\ell t\varepsilon_x - \sin\omega_\ell t\varepsilon_y) \tag{9.12}$$

ε_x, ε_y, and ε_z are the unit vectors of $\mathbf{\Omega}^{(\ell)}(t)$. $\mathbf{E}^{(\ell)}(t)$ represents the total field at the position of the ℓ-th atom. In general case, it is the summation of (9.4) and (9.5). Since the induced field part is comparatively weaker than the radiation part, we may neglect the induced field part and remain the radiation part, then the field can be approximately written as

$$\mathbf{E}^{(\ell)}(t) = \frac{2}{3c^3}\mathbf{n}\sum_m\omega_m^2\frac{d}{dt}D^{(m)}(t) \tag{9.13}$$

Inserting (9.13) into (9.10) gives

$$\frac{d}{dt}\sigma_1^{(\ell)} = -\frac{2}{\hbar}\mathbf{dn}\cdot\frac{2\mathbf{n}}{3c^3}\sum_m \omega_m^2 D^{(m)}\sin\omega_\ell t S_3^{(\ell)}$$
$$= -\frac{4d}{3\hbar c^3}S_3^{(\ell)}\sum_m \omega_m^2 d[\sigma_+^{(m)}\exp(i\omega_m t) - \sigma_-^{(m)}\exp(-i\omega_m t)]$$
$$\times\frac{1}{2i}[\exp(i\omega_\ell t) - \exp(-i\omega_\ell t)]i\omega_m$$
$$= -\frac{4d^2}{3\hbar c^3}S_3^{(\ell)}\sum_m \frac{\omega_m^3}{2}\{\sigma_+^{(m)}\exp[i(\omega_m+\omega_\ell)t]$$
$$-\sigma_-^{(m)}\exp[i(-\omega_m+\omega_\ell)t] - \sigma_+^{(m)}\exp[i(\omega_m-\omega_\ell)t]$$
$$+\sigma_-^{(m)}\exp[-i(\omega_m+\omega_\ell)t]\} \tag{9.14}$$

After adopting the rotating-wave approximation and replacing the operators $\sigma_i(t)$ by the corresponding expectation values $S_i(t)$, the above equation therefore reduces to

$$\frac{d}{dt}S_1^{(\ell)} = \frac{4d^2}{3\hbar c^3}S_3^{(\ell)}\sum_m \frac{\omega_m^3}{2}[S_1^{(m)}\cos(\omega_\ell-\omega_m)t + S_2^{(m)}\sin(\omega_\ell-\omega_m)t] \tag{9.15}$$

Following the same procedure, we have

$$\frac{d}{dt}S_2^{(\ell)} = \frac{4d^2}{3\hbar c^3}S_3^{(\ell)}\sum_m \frac{\omega_m^3}{2}[S_1^{(m)}\sin(\omega_\ell-\omega_m)t - S_2^{(m)}\cos(\omega_\ell-\omega_m)t]$$

$$\frac{d}{dt}S_3^{(\ell)} = \frac{4d^2}{3\hbar c^3}\left\{S_2^{(\ell)}\sum_m \frac{\omega_m^3}{2}[S_1^{(m)}\sin(\omega_\ell-\omega_m)t - S_2^{(m)}\cos(\omega_\ell-\omega_m)t]\right.$$
$$\left.-S_1^{(\ell)}\sum_m \frac{\omega_m^3}{2}[S_1^{(m)}\cos(\omega_\ell-\omega_m)t + S_2^{(m)}\sin(\omega_\ell-\omega_m)t]\right\}$$

If all the N atoms have the equal resonant frequency, i.e., $\omega_\ell = \omega_m$, then (9.15) reduce to

$$\frac{d}{dt}S_1^{(\ell)} = \frac{1}{2}\gamma S_3^{(\ell)}\sum_m S_1^{(m)}$$
$$\frac{d}{dt}S_2^{(\ell)} = \frac{1}{2}\gamma S_3^{(\ell)}\sum_m S_2^{(m)} \tag{9.16}$$
$$\frac{d}{dt}S_3^{(\ell)} = -\frac{1}{2}\gamma\left\{S_2^{(\ell)}\sum_m S_2^{(m)} + S_1^{(ell)}\sum_m S_1^{(m)}\right\}$$

with

$$\gamma = \frac{4d^2}{3\hbar c^3}\omega_0^3 \tag{9.17}$$

Summing eqs.(9.16) with respect to ℓ and introducing the collective atomic operator

$$\mathbf{S} = \sum_{\ell=1}^{N} \mathbf{S}^{(\ell)} \tag{9.18}$$

we have

$$\begin{aligned} \frac{d}{dt}S_1 &= \frac{1}{2}\gamma S_3 S_1 \\ \frac{d}{dt}S_2 &= \frac{1}{2}\gamma S_3 S_2 \\ \frac{d}{dt}S_3 &= -\frac{1}{2}\gamma(S_1^2 + S_2^2) \end{aligned} \tag{9.19}$$

here S_1, S_2, and S_3 are three components of the collective atomic operator $\mathbf{S}$. (9.19) describes the time evolution of the total Bloch vector of N atoms with same resonant frequency under the interaction of the field produced by the N atoms themselves. It is easily proved that $\mathbf{S}^{(\ell)} \cdot \mathbf{S}^{(m)}$ is a constant of motion. This is because

$$\begin{aligned} \frac{d}{dt}(\mathbf{S}^{(\ell)} \cdot \mathbf{S}^{(m)}) &= \frac{d}{dt}\mathbf{S}^{(\ell)} \cdot \mathbf{S}^{(m)} + \mathbf{S}^{(\ell)} \cdot \frac{d}{dt}\mathbf{S}^{(m)} \\ &= (\mathbf{\Omega}^{(\ell)} \times \mathbf{S}^{(m)}) \cdot \mathbf{S}^{m} + \mathbf{S}^{(\ell)} \cdot (\mathbf{\Omega}^{(m)} \times \mathbf{S}^{(m)}) \\ &= (\mathbf{\Omega}^{(\ell)} - \Omega^{(m)}) \cdot (\mathbf{S}^{(\ell)} \times \mathbf{S}^{(m)}) \end{aligned}$$

It is known from (9.12) and (9.13) that for $\omega_\ell = \omega_m$, $\mathbf{\Omega}^{(\ell)} = \mathbf{\Omega}^{(m)}$, so the above equation becomes

$$\frac{d}{dt}(\mathbf{S}^{(\ell)}(t) \cdot \mathbf{S}^{(m)}(t)) = 0 \tag{9.20}$$

Therefore, the $\mathbf{S}^{(\ell)} \cdot \mathbf{S}^{(m)}$ is a constant of motion. This means that

$$\mathbf{S}^2(t) = \sum_{\ell,m} \mathbf{S}^{(\ell)}(t) \cdot \mathbf{S}^{(m)}(t) = \sum_{\ell,m} \mathbf{S}^{(\ell)}(0) \cdot \mathbf{S}^{(m)}(0) = \mathbf{S}^2(0) \tag{9.21}$$

That is to say, the magnitude of the total Bloch vector describing the N two-level atoms with same resonant frequency is decided by the initial directions of the N independent Bloch vectors $\mathbf{S}^{(\ell)}(0)$. And the length of $\mathbf{S}$ can be assumed to be arbitrary in the range from 0 to N. If all the N atoms are initially assumed to be polarized and all the N Bloch vectors are assumed to be along the same direction, then at t=0, the magnitude of the total Bloch vector $\mathbf{S}$ is proportional to N, i.e., $S \sim N$. In this case, the intensity of the coherent decay is proportional to the square of the amplitude of the Bloch vector $\mathbf{S}$, there displays a maximum intensity of pulse, this phenomenon is termed as superradiance. It is worth to point out that as mentioned previously, the appearance of evident polarization in atomic system needs a delay self-revival time T_D. Bonifacio et al. call the cooperative spontaneous emission in which the atoms starting from the completely polarized state as superradiance, and the cooperative spontaneous emission with a delay time T_D as superfluorescence. Since they are referred to a same physical process in physical origin and the practical cooperative spontaneous emission system must has the delay time, we may unify to interpret these phenomena as superfluorescence.

For the case of $S = N$, we have

$$S_1^2 + S_2^2 = N^2 - S_3^2 \tag{9.22}$$

Inserting the above equation into (9.19) gives

$$\frac{d}{dt}S_3 = -\frac{\gamma}{2}(N^2 - S_3^2) \tag{9.23}$$

Because the radiation intensity $I(t)$ of the atomic system is proportional to the derivative of the atomic energy, then

$$I(t) = -\frac{d}{dt}\sum_{\ell=1}^{N}\langle H_A^{(\ell)}\rangle = -\frac{1}{2}\hbar\omega_0\frac{d}{dt}S_3 = \frac{1}{4}\hbar\omega_0\gamma(N^2 - S_3^2) \tag{9.24}$$

The solution of (9.24) is

$$S_3(t) = -N\tanh\left(\frac{t-t_0}{T_R}\right) \tag{9.25}$$

with

$$T_R = \left(\frac{N\gamma}{2}\right)^{-1} \tag{9.26}$$

Substituting (9.25) into (9.24), the radiation intensity becomes

$$I(t) = -\frac{1}{2}\hbar\omega_0\frac{d}{dt}S_3(t) = I_0\frac{N^2}{4}sech^2\left(\frac{t-t_0}{T_R}\right) \tag{9.27}$$

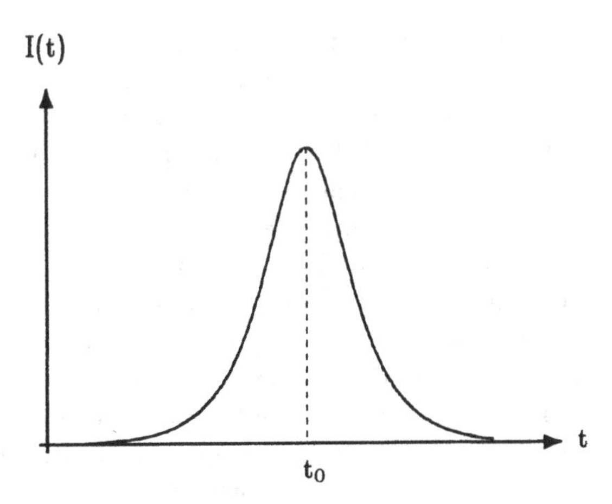

Figure 9.6: Time evolution of the emission intensity I described by (9.27)

Fig.(9.6) plots the curve for $I(t)$ versus t.

When $t = t_0$, the radiation intensity $I(t)$ reaches its maximum value

$$I(t_0) = \frac{N^2 I_0}{4} \tag{9.28}$$

$I(t_0)$ is proportional to N^2, so t_0 in (9.27) is usually called the delay time T_D of the superfluorescence pulse. The magnitude of T_D is much larger than the decay time T_R of the pulse, i.e., $T_R \ll T_D$.

The above main results obtained by the quasi-classical theory are coincident with the experimental results. However, it must be pointed out that the quasi-classical theory has its limitations when it is used to explain initial process of the superfluorescence radiation. Since

$$S^2(t) = S^2(0) = S_1^2(0) + S_2^2(0) + S_3^2(0) \tag{9.29}$$

is a conserving quantity independent of time, for the complete polarization system, $S^2(0) = N$. From (9.22) and (9.25), the delay time obeys the following equation

$$S_1^2(0) + S_2^2(0) = N^2 - S_3^2(0) = N^2 \left[1 - \tanh^2\left(\frac{T_D}{T_R}\right)\right] \tag{9.30}$$

so the value of T_D is determined by the initial condition of the system, especially determined by the absolute value of the transverse component of the total Bloch vector. If the initial dipole moment is very small, i.e., $S_1^2(0)+S_2^2(0) \ll 1$, then from (9.30) we see that the value of the delay time T_D is very large. That is to say, the complete polarization of the atomic system needs a very long time. When $S_1^2(0) + S_2^2(0) = 0$, $T_D \to \infty$, this means that there exists a sub-stable state, in which the system does not decay and no superfluorescence pulse appears. Evidently this theoretical result is incorrect. As mentioned previously, when the atoms are excited sufficiently, the vacuum fluctuations and thermal field effects can lead to the atomic system to be polarized, and then the atomic system radiates the superfluorescence pulse. So there does not exist the undamped state with infinite delay time T_D. That is to say, the quasi-classical theory can not explain the quantum fluctuations in starting process well. To explain this phenomenon well, the quantum theory must be adopted. Next we introduce the quantum theory of superfluorescence.

9.3 Quantum theoretical description of superfluorescence

9.3.1 Heisenberg equation of the system

In the quasi-classical theory, the radiation field is treated as a classical quantity. Now we turn to the quantum description of superfluorescence, so the radiation field must be treated as a quantized one. The Hamiltonian describing the system of N two-level atoms with different resonant frequencies ω_ℓ under the interaction of radiation field is written in the form

$$H = \sum_\ell \hbar\omega_\ell S_z^{(\ell)} + \hbar \sum_k \omega_k a_k^\dagger a_k + \sum_k \varepsilon_k (a_k S_+^{(\ell)} + a_k^\dagger S_-^{(\ell)}) \tag{9.31}$$

Here the rotating-wave approximation is used and the atom-field coupling constant is assumed to be independent of the different atom. If we assume that the atomic resonant frequencies are distributed in the vicinity of ω_0, then the above equation can be rewritten as

$$H = H_0 + \Delta + V \tag{9.32}$$

with

$$\begin{aligned} H_0 &= \hbar \sum_k \omega_k a_k^\dagger a_k + \hbar\omega_0 \sum_\ell S_z^{(\ell)} \\ \Delta &= \hbar \sum_\ell (\omega_\ell - \omega_0) S_z^{(\ell)} \end{aligned} \tag{9.33}$$

$$V = \sum_k \varepsilon_k \left(a_k S_+^{(\ell)} + a_k^\dagger S_-^{(\ell)} \right)$$

From (9.32), we can discuss the time evolution of atomic operators. Here we adopt the normal ordering for the operators. The normal ordering is referred to that the creation operator is placed in the left-side of the product of operators, and the annihilation operator is placed in the right-side of the product of operators. The order between operators can be changed through the commutation relation of operators. Generally, the atomic operators and the field operators commutate with each other, so the ordering between atomic operators and field operators can be changed arbitrarily. However, if the field operator are separated into the vacuum part $a_k(0)$ and the radiation part $a_k(t)$, because the radiation one is related to atomic operators, so in the calculating process, the ordering between atomic operators and field operators cannot be changed. Once the ordering has to be changed, the commutation relation must be used. We will see that the different permutations of ordering operators (normal ordering and antinormal ordering) will exhibit different physical p ictures for same quantum effect.

Now we discuss the equation of motion of operators for a single atom in the normal ordering. In the Heisenberg picture, from (9.32) we have

$$\frac{d}{dt} S_z^{(\ell)} = \frac{i}{\hbar}[H, S_z^{(\ell)}] = -\frac{i}{\hbar} \sum_k \varepsilon_k (S_+^{(\ell)} a_k - a_k^\dagger S_-^{(\ell)}) \tag{9.34}$$

here the commutation relations between $S_z^{(\ell)}$ and $S_\pm^{(\ell)}$ have been used. Similarly,

$$\frac{d}{dt} S_-^{(\ell)} = \frac{i}{\hbar}[H, S_-^{(\ell)}] = -i\omega_\ell S_-^{(\ell)} - 2\frac{i}{\hbar} \sum_k \varepsilon_k S_z^{(\ell)} a_k \tag{9.35}$$

$$\frac{d}{dt} S_+^{(\ell)} = \frac{i}{\hbar}[H, S_+^{(\ell)}] = i\omega_\ell S_+^{(\ell)} - 2\frac{i}{\hbar} \sum_k \varepsilon_k a_k^\dagger S_z^{(\ell)} \tag{9.36}$$

In order to obtain the solutions of atomic operators $S_i^{(\ell)}(t)$ $(i = \pm, z)$, the time-dependence of the field operator must be given. From the Heisenberg equation

$$\frac{d}{dt} a_k = \frac{i}{\hbar}[H, a_k] = -i\omega_k a_k - \frac{i}{\hbar}\varepsilon_k \sum_{\ell'} S_-^{(\ell')} \tag{9.37}$$

The solution of the above equation gives

$$a_k(t) = a_k(0)\exp(-i\omega_k t) - \frac{i}{\hbar}\varepsilon_k \sum_k \int_0^t S_-^{(\ell')}(t')\exp[-i\omega_k(t-t')]dt' \quad (9.38)$$

The integrand $S_-^{(\ell')}(t)$ in eq.(9.38) can be replaced by its zero-order approximation solution. The zero-order approximation equation of $S_-^{(\ell')}(t)$ is

$$\frac{d}{dt}S_-^{(\ell')}(t) = \frac{i}{\hbar}[H_0, S_-^{(\ell')}] = -i\omega_{\ell'}S_-^{(\ell')} \quad (9.39)$$

its solution gives immediately

$$S_-^{(\ell')}(t') = S_-^{(\ell')}(t)\exp[-i\omega_{\ell'}(t'-t)] \quad (9.40)$$

Inserting (9.40) into (9.38) we have

$$\begin{aligned}
a_k(t) &= a_k(0)\exp(-i\omega_k t) \\
&\quad -\frac{i}{\hbar}\varepsilon_k \sum_k \int_0^t S_-^{(\ell')}(t')\exp[-i\omega_{\ell'}(t'-t)]\exp[-i\omega_k(t-t')]dt' \\
&= a_k(0)\exp(-i\omega_k t) \\
&\quad -\frac{i}{\hbar}\varepsilon_k \sum_{\ell'} S_-^{(\ell')}(t)\int_0^t \exp[i(\omega_k-\omega_{\ell'})(t'-t)]dt' \\
&= a_k(0)\exp(-i\omega_k t) \\
&\quad -\frac{i}{\hbar}\varepsilon_k \sum_{\ell'}\left\{i\pi\delta(\omega_k-\omega_{\ell'}) + P\left(\frac{1}{\omega_k-\omega_{\ell'}}\right)\right\}S_-^{(\ell')}(t)
\end{aligned} \quad (9.41)$$

The above equation indicates that the field is divided into the vacuum part and the radiation part, but the radiation part is related to the atomic operators. Substituting (9.41) into (9.34) yields

$$\begin{aligned}
\frac{d}{dt}S_z^{(\ell)} &= -\frac{i}{\hbar}\sum_k \varepsilon_k(S_+^{(\ell)}a_k + a_k^\dagger S_-^{(\ell)}) = -\frac{i}{\hbar}\sum_k \varepsilon_k\left\{S_+^{(\ell)}\{a_k(0)\exp(-i\omega_k t)\right. \\
&\quad \left. -\frac{\varepsilon_k}{\hbar}\sum_{\ell'}\left[i\pi\delta(\omega_k-\omega_{\ell'}) + P\left(\frac{1}{\omega_k-\omega_{\ell'}}\right)\right]S_-^{(\ell')}(t)\right\}
\end{aligned}$$

$$+\left\{a_k^\dagger(0)\exp(i\omega_k t) - \frac{\varepsilon_k}{\hbar}\sum_{\ell'}\left[-i\pi\delta(\omega_k - \omega_{\ell'})\right.\right.$$

$$\left.\left.+P\left(\frac{1}{\omega_k - \omega_{\ell'}}\right)\right] S_+^{(\ell')}(t)\right\} S_-^{(\ell)}(t)\Big\} \tag{9.42}$$

The principal function $P(\frac{1}{\omega_k - \omega_{\ell'}})$ corresponds to the Lamb shift, which does not affect the decay rate of atomic operators, so as an approximation it can be ignored. Therefore, the Langevin equation for the operator $S_z^{(\ell)}$ is

$$\frac{d}{dt}S_z^{(\ell)} = -\gamma S_+^{(\ell)}\sum_{\ell'} S_-^{(\ell')} - [S_+^{(\ell)}F(t)\exp(-i\omega_0 t) + \text{h.c.}] \tag{9.43}$$

where

$$\gamma = \frac{2\pi}{\hbar^2}\sum_k \varepsilon_k^2 \delta(\omega_k - \omega_\ell) \tag{9.44}$$

$$F(t) = \frac{i}{\hbar}\sum_k \varepsilon_k a_k(0)\exp[-i(\omega_k - \omega_0)t] \tag{9.45}$$

here γ represents the lifetime of the spontaneous decay for the ℓ-th atom. Since all the N atoms are assumed to be identical. here γ is assumed to be independent of ℓ. Because $F(t)$ is proportional to the vacuum field part, then the last term in (9.43) describes the influence of the vacuum field on the ℓ-th atom. It is easy to prove that $F(t)$ and $F^\dagger(t)$ have the following statistical characteristics,

$$\langle F(t)\rangle = \langle F^\dagger(t)\rangle = 0 \tag{9.46}$$

$$\langle F^\dagger(t)F(t)\rangle = 0 \tag{9.47}$$

$$\langle F(t)F^\dagger(t')\rangle = \gamma\delta(t - t') \tag{9.48}$$

so $F(t)$ has the Gaussian statistical properties similar to the statistical force in the classical Brown motion. Using (9.46)-(9.48), it is easy to obtain correlation functions as

$$\langle F(t_1)F(t_2)\cdots F(t_n)F^\dagger(t_1')F^\dagger(t_2')\cdots F^\dagger(t_m')\rangle = 0 \quad (n \neq m)$$

$$\langle F(t_1)F(t_2)\cdots F(t_n)F^\dagger(t_1')F^\dagger(t_2')\cdots F^\dagger(t_n')\rangle$$

$$= \sum_p \langle F(t_1)F^\dagger(t_1')\rangle\langle F(t_2)F^\dagger(t_2')\rangle\cdots\langle F(t_n)F^\dagger(t_n')\rangle$$

$$= \gamma^n \sum_p \delta(t_1 - t_1')\delta(t_2 - t_2')\cdots\delta(t_n - t_n') \quad (n = m)$$

Similarly, substituting (9.41) into (9.35) and (9.36) yield

$$\frac{d}{dt}S_-^{(\ell)} = -i\omega_\ell S_-^{(\ell)} + \gamma S_z^{(\ell)} \sum_{\ell'} S_-^{\ell'} + 2S_z^{(\ell)} F(t)\exp(-i\omega_0 t) \qquad (9.49)$$

$$\frac{d}{dt}S_+^{(\ell)} = i\omega_\ell S_+^{(\ell)} + \gamma S_z^{(\ell)} \sum_{\ell'} S_+^{(\ell')} + 2S_z^{(\ell)} F^\dagger(t)\exp(i\omega_0 t) \qquad (9.50)$$

From (9.49) it can be directly obtained that

$$S_-^{(\ell)} = S_-^{(\ell)}(0)e^{-i\omega_\ell t} + 2S_z(0)\lim_{t\to 0+} e^{-i\omega_\ell t}\int_0^t \exp[i(\omega_\ell - \omega_0)t']F(t')dt' \quad (9.51)$$

The conjugation of the above equation is just the solution of (9.50). If we assume that the field is in the vacuum state $|\{0_k\}\rangle$ and the atom in $|\psi\rangle$ at time t=0, i.e., the initial state vector of the system is

$$|\ \rangle = |\psi, \{0_k\}\rangle \qquad (9.52)$$

then inserting (9.51) and its conjugation into (9.43), the expectation value of $\langle \frac{d}{dt}S_z^{(\ell)}\rangle$ in the state described by (9.52) is written as

$$\left\langle \frac{d}{dt}S_z^{(\ell)} \right\rangle = -\gamma \sum_{\ell'} \langle S_+^{(\ell)}(0)S_-^{(\ell')}(0)\rangle = -\gamma \langle S_+^{(\ell)}(0)S_-^{(\ell)}(0)\rangle \qquad (9.53)$$

here eqs.(9.46)-(9.48) are used. For a single atom

$$S_+^{(\ell)}(0)S_-^{(\ell)}(0) = S_z^{(\ell)}(0) + \frac{1}{2} \qquad (9.54)$$

thus

$$\left\langle \frac{d}{dt}S_z^{(\ell)}(0) \right\rangle = -\gamma \qquad (9.55)$$

Then for the decay of a single-atom described by the normal ordering method, the vacuum fluctuation does not affect the initial decay. In order to reveal the initial decay behavior, we must consider the dipole fluctuations. In the following we will reveal the effect of vacuum fluctuations on the initial decay by means of the antinormal ordering method.

9.3.2 Dicke model for superfluorescence

The time-dependence of atomic operators for a single atom has been discussed. Now we turn to discuss the collective behavior of N atoms. The collective atomic operators are defined as

$$S_i = \sum_{\ell=1}^{N} S_i^{(\ell)} \qquad (i = z, +, -, x, y) \tag{9.56}$$

and all the N atoms have the same frequency $\omega_\ell = \omega_0$. Summing (9.43) with respect to ℓ and averaging it in state $|\ \rangle$, then

$$\sum_{\ell} \left\langle \frac{d}{dt} S_z^{(\ell)} \right\rangle = -\gamma \sum_{\ell,\ell'} \langle S_+^{(\ell)} S_-^{(\ell')} \rangle \tag{9.57}$$

Calling (9.56), we have

$$\left\langle \frac{d}{dt} S_z \right\rangle = -\gamma \langle S_+ S_- \rangle \tag{9.58}$$

Note the relation

$$S^2 = S_z^2 + \frac{1}{2}(S_+S_- + S_-S_+) \tag{9.59}$$

and the commutation relation

$$[S_+, S_-] = 2S_z \tag{9.60}$$

i.e.,

$$S_-S_+ = S_+S_- - 2S_z \tag{9.61}$$

we have

$$S_+S_- = S^2 - S_z^2 + S_z \tag{9.62}$$

Substituting (9.62) into (9.58) gives

$$\left\langle \frac{d}{dt} S_z \right\rangle = -\gamma \langle (S^2 - S_z^2 + S_z) \rangle \tag{9.63}$$

In order to reveal the time evolution of the atomic energy operator, we must solve (9.63). Usually the state $|\ \rangle$ can be chosen arbitrarily, here we choose the Dicke state. The Dicke state is referred to the common eigenvectors of operators S^2 and S_z. For the N atoms system, when $\omega_k = \omega_0$, i.e., the term Δ

in (9.32) is equal to zero, the Hamiltonian of this atom-field coupling system can be rewritten in terms of the collective atomic operators (9.56) as

$$H = \sum_k \hbar\omega_k a_k^\dagger a_k + \hbar\omega_0 S_z + \sum_k \hbar\varepsilon_k (a_k S_+ + a_k^\dagger S_-) \tag{9.64}$$

It is easy to prove that the operator S^2 is a conserving quantity, i.e.,

$$\frac{d}{dt}S^2 = \frac{i}{\hbar}[H, S^2] = \frac{i}{\hbar}\left\{\omega_0 \left[S_z, \frac{1}{2}(S_+S_- + S_-S_+)\right]\right.$$
$$\left. + \sum_k \varepsilon_k \left[a_k S_+ + a_k^\dagger S_-, \frac{1}{2}(S_+S_- + S_-S_+) + S_z^2\right]\right\} = 0 \tag{9.65}$$

here (9.59) and the commutation relation among the atomic operators have been used. From (9.65) we know

$$\langle S^2(t)\rangle = \langle S^2(0)\rangle = \text{constant} \tag{9.66}$$

That S^2 is a conserving quantity indicates that with the time development, the eigenvalue ℓ of S^2 does not change in the process of N atoms damping from the initial excited state to the final ground state, the value ℓ is decided by the initial excitation case of atoms. With the time development, only the eigenvalue m of the z-component S_z changes. For the Dicke state $|\ell, m\rangle$, we have

$$S^2|\ell, m\rangle = \ell(\ell+1)|\ell, m\rangle \tag{9.67}$$
$$S_z^2|\ell, m\rangle = m|\ell, m\rangle \tag{9.68}$$

where the values of ℓ may be

$$\ell = \frac{N}{2}, \frac{N}{2} - 1, \cdots, 0 \tag{9.69}$$

For a given system, the value of ℓ is completely determined by the atomic initial excitation, its maximum value is $N/2$, which corresponds to all the N atoms initially in their excited states. However, for a given ℓ, the value of m may be chosen as

$$m = \ell, \ell - 1, \cdots, -(\ell - 1), -\ell \tag{9.70}$$

This means that with the time development, m varies from ℓ to $-\ell$.

The atomic transitions obey the selective rule $\Delta m = \pm 1$, therefore, for the

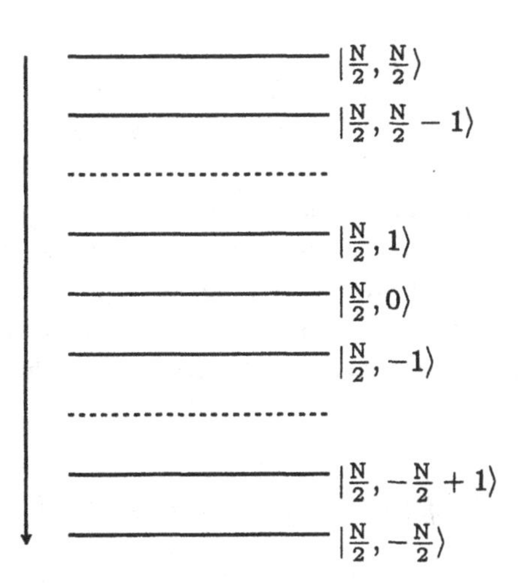

Figure 9.7: Collective atomic state for $\ell = N/2$. The arrow in left-side represents the time evolution direction for the states of the system when ℓ is conserved

fully excited atomic system, the time evolution of the system may be shown as the corresponding level-transition processes in Fig.(9.7).

Now we turn to calculate the derivate $\frac{d}{dt}\langle S_z\rangle$. Assuming that the field is initially in the vacuum state, and the atomic system in the Dicke state $|\ell, m\rangle$, i.e.,

$$|\ \rangle = |\ell, m, 0_k\rangle \tag{9.71}$$

then from (9.63) we obtain

$$\frac{d}{dt}\langle S_z\rangle = -\gamma[\ell(\ell+1) - m(m-1)] \tag{9.72}$$

The fundamental condition of exhibiting the superfluorescence demands that all the N atoms are initially in their excited states, so $\ell = N/2$, and the maximum value of m is also $N/2$. With the time development, $\ell = N/2$ does not change, but m can be vary from its maximum value $N/2$ to -$N/2$. From (9.72) we see that the values of m are different at different transition moment, the decay rates $\frac{d}{dt}\langle S_z\rangle$ are also different. For example, when $t = 0, m = N/2$,

(9.72) gives

$$\frac{d}{dt}\langle S_z\rangle = -\gamma\left[\frac{N}{2}\left(\frac{N}{2}+1\right)-\left(\frac{N}{2}\right)^2+\frac{N}{2}\right] = -\gamma N \tag{9.73}$$

Eq.(9.73) implies that at the initial time, the decay rate of the system is proportional to N. Because at t=0, all the N atoms do not couple to each other, this means that each atom decays separately, the total decay rate is the summation of all independent decay rates. Subsequently, the decay will lead to the reduction of m, when m decreases to m=0 as shown in the middle part of Fig.(9.7), (9.72) becomes

$$\frac{d}{dt}\langle S_z\rangle = -\gamma\left[\frac{N}{2}\left(\frac{N}{2}+1\right)\right] \approx \gamma\frac{N^2}{4} \tag{9.74}$$

In this case, the decay rate reaches its maximum value, which is proportional to the square of the number of atoms. At this time, the coupling between field and atoms makes the atoms to be collective decay, then the system can radiate the superfluorescence pulse. This state in which ℓ reaches its maximum value and m=0 is termed as the superradiance Dicke state $|N/2,0\rangle$. With the time development, m decreases to its minimum value $m = -N/2$, in this moment, (9.72) becomes

$$\frac{d}{dt}\langle S_z\rangle = 0 \tag{9.75}$$

This means that the decay rate decreases to zero and all the atom are in their ground states.

By choosing the special Dicke state (9.71), we have quantitatively analyzed the decay processes of the atomic energy operator for the system. Now we study the time evolution of the atomic energy operator for a general state vector $|\ \rangle$. For an arbitrary state $|\ \rangle$, (9.63) becomes

$$\frac{d}{dt}\langle S_z\rangle = -\gamma\{\langle S^2\rangle - \langle S_z^2\rangle + \langle S_z\rangle\} \tag{9.76}$$

Because S^2 is a conserving quantity, it can be replaced by $\langle S^2(t)\rangle = \langle S^2(0)\rangle = \frac{N}{2}(\frac{N}{2}+1)$. So (9.76) can be rewritten as

$$\frac{d}{dt}\langle S_z\rangle = -\gamma\left\{\frac{N}{2}\left(\frac{N}{2}+1\right) - \langle S_z^2\rangle + \langle S_z\rangle\right\} \tag{9.77}$$

Inserting $\langle S_z\rangle^2 - \langle S_z\rangle^2$ into (9.77), then (9.77) becomes

$$\frac{d}{dt}\langle S_z\rangle = -\gamma\left[\frac{N}{2}\left(\frac{N}{2}+1\right) - \langle S_z\rangle^2 + \langle S_z\rangle\right] \\ +\gamma\langle S_z^2 - \langle S_z\rangle^2\rangle \tag{9.78}$$

Obviously, the last term in (9.78) describes the fluctuations of the atomic operator S_z. And from the view of physical effects, these fluctuations only give minor modification to the time behavior of S_z. So as an approximation, we can neglect this term in (9.78). Then (9.78) reduces to

$$\frac{d}{dt}\langle S_z\rangle = -\gamma\left[\frac{N}{2}\left(\frac{N}{2}+1\right) - \langle S_z\rangle^2 + \langle S_z\rangle\right] \\ = -\left\{\left(\frac{N}{2}+\langle S_z\rangle\right)\left(\frac{N}{2}-\langle S_z\rangle+1\right)\right\} \tag{9.79}$$

The solution of the above equation gives

$$\langle S_z(t)\rangle = -\frac{N}{2}\tanh\left(\frac{t-T_D}{T_R}\right) \tag{9.80}$$

where

$$T_D = \frac{1}{2}T_R \ln N \tag{9.81}$$

$$T_R = \frac{2}{N\gamma} \tag{9.82}$$

The time behavior of $\langle S_z\rangle$ (9.80) is illustrated in Fig.(9.8). From eq.(9.80) we can obtain the intensity of the radiation field as

$$I(t) = -\hbar\omega_0\frac{d}{dt}\langle S_z\rangle = I_0\frac{N^2}{4} sech^2\left(\frac{t-T_D}{T_R}\right) \tag{9.83}$$

Eq.(9.83) shows that the intensity of the superfluorescence pulse is proportional to the square of the number of atoms, and the decay time T_R is smaller than the delay time T_D, i.e., $T_R \ll T_D$. Different from the result obtained by the quasi-classical theory, here the delay time T_D has certain limited value defined by (9.81).

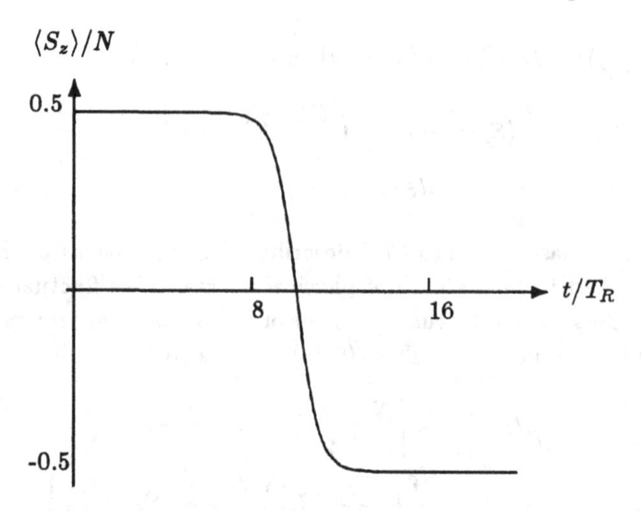

Figure 9.8: Time evolution of $\langle S_z\rangle$ described by (9.80)

9.3.3 Quantum statistical properties of superfluorescence

A. Langevin equation for collective atomic operators and the effects of vacuum fluctuations

Superfluorescence is the effect of cooperative spontaneous emission. In order to analyze this effect exactly, the quantum statistical property of the system and the effects of quantum fluctuations must be considered. Here we first derive the quantum Langevin equation of the collective atomic system, then discuss the effects of vacuum fluctuations.

As we know, the time evolutions of operators of the ℓ-th atom obey eqs.(9.43), (9.49), and (9.50). According to eqs.(9.49), (9.50) and (9.56), the collective atomic operators S_- and S_+ obey the Langevin equations

$$\frac{d}{dt}S_- = -i\omega_0 S_- + \gamma S_- S_z + 2F(t)S_z \exp(-i\omega_0 t) \tag{9.84}$$

$$\frac{d}{dt}S_+ = i\omega_0 S_+ + \gamma S_z S_+ + 2F^\dagger(t)S_z \exp(i\omega_0 t) \tag{9.85}$$

Here the random force operators $F(t)$ and $F^\dagger(t)$ are defined by (9.45). These operators describe the effects of vacuum field on the atomic dynamic behavior. Their statistical properties are given by eqs.(9.46)-(9.48). From (9.43), the

Langevin equation of the collective operator S_z is given as

$$\frac{d}{dt}S_z = -\gamma S_+ S_- - [F(t)S_+ \exp(-i\omega_0 t) + \text{h.c.}] \tag{9.86}$$

It must be mentioned that the antinormal ordering of the operators has been used when eqs.(9.84)-(9.86) are obtained. That is to say, in the above three equations, the creation operators are placed in the right hand side of the product of operators, and the annihilation operators are placed in the left hand side. To do this is useful to reveal the effects of vacuum fluctuations on the initial process of superfluorescence.

For (9.84) and (9.85), if we only consider the linear range of the initial stage, then we can assume

$$S_z(t) \approx S_z(0) \tag{9.87}$$

In this case, the solution of (9.84) gives

$$S_-(t) = \left\{ S_-(0) + 2S_z(0) \int_0^t e^{-\gamma S_z(0)t'} F(t')dt' \right\} \exp\{[-i\omega_0 - \gamma S_z(0)]t\} \tag{9.88}$$

Here the commutation relation

$$[S_z(0), F(t)] = 0 \tag{9.89}$$

has been used. Similarly, we have

$$S_+(t) = \left\{ S_+(0) + 2S_z(0) \int_0^t dt' e^{-\gamma S_z(0)t'} F^\dagger(t') \right\} \exp\{[i\omega_0 - \gamma S_z(0)]t\} \tag{9.90}$$

Replacing $S_-(t)$ and $S_+(t)$ in the right hand in (9.86) by eqs.(9.88) and (9.90), the time behavior of S_z at initial stage can be obtained. If all the N atoms are initially in their excited states, and the field in a vacuum state, i.e., the initial state of the system is $|\frac{N}{2}, \{0_k\}\rangle$, then the first term in the right hand of (9.86) gives

$$\langle S_-(t)S_+(t)\rangle = \langle S_-(0)S_+(0)\rangle$$
$$+4\langle S_z^2(0) \int_0^t \int_0^{t'} \exp[-\gamma S_z(0)(t' + t'')]F(t')F^\dagger(t'')\rangle dt' dt'' e^{2\gamma S_z(0)t}$$

$$= N^2 e^{N\gamma t} \int_0^t dt' \int_0^{t'} dt'' \exp\left[-\frac{N}{2}\gamma(t'+t'')\right] \langle F(t')F^\dagger(t'')\rangle$$

$$= N^2 e^{N\gamma t} \int_0^t dt' \int_0^{t'} dt'' \exp\left[-\frac{N}{2}\gamma(t'+t'')\right] \gamma\delta(t'-t'')$$

$$= N^2 e^{N\gamma t}\gamma \int_0^t e^{-N\gamma t'} dt' = N(e^{N\gamma t}-1) \tag{9.91}$$

The second term in the right hand of (9.86) becomes

$$\langle F(t)S_+ e^{-i\omega_0 t} + h.c.\rangle$$

$$= N \int_0^t \exp\left[\frac{N\gamma}{2}(t'-t'')\right] \langle F(t)F^\dagger(t') + \text{h.c.}\rangle dt'$$

$$= N \int_0^t \exp[-N\gamma(t-t')]\gamma\delta(t-t')dt' = N\gamma \tag{9.92}$$

Combining eqs.(9.91) and (9.92) with (9.86), then we have

$$\frac{d}{dt}\langle S_z\rangle = -N\gamma(e^{N\gamma t}-1) - N\gamma = -N\gamma e^{N\gamma t} \tag{9.93}$$

We can see that when the influence of vacuum fluctuations ($F(t)$) is taken into account at the initial stage, the intensity of the field radiated by the collective atoms can be written in the form

$$I(t) \propto I_0 N e^{N\gamma t} \tag{9.94}$$

It is different from (9.45) and (9.73) obtained by means of the normal ordering of operators. So we see that using the antinormal ordering of operators can reveal the important role of the vacuum fluctuations in the initial decay stage of collective atoms. If we neglect the terms proportional to $\langle F(t)F^\dagger(t)\rangle$ which reflect the vacuum field effects, then from (9.91) we find, at the initial stage the decay rate becomes

$$\frac{d}{dt}\langle S_z\rangle|_{t=0^+} = 0 \tag{9.95}$$

That is to say, the excited atomic system becomes stable, which is similar to the result in the quasi-classical theory. This result is disagreement with the experimental result. So the influence of the vacuum field can not be neglected.

However, if at t=0, all the N atoms are in their ground state, i.e., the initial state vector is $|-\frac{N}{2},\{0_k\}\rangle$, then the right hand of (9.91) is equal to $-N\gamma$. So we find that in (9.86) the first term which corresponding to the contribution of atomic dipole fluctuations, and the second term which reflecting the influence of vacuum fluctuations, cancel out each other. Therefore, the decay rate $\frac{d}{dt}\langle S_z\rangle_{t=0}=0$. This means that the system initial in the ground state is forever in its ground state even though the dipole fluctuations and vacuum fluctuations are existed.

B. P representation of the atom

The quantum statistical properties of atomic system can also described by means of the probability distribution function. Generally, for an arbitrary function $Q(S_-,S_z,S_+)$ which contains the atomic operators S_-, S_z, and S_+, it can be expanded as power series

$$Q^{(a)}(S_-,S_z,S_+)=\sum_{r_1,r_2,r_3}Q^{(a)}_{r_1,r_2,r_3}S_-^{r_1}S_z^{r_2}S_+^{r_3} \tag{9.96}$$

The upper index (a) stands for the antinormal ordering. If the upper index is denoted by (o), this means that the normal ordering is adopted. Correspondingly, a C-number function associated with (9.96) can also be defined as

$$Q^{(a)}(\beta,m,\beta^*)=\sum_{r_1,r_2,r_3}Q^{(a)}_{r_1,r_2,r_3}\beta^{r_1}m^{r_2}\beta^{*r_3} \tag{9.97}$$

here β and β^* are complex parameters corresponding to the non-Hermitian operators S_- and S_+, m is a real parameter related to the Hermitian operator S_z. We can also formally rewrite (9.96) in terms of the δ function as

$$\begin{aligned}Q^{(a)}(S_-,S_z,S_+)=&\int\int\int d\beta dm d\beta^*\overline{Q}^{(a)}(\beta,m,\beta^*)\delta(\beta-S_-)\\&\times\delta(m-S_z)\delta(\beta^*-S_+)\end{aligned} \tag{9.98}$$

here the δ function can be defined as

$$\delta(\beta_i-S_i)=\frac{1}{2\pi}\int_{-\infty}^{\infty}\exp[-i\xi_i(\beta-S_i)]d\xi_i \tag{9.99}$$

If S_i is a non-Hermitian operator, then the δ function is defined as

$$\delta(\beta_i - S_i)\delta(\beta_i - S_i^\dagger) = \left(\frac{1}{2\pi}\right)^2 \int\int d^2\xi_i e^{-i\xi_i(\beta - S_i)} e^{-i\xi_i(\beta^* - S_i^\dagger)} \tag{9.100}$$

with

$$d^2\xi_i = d(\mathrm{Re}\xi_i)d(\mathrm{Im}\xi_i) \tag{9.101}$$

Therefore, (9.98) can be written as

$$Q^{(a)}(S_-, S_z, S_+) = \left(\frac{1}{2\pi}\right)^3 \int\int\int \overline{Q}^{(a)}(\beta, m, \beta^*) d\beta dm d\beta^*$$
$$\times \int\int\int d\xi_1 d\xi_2 d\xi_3 e^{-i\xi_1(\beta - S_-)} e^{-i\xi_2(m - S_z)} e^{-i\xi_3(\beta^* - S_+)} \tag{9.102}$$

Changing the integrating order, (9.102) can also be written as

$$Q^{(a)}(S_-, S_z, S_+) = \int\int\int d\xi_1 d\xi_2 d\xi_3 e^{i\xi_1 S_-} e^{i\xi_2 S_z} e^{i\xi_3 S_+} F(\xi_1, \xi_2, \xi_3) \tag{9.103}$$

where $F(\xi_1, \xi_2, \xi_3)$ is the Fourier transformation of the C-number function $\overline{Q}^{(a)}(\beta, m, \beta^*)$, i.e.,

$$F(\xi_1, \xi_2, \xi_3) = \left(\frac{1}{2\pi}\right)^3 \int\int\int d\beta dm d\beta^* \overline{Q}^{(a)}(\beta, m, \beta^*)$$
$$\times \exp[-i(\xi_1\beta + \xi_2 m + \xi_3\beta^*)] \tag{9.104}$$

Its inverse transformation gives the C-number function $\overline{Q}^{(a)}(\beta, m, \beta^*)$, i.e.,

$$\overline{Q}^{(a)}(\beta, m, \beta^*) = \int\int\int d\xi_1 d\xi_2 d\xi_3 e^{i(\xi_1\beta + \xi_2 m + \xi_3\beta^*)} F(\xi_1, \xi_2, \xi_3) \tag{9.105}$$

If the operator function Q is just the density operator $\rho(S_-, S_z, S_+, t)$ of the atomic system in the Schrödinger picture, then from (9.103) we have

$$\rho^{(a)}(S_-, S_z, S_+) = \int\int\int d\xi_1 d\xi_2 d\xi_3 e^{i\xi_1 S_-} e^{i\xi_2 S_z} e^{i\xi_3 S_+} F(\xi_1, \xi_2, \xi_3) \tag{9.106}$$

where

$$F(\xi_1, \xi_2, \xi_3) = \left(\frac{1}{2\pi}\right)^3 \int\int\int d\beta dm d\beta^* \rho^{(a)}(\beta, m, \beta^*) e^{-i(\xi_1\beta + \xi_2 m + \xi_3\beta^*)} \tag{9.107}$$

By use of the density operator $\rho(t)$, one can obtain the expectation value of an arbitrary operator Q, i.e.,

$$\langle Q^{(a)}(S_-, S_z, S_+, t)\rangle = \mathrm{Tr}[\rho(t)Q^{(a)}(S_-, S_z, S_+)] \tag{9.108}$$

Inserting (9.108) into (9.103), we have

$$\begin{aligned}
&\langle Q^{(a)}(S_-, S_z, S_+, t)\rangle \\
&= \left\langle \int\int\int d\xi_1 d\xi_2 d\xi_3 e^{i\xi_1 S_-} e^{i\xi_2 S_z} e^{i\xi_3 S_+} F(\xi_1, \xi_2, \xi_3) \right\rangle \\
&= \int\int\int d\xi_1 d\xi_2 d\xi_3 C^{(a)}(\xi_1, \xi_2, \xi_3) F(\xi_1, \xi_2, \xi_3)
\end{aligned} \tag{9.109}$$

where

$$C^{(a)}(\xi_1, \xi_2, \xi_3) = \langle e^{i\xi_1 S_-} e^{i\xi_2 S_z} e^{i\xi_3 S_+}\rangle = \mathrm{Tr}[\rho(t) e^{i\xi_1 S_-} e^{i\xi_2 S_z} e^{i\xi_3 S_+}] \tag{9.110}$$

is called the characteristic function of the atomic operators. $F(\xi_1, \xi_2, \xi_3)$ is the Fourier transformation of the C-number function given by (9.104). Eq.(9.109) shows that calculating the expectation value of the atomic operator function $Q^{(a)}(S_-, S_z, S_+)$ may be transferred to evaluate the integration of a C-number function.

We can also adopt another method to transform calculating the expectation value of the atomic operator into calculating the integration of C-number function. In fact, (9.98) can be rewritten as

$$\begin{aligned}
&\langle Q^{(a)}(S_-, S_z, S_+)\rangle \\
&= \int\int\int d\beta dm d\beta^* \overline{Q}^{(a)}(\beta, m, \beta^*) \langle \delta(\beta - S_-)\delta(m - S_z)\delta(\beta^* - S_+)\rangle \\
&= \int\int\int d\beta dm d\beta^* \overline{Q}^{(a)}(\beta, m, \beta^*) P^{(a)}(t)
\end{aligned} \tag{9.111}$$

here $P^{(a)}(t)$ is called the quasi-probability distribution function of the atomic system, this representation is also termed as P-representation of atomic operators. $P^{(a)}(t)$ is defined as

$$\begin{aligned}
P^{(a)}(t) &= \langle \delta(\beta - S_-(t))\delta(m - S_z(t))\delta(\beta^* - S_+(t))\rangle \\
&= \mathrm{Tr}[\rho(t)\delta(\beta - S_-(t))\delta(m - S_z(t))\delta(\beta^* - S_+(t))]
\end{aligned} \tag{9.112}$$

Calling eqs.(9.99) and (9.110), the above equation becomes

$$P^{(a)}(t) = \left(\frac{1}{2\pi}\right)^3 \int\int\int d\xi_1 d\xi_2 d\xi_3 \langle e^{i\xi_1 S_-} e^{i\xi_2 S_z} e^{i\xi_3 S_+}\rangle$$
$$\times e^{-i(\xi_1\beta+\xi_2 m+\xi_3\beta^*)}$$
$$= \left(\frac{1}{2\pi}\right)^3 \int\int\int d\xi_1 d\xi_2 d\xi_3 e^{-i(\xi_1\beta+\xi_2 m+\xi_3\beta^*)} C^{(a)}(\xi_1,\xi_2,\xi_3,t) \quad (9.113)$$

Evidently, the quasi-probability distribution function $P^{(a)}(t)$ is just the Fourier transformation of the characteristic function (9.110). That is to say, by calculating the Fourier transformation of the characteristic function (9.110) of the atomic operator, one can obtain the P-representation function $P^{(a)}(t)$.

If expanding the characteristic function

$$C^{(a)}(\xi_1,\xi_2,\xi_3,t) = \langle \exp(i\xi_1 S_-(t)) \exp(i\xi_2 S_z(t)) \exp(i\xi_3 S_+(t))\rangle$$
$$= \sum_{p,q,r} \frac{1}{p!q!r!} \langle (i\xi_1 S_-(t))^p (i\xi_2 S_z(t))^q (i\xi_3 S_+(t))^r\rangle \quad (9.114)$$

and also considering the initial linear range, i.e., eqs.(9.87) and (9.89) are reasonable, then inserting (9.88) into the above equation, we find that only the terms proportional to the random force is not equal to zero, and (9.114) becomes

$$C^{(a)}(\xi_1,\xi_2,\xi_3,t) = \exp\left(\frac{i}{2}\xi_2 N\right) \exp[-|\xi_1|^2 \langle S_-(t) S_+(t)\rangle]$$
$$= \exp\left[\frac{i}{2}\xi_2 N - |\xi_1|^2 \langle S_-(t) S_+(t)\rangle\right] \quad (9.115)$$

here we have considered that $\xi_3 = \xi_1^*$. Making the Fourier transformation for (9.115), we obtain the quasi-probability distribution function as

$$P^{(a)}(t) = \left(\frac{1}{2\pi}\right)^3 \int\int\int d\xi_2 d^2\xi_1 \exp[-i\xi_2 m - i(\xi_1\beta + \xi_1^*\beta^*)]$$
$$\times \exp\left[\frac{i}{2}\xi_2 N - |\xi_1|^2 \langle S_-(t) S_+(t)\rangle\right]$$

Using the identify relation

$$\int d\xi_2 \exp\left[-i\xi_2\left(\frac{N}{2} - m\right)\right] = \delta\left(m - \frac{N}{2}\right)$$

we have

$$
\begin{aligned}
P^{(a)}(t) &= \left(\frac{1}{2\pi}\right)^2 \int\int d^2\xi_1 \exp[-i(\xi_1\beta + \xi_1^*\beta^*) - |\xi|^2 \langle S_-(t)S_+(t)\rangle] \\
&= \frac{1}{\pi}\int_0^\infty r dr \exp[-2ir|\beta| - r^2\langle S_-(t)S_+(t)\rangle] \\
&= \frac{1}{\pi}\int_0^\infty r dr \exp\left[-\langle S_-S_+\rangle\left(r - \frac{i|\beta|}{\langle S_-S_+\rangle}\right)^2\right]\exp\left(-\frac{|\beta|^2}{\langle S_-S_+\rangle}\right) \\
&= \frac{1}{2\pi\langle S_-S_+\rangle}\exp\left(-\frac{|\beta|^2}{\langle S_-S_+\rangle}\right) \qquad (9.116)
\end{aligned}
$$

here we have assumed $r = |\xi_1|$ and used the relation

$$\int_0^\infty r \exp(-\alpha r) dr = \frac{1}{2\alpha} \qquad (9.117)$$

Eq.(9.116) shows that at the beginning of the decay of the excited atomic system, the quasi-probability distribution function has the Gaussian distribution properties. Note that the expectation value of the product for the transverse components of atomic operators obeys (9.91), so the value of $P^{(a)}(t)$ will decrease with the time development, that is to say, the Gaussian characteristic will be reduced, the system will evolve toward the classical behavior. So the influence of vacuum fluctuations is not only the physical origin of triggering the initial decay of the fully excited atomic system, but also the fundamental cause of leading to the system to exhibit the Brown motion feature. However, with the enhancing of the cooperative radiation, the system will exhibit the quasiclassical-like coherence radiation behavior.

C. Triggering effect of the thermal radiation field

If the initial radiation field is not in the vacuum state but in a thermal radiation field, which contains the mode whose frequency is resonant to the transition frequency of a two-level atom. This case is related to the superradiance of Rydberg atoms. The equations of motion of the collective atomic operators $S_\pm$ and S_z of the two-level atoms interacting with the thermal field are given by eqs.(9.34)- (9.36). The effect of thermal field on atomic system is also characterized by the random force $F(t)$. The density operator of the thermal radiation field is given in Chapter 3 as

$$\rho = \sum_{n_k} P_{(n_k)} |\{n_k\}\rangle\langle\{n_k\}| \tag{9.118}$$

here the probability $P_{(n_k)}$ is defined as

$$\begin{aligned} P_{(n_k)} &= \prod_k \frac{\overline{n_k}^{n_k}}{(1+\overline{n_k})^{1+n_k}} \\ \overline{n_k} &= [\exp(\hbar\omega_k/KT) - 1]^{-1} \end{aligned} \tag{9.119}$$

From eq.(9.118) we obtain the statistical properties of the random force as

$$\begin{aligned} \langle F(t)\rangle &= \langle F^\dagger(t)\rangle = 0 \\ \langle F(t)F^\dagger(t)\rangle &= (\overline{n}_0 + 1)\gamma\delta(t-t') \\ \overline{n}_0 &= [\exp(\hbar\omega_0/KT) - 1]^{-1} \end{aligned} \tag{9.120}$$

Comparing (9.120) with (9.48) we see that the γ in (9.48) is replaced by $(\overline{n_0} + 1)\gamma$, but the Gaussian property of the random force $F(t)$ does not change, so the previous discussion needs minor modifications. Similar to the effect of the vacuum field, the influence of the thermal radiation field only affect the initial decay of the atomic system. First, it triggers the decay of the fully excited atomic system. In addition the quasi-probability distribution function $P^{(a)}(t)$ has the same formation as (9.116) except that the correlation of the random force is described by (9.120). Moreover, the expectation value $\langle S_- S_+\rangle$ is different from (9.91), it becomes

$$\langle S_-(t)S_+(t)\rangle = N(\overline{n}_0 + 1)(e^{N\gamma t} - 1) \tag{9.121}$$

Evidently (9.121) is similar to (9.91), so the statistical characteristic of the system does not change. Thus the thermal radiation can accelerate the Brown motion of the atomic system.

9.4 Superfluorescent beats

9.4.1 Basic characteristics of the superfluorescent beats

In the above discussion for the superfluorescence, we ever assumed that the transition frequencies of the excited atoms are identical, however, if the atomic level with fine structure is split by a very weak external field, the atoms may be distributed in different energy levels. Suppose that the excited atoms are populated in two high-level states as shown in Fig.(9.9a), then the atoms decay independently to the same or the different low levels as shown in Fig.(9.9a,b). The atoms in an excited state may also decay to two low-levels as shown in Fig.(9.9c). In these cases, the system may be considered as that two groups

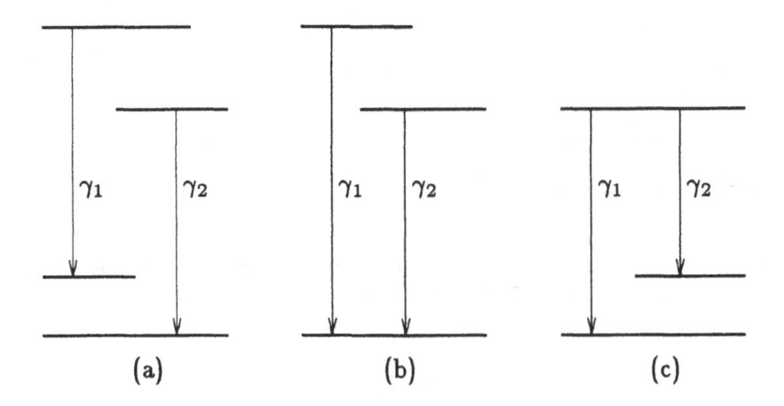

Figure 9.9: Three kinds of transitions resulting in superfluorescent beat, (a). Mixed-type transition; (b) V-type transition; (c) Λ-type transition

of excited atoms with two transition frequencies (ω_1 and ω_2) decay independently. But this kind of decay may present instantaneous interference between two transitions, and induce the output superfluorescences to exhibit quantum beats, which is called the superfluorescent beats.

Superfluorescent beats was first observed by Verhen et al in 1976. They found that C_S atoms which are stimulated by a laser field may decay from hyperfine state $7^2P_{3/2}$ to $7^2P_{1/2}$ as shown in Fig.(9.10), and displays the superfluorescent beats. Later, Marek and Ryschka observed a superfluorescent decay of C_s atoms from hyperfine state $5^2D_{3/2}$ to $6^2P_{1/2}$, and obtained the superfluorescent beats as shown in Fig.(9.11). These experimental results show that the superfluorescent decay of two groups of excited atoms with two transition frequencies (ω_1 and ω_2) induces instantaneous interference effect, and display the superfluorescent beats, the amplitude of the output pulse is modulated by the detuning $\delta\omega = \omega_1 - \omega_2$.

The quantum beats of a single-atom spontaneous emission have observed

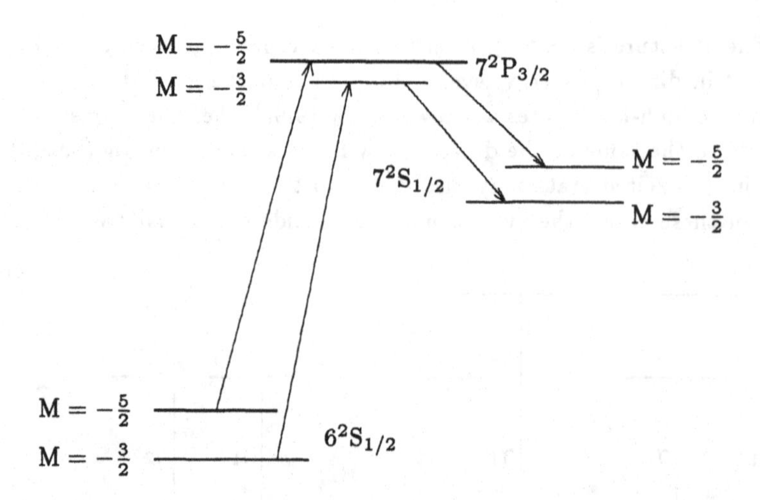

Figure 9.10: The transitions of Cesium atom for superfluorescent beat, the hyperfine structure decay for $7^2P_{3/2} \rightarrow 7^2S_{1/2}$.

in experiments for long time, it resulted from that the excited atomic state is initially in a superposition state of the split states, and then decays to one final state. Therefore, the quantum beats can reveal the splitting structure of the initial energy level of the atom. Since the superfluorescent beats is resulted from the interference between two independent superfluorescence decays of two groups of atoms with different transition frequencies (ω_1 and ω_2), it can be used to reveal not only the splitting structure for the initial energy levels, but also for the final energy levels. On the other hand, because the superfluorescence pulse is a very intensive signal with fine orientation, so that the superfluorescent beats may be utilized in the technology of the high resolution spectrum analysis.

9.4.2 Superfluorescent beats in the Dicke model

We now consider a system with N two-level atoms divided into two groups according to their resonance frequencies ω_1 and ω_2. Suppose that both groups have the same number of atoms, and at t=0 all of them are in their excited states and the radiation field is in vaccum state. For the two groups of atoms,

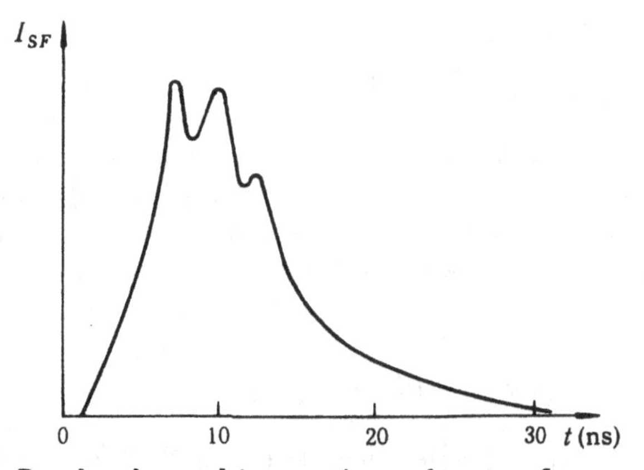

Figure 9.11: Results observed in experiment for superfluorescent beat

they can be described by the collective atomic operators S_1 and S_2

$$S_\eta = \sum_{\ell=1}^{\frac{N}{2}} S^{(\eta,\ell)} \qquad (\eta = 1, 2) \tag{9.122}$$

Their components are defined as

$$S_{\eta,q} = \sum_{\ell=1}^{\frac{N}{2}} S_q^{(\eta,\ell)} \qquad (q = x, y, x, +, -) \tag{9.123}$$

here $S_q^{(\eta,\ell)}$ represents the q component of the ℓ-th atom with resonance frequency ω_η. Therefore, the Hamiltonian of the system in the rotating-wave approximation is written by

$$H = H_0 + \Delta + V \tag{9.124}$$

where

$$H_0 = \sum_k \hbar\omega_k a_k^\dagger a_k + \frac{1}{2}\hbar(\omega_1 + \omega_2)\sum_\ell (S_z^{(1,\ell)} + S_z^{(2,\ell)}) \tag{9.125}$$

$$\Delta = \frac{\hbar}{2}(\omega_1 - \omega_2)\sum_\ell (S_z^{(1,\ell)} - S_z^{(2,\ell)}) \tag{9.126}$$

$$V = \sum_k [e_1(k,\ell) + e_2(k,\ell)] \tag{9.127}$$

$$
\begin{aligned}
e_1(k,\ell) &= \varepsilon_k a_k S_+^{(1,\ell)} + \text{h.c.} \\
e_2(k,\ell) &= \varepsilon_k a_k S_+^{(2,\ell)} + \text{h.c.}
\end{aligned} \tag{9.128}
$$

In order to realize the time behavior of the total system, we first discuss the time evolution of the photon number operator $P(k) = a_k^\dagger a_k$ for each k mode, and the interaction energies $e_1(k,\ell)$ and $e_2(k,\ell)$ for an atom interacting with the mode k. Because of the right-left symmetry, it is convenient to define the operators as

$$
p(|k|) = \frac{1}{2}[P(k) + P(-k)] \tag{9.129}
$$

$$
e(|k|,\eta,\ell) = \frac{1}{2}[e(k,\eta,\ell) + e(-k,\eta,\ell)] \quad (\eta = 1,2) \tag{9.130}
$$

The Heisenberg equation of motion for the photon number operator $P(|k|)$ gives

$$
\hbar\frac{d}{dt}P(|k|) = -i[P(|k|),H] = \sum_{\eta,\ell}\bar{e}(|k|,\eta,\ell) \tag{9.131}
$$

where

$$
\begin{aligned}
\bar{e}(|k|,\eta,\ell) &= \frac{1}{2}[\bar{e}(k,\eta,\ell) + \bar{e}(-k,\eta,\ell)] \\
\bar{e}(|k|,\eta,\ell) &= -i(\varepsilon_k a_k S_+^{(\eta,\ell)} - \text{h.c.})
\end{aligned} \tag{9.132}
$$

The equations for $e(|k|,\eta,\ell)$ and $\bar{e}(|k|,\eta,\ell)$ are, respectively,

$$
\hbar\frac{d}{dt}e(|k|,\eta,\ell) = -\hbar(\Omega_k - \delta\omega_\eta)\bar{e}(|k|,\eta,\ell) - |\varepsilon_k|^2\sum_{m,\xi}S(\eta,\ell,\xi,m) \tag{9.133}
$$

$$
\begin{aligned}
\hbar\frac{d}{dt}\bar{e}(|k|,\eta,\ell) &= -\hbar(\Omega_k - \delta\omega_\eta)e(|k|,\eta,\ell) + 2|\varepsilon_k|^2(S_z^{(\eta,\ell)} + 1/2) \\
&\quad + |\varepsilon_k|^2\sum_{m,\xi}(1 - \delta_{\eta,\ell;\xi,m})S(\eta,\ell,\xi,m)
\end{aligned} \tag{9.134}
$$

with

$$
\begin{aligned}
&\Omega_k = \omega_k - \omega_0, \quad \omega_0 = \frac{1}{2}(\omega_1 + \omega_2), \quad \delta\omega_\eta = \omega_\eta - \omega_0 \\
&\xi = 1,2; \quad m = 1,2,3,\ldots,\frac{N}{2}
\end{aligned} \tag{9.135}
$$

The two spin operators $S(\eta, \ell, \xi, m)$ and $\overline{S}(\eta, \ell, \xi, m)$ which appear in the above equations are, respectively, defined as

$$\begin{aligned} S(\eta, \ell, \xi, m) &= S_+^{(\eta,\ell)} S_-^{(\xi,m)} + h.c. \\ \overline{S}(\eta, \ell, \xi, m) &= i(S_+^{(\eta,\ell)} S_-^{(\xi,m)} - h.c.) \end{aligned} \tag{9.136}$$

In addition, the Heisenberg equation of motion for the operator $S_z^{(\eta,\ell)}$ gives

$$\frac{d}{dt} S_z^{(\eta,\ell)} = -2 \sum_{k>0} \overline{e}(|k|, \eta, \ell) \tag{9.137}$$

In fact, we have to know the time evolution of the emitted field, or equivalently, the time behavior of

$$\langle S_z \rangle = \sum_{\eta,\ell} \langle S_z^{(\eta,\ell)} \rangle \tag{9.138}$$

So we must solve (9.137) to obtain $S_z^{(\eta,\ell)}(t)$. This demands to obtain $\overline{e}(|k|, \eta, \ell)$ first. Derivating (9.134) with respect to time t gives

$$\begin{aligned} \hbar \frac{d^2}{dt^2} \overline{e}(|k|, \eta, \ell) &= -(\Omega_k - \delta\omega_\eta) \frac{d}{dt} e(|k|, \eta, \ell) + 2|\varepsilon_k|^2 \frac{d}{dt} \left(S_z^{(\eta,\ell)} + \frac{1}{2} \right) \\ &\quad + |\varepsilon_k|^2 \sum_{\ell,m} \left(1 - \delta_{\eta,\ell;\xi,m} \frac{d}{dt} S(\eta, \ell; \xi, m) \right) \\ &= \hbar(\Omega_k - \delta\omega_\eta)^2 \overline{e}(|k|, \eta, \ell) - |\varepsilon_k|^2 (\Omega_k - \delta\omega_\eta) \sum_{\xi,m} \overline{S}(\eta, \ell; \xi, m) \\ &\quad + 2|\varepsilon_k|^2 \frac{d}{dt} \left[S_z^{(\eta,\ell)} + \frac{1}{2} + \sum_{\xi,m} (1 - \delta_{\eta,\ell;\xi,m}) S(\eta, \ell; \xi, m) \right] \end{aligned} \tag{9.139}$$

The terms containing $|\varepsilon_k|^2$ can be considered as the perturbation parts. For the main parts in (9.139)

$$\frac{d^2}{dt^2} \overline{e}(|k|, \eta, \ell) = (\Omega_k - \delta\omega_\eta)^2 \overline{e}(|k|, \eta, \ell) \tag{9.140}$$

its solution gives

$$\overline{e}(|k|, \eta, \ell) = A \sin(\Omega_k - \delta\omega_\eta) t \tag{9.141}$$

Considering the initial condition

$$\langle \bar{e}(|k|,\eta,\ell)\rangle|_{t=0} = \left\langle \sum_{\ell} \bar{e}(|k|,\eta,\ell) \right\rangle\Bigg|_{t=0} = 0 \tag{9.142}$$

$$\langle e(|k|,\eta,\ell)\rangle|_{t=0} = \frac{2}{\hbar}|\varepsilon_k|^2 \tag{9.143}$$

then the expectation value of the solution of (9.139) gives

$$\begin{aligned}\langle \bar{e}(|k|,\eta,\ell)\rangle = {} & \frac{1}{\hbar}|\varepsilon_k|^2 \int_0^t dt' \frac{\sin(\Omega_k - \delta\omega_\eta)(t-t')}{\Omega_k - \delta\omega_\eta} \\ & \times \left[2\langle S_z^\eta\rangle + 1 + \left(\frac{N}{2} - 1\right)\langle S(\eta,\eta)\rangle + \frac{N}{2}\langle S(\eta,\xi)\rangle\right] \\ & - \frac{1}{\hbar}|\varepsilon_k|^2 \int_0^t dt' [\sin(\Omega_k - \delta\omega_\eta)(t-t')]\langle S(\eta,\xi)\rangle \\ & + 2\frac{1}{\hbar}|\varepsilon_k|^2 \frac{\sin(\Omega_k - \delta\omega_\eta)t}{\Omega_k - \delta\omega_\eta} \qquad (\eta \neq \xi)\end{aligned} \tag{9.144}$$

Replacing the summation to all modes k of the radiation field by the integration in continuum limit, (9.137) becomes

$$\begin{aligned}T_{1,\ell}\frac{d}{dt}\langle S_z^{(\eta)}\rangle = {} & -\left\langle S_z^{(\eta)} + \frac{1}{2}\right\rangle \\ & - \frac{1}{2}\left[\left(\frac{N}{2} - 1\right)\langle S(\eta,\eta)\rangle + \frac{N}{2}\langle S(\eta,\xi)\rangle\right] \quad (\eta \neq \xi)\end{aligned} \tag{9.145}$$

where $T_{1\ell}$ is the single-atom relaxation lifetime.

When $\omega_1 = \omega_2$, the two groups of atoms are degenerated to one group with N atoms. In this case, the system is just the cooperative emission of the N atoms with the same transition frequency and the superfluorescence pulse appears. In fact, the total spin operator S of the system is a conserving constant of motion, the expectation value of the transverse component of the total spin operator can be defined as

$$\langle \mathbf{S}_\perp^2 \rangle = \sum_{\eta,\ell,\xi,m} \langle \mathbf{S}_\perp^{(\eta,\ell)} \cdot \mathbf{S}_\perp^{(\xi,m)} \rangle \tag{9.146}$$

and can be rewritten as

$$\mathbf{S}_{\perp}^2 = \frac{1}{2}N\left(\frac{1}{2}N+1\right) - \langle S_z^2\rangle \tag{9.147}$$

Making the summation in (9.145) with respect to all atoms and applying eqs.(9.136) and (9.147), then the well-known equation (9.63) for the superfluorescence can be obtained as

$$T_{1,\ell}\frac{d}{dt}\langle S_z\rangle = -\left(\langle S_z\rangle + \frac{N}{2}\right) - \left(\frac{N^2}{4} - \langle S_z\rangle^2\right) \tag{9.148}$$

here the approximation $\langle S_z^2\rangle \approx \langle S_z\rangle^2$ has been used.

When $\omega_1 \neq \omega_2$, from (9.124) we know that the square of the total spin

$$\mathbf{S}^2 = \mathbf{S}_1^2 + \mathbf{S}_2^2 + 2\mathbf{S}_1 \cdot \mathbf{S}_2 \tag{9.149}$$

is not a conserving quantity. This can be proved as follows

$$[H, \mathbf{S}^2] = [\Delta, 2\mathbf{S}_1 \cdot \mathbf{S}_2] \neq 0 \tag{9.150}$$

However, the square of the total spin operator for the i-th (i=1,2) group of atoms remains a conserving constant, i.e., $[H, \mathbf{S}_i^2] = 0$. Therefore, (9.147) can be rewritten as

$$\langle \mathbf{S}_{\perp}^2\rangle = \langle \mathbf{S}^2\rangle - \langle S_z^2\rangle \tag{9.151}$$

In this case, $\langle \mathbf{S}^2\rangle$ is time-dependent.

Because $\mathbf{S}_1^2$ and $\mathbf{S}_2^2$ are still the constants of motion when $\omega_1 \neq \omega_2$, the atomic states which the system evolves are of the form $|S_1, S_2, m_1, m_2\rangle$, where S_1 and S_2 are the quantum numbers of the quadratic angular momenta of the two groups ($S_1 = S_2 = \frac{N}{4}$), m_1 and m_2 are the quantum numbers of the two z-components. If one looks at Fig.(9.7) in which show the time evolution of the atomic states of the superfluorescence system, one can find that in this case the trajectory in the diagram is much more complex. As is well known from the rules of addition of the angular momenta, the states of the system can also be written in the form $|S, m_1, m_2\rangle$, where the total quantum number of angular momenta may be $|S_1 - S_2|$, $|S_1 - S_2 + 1|$, $\cdots$, $S_1 + S_2 - 1$, $S_1 + S_2$, i.e., may be equal to $0,1,2,\cdots,\frac{N}{2}$. Therefore, the system in its evolution may occupy states having S smaller than its initial value $\frac{N}{2}$ and even land temporarily in the state belonging to S=0. Inasmuch as the emission rate strongly depends on the value of S, the emitted intensity of the radiation field of the system has more complex variation than the superfluorescence pulse.

Making a summation with respect to the number ℓ of atoms in (9.145), and using (9.147), then we have

$$T_{1\ell}\frac{d}{dt}\langle S_z\rangle = -\left(\langle S_z\rangle + \frac{N}{2}\right) - \left(\langle \mathbf{S}^2\rangle - \langle S_z\rangle^2 - \frac{N}{2}\right) \tag{9.152}$$

or

$$\frac{d}{dt}\langle S_z\rangle = -\gamma(\langle \mathbf{S}^2\rangle - \langle S_z\rangle^2 + \langle S_z\rangle) \tag{9.153}$$

with

$$\gamma = \frac{1}{T_{1\ell}}$$

Evidently, obtaining the solution of (9.153) requires the knowledge of $\langle \mathbf{S}^2\rangle$ as a function of time.

Since $\mathbf{S}_1^2$ and $\mathbf{S}_2^2$ are the constants of motion, as suggested by (9.149) the only part of $\langle \mathbf{S}^2\rangle$ which has to be determined is the scalar product $\langle \mathbf{S}_1 \cdot \mathbf{S}_2\rangle$, i.e., the time evolution of the operator $S(\eta, \xi)$ (9.136). Using the Heisenberg equation and proceeding a complicated calculation, we obtain the second-order differential equation for $\langle \mathbf{S}^2\rangle$ as

$$\begin{aligned}&\frac{d^2}{dt^2}\langle \mathbf{S}^2\rangle - \gamma\langle S_z\rangle\frac{d}{dt}\langle \mathbf{S}^2\rangle + (\omega_1 - \omega_2)^2\left[\mathbf{S}^2 - \frac{N}{2}\left(\frac{N}{2}+1\right)\right]\\ &= \frac{1}{2}(\omega_1 - \omega_2)^2\left[\langle S_z\rangle^2 - \left(\frac{N}{2}\right)^2\right]\end{aligned} \tag{9.154}$$

Solving the two coupled eqs.(.153) and (9.154), one can obtain the solution of $\langle S_z(t)\rangle$, furthermore the emitted intensity I(t) ($\approx \frac{d}{dt}\langle S_z\rangle$) may be determined. Here we give the numerical solutions of $I(t)$. As shown in Fig.(9.12), the curve (a) represents the time evolution of the superfluorescent beats for small detuning $\delta\omega = \omega_1 - \omega_2$. With the increasing of detuning $\delta\omega$, curves (b) and (c) shows that the system exhibits clear superfluorescent beats, the frequency of the beats also increases, and the variation of the beats amplitude decreases. For large detuning $\delta\omega$, the beats phenomena disappear as shown in the cure (d), in this case, the system displays the non-coherence independent radiation. It is worthy pointing out that the theoretical results about the emitted intensity, the variation of beat frequency and the beat amplitude are

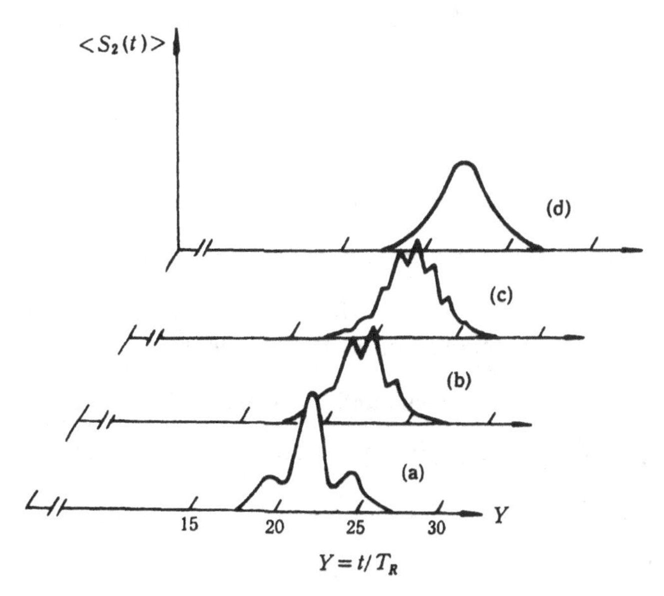

Figure 9.12: Time evolution of the emission intensity I obtained by solving (9.153) and (9.154). (a) $A = 2$, (b) $A = 4$, (c) $A = 6$, and (d) $A = \infty$

coincident with the main experimental results but not precise. For further investigation of the superfluorescence beats, some physical processes such as the quantum fluctuations must be taken into account in detail, then the theoretical results probably may be more coincident with the experimental results.

References

1. R.H.Dicke, Phys. Rev. **93** (1954) 99.
2. C.Leonardi, F.Persico, and G.Vetri, *Dicke model and theory of driven and spontaneous emission*, (La Rivista de Nuovo Cimento, 1986).
3. Jin-sheng Peng, *Resonance fluorescence and superfluorescence* (Science press, 1993, Beijing).
4. N.E.Rehler and J.H.Eberly, *Phys.Rev.* **A3** (1971) 1735.
5. N.Skrianowitz, E.P.Herman, J.C.Mcgillivrag, and M.S.Feld, *Phys. Rev. Lett.* **30** (1973) 309.

6. R.Banifacio and L.A.Lugiato, *Phys.Rev.* **A11** (1975) 1507; **A12** (1975) 587.

7. Q.H.F.Verhen, H.J.Hispoors, and H.M.Gibbs, *Phys.Rev.Lett.* **38** (1977) 764.

8. M.Gross, J.Raimond, and S.Haroche, *Phys.Rev.Lett.* **40** (1978) 1711

9. M.Gross and S.Haroche, *Phys.Rep.* **93** (1982) 301.

10. Q.H.F.Verhen and H.M.Gibbs, in *Dissipative system in quantum optics*, edited by R.Bonifacio (Springer, Berlin, 1982).

11. C.Leonardi, Jin-sheng Peng, and A.Vaglica, *J.Phys.* **B15** (1982) 4017.

12. Jin-sheng Peng, *J. of C.C.N.U.* **18** (1984) 24.

13. Jin-sheng Peng, *ACTA Physica Sinica* **33** (1984) 884.

14. Jin-sheng Peng, *J. of C.C.N.U.* **20** (1986) 36.

15. C.H.Keitel, M.O.Scully, and G.Sussman, *Phys.Rev.* **A45** (1992) 3242. xlatex plb9

16. R.Michalska-Trautman, *Phys.Rev.* **A46** (1992) 7270.

17. F.Haake and M.I.Kolobov, *Phys.Rev.Lett.* **71** (1993) 995.

CHAPTER 10

OPTICAL BISTABILITY

One of the most important subjects in quantum optics is the optical bistability which was first discussed theoretically by Szoke et al. in 1969, and observed experimentally by Gibbs et al. in 1976. Inasmuch as the optical bistability devices may be extensively applied in the optical communication and optical computer, it has extensively investigated both theoretically and experimentally over the recent years.

This chapter is devoted to discussing the basic features of the optical bistability by use of the quantum theory. First, we introduce the general characteristics and the production mechanism of the optical bistability. Then, we present a statement of quantum description for dispersive optical bistability.

10.1 Basic characteristics and the production mechanism of optical bistability

When a laser field passing through a medium which is consisted of a large number of atoms or molecules, the intensity of the output light changes nonlinearly owing to the interaction between the light and the medium. The term optical bistability refers to the situation in which two different output intensities of light are possible for a given input intensity of light. For example, when the laser field with intensity I_i incidents into the Fabry-Perot optical cavity which is full of the nonlinear medium as shown in Fig.(10.1), the relation between the intensity I_T of output light and the input intensity I_i of light can exhibit the characteristic as shown in Fig.(10.2).

Fig.(10.2) indicates that, if the input intensity I_i of light is small, the output intensity I_T of light is smaller than I_i due to the absorption and dispersion of medium. With the increasing of I_i, I_T also increases as shown in OA. But when the input intensity I_i reaches to I_B, the output intensity I_T jumps to

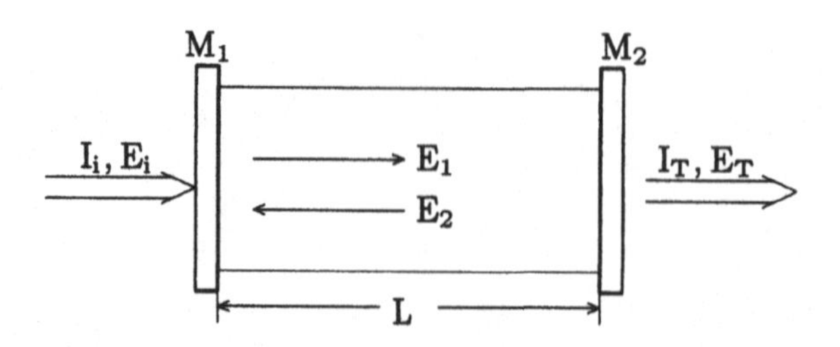

Figure 10.1: Schematic illustration for device generating optics bistability

I_C suddenly. Subsequently, I_T increases along the line CE continuously. Evidently because the saturation of the absorption and dispersion of medium, I_T increases linearly with the increasing of I_i and the slope of the line CE is larger than that of OB. In this moment, if I_T decreases slowly, then I_T decreases linearly along the line EC. When I_i decreases to I_B, I_T does not go back to I'_B and remains on the upper branch given by the curve segment ECD. As the input intensity I_i passes through the value I_A, the output intensity makes a transition from I'_D to I'_A suddenly, and traces out the curve AO. So we see, in the range ABCD, there exist two possible output intensities of light, that is to say, the optical bistability occurs. Which branch the output signal is in depends on how the input signal injects. Evidently, in the bistability range, the system can be forced to make a transition to the upper branch CE by injecting a pulse of light so that the total input intensity exceeds I_B, or the system can also be forced to make a transition to the lower state by injecting an opposed pulse of light.

Now we analyze the input and output characteristics of the system in which an input laser passing through the Fabry-Perot cavity. When a beam of the continuous wave laser injected in a optical cavity whose resonant frequency is resonant or near resonant to the frequency of laser, the incident field can be reflected and transmitted by the cavity mirrors placed in the two sides of the cavity. If the cavity is an empty one, then the intensity of output field is proportional to the intensity of input field, i.e., $I_T = |E_T|^2 \propto I_i$, here E_T is the amplitude of the output field. But when the cavity is full of medium, the interaction between the medium and the field leads to the absorption and dispersion of the field, so the amplitude E_T of output field becomes a non-

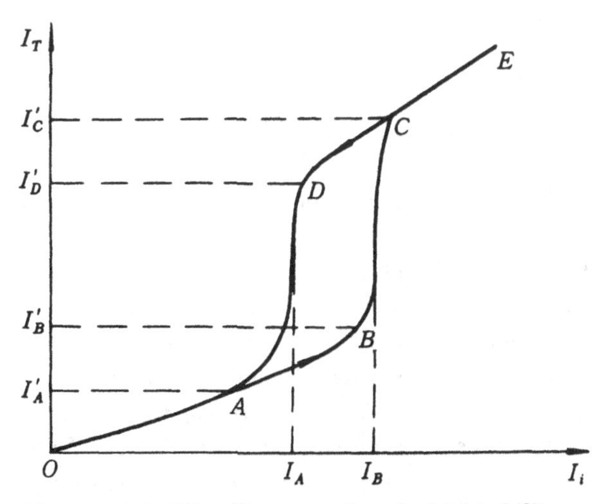

Figure 10.2: The diagram of optical bistability

linear function of the amplitude E_i, therefore, the intensity I_T of output field varies nonlinearly with the intensity I_i of input field. The cavity mirrors are assumed to be identical and lossless, and the reflectance coefficient R and the transmittance one T are defined as

$$R = |\rho|^2, \quad T = |\tau|^2 \tag{10.1}$$

where ρ and τ are the reflectance and transmittance amplitudes corresponding to R and T, respectively. E_1 and E_2 denote the amplitudes of the forward-going and backforward-going waves within the cavity, then according to the boundary conditions, E_1, E_2, E_T, and E_i satisfy

$$E_1 = \tau E_i + \rho E_2 \tag{10.2}$$

$$E_T = \tau E_1 \tag{10.3}$$

$$E_2 = \rho E_1 \exp(2ikL - \alpha L) \tag{10.4}$$

Here we have assumed that the atoms filled in the cavity are identical and are homogeneous. And k is the propagation constant of the field in the cavity, which obeys $k = n\omega_L/c$, ω_L is the frequency of incident field, n is the transmittance rate of medium, the term 2kL represents the difference of optical paths between E_1 and E_2, and α stands for the intensity absorption coefficient. Because E_T varies linearly with E_1, we only concern on the relation between I_1

and I_i in the following.

Combining (10.2) with (10.4) gives

$$E_1 = \frac{\tau E_i}{1 - \rho^2 \exp(2ikL - \alpha L)} \tag{10.5}$$

Clearly, if α and k are the nonlinear function of the light intensity in the cavity, then I_1 depends on I_i nonlinearly.

Assuming that the medium filled in the cavity is a pure absorption one, this means that there is no dispersion effect of medium on the field, that is to say, the propagation coefficient k is independent of the intensity of the field in the cavity, and the absorption coefficient α depends nonlinearly on the intensity of the field. To simplify the following analysis, we assume that the mirror separation L is adjusted so that $\rho^2 \exp(2ikL) = |\rho|^2 = R$. We also assume that $\alpha L \ll 1$, so that we can ignore the spatial variation of the intensity of the field inside the cavity. Then (10.5) reduces to

$$E_1 = \frac{\tau E_1}{1 - R(1 - \alpha L)} \tag{10.6}$$

And the relation between I_1 and I_i obeys

$$I_1 = \frac{T I_i}{[1 - R(1 - \alpha L)]^2} \tag{10.7}$$

This equation can be simplified by introducing the dimensionless parameter C

$$C = \frac{R\alpha L}{1 - R} \tag{10.8}$$

then becomes

$$I_1 = \frac{1}{T} \frac{I_i}{(1 + C)^2} \tag{10.9}$$

here the relation $R + T = 1$ has been used.

For simplicity, we assume that the atoms (for example, the two-level atoms) in medium interacting with the field via one-photon process, according to Chapter 8, the unsaturated absorption coefficient α obeys the relation

$$\alpha = \frac{\alpha_0}{1 + (I_1 + I_2)/I_s} \tag{10.10}$$

where α_0 denotes the saturated absorption coefficient, I_s is the saturation intensity corresponding to α_0. In the long-time limit, the intensities of the forward-going and backward-going fields are approximately equal, i.e., $I_1 \approx I_2$. Therefore, the dimensionless parameter C is given by

$$C = \frac{C_0}{1 + 2I_2/I_s} \tag{10.11}$$

with $C_0 = R\alpha_0 L/(1-R)$. Substituting eq.(10.11) into (10.9) gives

$$I_i = TI_1\left(1 + \frac{C_0}{1 + 2I_1/I_s}\right)^2 \tag{10.12}$$

Noticing (10.3), i.e., $I_T = TI_1$, then the relation between the output intensity I_T and the input intensity I_i obeys

$$I_i = I_T\left(1 + \frac{C_0}{1 + 2I_T/(TI_s)}\right)^2 \tag{10.13}$$

The input-output relation implied by (10.13) is illustrated graphically in Fig.(10.3). As we see, when C_0 is small, i.e., the saturated absorption coefficient is weak, the output intensity I_T depends linearly on the input intensity

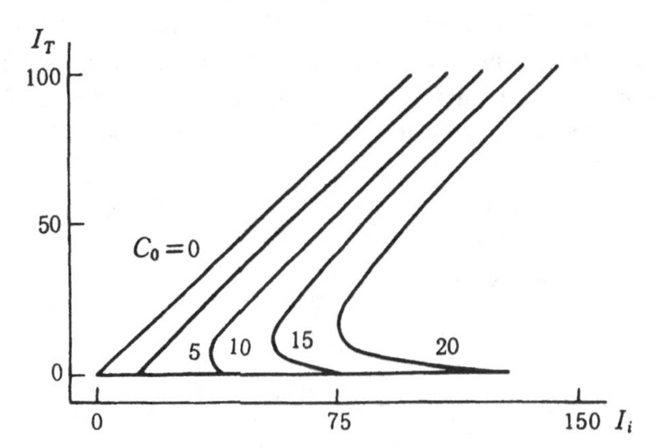

Figure 10.3: Relation between the outpout intensity I_T and the input one I_i for the absorption bistability

I_i. But for C_0 greater than a given value (for example, $C_0 \geq 10$ as shown in Fig.(10.3)), this input-output curve becomes S-shaped. Evidently, in the two ends of the S-shaped curve, i.e., the parts of $\frac{dI_T}{dI_i} > 0$, I_T has two stable branches. That is to say, for a certain range of the input intensity I_i, the output intensity I_T can have two different stability states, i.e., the optical bistability is exhibited. In the middle part of the S-shaped curve, $\frac{dI_T}{dI_i} < 0$, so the system in this part is unstable. If the system is initially in this state, it will rapidly switch to one of the stable branches due to the growth of a small perturbation. Since the medium we discuss here only has the absorption effect of the field, we interpret this bistability as the absorptive bistability.

From the view of physical origin, the appearance of the absorptive bistability arises from the nonlinear absorption of medium and the feedback effect. When the incident field is weak, the unsaturated absorption of medium leads to the output field linearly dependent on the input field. With the increasing of intensity of the incident field, the absorption of medium is saturated, so the cavity field becomes stronger. The stronger the cavity field is, the weaker the effect of absorption. This is just a positive feedback effect. When the intensity of the incident field reaches to a certain value, the slope becomes very large, this make a transition from the lower branch to the upper one.

Up to now, we have discussed the basic mechanism of the absorption optical bistability. If the medium in the Fabry-Perot cavity has the dispersive effect on the cavity field, what is the relation between the output intensity I_T and the input intensity I_i of light ? Now we discuss this in the following.

If the medium in the cavity only has the dispersion effect and no absorption effect on the cavity field, i.e., $\alpha = 0$, and the refractive index n depends nonlinearly on the field intensity. In this case, (10.5) becomes

$$E_1 = \frac{\tau E_i}{1 - \rho^2 \exp(2ikL)} = \frac{\tau E_i}{1 - Re^{i\delta}} \tag{10.14}$$

here we have assumed that

$$\rho^2 = Re^{i\phi} \tag{10.15}$$

$$\delta = \phi + \frac{2n(I)\omega_L L}{c} \tag{10.16}$$

where the refractive index $n(I)$ is the function of the intensity I. For a typical dispersive medium–Kerr medium, the refractive index is

$$n(I) = n_0 + n_2 I \tag{10.17}$$

here n_0 and n_2 are the parameters unrelated to I. Inserting (10.16) and (10.17) into (10.14) gives

$$T' = \frac{I_1}{I_i} = \frac{T^{-1}}{1 + (4R^2/T^2)\sin^2(\delta/2)} \tag{10.18}$$

The relation $I = I_1 + I_2 \approx 2I_1$ has been used. And the phase shift δ is

$$\delta = \phi + \frac{2n_0\omega_L L}{c} + \frac{4n_2\omega_L L I_1}{c} = \delta_0 + \left(\frac{4n_2\omega_L L}{c}\right) I_1 \tag{10.19}$$

It is evident that T' is the transmittance coefficient with considering the effect of the Kerr medium in the cavity, it reaches to the maximum values for $\delta = 2m\pi$ (m=0,1,2,$\cdots$), as shown in Fig.(10.4). We can choose appropriate parameters such as R, T, n_2 and L so that the width of peaks may be very sharp. In this

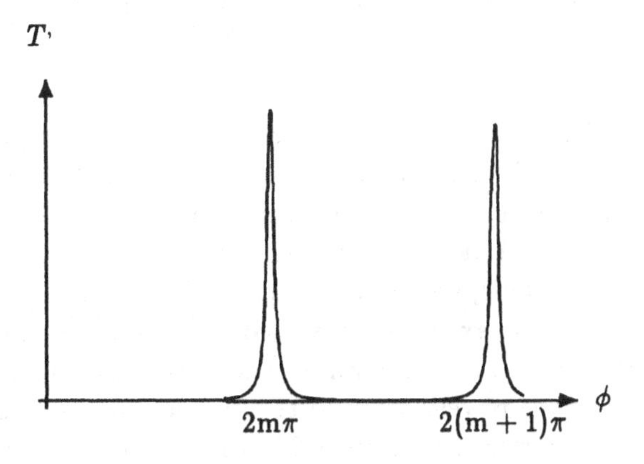

Figure 10.4: Curve for $T' - \phi$ relationship

case, $\sin(\delta/2)$ can be expanded as a series of power terms at $\delta = 2m\pi$ and only retain to its first-order term, i.e.,

$$\sin(\delta/2) = \frac{\delta - 2m\pi}{2} = \phi_0 + \left(\frac{2n_2\omega_L L}{c}\right) I_1 \tag{10.20}$$

with $\phi_0 = (\delta_0 - 2m\pi)/2$. Inserting (10.20) into (10.18) gives the relation between I_1 and I_i as

$$I_i = I_1 \left\{1 + 4R^2 \frac{[\phi_0 + (2n_2\omega_L L/c)I_1]^2}{T^2}\right\} \tag{10.21}$$

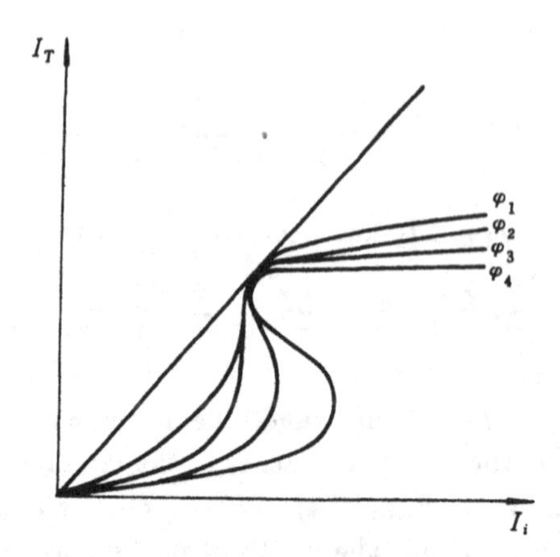

Figure 10.5: Characteristic curve of the dispersive optical bistability for different ϕ_0

Noticing $I_T = TI_1$, the output-input relation can also obtained from eq.(10.21). As shown in Fig.(10.5), the output-input curve exhibits bistability property for different initial phase ϕ_0. Since the medium in cavity is a dispersive one, we call this kind of optical bistability as the dispersive optical bistability.

The generation of dispersive optical bistability can be explained as follows. Since the refractive index n depends nonlinearly on the internal intensity I of the field, the resonant wavelength of cavity is related to I. The difference between the resonant wavelength and the wavelength λ_L of incident laser affects on the transimittance coefficient T', which leads to the internal intensity I to be changed. This is just a feedback process. So similar to the case in the absorptive bistability, the physics origin of generating dispersive bistability is also due to the nonlinear interaction between the field and the medium, and to the feedback effect.

Except these two types, there is a kind of absorptive-dispersive mixture optical bistability. In this bistability system, both the absorption and dispersion effects of medium are considered. Since the basic generation mechanism is similar to the above discussions, here we no longer give detail discussions.

10.2 Quantum description of the dispersive optical bistability

In Section 1, we have generally discussed the basic features of the optical bistability and analyzed its generation mechanism. In this section, our aim is to discuss how to describe optical bistability by means of quantum theory. Here we only concentrate our attention on the dispersive bistability.

10.2.1 Hamiltonian describing the optical bistability system

Considering that the cavity whose resonant frequency is ω_c contains a non-linear dispersive medium. From the theory of nonlinear optics, the polarization of the medium interacting with a single-mode field is expressed as

$$\mathbf{P} = \chi^{(1)} \cdot \mathbf{E} + \chi^{(2)} \cdot \mathbf{EE} + \chi^{(3)} \cdot \mathbf{EEE} + \cdots \tag{10.22}$$

where $\mathbf{P}$ is the polarization of the medium, $\mathbf{E}$ is the electric field, $\chi^{(n)}$ is a $(n+1)$th rank susceptibility tensor. According to the Maxwell equations

$$\begin{aligned}
\nabla \times \mathbf{E} &= -\frac{1}{c}\frac{\partial}{\partial t}\mathbf{B} \\
\nabla \times \mathbf{B} &= \frac{1}{c}\frac{\partial}{\partial t}(\varepsilon_0 \mathbf{E} + \mathbf{P}) \\
\nabla \cdot (\varepsilon_0 \mathbf{E} + \mathbf{P}) &= 0 \\
\nabla \cdot \mathbf{B} &= 0
\end{aligned} \tag{10.23}$$

The energy density of the radiation field interacting with the nonlinear medium obeys

$$c\nabla \cdot (\mathbf{E} \times \mathbf{B}) = -\frac{1}{2}\frac{\partial}{\partial t}(\varepsilon_0 \mathbf{E}^2 + \frac{\mathbf{B}^2}{\mu_0}) - \mathbf{E} \cdot \frac{\partial}{\partial t}\mathbf{P} \tag{10.24}$$

Since $\mathbf{E} \times \mathbf{B}$ is the Poynting vector, (10.24) shows that the rate of the electromagnetic energy flowing from unit volume is equal to the decreased rate of the electromagnetic energy density stored in the cavity. Inserting (10.22) into (10.24) gives

$$c\nabla \cdot (\mathbf{E} \times \mathbf{B}) = -\frac{\partial}{\partial t}U(r,t) \tag{10.25}$$

where

$$\begin{aligned}
U(\mathbf{r},t) = &\frac{1}{2}(\varepsilon_0 \mathbf{E}^2 + \mathbf{B}^2) + \frac{1}{2}\left(\mathbf{E} \cdot \chi^{(1)} \cdot \mathbf{E} + \frac{2}{3}\mathbf{E} \cdot \chi^{(2)} \cdot \mathbf{EE}\right. \\
&\left. + \frac{1}{2}\mathbf{E} \cdot \chi^{(3)} \cdot \mathbf{EEE} + \cdots\right)
\end{aligned} \tag{10.26}$$

is the instantaneous electromagnetic energy density. Therefore, the Hamiltonian of the nonlinear medium-field coupling system is written as

$$H =: \int d^3\mathbf{r} \left\{ \frac{1}{2\mu_0}|\mathbf{B}|^2 + \mathbf{E} \cdot \left[\frac{1}{2}(\varepsilon_0 + \chi^{(1)}) \cdot \mathbf{E} + \frac{1}{3}\chi^{(2)} \cdot \mathbf{EE} \right.\right.$$
$$\left.\left. + \frac{1}{4}\chi^{(3)} \cdot \mathbf{EEE} + \cdots \right] \right\} : \tag{10.27}$$

where :: denotes the normal ordering. If the third-order polarization effect plays an dominant role, then the terms containing $\chi^{(2)}$ and $\chi^{(n)}$ can be ignored. The above equation therefore reduces to

$$H =: \int d^3\mathbf{r} \left\{ \frac{1}{2\mu_0}|\mathbf{B}|^2 + \mathbf{E} \cdot [(\varepsilon_0 + \chi^{(1)}) \cdot \mathbf{E}/2 + \chi^{(3)}\mathbf{EEE}/4] \right\} : \tag{10.28}$$

Now we derive the quantized expression of the Hamiltonian (10.28). Since the electric operator E can be expressed as

$$\mathbf{E} = i \left(\frac{\hbar\omega_c}{2\varepsilon_0} \right)^{1/2} [a\mathbf{u}(\mathbf{r}) \exp(-i\omega_c t) - a^\dagger \mathbf{u}^*(\mathbf{r}) \exp(i\omega_c t)] \tag{10.29}$$

where the mode function $\mathbf{u}$ is defined to satisfy

$$\int \mathbf{u}^*(\mathbf{r}) \cdot (1 + \chi^{(1)}/\varepsilon_0) \cdot \mathbf{u}(\mathbf{r}) d^3\mathbf{r} = 1 \tag{10.30}$$

Inserting (10.29) into (10.28), using (10.30), and considering the rotating-wave approximation, i.e., neglecting the terms such as $a^4 e^{-i4\omega_c t}$ and $a^{\dagger 3} a e^{2i\omega_c t}$, then (10.28) becomes

$$H = \hbar\omega_c a^\dagger a + \hbar\chi'' a^{\dagger 2} a^2 \tag{10.31}$$

where

$$\chi'' = \frac{3\hbar\omega_c^2}{8\varepsilon_0^2} \int d^3\mathbf{r} \chi^{(3)} |\mathbf{u}(\mathbf{r})|^4 \tag{10.32}$$

Eq.(10.28) is the quantized expression of the Hamiltonian describing a cavity field interacting with a nonlinear medium.

For the system shown in Fig.(10.1), if the incident field is a strong laser field, then it can be characterized by a classical electromagnetic field, so the interaction of the cavity field with the incident one can be characterized by the Hamiltonian

$$V = i\hbar[a^\dagger E(t) \exp(-i\omega_L t) - aE^*(t) \exp(i\omega_L t)] \tag{10.33}$$

Considering the field damped by the cavity, then the total Hamiltonian of this system may be written as

$$H = \hbar\omega_c a^\dagger a + \hbar\chi'' a^{\dagger 2} a^2 + i\hbar[a^\dagger E(t)\exp(-i\omega_L t) - aE^*(t)\exp(i\omega_L t)] + \sum_k \omega_k b_k^\dagger b_k + \sum_k \varepsilon_k (a^\dagger b_k + a b_k^\dagger) \tag{10.34}$$

Here $E(t)$ and ω_l are the amplitude and the frequency of the laser field respectively, b_k and $b_k^\dagger$ are the reservoir operators for the cavity damping. Next we discuss the optical bistability property of the system starting from (10.34).

10.2.2 Optical bistability properties of the system

If the mirror M_2 in Fig.(10.1) is a full-transmittance one, then the expectation value of the operator a is just the transmittance field amplitude. In order to obtain $\langle a\rangle$, we need to know the density operator of the system. This demands to solve the master equation of the system. In the interaction picture, from (5.89) we get

$$\frac{d}{dt}\rho = -i\Delta\omega[a^\dagger a, \rho] - i\chi''[a^{\dagger 2} a^2, \rho] + [E(t)a^\dagger - E^*(t)a, \rho] + \kappa'(2a\rho a^\dagger - \rho a^\dagger a - a^\dagger a\rho) \tag{10.35}$$

here we have assumed that the reservoir field is in a vacuum state, and $\Delta\omega = \omega_c - \omega_l$ is the frequency detuning between the incident field and the cavity one. Denoting the mean amplitude of the transmittance field is

$$\alpha = \langle a\rangle = \mathrm{Tr}(\rho a) \tag{10.36}$$

then from (10.35) we get

$$\frac{\partial}{\partial t}\alpha = -i\Delta\omega\alpha - 2i\chi''\langle a^\dagger a^2\rangle + E(t) - \kappa'\alpha \tag{10.37}$$

Similar to (9.78)-(9.79), here we have neglected the quantum fluctuations $\langle a^\dagger a^2\rangle - \langle a^\dagger\rangle\langle a\rangle^2$, i.e., we have assumed

$$\langle a^\dagger a^2\rangle = \langle a^\dagger\rangle\langle a\rangle^2 = \alpha^*\alpha^2 \tag{10.38}$$

then eq.(10.37) becomes

$$\frac{\partial}{\partial t}\alpha = -i\Delta\omega\alpha - 2i\chi''\alpha^*\alpha^2 + E(t) - \kappa'\alpha \tag{10.39}$$

Similarly, $\alpha^* = \langle a^\dagger \rangle$ obeys

$$\frac{\partial}{\partial t}\alpha^* = i\Delta\omega\alpha^* + 2i\chi''\alpha\alpha^{*2} + E^*(t) - \kappa'\alpha^* \tag{10.40}$$

Since α and α^* are two independent variables, (10.39) and (10.40) can be rewritten as

$$\frac{\partial}{\partial t}\begin{pmatrix}\alpha \\ \alpha^*\end{pmatrix} = \begin{pmatrix} E(t) - (\kappa' + i\Delta\omega)\alpha - 2i\chi''\alpha^*\alpha^2 \\ E^*(t) - (\kappa' - i\Delta\omega)\alpha^* + 2i\chi''\alpha\alpha^{*2}\end{pmatrix} \tag{10.41}$$

Usually, we only focus our attention on the case of the steady output of transmittance field, i.e., on the steady-state solution of (10.41). The steady-state solutions of (10.41) give

$$\begin{aligned} E &= (\kappa' + i\Delta\omega)\alpha + 2i\chi''\alpha^*\alpha^2 \\ E^* &= (\kappa' - i\Delta\omega)\alpha^* - 2i\chi''\alpha\alpha^{*2} \end{aligned} \tag{10.42}$$

If we define that the intensity of transmittance field is proportional to the mean photon number as

$$|E_T|^2 = \alpha^*\alpha \tag{10.43}$$

then from (10.42) the relation between the intensity of this incident field and that of the transmittance field gives

$$|E|^2 = |E_T|^2[\kappa' + (\Delta\omega + 2|E_T|^2\chi'')^2] \tag{10.44}$$

Evidently, this equation is similar to (10.21). Fig.(10.6) plots the relation between $|E_T|^2$ and $|E|^2$ for different κ', χ'' and $\delta\omega$. The properties of the transmittance intensity depend strongly on the different choices of κ', χ'', and $\Delta\omega$. Fig.(10.6) indicates that for $\kappa' = 1$, $\chi'' = 0.5$, and $\Delta\omega = -10$, the optical bistability of the transmittance intensity occurs. But for $\kappa' = 0.1$, $\chi'' = 0.2$, and $\Delta\omega = 0.5$, there is no optical bistability in the system. Under what conditions does the optical bistability occur ? Next we investigate the regions of the stability by use of a linearized analysis method.

Due to the quantum fluctuations of the system, there will display a small additional variation in the steady-state solution. The basic idea of the linearized analysis method is to introduce a small variation in the neighbourhood of the steady-state solution

$$\alpha(t) = \alpha_0 + \alpha_1(t) \tag{10.45}$$

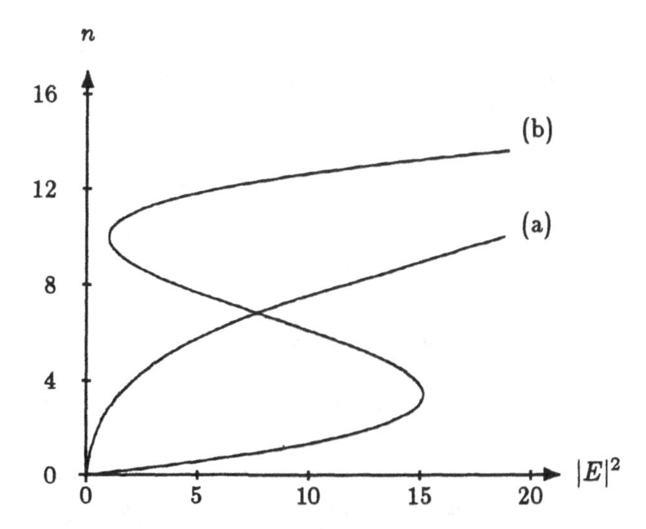

Figure 10.6: n versus $|E|^2$ for different $\Delta\omega$, κ' and χ'', (a) $\Delta\omega = 0.5$, $\kappa' = 1$, $\chi'' = 0.2$, (b) $\Delta\omega = -10$, $\kappa' = 1$, $\chi'' = 0.5$

where α_0 is the steady-state solution, $\alpha_1(t)$ represents the small variation. Inserting (10.45) and its conjugation into (10.41) and only retaining the first-order terms of $\alpha_1(t)$, then we obtain the following linearized equations for the variation $\alpha_1(t)$ as

$$\frac{\partial}{\partial t}\begin{pmatrix}\alpha_1(t)\\ \alpha_1^*(t)\end{pmatrix} = -\begin{pmatrix}\kappa + 4\chi|E_T|^2 & 2\chi\alpha_0^2\\ 2\chi^*\alpha_0^{*2} & \kappa^* + 4\chi^*|E_T|^2\end{pmatrix}\begin{pmatrix}\alpha_1(t)\\ \alpha_1^*(t)\end{pmatrix}$$
$$= -A\begin{pmatrix}\alpha_1(t)\\ \alpha_1^*(t)\end{pmatrix} \tag{10.46}$$

where $\chi = i\chi''$, $\kappa = \kappa' + i\Delta\omega$. From the theory of differential equations we know that in order to have a steady-state solution in (10.41), the matrix A must obey

$$\mathrm{Tr}(A) > 0, \quad \mathrm{Det}(A) > 0 \tag{10.47}$$

From (10.46) we know

$$\mathrm{Tr}(A) = 2\kappa' \tag{10.48}$$
$$\mathrm{Det}(A) = 12|\chi|^2|E_T|^2 + 4|E_T|^2(\chi\kappa^* + \chi^*\kappa) + |\kappa|^2 \tag{10.49}$$

Because κ' represents the damping of the field in the cavity, its value is always larger than zero, i.e., $\kappa' > 0$, so that $\mathrm{Tr}(A) > 0$. In order that (10.49) obeys (10.47b), $|E_T|^2$ must obey

$$|E_T|^2 > n_+ \qquad or \qquad |E_T|^2 < n_- \tag{10.50}$$

where

$$n_\pm = \frac{-2\Delta\omega \pm \sqrt{\Delta\omega)^2 - 3\kappa'}}{6\chi''} \tag{10.51}$$

Since $n_\pm$ must be positive number, (10.51) shows that κ', χ'', and $\Delta\omega$ satisfy

$$(\Delta\omega)^2 > 3\kappa'^2 \tag{10.52}$$
$$\Delta\omega\chi'' < 0$$

That is to say, when the parameters κ', χ'', and $\Delta\omega$ satisfy (10.52), the transmittance intensity of the field can display the bistability. Evidently, the parameters $\kappa' = 1$, $\chi'' = 0.5$, and $\Delta\omega = -10$ obey (10.52), so there are two stable transmittance branches of $|E_T|^2$ in Fig.(10.5).

From (10.3) we know that the transmittance intensity $|E_T|^2$ is proportional to the output intensity, thus there are two stable branches for the output intensity. Consequently, the above result shows that if the Fabry-Perot cavity contains nonlinear dispersive medium, the optical bistability can occur under certain conditions.

References

1. A.Szoke, V.Daneu, J.Goldhar, and N.A.Kurnit, *Appl.Phys.Lett.* **15** (1969) 376.

2. H.M.Gibbs, S.L.McCall, and T.N.C.Venkatesan, *Phys.Rev.Lett.* **36** (1976) 113.

3. P.D.Drummond and D.F.Walls, *J.Phys.* **A13** (1980) 725.

4. L.A.Lugiato, *Theory of Optical Bistability* in Progress in Optics XXI, edited by E.Wolf (North-Holland, 1984).

5. Y.Shen, *The Principles of Nonlinear Optics*, Chap.2 (John Wiley and Sons, Inc., 1984).

6. F.A.M.de Oliveira and P.L.Knight, *Phys.Rev.* **A39** (1989) 3417.

7. R.W.Boyd, *Nonlinear Optics*, Chap.6 (Academic Press, Inc. 1992).

CHAPTER 11

EFFECTS OF VIRTUAL PHOTON PROCESSES

In the previous chapters, the rotating-wave approximation has been adopted when we discuss the interaction between a two-level atom and a radiation field, then the effects of the non-rotating-wave terms such as

$$\varepsilon(a^{\dagger}S_{+}+aS_{-}) \quad or \quad \sum_{k}\varepsilon_{k}(a_{k}^{\dagger}S_{+}+a_{k}S_{-}) \tag{11.1}$$

are ignored. This ignorance is on the basis of the fact that the energy variation of the atom-field coupling system due to the transition processes describing by (11.1) is large, so that the corresponding transitions are much less than the processes described by the rotating-wave terms. Eq.(11.1) represents two kinds of transition processes which the atom in its lower state returns to its upper state with emitting a photon simultaneously (shown in Fig.11.1a), and the atom in its upper state goes to its lower state with absorbing a photon from the field simultaneously (shown in Fig.11.1b). Meanwhile, the energy variation of the system is $\Delta E = E_2 - E_1 + \hbar\omega = \hbar\omega_0 + \hbar\omega \gg 0$, this means that these transition processes are not energy conservation over the transition duration in the atom-field coupling system, so that these processes can usually be neglected.

On the other hand, according to the Heisenberg uncertainty relation for the energy and time

$$\Delta E \cdot \Delta\tau = \hbar \tag{11.2}$$

The lifetime of the photons created in the transition processes described by the non-rotating-wave terms is very short because of $\Delta E \gg 0$, we call these photons as virtual photons. Since the energy variation of the system due to the transition processes corresponding to the rotating-wave approximation terms $\Delta E = \hbar\omega_0 - \hbar\omega$ is nearly equal to zero, the lifetime $\Delta\tau$ of the photons created

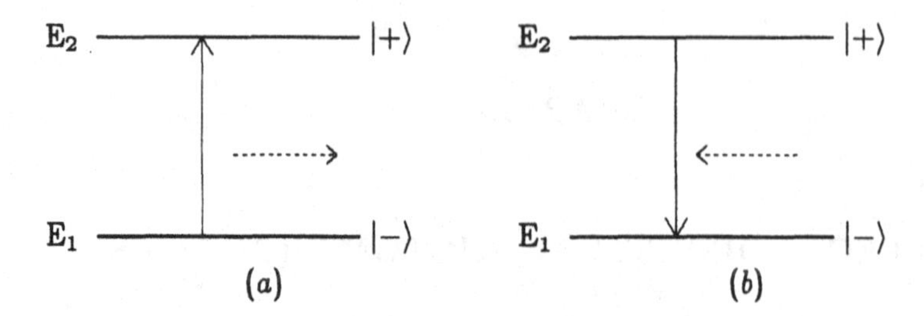

Figure 11.1: Diagrammatic illustration of the virtual photon transition processes

in these processes is very long, then these photons are just the real photons as usually named. However, the virtual photons can be reabsorbed by the atom in very short time, in this moment, the atom in its upper state returns to its lower state again. So the energy conservation law holds in a relative long time. As usual, the transition processes are named the virtual photon processes, the field made up of virtual photons is termed as the virtual field, and the effect due to the virtual photon processes is called the virtual field effect.

In view of the whole atom-field system, except the real photons with long lifetime which can be measured, there are many virtual photons which are created and absorbed quickly. These virtual photons form a cloud surrounding the field source. Even in a neutral atom the effects of the quantum fluctuations make certain energy variation of the atomic system, then induce the emittion and reabsorption processes of the virtual photons. So the neutral atom is surrounded by the virtual photon cloud. Since the virtual photon processes exist certainly in the atom-field coupling system, it is necessary to investigate the influences of the virtual field on the characteristics of the system.

In the present chapter, we first discuss the correction of the virtual field to the ground energy of a Hydrogen atom to reveal the physical origin of the Lamb shift. Subsequently, the effects of virtual photon processes on the phase features of the field in the system of a two-level atom interacting with a single-mode field are also analyzed. Finally, we discuss the influences of the virtual photon processes on the squeezing of light.

11.1 Relation between the Lamb shift of a Hydrogen atom and the virtual photon field

As we know, the Hamiltonian describing a bare Hydrogen atom is expressed as

$$H_a = \sum_n E_n \sigma_{nn}, \qquad \sigma_{nm} = |n\rangle\langle m| \tag{11.3}$$

here $|n\rangle$ is the eigenstate of the Hamiltonian of the bare Hydrogen atom corresponding to the eigenvalue E_n, σ_{nm} is the generalized atomic operator of the Hydrogen atom. In the dipole approximation, the Hydrogen atom interacting with the radiation field can be characterized by the Hamiltonian as

$$H = H_a + H_f + H_i \tag{11.4}$$

where $H_f = \sum_k \omega_k a_{kj}^\dagger a_{kj}$ is the Hamiltonian of the radiation field, H_i describes the interaction between the radiation field and the Hydrogen atom, it can be written as

$$H_i = -\frac{e}{mc}\mathbf{A}\cdot\mathbf{P} = -\frac{e}{m}\sum_{\ell,n}\sum_{kj}\left(\frac{2\pi\hbar}{V\omega_k}\right)^{1/2}(\mathbf{e}_{kj}\cdot\mathbf{P}_{\ell n})\sigma_{\ell n}(a_{kj}+a_{kj}^\dagger) \tag{11.5}$$

where $\mathbf{P}_{\ell n}$ describes the dipole moment strength between $|\ell\rangle$ and $|n\rangle$.

If the Hydrogen atom is in its bare ground state $|1s\rangle$ and the field in the vacuum state $|\{0_k\}\rangle$, the whole system (atom+field) is in the ground state

$$|\phi\rangle_0 = |1s, \{0_k\}\rangle \tag{11.6}$$

In this case, the system can not emit real photons, but due to the effects of fluctuations of the vacuum field, the energy variation ΔE of the system can be changed, thus the virtual photon processes as shown in Fig.(11.1) can also take place. That is to say, under the interaction of the vacuum field, the atom in its bare ground state goes to excited states and returns to ground state with fast creation and reabsorption of the virtual photons. These transition processes take place continuously, so the atom in its bare ground state is surrounded by the virtual photon cloud. We call this ground state surrounded by the virtual photon cloud as the dressed ground state.

Now, we discuss the influences of virtual photon field on the ground energy of the Hydrogen atom. Since the interaction energy H_i related to the virtual photon processes is much less than H_a and H_f , H_i can be regarded as a perturbation term. By use of the perturbation theory, the first-order perturbation correction to the ground state function (11.6) gives

$$|\phi\rangle_1 = \sum_i \sum_{kj} \frac{\langle i, 1(kj)|H_i|1s, \{0_k\}\rangle}{E_{1s} - E_i - \hbar\omega_k} |i, 1(kj)\rangle \tag{11.7}$$

where the state $|i, 1(kj)\rangle$ represents the atom in its i-th excited state and one photon in the k_j-th mode of the radiation field. Explicitly, $\langle i, 1(kj)|H_i|1s, \{0_k\}\rangle$ is the interaction matrix element, which represents that the atom goes to its excited state $|i\rangle$ from its ground state $|1s\rangle$ with emitting a photon simultaneously, or the atom returns to its ground state $|1s\rangle$ from excited state $|i\rangle$ with absorbing a photon simultaneously.

The second-order perturbation correction is

$$\begin{aligned} |\phi\rangle_2 = & -\frac{1}{2} \sum_i \sum_{kj} \frac{|\langle i, 1(kj)|H_i|1s, \{0_k\}\rangle|^2}{(E_{1s} - E_i - \hbar\omega_k)^2} |1s, \{0_k\}\rangle \\ & + \sum_{i\ell} \sum_{kj,k'j'} D^{i\ell}_{kjk'j'} |l, 1(kj), 1(k'j')\rangle \end{aligned} \tag{11.8}$$

where $|l, 1(kj), 1(k'j')\rangle$ is a state which the atom is in the state characterized by quantum number ℓ and two photons in the kj and $k'j'$ normal modes. The second term related to two-photon processes may be neglected, because the two-photon processes do not contribute to the second-order energy level shift of the ground state. Thus, the modified function of the ground state gives

$$|\phi\rangle = |\phi\rangle_0 + |\phi\rangle_1 + |\phi\rangle_2 \tag{11.9}$$

This is the dressed ground state of the system within second-order perturbation correction.

By means of (11.3) and (11.9) we obtain the expectation value of the bare atomic Hamiltonian H_a as follows

$$\begin{aligned} \langle H_a\rangle = \langle\phi|H_a|\phi\rangle = {}_0\langle\phi|H_a|\phi\rangle_0 + {}_1\langle\phi|H_a|\phi\rangle_1 \\ + {}_0\langle\phi|H_a|\phi\rangle_2 + {}_2\langle\phi|H_a|\phi\rangle_0 \end{aligned} \tag{11.10}$$

where

$${}_0\langle\phi|H_a|\phi\rangle_0 = \sum_n E_n \langle 1s, \{0_k\}|\sigma_{nn}|1s, \{0_k\}\rangle = E_{1n}$$

$$
{}_1\langle\phi|H_a|\phi\rangle_1 = \sum_i \sum_{kj} \frac{E_i 2\pi e^2 |\mathbf{e}_{kj}\cdot\mathbf{P}_{0i}|^2}{\hbar V m^2 \omega_k(\omega_{0i}+\omega_k)^2} \tag{11.11}
$$

$$
{}_0\langle\phi|H_a|\phi\rangle_2 = -\sum_i \sum_{kj} \frac{E_0 2\pi e^2 |\mathbf{e}_{kj}\cdot\mathbf{P}_{0i}|^2}{\hbar V m^2 \omega_k(\omega_{0i}+\omega_k)^2} = {}_2\langle\phi|H_a|\phi\rangle_0
$$

Substituting (11.11) into (11.10), we have

$$
\langle H_a\rangle = E_{1s} + \frac{2\pi e^2}{V m^2}\Sigma_i \Sigma_{kj} \frac{|\mathbf{e}_{kj}\cdot\mathbf{P}_{0i}|^2}{\omega_k(\omega_{0i}+\omega_k)^2}\omega_{0i} \tag{11.12}
$$

In the continuum limit, for the fixed polarization vector j, the summation of all k may be replaced by integral, i.e.,

$$
\sum_k \{\ \} \longrightarrow \left(\frac{V^{1/3}}{2\pi c}\right)^3 \int \omega_k^2 d\omega_k d\Omega \tag{11.13}
$$

Since the photons are produced in the processes of the atomic transition from excited state to ground state, the frequencies of the photons are restricted. In the non-relativistic approximation, the frequency of photon emitted by the atom must satisfy the equation $\hbar\omega \leq mc^2$, so that the maximum frequency of radiation field is $\omega_M = mc^2/\hbar$, which may be regarded as the upper limit in the integration (11.13). Thus,

$$
\langle H_a\rangle = E_{1s} + \frac{e^2}{m^2 c^3 4\pi^2}\sum_i \sum_{j=1}^{2} \int_0^{\omega_M} d\omega_k \frac{\omega_k\omega_{0i}}{(\omega_{0i}+\omega_k)} \int d\Omega |\mathbf{e}_{kj}\cdot\mathbf{p}_{0i}|^2 \tag{11.14}
$$

Using the summation relation for polarization vectors

$$
\sum_{j=1}^{2} |\mathbf{e}_{kj}\cdot\mathbf{P}_{0i}|^2 = |\mathbf{P}_{0i}|^2(\cos^2\alpha + \cos^2\beta) \tag{11.15}
$$

where α is the angle between the dipole moment vector $\mathbf{P}_{0i}$ and the polarization vector $\mathbf{e}_{kj}$, and β stands for the angle between $\mathbf{P}_{0i}$ and $\mathbf{e}_{kj}$ (shown in Fig.2). If θ is the angle between $\mathbf{P}_{0i}$ and the wave vector $\mathbf{k}$, then by the law of cosines, we have

$$
\sum_{j=1}^{2} |\mathbf{e}_{kj}\cdot\mathbf{P}_{0i}|^2 = |\mathbf{P}_{0i}|^2(1-\cos^2\theta) \tag{11.16}
$$

When we take the z-axis for the angular integration in (11.14) to be along $\mathbf{P}_{0i}$, then

$$d\Omega = \sin\theta d\theta d\phi \tag{11.17}$$

and

$$\int_0^{2\pi} d\phi \int_0^{\pi} \sin\theta(1-\cos^2\theta)d\theta = \frac{8\pi}{3} \tag{11.18}$$

Therefore, (11.14) becomes

$$\begin{aligned}\langle H_a\rangle &= E_{1s} + \frac{2e^2}{3m^2c^3\pi}\sum_i \int_0^{\omega_M} d\omega_k \frac{\omega_k\omega_{0i}}{(\omega_k+\omega_{0i})}|\mathbf{P}_{0i}|^2 \\ &= E_{1s} + \frac{2e^2}{3\pi m^2c^3}\sum_i \omega_{0i}|\mathbf{P}_{0i}|^2\left[-\frac{\omega_M}{\omega_M+\omega_{0i}} + \ln\left(\frac{\omega_M+\omega_{0i}}{\omega_{0i}}\right)\right] \\ &= E_{1s} + \frac{2e^2}{3\pi m^2c^3}\sum_i \omega_{0i}|\mathbf{P}_{0i}|^2 \ln\left(\frac{\omega_M}{\omega_{0i}}\right)\end{aligned} \tag{11.19}$$

here we have assumed $\omega_M \gg \omega_{0i}$.

Eq.(11.19) is the expectation value of H_a in the dressed state. Within the same approximations we may obtain

$$\begin{aligned}\langle H_f\rangle = \langle\phi|H_f|\phi\rangle &= \frac{2\pi e^2}{Vm^2}\sum_i\sum_{kj}\frac{|\mathbf{e}_{kj}\cdot\mathbf{P}_{0i}|^2}{(\omega_k+\omega_{0i})^2} \\ &\approx \frac{2e^2}{3\pi m^2c^3}\sum_i|\mathbf{P}_{0i}|^2\left[\omega_M - 2\omega_{0i}\ln\left(\frac{\omega_M}{\omega_{0i}}\right)\right]\end{aligned} \tag{11.20}$$

$$\begin{aligned}\langle H_i\rangle = \langle\phi|H_i|\phi\rangle &= -\frac{4\pi e^2}{Vm^2}\sum_i\sum_{kj}\frac{|\mathbf{e}_{kj}\cdot\mathbf{P}_{0i}|^2}{\omega_k(\omega_k+\omega_{0i})} \\ &\approx -\frac{4e^2}{3\pi m^2c^3}\sum_i|\mathbf{P}_{0i}|^2\left[\omega_M - \omega_{0i}\ln\left(\frac{\omega_M}{\omega_{0i}}\right)\right]\end{aligned} \tag{11.21}$$

The energy shift of the total system can be obtained as

$$\begin{aligned}&\langle H_a\rangle + \langle H_f\rangle + \langle H_i\rangle - E_{1s} \\ &= \frac{2e^2}{3\pi m^2c^3}\sum_i|\mathbf{P}_{0i}|^2\left[-\omega_M + \omega_{0i}\ln\left(\frac{\omega_M}{\omega_{0i}}\right)\right]\end{aligned} \tag{11.22}$$

The first term in the right-hand side of (11.22) can be written as

$$\Delta = -\frac{2e^2}{3\pi m^2 c^3}\sum_i |\mathbf{P}_{0i}|^2 \omega_M = -\frac{2\omega_M e^2}{3\pi m^2 c^3}|\langle 1s,\{0_k\}|P^2|1s,\{0_k\}\rangle|^2 \quad (11.23)$$

Evidently, $\langle H_a\rangle$ does not contain Δ, which only appears in $\langle H_f\rangle$ and $\langle H_i\rangle$. That is to say, the electron moving around the Hydrogen nucleus can create a cloud of virtual photons through the virtual photon transition processes due to the vacuum fluctuations, the interaction between this virtual photon field and the electron contributes an energy variation Δ to $\langle H_i\rangle$ and $\langle H_f\rangle$. This permits us to interpret the term Δ as the mass renormalization due to the self reaction of the virtual photon field.

The second term in the right-hand side of (11.22)

$$\delta = \frac{2e^2}{3\pi m^2 c^3}\sum_i \omega_{0i}|\mathbf{P}_{0i}|^2 \ln\left(\frac{\omega_M}{\omega_{0i}}\right) \quad (11.24)$$

is interpreted as the energy shift of the Hydrogen ground level due to the interaction of the virtual field, this is just the Lamb shift. As we know, for all i in (11.24), there is a relation $\omega_M \gg \omega_{0i}$, so the argument of the logarithm in (11.24) is very large, its ith-dependence is negligible, therefore, (11.24) may be approximated as

$$\delta = \frac{2e^2}{3\pi m^2 c^3}\ln\left(\frac{\omega_M}{\omega_{0i}}\right)\sum_i \omega_{0i}|\mathbf{P}_{0i}|^2 \quad (11.25)$$

In order to evaluate the Lamb shift δ of the Hydrogen atom, it is necessary to simplify the expression

$$\sum_i \omega_{0i}|\mathbf{P}_{0i}|^2 \quad (11.26)$$

The sum in (11.26) may be evaluated as follows. Notice the Hamiltonian of the Hydrogen atom to be

$$H_a = \frac{\mathbf{P}^2}{2m} - \frac{e^2}{4\pi\varepsilon_0 r} = \frac{\mathbf{P}^2}{2m} + V(r) \quad (11.27)$$

and for the arbitrary state $|\Psi\rangle$ of the system, there exists the relation

$$[P_j, H_a]|\Psi\rangle = \left[P_j, \frac{\mathbf{P}^2}{2m} + V(r)\right]|\Psi\rangle$$
$$= [P_j V(r) - V(r)P_j]|\Psi\rangle = -i\hbar\frac{\partial V(r)}{\partial r}|\Psi\rangle$$

so that

$$[P_j, H_a] = -i\hbar \frac{\partial V(r)}{\partial r} \tag{11.28}$$

Similarly, it is easy to verify

$$\sum_j [P_j, [P_j, H_a]] = -\hbar^2 \frac{e^2}{\varepsilon_0} \delta(r) \tag{11.29}$$

here we have adopted the relation

$$\delta(r) = -\frac{1}{4\pi} \nabla^2 \left(\frac{1}{r}\right)$$

If we take the diagonal matrix elements of right side of equation (11.29) for state $|k\rangle$, then we may have

$$-\frac{1}{2}\sum_j \langle k|[P_j, [P_j, H_a]]|k\rangle = -E_k\langle k|\mathbf{P}^2|k\rangle + \Sigma_j\langle k|P_j H_a P_j|k\rangle$$

$$= -E_k \sum_i \langle k|\mathbf{P}|i\rangle\langle i|\mathbf{P}|k\rangle + \sum_i \langle k|\mathbf{P}H_a|i\rangle\langle i|\mathbf{P}|k\rangle$$

$$= \sum_i E_{ki}|\mathbf{P}_{ki}|^2 \tag{11.30}$$

Therefore, from (11.29) we may obtain

$$\sum_i E_{ki}|\mathbf{P}_{ki}|^2 = \frac{\hbar^2 e^2}{2\varepsilon_0}\langle k|\delta(r)|k\rangle = \frac{\hbar^2 e^2}{2\varepsilon_0}|\Psi_k(0)|^2 \tag{11.31}$$

here $\Psi_k(r)$ is the k-state function of the Hydrogen atom. By use of the above equation, (11.25) reduces to be

$$\delta = \frac{e^4 \hbar}{3\pi\varepsilon_0 m^2 c^3} \ln\left(\frac{\omega_M}{\omega_{0i}}\right) |\Psi_{1s}(0)|^2 \tag{11.32}$$

Noticing for the 1s-state of the Hydrogen atom

$$|\Psi_{1s}(0)|^2 = \frac{1}{2}\left(\frac{1}{4\pi\varepsilon_0 a_0}\right)^3 \tag{11.33}$$

here a_0 is the Bohr radius, and taking $\omega_M = mc^2$ and estimating $\omega_{02} = (E_{2s} - E_0)/\hbar$ for the 2s-state, then δ turns out to be 1040 MHz which is in very good agreement with the measured value of 1057 MHz observed by Lamb and Retherford. So one can conclude that the effects of virtual field are the physical origin of the Lamb shift of the Hydrogen atom.

11.2 Influence of the virtual photon field on the phase fluctuations of the radiation field

From Section 3.3, we know that the phase property is an important characteristic for the radiation field, so it is valuable to investigate the effects of virtual photon processes on the phase properties of the field in the atom-field coupling system. Here we discuss the influence of the virtual field on the phase properties of the field in the system of a two-level atom interacting with the radiation field.

11.2.1 Time evolution of the phase operator in the atom-field coupling system with the rotating-wave approximation

Before investigating the influence of virtual field on the phase properties, we discuss the time evolution of the phase operator in the atom-field coupling system in which the virtual photon processes are neglected. We consider a system of a two-level atom coupled to a single-mode field. Within the rotating-wave approximation, this atom-field coupling system may be described by the Hamiltonian of the Jaynes-Cummings model

$$H = H_0 + V = \omega a^\dagger a + \omega_0 S_z + g(a^\dagger S_- + aS_+) \tag{11.34}$$

here g is the atom-field coupling constant. For simplicity, we take the exact-resonance case $\omega_0 = \omega$.

If at time t=0 the state vector of the atom-field coupling system is

$$|\Psi(0)\rangle = \sum_{n=0}^{\infty} F_n |e, n\rangle \tag{11.35}$$

it means that the atom is initially in its excited state $|e\rangle$, and the field is in the arbitrary superposition of number states $\sum_n F_n|n\rangle$. From (7.178), we know that with the time development, the state vector in the interaction picture gives

$$|\Psi^I(t)\rangle = \sum_{n=0}^{\infty} [a_n(t)|e, n\rangle + b_{n+1}(t)|g, n+1\rangle] \tag{11.36}$$

with

$$\begin{aligned} a_n(t) &= \frac{F_n}{2}\{\exp[-i\gamma(n)t] + \exp[i\gamma(n)t]\} \\ b_{n+1}(t) &= \frac{F_n}{2}\{\exp[-i\gamma(n)t] - exp[i\gamma(n)t]\} \\ \gamma(n) &= g\sqrt{n+1} \end{aligned} \tag{11.37}$$

Considering that the state vector (11.36) of the system at time t is decided in the interaction picture, we may change it into the Schrödinger picture as

$$|\Psi(t)\rangle = \sum_{n=0}^{\infty}[a_n(t)|e,n\rangle + b_{n+1}(t)|g,n+1\rangle]\exp[-i(n+1/2)\omega t] \tag{11.38}$$

In order to discuss the phase properties of the field, it is useful to expand the state vector $|\Psi(t)\rangle$ by the phase eigenkets $\{|\theta_m\rangle\}$ (3.131) defined by Pegg and Barnett in Chapter 3. So

$$|\Psi(t)\rangle = \sum_{n=0}^{\infty}\langle e,\theta_m|\Psi(t)\rangle|e,\theta_m\rangle + \langle g,\theta_m|\Psi(t)\rangle|g,\theta_m\rangle \tag{11.39}$$

where

$$\langle e,\theta_m|\Psi(t)\rangle = (s+1)^{-\frac{1}{2}}\sum_{n=0}^{\infty} F_n\cos(\gamma t)e^{-i(n+\frac{1}{2})\omega t}e^{-in\theta_m} \tag{11.40}$$

$$\langle g,\theta_m|\Psi(t)\rangle = -i(s+1)^{-\frac{1}{2}}\sum_{n=0}^{\infty} F_n\sin(\gamma t)e^{-i(n+\frac{1}{2})\omega t}e^{-in\theta_m} \tag{11.41}$$

Explicitly, here

$$P(\theta_m,t) = |\langle e,\theta_m|\Psi(t)\rangle|^2 + |\langle g,\theta_m|\Psi(t)\rangle|^2 \tag{11.42}$$

represents the phase-probability distribution. Thus the expectation values of the phase operator functions give

$$\langle\Phi\rangle = \sum_{m=0}^{s}\theta_m P(\theta_m,t) \tag{11.43}$$

$$\langle\Phi^2\rangle = \sum_{m=0}^{s}\theta_m^2 P(\theta_m,t) \tag{11.44}$$

If the radiation field is initially in a coherent state, i.e.,

$$F_n = \exp\left(-\frac{\overline{n}}{2}\right)\frac{\alpha^n}{\sqrt{n!}} \tag{11.45}$$

where

$$\alpha = \overline{n}^{1/2}\exp(i\xi) \tag{11.46}$$

Here $\overline{n}$ is the mean photon number, and ξ is the phase angle of α. For $\overline{n} \gg 1$, the photon distribution of the field may be approximated in Gaussian form according to (3.164)

$$F_n = (2\pi\overline{n})^{-1/4}\exp\left[-\frac{(n-\overline{n})^2}{4\overline{n}}\right]\exp(in\xi) \tag{11.47}$$

Substituting (11.47) into (11.40), we have

$$\begin{aligned}\langle e, \theta_m|\Psi(t)\rangle &= \frac{1}{2}(s+1)^{-1/2}(2\pi\overline{n})^{-1/4}\exp(-i\omega t/2)\int_0^\infty \exp\left[-\frac{(n-\overline{n})^2}{4\overline{n}}\right] \\ &\times[\exp(-i\gamma(n)t)+\exp(i\gamma(n)t)]\exp[in(\xi-\theta_m-\omega t)]dn\end{aligned} \tag{11.48}$$

Considering the Gaussian distribution properties, $\gamma(n)$ can be approximated as

$$g\sqrt{n+1} = g\left(\sqrt{\overline{n}+1}+\frac{n-\overline{n}}{2\sqrt{\overline{n}+1}}\right) \approx g\sqrt{\overline{n}}+g\frac{n-\overline{n}}{2\sqrt{\overline{n}}} \tag{11.49}$$

and noticing the integral formula

$$\int_{-\infty}^{\infty}\exp(-\alpha x^2)dx = \sqrt{\frac{\pi}{\alpha}}$$

then (11.48) is simplified as

$$\begin{aligned}\langle e, \theta_m|\Psi(t)\rangle = \frac{1}{2}&\left(\frac{2\pi}{s+1}\right)^{1/2}\left(\frac{4\overline{n}}{2\pi}\right)^{1/4} e^{-\frac{i}{2}\omega t}e^{i\overline{n}(\xi-\theta_m-\omega t)} \\ &\times\left\{\exp(ig\sqrt{\overline{n}}t)\exp\left[-\overline{n}\left(\xi-\theta_m-\omega t+\frac{gt}{2\sqrt{\overline{n}}}\right)^2\right]\right. \\ &\left.+\exp(-ig\sqrt{\overline{n}}t)\exp\left[-\overline{n}\left(\xi-\theta_m-\omega t-\frac{gt}{2\sqrt{\overline{n}}}\right)^2\right]\right\}\end{aligned} \tag{11.50}$$

Following the same procedure, we obtain

$$\langle g,\theta_m|\Psi(t)\rangle = \frac{1}{2}\left(\frac{2\pi}{s+1}\right)^{1/2}\left(\frac{4\overline{n}}{2\pi}\right)^{1/4} e^{-\frac{i}{2}\omega t}e^{i\overline{n}(\xi-\theta_m-\omega t)}$$
$$\times\left\{\exp(-ig\sqrt{\overline{n}}t)\exp\left[-\overline{n}\left(\xi-\theta_m-\omega t-\frac{gt}{2\sqrt{\overline{n}}}\right)^2\right]\right.$$
$$\left.-\exp(ig\sqrt{\overline{n}}t)\exp\left[-\overline{n}\left(\xi-\theta_m-\omega t+\frac{gt}{2\sqrt{\overline{n}}}\right)^2\right]\right\} \tag{11.51}$$

Substituting (11.50) and (11.51) into (11.42), the phase-probability distribution can be obtained as

$$P(\theta_m,t) = \frac{1}{2}\frac{2\pi}{s+1}\left(\frac{4\overline{n}}{2\pi}\right)^{\frac{1}{2}}\left\{\exp\left[-2\overline{n}\left(\xi-\theta_m-\omega t+\frac{gt}{2\sqrt{\overline{n}}}\right)^2\right]\right.$$
$$\left.+\exp\left[-2\overline{n}\left(\xi-\theta_m-\omega t-\frac{gt}{2\sqrt{\overline{n}}}\right)^2\right]\right\} \tag{11.52}$$

In the continuum limit, i.e., $s\to\infty$, θ_m is a continuum variable. The phase-probability distribution is to be normalized

$$\int P(\theta,t)\frac{s+1}{2\pi}d\theta = 1 \tag{11.53}$$

where $\frac{s+1}{2\pi}$ is the density of the phase states.

If considering $\frac{g}{2\sqrt{\overline{n}}} \ll \omega$ in (11.52) for $\overline{n}\gg 1$, we can neglect the term $\frac{g}{2\sqrt{\overline{n}}}$ associated with the atom-field coupling. Using eqs.(11.43), (11.44) and (11.52), we obtain the time evolution of the phase operator to be

$$\langle\Phi\rangle = \int \theta P(\theta,t)\frac{s+1}{2\pi}d\theta = \left(\frac{4\overline{n}}{2\pi}\right)^{\frac{1}{2}}\int\theta\exp[-2\overline{n}(\xi-\theta-\omega t)^2]d\theta$$
$$=\xi-\omega t \tag{11.54}$$
$$\langle\Phi^2\rangle = \int\theta^2 P(\theta,t)\frac{s+1}{2\pi}d\theta = (\xi-\omega t)^2+\frac{1}{4\overline{n}} \tag{11.55}$$

From (11.54) we see that the phase of the field varies within the ratio of ω. The phase fluctuation is expressed as

$$\langle\Delta\Phi\rangle^2 = \langle\Phi^2\rangle - \langle\Phi\rangle^2 = \frac{1}{4\overline{n}} \tag{11.56}$$

And according to (11.38) and (11.45), we know that the photon-number distribution at time t gives

$$P_n(t) = |a_n(t)|^2 + |b_n(t)|^2 = |F_n|^2 = \exp(-\overline{n})\frac{\overline{n}^n}{n!} \tag{11.57}$$

It is clear that the photon-number distribution is nearly in a Poissonian distribution, so the photon-number fluctuation is

$$\langle \Delta N \rangle^2 \approx \overline{n} \tag{11.58}$$

Thus the number-phase-uncertainty product gives

$$\langle \Delta\Phi \rangle^2 \cdot \langle \Delta N \rangle^2 = \frac{1}{4} \tag{11.59}$$

We see that the field, which is initially in a coherent states, retains its coherence with the time development if one neglects the influence of the atom-field coupling on the phase-probability distribution.

However, if we do not neglect $g/2\sqrt{\overline{n}}$ in (11.52), then

$$\langle \Phi^2 \rangle = (\xi - \omega t)^2 + \frac{1}{4\overline{n}} + \frac{g^2t^2}{4\overline{n}} \tag{11.60}$$

and $\langle \Phi \rangle$ still satisfies (11.54). The phase fluctuations can be shown as

$$\langle \Delta\Phi \rangle^2 = \frac{1}{4\overline{n}} + \frac{g^2t^2}{4\overline{n}} \tag{11.61}$$

Also, the number-phase-uncertainty product changes to be

$$\langle \Delta\Phi \rangle^2 \langle \Delta N \rangle^2 = \frac{1}{4} + \frac{g^2t^2}{4} \tag{11.62}$$

This means that the phase fluctuations are enhanced with the time development, and the field does not keep its phase-number-minimum-uncertainty product. From eqs.(11.56) and (11.60), we can see that the cause leading to the enhancement of the phase fluctuations is the atom-field coupling.

11.2.2 Time evolution of the phase operator without the rotating-wave approximation

Now we discuss the influence of virtual photon processes on the phase properties of the field. As we know, in the Jaynes-Cummings model the virtual photon processes are represented by the counterrotating wave terms $\varepsilon(a^\dagger S_+ + aS_-)$.

In order to investigate the role of virtual photon processes in the time development of the phase operator, we can not neglect the counterrotating wave terms in the Hamiltonian.

The Hamiltonian for a system of a two-level atom interacting with a single-mode field in the interaction picture is

$$H^I = H_0 + V^I(t) \tag{11.63}$$

where

$$H_0 = \omega a^\dagger a + \omega S_z \tag{11.64}$$

$$V^I = g[a^\dagger S_- + aS_+ + a^\dagger S_+ \exp(2i\omega t) + aS_- \exp(-2i\omega t)] \tag{11.65}$$

For simplicity, we also take the field to be resonant with the atomic transition frequency. Substituting (11.36) and (11.65) into the Schrödinger equation in the interaction picture

$$i\hbar\frac{\partial}{\partial t}|\Psi^I(t)\rangle = V^I(t)|\Psi^I(t)\rangle$$

we obtain

$$i\frac{d}{dt}a_n(t) = g[\sqrt{n+1}b_{n+1}(t) + \sqrt{n}\exp(2i\omega t)b_{n-1}(t)] \tag{11.66}$$

$$i\frac{d}{dt}b_{n+1}(t) = g[\sqrt{n+1}a_n(t) + \sqrt{n+2}\exp(-2i\omega t)a_{n+2}(t)] \tag{11.67}$$

It is easily seen that the last terms in (11.66) and (11.67) represent the influences of the virtual photon processes on $a_n(t)$ and $b_{n+1}(t)$. If we regard the last terms in (11.66) and (11.67) as perturbation terms, and define a perturbation parameter that is the ratio $g\sqrt{\bar{n}}/2\omega$, the $a_n(t)$ and $b_{n+1}(t)$ are then in the form of a perturbation series. The zeroth-order terms in $a_n(t)$ and $b_{n+1}(t)$ correspond to the solution with the rotating-wave approximation, and the terms including $g\sqrt{\bar{n}}/2\omega$ are the first-order correction due to the energy-nonconserving terms. We only retain terms up to the first order of $g/2\omega$ in $a_n(t)$ and in $b_{n+1}(t)$, then the solutions of (11.66) and (11.67) are derived to be

$$a_n(t) = \frac{F_n}{2}[\exp[i\gamma(n)t] + exp[-i\gamma(n)t]] + \frac{g\sqrt{n-2}}{2}F_n$$

$$\times \left[\frac{\exp[i(2\omega + g\sqrt{n-1})t] - 1}{2\omega + g\sqrt{n-1}} - \frac{\exp[i(2\omega - g\sqrt{n-1})t] - 1}{2\omega - g\sqrt{n-1}} \right] \tag{11.68}$$

$$b_{n+1}(t) = \frac{F_n}{2}[\exp[-i\gamma(n)t] - \exp[i\gamma(n)t]] + \frac{g\sqrt{n+2}}{2}F_n$$
$$\times \left[\frac{\exp[-i(2\omega - g\sqrt{n+3})t] - 1}{2\omega - g\sqrt{n+3}} + \frac{\exp[-i(2\omega + g\sqrt{n+3})t] - 1}{2\omega + g\sqrt{n+3}} \right] \tag{11.69}$$

Clearly, the first two terms in the above equations represent the contributions of the real photon processes to the amplitudes $a_n(t)$ and $b_{n+1}(t)$, the last two terms describe the modifications to the amplitudes due to the virtual photon processes. Meanwhile we find that $g\sqrt{n}/(2\omega \pm g\sqrt{n-1})$ and $g\sqrt{n+2}/(2\omega \pm g\sqrt{n+3})$ are no longer small when the field is intensive, then the last two terms in (11.68) and (11.69) can not be neglected. So that for very intensive field, the rotating-wave approximation in the Jaynes-Cummings model is destroyed. Thus the condition to adopt the rotating-wave approximation requires the radiation field to be not very intensive. To obtain the state function (11.36) in non-rotating-wave approximation, here we suppose that the terms $g\sqrt{n}/(2\omega \pm g\sqrt{n-1})$ and $g\sqrt{n+2}/(2\omega \pm g\sqrt{n+3})$ are relatively small, so the perturbation theory can be adopted. Substituting (11.68) and (11.69) into (11.38), the state vector of the system without rotating-wave approximation in the Schrödinger picture is determined.

As before, the phase-probability distribution of which the field initially in a coherent state for $\bar{n} \gg 1$ gives

$$\begin{aligned} P(\theta, t) = & \frac{1}{2}\frac{2\pi}{s+1}\left(\frac{4\bar{n}}{2\pi}\right)^{1/2} \{\exp(-2\bar{n}x^2) + \exp(-2\bar{n}y^2) \\ & -(g\sqrt{\bar{n}}/\omega)\{\exp[-\bar{n}(x^2+z^2)]\cos(2\xi + g\sqrt{\bar{n}})t \\ & +\exp[-\bar{n}(y^2+z^2)]\cos(2\xi - g\sqrt{\bar{n}})t \\ & -2\exp[-\bar{n}(x^2+y^2)]\sin[2(\omega t - \xi)]\sin(2g\sqrt{\bar{n}}t)\}\Big\} \end{aligned} \tag{11.70}$$

where

$$\begin{aligned} x &= \xi - \theta - \omega t + \frac{gt}{2\sqrt{\bar{n}}} \\ y &= \xi - \theta - \omega t - \frac{gt}{2\sqrt{\bar{n}}} \\ z &= \xi - \theta - \omega t \end{aligned} \tag{11.71}$$

Noticing the state vector described by (11.68) and (11.69) is not normalized, so it is necessary to normalize the phase-probability distribution function $P(\theta,t)$. Using the relation

$$A\int P(\theta,t)\frac{s+1}{2\pi}d\theta = 1 \tag{11.72}$$

we obtain the normalized $P(\theta,t)$ exact to the first order in $g/2\omega$ as

$$\begin{aligned}P(\theta,t) = & A\frac{1}{2}\frac{2\pi}{s+1}\left(\frac{4\bar{n}}{2\pi}\right)^{1/2}\{\exp(-2\bar{n}x^2)+\exp(-2\bar{n}y^2)\\ & -(g\sqrt{\bar{n}}/\omega)\{\exp[-\bar{n}(x^2+y^2)]\cos(2\xi+g\sqrt{\bar{n}}t)\\ & -\exp[-\bar{n}(y^2+z^2)]\cos(2\xi-g\sqrt{\bar{n}}t)\\ & +2\exp[-\bar{n}(x^2+y^2)]\sin[2(\omega t-\xi)]\sin(2g\sqrt{\bar{n}}t)\}\}\end{aligned} \tag{11.73}$$

where

$$\begin{aligned}A = & 1-\frac{g\sqrt{\bar{n}}}{\omega}\exp(-g^2t^2/8)\sin(2\xi)\sin(g\sqrt{\bar{n}}t)\\ & +\frac{g\sqrt{\bar{n}}}{\omega}\exp(-g^2t^2/2)\sin[2(\omega t-\xi)]\sin(2g\sqrt{\bar{n}}t)\end{aligned} \tag{11.74}$$

A is a probability normalization parameter.

Using eqs.(11.43),(11.44) and (11.73), and retaining terms up to the first order in $g/2\omega$, we obtain the time evolution of the phase operators as

$$\langle\Phi\rangle = \xi-\omega t+\frac{g^2t}{4\omega}\cos(2\xi)\cos(g\sqrt{\bar{n}}t)\exp(-g^2t^2/8) \tag{11.75}$$

$$\begin{aligned}\langle\Phi^2\rangle = & (\xi-\omega t)^2+\frac{1}{4\bar{n}}+\frac{g^2t^2}{4\bar{n}}\\ & +\frac{g^2t}{2\omega}(\xi-\omega t)\exp(-g^2t^2/8)\cos(2\xi)\cos(g\sqrt{\bar{n}}t)\\ & +\frac{g^2t^2}{4\bar{n}}\frac{3}{4}\frac{g\sqrt{\bar{n}}}{\omega}\{\exp(-g^2t^2/2)\sin[2(\omega t-\xi)]\sin(2g\sqrt{\bar{n}}t)\\ & -\sin(2\xi)\sin(g\sqrt{\bar{n}}t)\exp(-g^2t^2/8)\}\end{aligned} \tag{11.76}$$

So the phase fluctuation of the radiation field gives

$$\begin{aligned}\langle\Delta\Phi\rangle^2 = & \frac{g^2t^2}{4\bar{n}}+\frac{g^2t^2}{4\bar{n}}\frac{3g\sqrt{\bar{n}}}{4\omega}\{\sin[2(\omega t-\xi)]\sin(2g\sqrt{\bar{n}}t)\exp(-g^2t^2/2)\\ & -\sin(2\xi)\sin(g\sqrt{\bar{n}}t)\exp(-g^2t^2/8)\}+\frac{1}{4\bar{n}}.\end{aligned} \tag{11.77}$$

From (11.76)-(11.77), we can find that the terms of the first order in $g/2\omega$ appear in the expectation value of the phase operator. The appearance of $g/2\omega$ indicates the role of virtual-photon processes in the system. The effects of virtual field induce small-amplitude fluctuations in the expectation value of the phase operator, these fluctuations reflect the quantum noises of the atom-field coupling system. That is to say, the system without the rotating-wave approximation can explicitly exhibit the quantum noises of the system, but these quantum noises are not exhibited in the rotating-wave approximation. The amplitudes of quantum fluctuations are associated with the coupling constant g, the mean photon number $\overline{n}$, the resonant frequency ω and the phase angle ξ. Comparing eqs.(11.76) and (11.77) with (11.60) and (11.61), we can see that the phase fluctuations are associated with not only the coupling constant g and the mean photon number $\overline{n}$, but also the frequency ω and the initial phase ξ of the field. So the different frequency ω and the initial phase ξ of the field induce different fluctuations due to the effects of virtual field, even if the coupling constant g and the mean photon number $\overline{n}$ are the same. However, in the rotating-wave approximation, the phase fluctuation depends only on the coupling constant g and the mean photon number $\overline{n}$. That is to say, the phase fluctuations in the Jaynes-Cummings model without the rotating-wave approximation are associated with not only the intensity of the field, but also the frequency and the phase of the field.

Comparing (11.75) with (11.54), we find

$$\frac{d}{dt}\langle\Phi\rangle \neq -\omega \tag{11.78}$$

This means that the frequency of the field is shifted in the non-rotating-wave approximation. the value of the shift is determined by the atom-field coupling constant, the intensity and frequency of the field, also related to the time development. The cause leading this shift is that the effects of virtual field induce the atomic level to be shifted (namely, the Lamb shift). Thus the frequency of the field radiated by the atom is shifted. The above result shows that the role of virtual field induces not only the enhancement of phase fluctuations but also the shift of the frequency of field.

11.3 Influences of the virtual photon processes on the squeezing of light

In Section 7.2 the effect of virtual field is neglected when we discuss the squeezing of light in the Jaynes-Cummings model. In order to investigate the squeezing properties of the field in the atom-field coupling system accurately, it is necessary to discuss the influences of the virtual photon processes in detail.

The system we consider here is consisted of a two-level atom with the same parity interacting with a single-mode field via two-photon processes (termed as the two-photon Jaynes-Cummings model). The Hamiltonian for this system is

$$H = H_0 + V \tag{11.79}$$

where

$$H_0 = \omega a^\dagger a + \omega_0 S_z \qquad (\hbar = 1) \tag{11.80}$$

$$V = \varepsilon(a^{\dagger 2} + a^2)(S_+ + S_-) = \varepsilon(a^{\dagger 2} S_- + a^2 S_+ + a^{\dagger 2} S_+ + a^2 S_-) \tag{11.81}$$

Here $a^{\dagger 2} S_-$ describes the transition process which the atom goes from its excited state $|e\rangle$ to its ground state $|g\rangle$ by emitting two photons with frequency ω simultaneously. Conversely, $a^2 S_+$ represents the absorption of two photons and the excitation of the atom from its ground state $|g\rangle$ to its excited state $|e\rangle$. These two transition processes are the energy-conserving processes, namely, the real photon processes. However, $a^2 S_-$ corresponds to the absorption of two photons and the excitation of the atom from state $|g\rangle$ to state $|e\rangle$, and $a^{\dagger 2} S_+$ describes the emission of two photons and the deexcitation of the atom. In these two processes, the energy-conserving is violated instantaneously, that is to say, the meaning of $a^{\dagger 2} S_+$ and $a^2 S_-$ represents the virtual photon processes in the atom-field coupling system. In order to discuss the influences of the virtual photon processes, we first investigate the time evolution of the squeezing of field in the two-photon Jaynes-Cummings model with the rotating-wave approximation.

11.3.1 Squeezing of the field in the two-photon Jaynes-Cummings model with the rotating-wave approximation

In the rotating-wave approximation, the interaction Hamiltonian of the atom-field coupling system is

$$V = \varepsilon(a^{\dagger 2} S_- + a^2 S_+) \tag{11.82}$$

For simplicity, we only consider the field to be resonant with the atomic transition frequency via a two-photon process, i.e., $\omega_0 = 2\omega$. In the interaction picture, (11.82) becomes

$$V^I(t) = \varepsilon(a^{\dagger 2}S_- + a^2 S_+) \tag{11.83}$$

and the annihilation operator of the field satisfies

$$a^I(t) = a^S \exp(-i\omega t) \tag{11.84}$$

If the initial state vector of the atom-field coupling system is

$$|\Psi(0)\rangle = \sum_n F_n |g, n\rangle \tag{11.85}$$

this means that the atom is in its ground state $|g\rangle$, and the field in the superposition state $\sum\limits_n F_n|n\rangle$. At time t, it evolves to

$$|\Psi^I(t)\rangle = \sum_n a_n(t)|e, n\rangle + b_{n+2}(t)|g, n+2\rangle \tag{11.86}$$

Bringing the Schrödinger equation into the interaction picture, we obtain

$$i\frac{d}{dt}a_n = \varepsilon\sqrt{(n+2)(n+1)}\,b_{n+2} \tag{11.87}$$

$$i\frac{d}{dt}b_{n+2} = \varepsilon\sqrt{(n+2)(n+1)}\,a_n \tag{11.88}$$

Considering the initial condition (11.85), the solutions of the above equations are

$$a_n(t) = -iF_{n+2}\sin(\gamma(n)t) \tag{11.89}$$

$$b_{n+2}(t) = F_{n+2}\cos(\gamma(n)t) \tag{11.90}$$

$$\gamma(n) = \varepsilon\sqrt{(n+2)(n+1)} \tag{11.91}$$

Substituting (11.89)-(11.91) into (11.86), we obtain the density matrix $\rho^I(t)$ of the atom-field coupling system as

$$\begin{aligned}\rho^I(t) &= |\Psi^I(t)\rangle\langle\Psi^I(t)| \\ &= \sum_{n,k}\begin{pmatrix} a_k^* a_n|n\rangle\langle k| & a_n^* b_{k+2}|n\rangle\langle k+2| \\ a_k^* b_{n+2}|n+2\rangle\langle k| & b_{k+2}^* b_{n+2}|n+2\rangle\langle k+2| \end{pmatrix}\end{aligned} \tag{11.92}$$

In order to discuss the squeezing of the field, we define two Hermitian quadrature operators d_1 and d_2 of the field as

$$d_1 = \frac{1}{2}[a \exp(i\omega t) + a^{\dagger} \exp(-i\omega t)] \tag{11.93}$$

$$d_2 = \frac{1}{2i}[a \exp(i\omega t) - a^{\dagger} \exp(-i\omega t)] \tag{11.94}$$

It is easy to get that

$$(\Delta d_1)^2 = \frac{1}{4}[2\langle a^{\dagger} a\rangle + 1 + \langle a^2 \exp(2i\omega t)\rangle + \langle a^{\dagger 2} \exp(-2i\omega t)\rangle - \langle a \exp(i\omega t) + a^{\dagger} \exp(-i\omega t)\rangle^2] \tag{11.95}$$

$$(\Delta d_2)^2 = \frac{1}{4}[2\langle a^{\dagger} a\rangle + 1 - \langle a^2 \exp(2i\omega t)\rangle - \langle a^{\dagger 2} \exp(-2i\omega t)\rangle - \langle a \exp(i\omega t) - a^{\dagger} \exp(-i\omega t)\rangle^2] \tag{11.96}$$

If one of the uncertainties $(\Delta d_1)^2$ or $(\Delta d_2)^2$ satisfies the relation

$$Q_i = (\Delta d_i)^2 - \frac{1}{4} < 0 \quad (i = 1 \text{ or } 2) \tag{11.97}$$

we call the d_i component of the field to be squeezed.

If the field is initially in a coherent state, i.e.,

$$F_n = \exp(-\overline{n}/2)\frac{\overline{n}^{n/2} e^{in\phi}}{\sqrt{n!}} \tag{11.98}$$

then we may obtain the density matrix $\rho(t)$ (11.92) of the field, furthermore we have

$$\langle a\, e^{i\omega t}\rangle = e^{-\overline{n}} \sum_{n=-2}^{\infty} \frac{\overline{n}^{(n+5/2)} e^{i\phi}}{(n+2)!} \times \left[\sqrt{\frac{n+1}{n+3}} \sin(C_1 t)\sin(C_2 t) + \cos(C_1 t)\cos(C_2 t)\right] \tag{11.99}$$

$$\langle a^2\, e^{i2\omega t}\rangle = e^{-\overline{n}} \sum_{n=-2}^{\infty} \frac{\overline{n}^{(n+3)} e^{i2\phi}}{(n+2)!} \times \left[\sqrt{\frac{(n+2)(n+1)}{(n+3)(n+4)}} \sin(C_1 t)\sin(C_3 t) + \cos(C_1 t)\cos(C_3 t)\right] \tag{11.100}$$

$$\langle a^{\dagger}\, a\rangle = \overline{n} - 1 + e^{-\overline{n}} \sum_{n=-2}^{\infty} \frac{\overline{n}^{(n+2)}}{(n+2)!} \cos(2C_1 t) \tag{11.101}$$

where

$$C_1 = \varepsilon\sqrt{(n+2)(n+1)},\ C_2 = \varepsilon\sqrt{(n+2)(n+3)},\ C_3 = \varepsilon\sqrt{(n+3)(n+4)}$$

Substituting eqs.(11.99)-(11.101) into (11.97), we can discuss the squeezing properties of the field with the aid of the numerical method.

In Fig.(11.2), we show the time evolution of Q_i in the case $\phi = 0$, $\overline{n} = 1, 4, 6$ and ε=0.1, 0.05. Evidently, there appears the phenomenon of $Q_1 < 0$ in Figs.(11.2-a-d), which means that the squeezing of the field occurs in the corresponding time ranges. With the increasing of $\overline{n}$, the squeezing degree of the first squeezing phenomenon decreases, and the squeezing phenomenon becomes periodic characteristic related to the coupling constant ε gradually. In addition, it is valuable to point out that from eqs.(11.99)-(11.101) the squeezing degree of the field is only related to the mean photon number $\overline{n}$ and the phase angle ϕ, but independent of the frequency ω and the atom-field coupling constant ε (shown in Figs.(11.2-c-d)).

11.3.2 Influences of the virtual photon processes on the squeezing of light

Now we turn to discuss the influences of the virtual photon processes on the squeezing of the radiation field, in this case, the counterrotating-wave terms in (11.83) can not be neglected. In the interaction picture, the interaction Hamiltonian becomes

$$V^I(t) = \varepsilon(a^2 S_+ + a^{\dagger 2} S_- + a^2 S_- \exp(-4i\omega t) + a^{\dagger 2} S_+ \exp(4i\omega t)) \quad (11.102)$$

Similarly, substituting (11.86) and (11.102) into the Schrödinger equation, we get

$$\begin{aligned} i\frac{d}{dt}a_n(t) &= \varepsilon\sqrt{(n+2)(n+1)}b_{n+2}(t) \\ &\quad + \varepsilon\sqrt{(n-1)n}b_{n-2}(t)\exp(i4\omega t) \quad (11.103) \\ i\frac{d}{dt}b_{n+2}(t) &= \varepsilon\sqrt{(n+2)(n+1)}a_n(t) \\ &\quad + \varepsilon\sqrt{(n+4)(n+3)}a_{n+4}(t)e^{-i4\omega t} \quad (11.104) \end{aligned}$$

It is easy to see that the first terms on the right-hand side in (11.103) and (11.104) represent the contributions of the real photon processes to the probability amplitude $a_n(t)$, $b_{n+2}(t)$, and the second terms imply the influences of

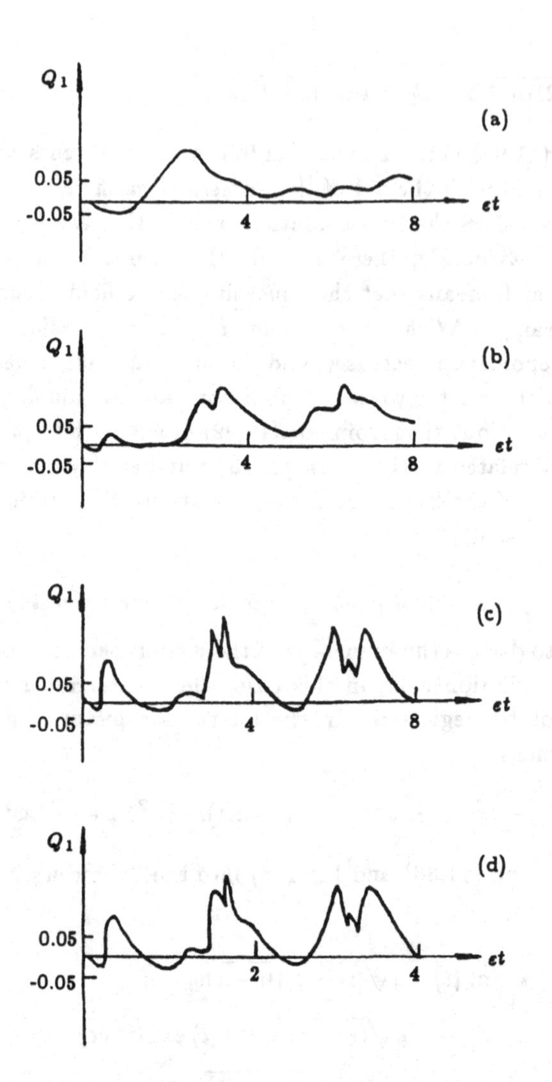

Figure 11.2: Time evolution of Q_1 with RWA, (a) $\overline{n} = 1$, $\varepsilon = 0.1$, (b) $\overline{n} = 4$, $\varepsilon = 0.1$, (c) $\overline{n} = 6$, $\varepsilon = 0.1$, (d) $\overline{n} = 6$, $\varepsilon = 0.05$

the virtual photon processes on $a_n(t)$ and $b_{n+2}(t)$.

In order to solve the above equations, we must mention that the influences of the virtual photon processes on $a_n(t)$, $b_{n+2}(t)$ are smaller than that of the real photon processes when the intensity of the initial field is not very intensive. So we can regard the solution of $a_n(t)$, $b_{n+2}(t)$ in the rotating-wave approximation (eqs.(11.89) and (11.90)) as the solution of the zero-order approximation of (11.103) and (11.104). Substituting eqs.(11.89) and (11.90) into eqs.(11.103) and (11.104) and only retaining terms up to the first order in ε/ω, we obtain

$$a_n(t) = A\left\{\frac{n(n-1)}{2\bar{n}^2}C_1 F_{n+2}\left[\frac{\exp[i(B_1 t - 4\phi)] - \exp(-i4\phi)}{B_1}\right.\right.$$
$$\left.\left.+\frac{\exp[i(B_2 t - 4\phi)] - \exp(-4i\phi)}{B_2}\right] - iF_{n+2}\sin(C_1 t)\right\} \tag{11.105}$$

$$b_{n+2}(t) = A\left\{-\frac{\varepsilon\bar{n}^2}{2\sqrt{(n+5)(n+6)}}F_{n+2}\left[\frac{\exp[-i(A_1 t - 4\phi)] - \exp(i4\phi)}{A_1}\right.\right.$$
$$\left.\left.-\frac{\exp[-i(A_2 t - 4\phi)] - \exp(i4\phi)}{A_2}\right] + F_{n+2}\cos(C_1 t)\right\} \tag{11.106}$$

where

$$A = \frac{\varepsilon\bar{n}^2}{2\sqrt{(n+5)(n+6)}}\left[\frac{\cos(A_1 t - 4\phi) - \cos(4\phi)}{A_1}\right.$$
$$\left.-\frac{\cos(A_2 t - 4\phi) - \cos(4\phi)}{A_2}\right]\cos(C_1 t) - \frac{n(n-1)C_1}{2\bar{n}^2}\left[\frac{\sin(B_1 t - 4\phi) - \sin(4\phi)}{B_1}\right.$$
$$\left.+\frac{\sin(B_2 t - 4\phi) + \sin(4\phi)}{B_2}\right]\sin(C_1 t) + 1$$
$$A_1 = 4\omega - \varepsilon\sqrt{(n+5)(n+6)}, \quad A_2 = 4\omega + \varepsilon\sqrt{(n+5)(n+6)}$$
$$B_1 = 4\omega - \varepsilon\sqrt{(n-2)(n-3)}, \quad B_2 = 4\omega + \varepsilon\sqrt{(n-2)(n-3)}$$

here the state vector $|\Psi(t)\rangle$ has been normalized according to the relation

$$|a_n(t)|^2 + |b_{n+2}(t)|^2 = |F_{n+2}|^2 \tag{11.107}$$

From (11.105) and (11.106), we can easily see that $\frac{\varepsilon\sqrt{n(n-1)}}{B_1}$, $\frac{\varepsilon\sqrt{n(n-1)}}{B_2}$, $\frac{\varepsilon\sqrt{(n+3)(n+4)}}{A_1}$, and $\frac{\varepsilon\sqrt{(n+3)(n+4)}}{A_2}$ are not infinitesimal if the field is intensive. In this case the rotating-wave approximation is invalid. For simplicity, we only consider that the field is not intensive so that $\varepsilon\sqrt{n(n-1)}/B_1$, $\varepsilon\sqrt{n(n-1)}/B_2$,$\varepsilon\sqrt{(n+3)(n+4)}/A_1$, and $\varepsilon\sqrt{(n+3)(n+4)}/A_4$ are not very large. Then we can substitute $a_n(t)$ (11.105), $b_{n+2}(t)$ (11.106) into the density matrix $\rho(t)$ by using the perturbation theory and retaining the terms up to ε/ω, therefore obtain

$$\begin{aligned}
\langle a\exp(i\omega t)\rangle &= \langle a\exp(i\omega t)\rangle_{RWA} + \exp(-\overline{n})\sum_{n=-2}^{\infty}\frac{\overline{n}^{n+5/2}}{(n+2)!} \\
&\times\left\{\frac{\varepsilon\overline{n}^2}{4\sqrt{(n+5)(n+6)}}\left(\frac{\cos(A_1t)-1}{A_1}-\frac{\cos(A_2t)-1}{A_2}\right)\right. \\
&\times\left[\sqrt{\frac{n+1}{n+3}}\sin(2C_1t)\sin(C_2t)-2\sin^2(C_1t)\cos(C_2t)\right] \\
&+\frac{n(n-1)C_1}{4\overline{n}^2}\left(\frac{\sin(B_1t)}{B_1}+\frac{\sin(B_2t)}{B_2}\right) \\
&\times\left.\left[\sqrt{\frac{n+1}{n+3}}\cos^2(C_1t)\sin(C_2t)-\sin(2C_1t)\cos(C_2t)\right]\right\} \qquad (11.108)
\end{aligned}$$

$$\begin{aligned}
\langle a^2\exp(2i\omega t)\rangle &= \langle a^2\exp(2i\omega t)\rangle_{RWA} + \exp(-\overline{n})\sum_{n=-2}^{\infty}\frac{\overline{n}^{n+3}}{(n+2)!} \\
&\times\left\{\frac{n(n-1)C_1}{4\overline{n}^2}\left(\frac{sin(B_1t)}{B_1}+\frac{\sin(B_2t)}{B_2}\right)\right. \\
&\times\left[\sqrt{\frac{(n+1)(n+2)}{(n+3)(n+4)}}2\cos^2(C_1t)\sin(C_3t)-\sin(2C_1t)\cos(C_3t)\right] \\
&+\frac{\varepsilon\overline{n}^2}{4\sqrt{(n+5)(n+6)}}\left(\frac{\cos(A_1t)-1}{A_1}-\frac{\cos(A_2t)-1}{A_2}\right) \\
&\times\left.\left[\sqrt{\frac{(n+1)(n+2)}{(n+3)(n+4)}}\sin(2C_1t)\sin(C_3t)-2\sin^2(C_1t)\cos(C_3t)\right]\right\} \qquad (11.109)
\end{aligned}$$

$$\begin{aligned}
\langle a^\dagger a> &=< a^\dagger a\rangle_{RWA} \\
&+e^{-\overline{n}}\sum_{n=-2}^{\infty}\frac{\overline{n}^{n+2}}{(n+2)!}\left\{\frac{\varepsilon\overline{n}^2}{\sqrt{(n+5)(n+6)}}\left[\frac{\cos(A_1t)-1}{A_1}-\frac{\cos(A_2t)-1}{A_2}\right]\right. \\
&+\left.\frac{\varepsilon n(n-1)}{\overline{n}^2}\sqrt{(n+1)(n+2)}\left[\frac{\sin(B_1t)}{B_1}+\frac{\sin(B_2t)}{B_2}\right]\right\} \qquad (11.110)
\end{aligned}$$

where

$$C_1 = \varepsilon\sqrt{(n+1)(n+2)}, \quad C_2 = \varepsilon\sqrt{(n+2)(n+3)},$$
$$C_3 = \varepsilon\sqrt{(n+3)(n+4)} \tag{11.111}$$

The terms such as $\langle a\exp(i\omega t)\rangle_{RWA}$, $\langle a^2\exp(2i\omega t)\rangle_{RWA}$ and $\langle a^\dagger a\rangle_{RWA}$ in the above equations are the expectation values of $a\exp(i\omega t)$, $a^2\exp(2i\omega t)$ and $a^\dagger a$ in the rotating-wave approximation case (eqs.(11.99)-(11.101)), these terms represent the effect of the real photon processes. The residual terms in the right-hand side of eqs.(11.108)-(11.110) reflect the modification of the virtual photon processes to $a\exp(i\omega t)$, $a^2\exp(2i\omega t)$ and $a^\dagger a$. Substituting eqs.(11.108)-(11.110) into (11.97), we can investigate the squeezing properties of the field. Fig.(11.3) presents the time evolution of Q_1 without the rotating-wave approximation by numerical method, the parameters ($\phi, \bar{n}$ and ε) are specified as in Fig.(11.2). Here the value of the mean photon number $\bar{n}$ and the coupling constant ε ensure that the amplitudes of $\varepsilon\bar{n}/A_1$, $\varepsilon\bar{n}/A_2$, $\varepsilon\sqrt{(n-1)n}/B_1$ and $\varepsilon(n-1)n/B_2$ are not large. Comparing Fig.(11.3) with Fig.(11.2), we find that the squeezing degree of the component d_1 of the field increases , and the time of the continuous squeezing phenomenon is lengthened due to the virtual photon processes. We also note a particular time interval of the squeezing which is absent in the two-photon Jaynes-Cummings model with the rotating-wave approximation, this is because of the influence of the virtual photon processes. Comparing Fig.(11.3) with Fig.(11.2) we can also see that the change of Q_1 is not evident when $\bar{n}$ is small ($\bar{n} = 1$) due to the virtual photon processes. But with the increasing of the mean photon number $\bar{n}$ ($\bar{n}$=4,6), the changes in the squeezing degree and squeezing time region are evident. This means that the rotating-wave approximation is legitimate only when the field is weak.

From Figs.(11.3-c)-(11.3-d), it is easily seen that the appearance of squeezing is also periodic. The periodic property is the same as that described by Figs.(11.2-c)-(11.2-d). The reason is that we only restrict the field to be not intensive. So we can neglect the influence of the virtual photon processes on the oscillating periodic time. If the field is very intensive, the rotating-wave approximation is unsuitable, and the oscillating periodic property can markedly change. Furthermore, we can find that not only the oscillating frequency of Q_1 but also the amplitude of Q_1 changes significantly in the two-photon Jaynes-Cummings model without the rotating-wave approximation for different ε. But

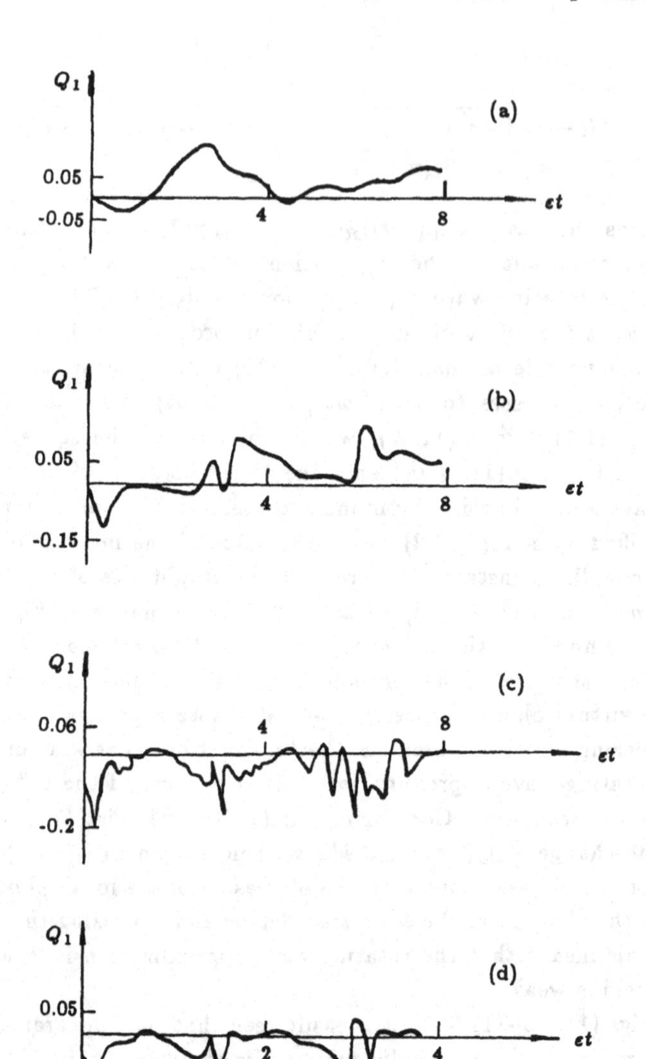

Figure 11.3: Q_1 versus without RWA, (a) $\overline{n} = 1$, $\epsilon = 0.1$, (b) $\overline{n} = 4$, $\epsilon = 0.1$, (c) $\overline{n} = 6$, $\epsilon = 0.1$, (d) $\overline{n} = 6$, $\epsilon = 0.05$

in the rotating-wave approximation case as shown in Fig.(11.2-c) and (11.2-d), the squeezing degree only changes with $\bar{n}$ and ϕ, and is not related to ε and ω. Here, the present results show that the influence of the virtual photon processes on the degree of squeezing is related to not only the intensity of the field but also the optical frequency ω and the atom-field coupling constant ε. So, considering the effect of virtual photon field, the different optical frequencies and the different coupling constants will induce different squeezing degree of the field in the atom-field coupling system, even if the intensity is the same. The degree of squeezing is completely determined by the characteristic of the atom-field coupling system.

References

1. D.P.Craig, *Molecular quantum electrodynamics—- An introduction to radiation-molecular interaction* (Academic, New York, 1984).
2. H.Haken, *Light* **Vol.1** (North Holland, Amsterdam, 1985).
3. G.Compagno, R.Passnate and F.Persico, *Phys.Lett.* **A93** (1983) 256.
4. R.Passnate, G.Compagno and F.Persico, *Phys.Rev.* **A31** (1985) 2827.
5. F.Persico and E.A.Power, *Phys.Rev.* **A36** (1987) 475.
6. G.Compagno, R.Passnate and F.Persico, *Phys.Scr.* **T21** (1988) 33.
7. G.Compagno and G.M.Palma, *Phys.Rev.* **A37** (1988) 2979.
8. G.Compagno, G.M.Palma, R.Passnate and F.Persico, *Europhys.Lett.* **9** (1989) 215.
9. A.K.Biswas, G.Compagno, G.M.Palma, R.Passnate and F.Persico, *Phys. Rev.* **A42** (1990) 4291.
10. G.Compagno, S.Vivirto and F.Persico, *Phys.Rev.* **A46** (1992) 7303.
11. X.Y.Huang and Jin-sheng Peng, *Phys.Scr.* **T21** (1988) 100.
12. Gao-xiang Li and Jin-sheng Peng, *Chin.Quantum Electonics* **7** (1990) 175.
13. Jin-sheng Peng and Gao-xiang Li, *Phys.Rev.* **A45** (1992) 3289.

14. Jin-sheng Peng, Gao-xiang Li and Peng Zhou, *ACTA Physica Sinica* **40** (1991) 1042.

15. Gao-xiang Li and Jin-sheng Peng, *ACTA Phys.Sin.* **41** (1992) 766.

16. Jin-sheng Peng and Gao-xiang Li, *Phys.Rev.* **A47** (1993) 3167.

17. Jin-sheng Peng and Gao-xiang Li, *ACTA Phys.Sin.* **42** (1993) 568.

PART III

QUANTUM PROPERTIES OF ATOMIC BEHAVIOR UNDER THE INTERACTION OF THE RADIATION FIELD

In the previous volume, we have discussed the quantum properties of light fields in a system of atoms interacting with a radiation field. This volume will investigate the quantum properties of atomic behavior in the same system. When an atom is driven by a laser field, the evolution of the atomic system exhibits a number of new quantum properties which are not exhibited in the absence of laser fields. To study these new quantum properties of atomic behavior both theoretically and experimentally is a very important task in modern quantum optics.

One of the nonclassical effects of the atom interacting with a laser field is the periodic collapses and revivals of atomic populations which is treated in Chapter 12. After discussing the time evolution of atomic operators, we concentrate on the conditions for periodic collapses and revivals of the population inversion. The role of nonclassical effects are also discussed. Chapter 13 is devoted to studying the squeezing effect of atomic operators. The conditions inducing this squeezing are investigated within the Jaynes-Cummings model. The squeezing effect of the dipole moment in resonance fluorescence is also discussed. Chapter 14 deals with the coherent trapping of atomic populations. Here we discuss the conditions leading to such a trapping. The influence of atomic initial states on the coherent population trapping is also analyzed in detail. As for the quantum properties of a system of two atoms interacting with the radiation field, this is examined in Chapter 15. Here the effects of dipole-dipole interaction on the time evolution of the system are investigated. The periodic collapses and revivals, the squeezing of atomic operators and the coherent trapping of atomic populations for the two-atom system are analyzed comprehensively. Chapter 16 investigates the autoionization of an atom induced by laser fields. The quantum theoretical treatments of atomic autoionization due to the interaction with a weak and with an intensive laser field respectively are discussed. The autoionization of the atom in the above-threshold region is also analyzed. Chapter 17 is devoted to investigating the motion of atom in the laser field. We discuss the theory of atomic diffraction and deflection in the standing-wave field, and describe the force exerted on an atom by the radiation field. The last chapter (Chapter 18) deals with laser cooling which has been extensively studied both theoretically and experimentally over the past few years. Here we introduce the general principle of laser cooling within a quantum theoretical framework. Finally, the temperature limits of laser cooling are also analyzed.

CHAPTER 12

COLLAPSES AND REVIVALS OF ATOMIC POPULATIONS

One of the most important quantum features in the atom-field coupling system is that the atomic dynamic behavior can exhibit the periodic collapse and revival phenomenon which cannot be revealed by means of the quasi-classical theory. In this chapter we focus our attention on the atomic dynamic behavior by taking into account the time evolution of the atomic operators.

12.1 Time evolution of the atomic operator of a two-level atom under the interaction of a classical electromagnetic field

In Chapter 2 we have shown that the Hamiltonian describing the system of a two-level atom interacting with a classical field $E(t)$ is given in the dipole approximation by (2.34)

$$H = \frac{\omega_0}{2}\sigma_3 - (\mathbf{d}_r \cdot \mathbf{E(t)})\sigma_1 = \omega_0 S_3 + 2\lambda E(t)(S_+ + S_-) \quad (\hbar = 1) \tag{12.1}$$

where the Pauli operators σ_1 and σ_2 are replaced by the pseudo-spin operators S_z and $S\pm$, the coupling constant λ is defined as

$$\mathbf{d}_r \cdot \mathbf{E}(t) = -2\lambda E(t) \tag{12.2}$$

It is clear that $\omega_0 S_3$ is the energy operator of a two-level atom and $2\lambda E(t)(S_+ + S_-)$ represents the interaction energy between the atom and the classical field. If the amplitude of the field is E_0 and the frequency is ω, namely

$$E(t) = E_0 \cos\omega t = \frac{E_0}{2}(e^{-i\omega t} + e^{i\omega t}) \tag{12.3}$$

then the Hamiltonian (12.1) becomes

$$H = \omega_0 S_3 + \lambda E_0(e^{-i\omega t} + e^{i\omega t})S_+ + \lambda E_0(e^{-i\omega t} + e^{i\omega t})S_- \tag{12.4}$$

In the interaction picture, the interaction Hamiltonian is written as

$$V^I(t) = \lambda E_0\{\exp[i(\omega-\omega_0)t] + \exp[-i(\omega+\omega_0)t]\}S_+ \\ +\lambda E_0\{\exp[i(\omega+\omega_0)t] + \exp[-i(\omega-\omega_0)t]\}S_- \tag{12.5}$$

Clearly, the terms containing the fast oscillating factors $\exp[\pm i(\omega+\omega_0)t]$ are the counter rotating-wave terms, which describe the transition processes of energy non-conservation corresponding to the virtual photon process terms such as aS_-, $a^\dagger S_+$, as pointed out in Chapter 11. Thus in the rotating-wave approximation, (12.5) is reduced to

$$V^I(t) = \lambda E_0 \exp[-i(\omega-\omega_0)t]S_- + \lambda E_0 \exp[i(\omega-\omega_0)t]S_+ \tag{12.6}$$

The right side of the above equation is similar to the term $\lambda\{aS_+\exp[i(\omega-\omega_0)t] + a^\dagger S_-\exp[-i(\omega-\omega_0)t]\}$ in the Jaynes-Cummings model, the only difference is that the field is described by the operators a and $a^\dagger$, but here the field is represented by the C-number E_0.

Supposing that the two-level atom is initially in the superposition of its ground state $|1\rangle$ and excited state $|2\rangle$

$$|\Psi(0)\rangle = \gamma_1|1\rangle + \gamma_2|2\rangle \tag{12.7}$$

where$|\gamma_1|^2 + |\gamma|^2 = 1$. At time t, the state vector in the interaction picture evolves into

$$|\Psi^I(t)\rangle = C_2(t)|2\rangle + C_1(t)|1\rangle \tag{12.8}$$

Substituting (12.6) and (12.8) into the Schrödinger equation in the interaction picture, we obtain

$$i\frac{d}{dt}C_2(t) = \lambda E_0 \exp[i(\omega-\omega_0)t]C_1(t) \\ i\frac{d}{dt}C_1(t) = \lambda E_0 \exp[-i(\omega-\omega_0)t]C_2(t) \tag{12.9}$$

Considering (12.7), the solutions of above equations in the resonance case (i.e., $\omega=\omega_0$) read as

$$C_1(t) = \gamma_1\cos(\lambda E_0 t) - i\gamma_2\sin(\lambda E_0 t) \tag{12.10}$$

$$C_2(t) = \gamma_2\cos(\lambda E_0 t) - i\gamma_1\sin(\lambda E_0 t) \tag{12.11}$$

As we know, $\langle S_3 \rangle$ describes the expectation value of the atomic energy operator and represents the atomic population inversion between the excited state and the ground state, so we can realize the atomic dynamic behavior by means of studying the time evolution of $\langle S_3 \rangle$. From eqs.(12.8), (12.10) and (12.11), we obtain $\langle S_3(t) \rangle$ as

$$\begin{aligned}\langle S_3(t)\rangle = \langle \Psi^I(t)|S_3^I|\Psi^I(t)\rangle &= \frac{1}{2}(|C_2(t)|^2 - |C_1(t)|^2) \\ &= \frac{1}{2}(|\gamma_2|^2 - |\gamma_1|^2)\cos(2\lambda E_0 t)\end{aligned} \tag{12.12}$$

The above equation shows that $\langle S_3(t) \rangle$ exhibits a cosine oscillation against t with frequency λE_0. Usually, this cosine oscillation is called the Rabi oscillation and the frequency

$$\Omega_R = 2\lambda E_0 \tag{12.13}$$

is called the Rabi oscillating frequency. Evidently, Ω_R is proportional to the strength E_0 of the classical field. It is clear that under the interaction of a classical field, the two-level atom displays the Rabi oscillation with the frequency Ω_R between the excited state and the ground state, and the oscillating amplitude $(|\gamma_2|^2 - |\gamma_1|^2)$ depends on the atomic initial state. When $\gamma_1 = 0$, corresponding to the atom in its ground state at time t=0, (12.12) is reduced to

$$\langle S_3(t)\rangle = -\frac{1}{2}\cos(2\lambda E_0 t) \tag{12.14}$$

If the strength E_0 of the classical field is equal to zero, which represents the field is a vacuum field, then $\langle S_3(t) \rangle = -\frac{1}{2}$. This means that the atom initially in its ground state interacting with the vacuum field maintains in its ground state. However, when the atom is initially in its excited state, i.e., $\gamma_1 = 0$, $\gamma_2 = 1$, then

$$\langle S_3(t)\rangle = \frac{1}{2}\cos(2\lambda E_0 t) \tag{12.15}$$

Obviously, equation (12.15) is similar to equation (12.14). But for $E_0 = 0$ in eq.(12.15), $\langle S_3 \rangle = 1/2$, which means that the atom initially in its excited state does not decay due to the interaction of the vacuum field. From the physical point of view, this result is unreasonable, because the atom in its excited state can spontaneously decay to its ground state due to the effect of vacuum fluctuations. Therefore, we can see that the quasi-classical theory has its limitations in revealing the time evolution of atomic variables in the atom-field coupling system. In fact, the exact descriptions of the time development of the atomic variable may be obtained in the quantum electrodynamics theory. In the following sections, we will discuss the dynamic characteristics of the atom under the interaction of a quantized field.

12.2 Periodic collapses and revivals of an atom interacting with a quantized field

Here we still concentrate on the system of a two-level atom coupled to a single-mode radiation field, but the field is quantized. That is to say, this system may be characterized by the Jaynes-Cummings model as

$$H = H_0 + V \qquad (\hbar = 1) \tag{12.16}$$

$$H_0 = \omega_0 S_3 + \omega a^\dagger a \tag{12.17}$$

$$V = g(a^\dagger S_- + a S_+) \tag{12.18}$$

In order to discuss the time evolution of the atomic dynamic variable, we must give the state vector of the system at time t.

We assume that the initial atomic state $|\Psi_A(0)\rangle$ is described by (12.7), and the initial field is represented by

$$|\Psi_F(0)\rangle = \sum_n F_n |n\rangle \tag{12.19}$$

here F_n is the probability amplitude of the field in the number state $|n\rangle$ and satisfies $\sum_n |F_n|^2 = 1$. For different fields, the expressions of F_n are different. At time t=0, the state vector of the atom-field coupling system gives

$$|\Psi(0)\rangle = \sum_n F_n(\gamma_1|1,n\rangle + \gamma_2|2,n\rangle) \tag{12.20}$$

Moreover, the state vector of the system at time t in the interaction picture can be expanded as

$$|\Psi^I(t)\rangle = \sum_n a_n(t)|2,n\rangle + b_{n+1}|1,n+1\rangle \tag{12.21}$$

Utilizing the Schrödinger equation in the interaction picture, we obtain in the resonant case

$$i\frac{d}{dt}a_n(t) = g\sqrt{n+1}\,b_{n+1}(t)$$

$$i\frac{d}{dt}b_{n+1}(t)=g\sqrt{n+1}a_n(t) \tag{12.22}$$

Considering the initial condition (12.20), the exact solutions of the above equations give

$$\begin{aligned} a_n(t) &= \gamma_2 F_n \cos(g\sqrt{n+1}t) - i\gamma_1 F_{n+1}\sin(g\sqrt{n+1}t) \\ b_{n+1}(t) &= \gamma_1 F_{n+1}\cos(g\sqrt{n+1}t) - i\gamma_2 F_n \sin(g\sqrt{n+1}t) \end{aligned} \tag{12.23}$$

Substituting (12.23) into (12.21), we may obtain the state vector $|\Psi^I(t)\rangle$ of the system at arbitrary time t. By use of $|\Psi^I(t)\rangle$, we can discuss the time evolution of the atomic operators in detail.

12.2.1 Time development of atomic operators under the interaction of the field in a number state $|m\rangle$

If the field is initially in the number state $|m\rangle$, that is, $F_n = \delta_{n,m}$, and the atom prepared in its ground state $|1\rangle$, i.e., $\gamma_2 = 0$ and $\gamma_1 = 1$, then from eqs.(20)-(23), the state of the system evolves to

$$|\Psi^I(t)\rangle = -i\sin(g\sqrt{m}t)|2, m-1\rangle + \cos(g\sqrt{m}t)|1, m\rangle \tag{12.24}$$

So the expectation value of the atomic inversion operator S_3 reads as

$$\langle S_3(t)\rangle = -\frac{1}{2}\cos(2g\sqrt{m}t) \tag{12.25}$$

Comparing the above equation with (12.14), we can see that under the interaction of a number state field, the atom exhibits a cosine oscillation between its excited state and the ground state, and the oscillating frequency is

$$\Omega = 2g\sqrt{m} \tag{12.26}$$

This equation explicitly shows that the different photon number states lead to different quantum Rabi oscillating frequencies. In particular, noticing the initial expectation value of the intensity of the electric field

$$\langle E^2\rangle \propto \langle a^\dagger a\rangle = m$$

and the intensity of the classical electric field satisfying $E^2 = E_0^2$, we find that the formalism of Ω is resemblance to that of Ω_R when the classical physical variable E_0^2 is replaced by the expectation $\langle E^2 \rangle$. In this sense, Ω is usually termed as the quantum Rabi oscillating frequency. For m=0 corresponding to the vacuum state, the meaning of (12.25) is identical to that of quasi-classical case, which means that under the interaction of the vacuum field, the atom initially in the ground state remains its ground state.

But if we assume the atom initially in its excited state, that is, $\gamma_1 = 0$ and $\gamma_2 = 1$, the state vector of the system may be simplified as

$$|\Psi^I(t)\rangle = \cos(g\sqrt{m+1}t)|2,m\rangle - i\sin(q\sqrt{m+1}t)|1,m+1\rangle \tag{12.27}$$

In this case, $\langle S_3(t)\rangle$ will be expressed as

$$\langle S_3(t)\rangle = \frac{1}{2}\cos(g\sqrt{m+1}t) \tag{12.28}$$

Obviously, with the time development, $\langle S_3 \rangle$ is also oscillating according to the cosine function, but the oscillating frequency is different from the case for the atom initially in its ground state. That is to say, the atomic initial state plays an important role in the time evolution of atomic operators. However, in the quasi-classical theory, we know from (12.14) and (12.15) that the oscillating frequency is the same. It is worthwhile to point out that when the field initially in the vacuum state $|0\rangle$, from (12.28) we obtain

$$\langle S_3(t)\rangle = \frac{1}{2}\cos(2gt) \tag{12.29}$$

This means that the atom initially in its excited state has the Rabi oscillation with frequency $2g$ due to the interaction of the vacuum field. Evidently, if $t = \frac{\pi}{2g}$, the atom will evolve to its ground state. This result is obviously different from that obtained by use of the quasi-classical theory. So it shows that by means of the quantum theory the influences of vacuum fluctuations on the atomic property can be revealed.

12.2.2 Periodic collapses and revivals of the atom under the interaction of a coherent field

The above results show that under the interaction of a number state field, the atom exhibits a cosine oscillation between its excited state and its ground state. However, if the initial field is in a coherent state, how is about the atomic dynamic behavior ? We discuss this problem in this section.

Here we only consider the initial conditions which the atom is in its ground state ($\gamma_1 = 1$, $\gamma_2 = 0$) and the field in a coherent state $|\alpha\rangle$. For the coherent field, the probability amplitude F_n in (12.19) is chosen as

$$F_n = \exp(-|\alpha|^2/2)\frac{\alpha^n}{\sqrt{n!}} \tag{12.30}$$

Then (12.22) can be rewritten as

$$\begin{aligned}|\Psi^I(t)\rangle = \sum_n e^{-|\alpha|^2/2}\frac{\alpha^n}{\sqrt{n!}}[-i\sin(g\sqrt{n}t)|2, n-1\rangle \\ + \cos(g\sqrt{n}t)|1, n\rangle]\end{aligned} \tag{12.31}$$

correspondingly, $\langle S_3(t)\rangle$ yields

$$\langle S_3(t)\rangle = -\frac{1}{2}\sum \exp(-\overline{n})\frac{\overline{n}^n}{n!}\cos(2g\sqrt{n}t) \tag{12.32}$$

here we have assumed $\overline{n} = |\alpha|^2$. The above equation shows that the expectation value of the atomic energy operator is seen to be a superposition of infinite number of oscillations with the frequency $2g\sqrt{n}$ and the Poisson weights $\exp(-\overline{n})\frac{\overline{n}^n}{n!}$. Evidently, this superposition induces that the time evolution of $\langle S_3\rangle$ is not a cosine oscillation. The time behavior of $\langle S_3(t)\rangle$ for $\overline{n} = 8$, 15 are shown in Fig.(12.1). The results show that in a short duration, $\langle S_3(t)\rangle$ exhibits the rapid oscillation with decreasing the amplitude quickly, this phenomenon is termed the collapse. And for a long time range, $\langle S_3(t)\rangle$ remains to be zero. In the subsequent time range, $\langle S_3(t)\rangle$ does the oscillations with increasing the amplitude rapidly and then decreasing repeatedly. As usual, we interpret the phenomenon which the amplitude of $\langle S_3(t)\rangle$ increases from zero rapidly as the quantum revival. From Fig.(12.1), we also see that the quantum revival of $\langle S_3(t)\rangle$ is periodic. So we find that the time evolution of the atomic operator in the system of an atom interacting with a coherent field exhibits the periodic collapses and revivals. In addition, Fig.(12.1) shows that the period of the quantum revival depends on the mean photon number $\overline{n}$, the decay time of the collapses independent of $\overline{n}$, and the maximum amplitude per revival decreases with the time development. In the following, we analyze these properties by use of the saddle-point techniques.

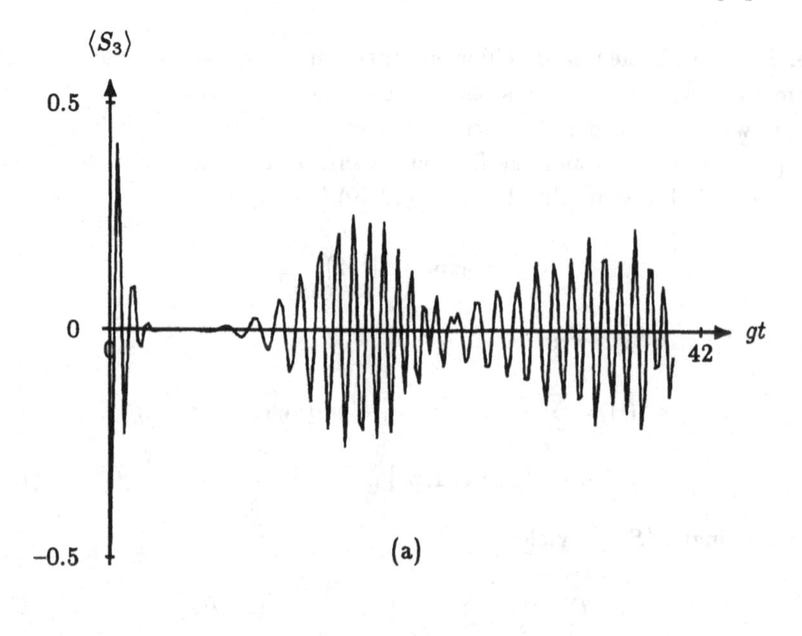

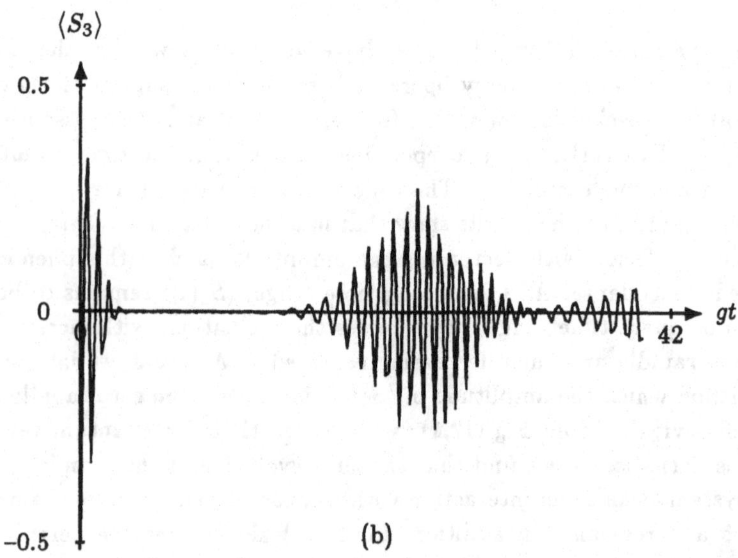

Figure 12.1: Time evolution of $\langle S_3(t) \rangle$, (a) $\overline{n} = 8$, (b) $\overline{n} = 15$

Noticing the Poisson weighting factor $P_n = \exp(-\overline{n})\frac{\overline{n}^n}{n!}$ will peak at a value $n = \overline{n}$ with relatively narrow dispersion $(\langle n^2\rangle - \overline{n}^2)^{1/2} = \sqrt{\overline{n}}$. This means that for the summation over n in (12.32), the dominant contributions to $\langle S_3(t)\rangle$ is the n located at the range $\overline{n} - \sqrt{\overline{n}} < n < \overline{n} + \sqrt{\overline{n}}$. When $\overline{n} \gg 1$, this range is very narrow. So the sum in (12.32) can be replaced by an integral, namely $\sum\limits_n \rightarrow \int\limits_0^\infty dn$. Utilizing the Stiring formula

$$\ln\Gamma(z) = \left(z - \frac{1}{2}\right)\ln z - z + \frac{1}{2}\ln(2\pi) + \frac{1}{12z} - \frac{1}{360z^3} + \frac{1}{1260z^5} + \cdots \quad (12.33)$$

$lnn!$ is approximated as

$$\ln n! \approx \left(n + \frac{1}{2}\right)\ln(n+1) - n + \frac{1}{2}\ln(2\pi) \approx n\ln n - n + \ln(2\pi n)^{1/2} \quad (12.34)$$

Therefore, P_n is approximated to be

$$\ln P_n = \ln\left[\exp(-\overline{n})\frac{\overline{n}^n}{n!}\right] = -\overline{n} + n + n\ln(\overline{n}/n) - \ln(2\pi n)^{1/2}$$

that is

$$P_n \approx (2\pi n)^{-1/2}\exp[-\overline{n} + n - n\ln(n/\overline{n})] \quad (12.35)$$

Consequently (12.32) can be changed as

$$\begin{aligned}\langle S_3(t)\rangle &= -\frac{1}{2}\int\limits_0^\infty dn(2\pi n)^{-1/2}\exp[-\overline{n} + n - n\ln(n/\overline{n})]\cos(2g\sqrt{n}t) \\ &= -\frac{1}{2}\mathrm{Re}\int\limits_0^\infty dn(2\pi n)^{-1/2}\exp[-\overline{n} + n - n\ln(n/\overline{n})]\exp(i2g\sqrt{n}t) \\ &= -\frac{1}{2}\mathrm{Re}\int\limits_0^\infty dy(2\pi n)^{-1/2}\overline{n}\exp[-\overline{n} + \overline{n}\psi(y)] \\ &\approx -\frac{1}{2}\overline{n}(2\pi\overline{n})^{-1/2}\mathrm{Re}\int\limits_0^\infty \exp[-\overline{n} + \overline{n}\psi(y)]dy \end{aligned} \quad (12.36)$$

where $y = n/\overline{n}$ and

$$\psi(y) = y(1 - \ln y) + i2g\overline{n}^{-1/2}yt \quad (12.37)$$

In order to evaluate the above integral, it is necessary to introduce the saddle-point techniques for calculating the integral such as

$$I = \int \exp[\alpha(x)]dx \tag{12.38}$$

First we expand $\alpha(x)$ as

$$\alpha(x) = \alpha(x_0) + \alpha'(x_0)(x - x_0) + \frac{1}{2}\alpha''(x_0)(x - x_0)^2 + \cdots \tag{12.39}$$

If x_0 is chosen as a steady point which corresponds to the maximum value of $\alpha(x)$, then the evaluation of (12.38) can be simplified. In fact, when x_0 is chosen as the value for $\alpha(x)$ being maximum, the integrating result in the neighbourhood centered at x_0 plays a dominant role in the integral (12.38) and the integral over the neighbourhood Δx of x_0 can be approximately regarded as the result of (12.38) over the all range of x . If $\alpha(x)$ reaches its maximum value at x_0, then $\alpha(x)$ must satisfy

$$\alpha'(x_0) = 0 \quad and \quad \alpha''(x_0) < 0 \tag{12.40}$$

So (12.38) can be rewritten at the neighbourhood Δx of x_0 as

$$I = \exp(\alpha(x_0)) \int dx \exp\left\{-\frac{1}{2}[-\alpha''(x_0)(x - x_0)^2 + \cdots\right\}$$

Clearly, since the integration function is a Gaussian-type formalism when the unimportant factors are neglected, the above integral is entirely integrable. We call x_0 obeying (12.40) as the saddle-point of the integral (12.38) and the method we discussed in the above as the saddle-point method.

Using the saddle-point method for evaluating the integral (12.36) we can get

$$\begin{aligned}\langle S_3(t)\rangle &= -\frac{1}{2}\mathrm{Re}\left(\frac{\overline{n}}{2\pi}\right)^{1/2} \exp(-\overline{n} + \overline{n}\psi(y_0)) \int_0^\infty e^{-\frac{1}{2}\psi''(y_0)\overline{n}(y-y_0)^2} dy \\ &= -\mathrm{Re}\frac{1}{2\sqrt{\psi''(y_0)}} \exp[-\overline{n} + \overline{n}\psi(y_0)]\end{aligned} \tag{12.41}$$

here y_0 is the saddle-point of $\psi(y)$ and satisfies

$$\psi'(y_0) = 0$$

That is to say, y_0 is the solution of the equation

$$y_0^{1/2}\ln(y_0) = ig\overline{n}^{-1/2}t \tag{12.42}$$

Thus y_0 depends on time evidently.

First we give a qualitative analysis on the dynamic behavior of the atom interacting with a coherent field by use of (12.41) and (12.42). For very short time, i.e.,

$$t \ll \tau_c = g^{-1} \tag{12.43}$$

Equation (12.42) has a trivial solution $y_0 \approx 1$ and

$$\psi(1) = 1 + i2g\overline{n}^{-1/2}t \tag{12.44}$$

Substituting (12.44) into (12.41) we obtain

$$\langle S_3(t)\rangle = -\frac{1}{2}\mathrm{Re}\frac{1}{\sqrt{\psi''(1)}}\exp(2ig\overline{n}^{1/2}t) \tag{12.45}$$

This means that the atom displays the Rabi oscillation with frequency $\Omega_R = 2g\overline{n}^{1/2}$ for $t \ll \tau_c$. But when the time develops to $t \sim \tau_c$, from (12.42) we see

$$y_0 - 1 \sim \overline{n}_0^{-1/2}, \quad \mathrm{Re}(1 - \psi(y_0)) \sim \overline{n}^{-1/2} \tag{12.46}$$

In this case, the amplitude of $\langle S_3(t)\rangle$ decreases exponentially. That is to say, the envelope of the Rabi oscillations collapses rapidly to zero. Therefore, τ_c can be considered as the natural "collapse time" for Rabi-type oscillations of a two-level atom interacting with a single-mode radiation field initially in a pure coherent state. Explicitly, τ_c only depends on the atom-field coupling constant g and does not related to the mean photon number $\overline{n}$ of the coherent field.

This collapse has a simple physical meaning. For time t=0, the atom-field system can be considered to be in the superposition of $\{|1,n\rangle\}$, so the term $-\frac{1}{2}\exp(-\overline{n})\frac{\overline{n}^n}{n!}\cos(2g\sqrt{n}t)$ in (12.32) can be regarded as the quantum Rabi oscillation due to the emission and absorption of photons for the system initially in state $|1,n\rangle$. And eq.(12.32) is exactly a superposition of these Rabi oscillations. At the moment t=0, the field is in the coherent state, so the system is prepared in a definite state. But since all terms in (12.32) oscillate with different Rabi frequencies, they will become decorrelated in time $t \approx \tau_c$. That is to say, for different n, the value of $\cos(2g\sqrt{n}t)$ may be positive or negative at certain time τ_c, which leading to these corresponding Rabi oscillations decorrelated and inducing that the value of $\langle S_3(t)\rangle$ collapses to zero. This decorrelation time τ_c is easily estimated as follows. Since for $\overline{n} \gg 1$, the main effects on $\langle S_3(t)\rangle$ result from the oscillations for n being in the range $\overline{n}-\sqrt{\overline{n}} \leq n \leq \overline{n}+\sqrt{\overline{n}}$. If the difference between the phases $2g\sqrt{\overline{n}+\sqrt{\overline{n}}}\tau_c$ and $2g\sqrt{\overline{n}-\sqrt{\overline{n}}}\tau_c$ is π, then for n in the dominant range of $\overline{n}-\sqrt{\overline{n}} \leq n \leq \overline{n}+\sqrt{\overline{n}}$, the values of $\cos(2g\sqrt{n}t)$ may be positive or negative, so τ_c must obey

$$2g(\sqrt{\overline{n}+\sqrt{\overline{n}}}-\sqrt{\overline{n}-\sqrt{\overline{n}}})\tau_c \sim 1 \tag{12.47}$$

Considering $\overline{n} \gg 1$ we get the decorrelation time to be

$$\tau_c \sim g^{-1} \tag{12.48}$$

Obviously, the decorrelation time τ_c is in agreement with the collapse time eq.(12.43).

We can realize the revival phenomenon as follows. When the time develops to a given value T_R, the phase of cosine function in (12.32) reaches to the difference of value 2π, then $\cos(2g\sqrt{n}t)$ recovers its value. The amplitude factor $\exp(-\overline{n})\frac{\overline{n}^n}{n!}$ for $\overline{n} \gg 1$ keeps almost steady, so that $\langle S_3\rangle$ revivals to its initial value. The interval T_R between two revivals can be found from the relation

$$2g(\sqrt{\overline{n}+1}-\sqrt{\overline{n}})T_R \approx 2\pi$$

thus

$$T_R = \frac{2\pi}{g}\overline{n}^{1/2} \tag{12.49}$$

Here T_R depends on the atom-field coupling constant g and the mean photon number $\overline{n}$ of the initial coherent field. This qualitative result (12.49) is in agreement with Fig.(12.1), which shows at the time points

$$t = kT_R \qquad (k = 0, 1, 2, \cdots) \tag{12.50}$$

the revivals occur. Substituting (12.50) into (12.42), the exact solutions of (12.42) give

$$y_0 = \exp(2ik\pi) \qquad (k = 0, 1, 2, \cdots) \tag{12.51}$$

Clearly, for t=0, $y_0 = 1$.

Here we have studied the time evolution of the atomic inversion in the neighborhood of $y_0 = 1$, however, for the neighborhood of $y_0 = \exp(i2\pi k)$, what will happen to $\langle S_3(t)\rangle$? That is to say, what is the time evolution of the atomic inversion near the revival points $t = kT_R$? In order to answer this question, we assume

$$t = kT_R + \tau, \quad \tau \ll 1 \tag{12.52}$$

correspondingly, $y_{0k}(t)$ can be written in the form

$$y_{0k}(t) = \eta_k \exp(i\phi_k), \quad \eta_k = 1 + \eta_k^{(1)}, \quad \phi_k = 2\pi k + \phi_k^{(1)} \tag{12.53}$$

under the condition that the local time τ is small. Here we have also supposed

$$\eta_k^{(1)} \ll 1, \quad \phi^{(1)} \ll 1, \quad g\overline{n}^{-1/2}\tau \ll 1 \tag{12.54}$$

Subsitituting (12.53) into (12.41) and only retaining the term as $\eta_k^{(1)}$ and $\phi_k^{(1)}$, we get

$$i\pi k\eta_k^{(1)} - \pi k\phi_k^{(1)} + \eta_k^{(1)} + i\phi_k^{(1)} = ig\overline{n}^{-1/2}\tau$$

Here we have considered eqs.(12.52) and (12.49). By means of the identical condition of complex, the above equation may be divided into two parts

$$\begin{aligned} \eta_k^{(1)} &= \pi k\phi_k^{(1)} \\ \pi k\eta_k^{(1)} + \phi_k^{(1)} &= g\overline{n}^{-1/2}\tau \end{aligned} \tag{12.55}$$

The solution of (12.55) is

$$\phi_k^{(1)} = g\overline{n}^{-1/2}\tau[1 + \pi^2k^2]^{-1} \tag{12.56}$$

$$\eta_k^{(1)} = g\overline{n}^{-1/2}\tau\pi k[1 + \pi^2k^2]^{-1} \tag{12.57}$$

Substituting (12.53) into (12.57) and considering (12.41), we obtain

$$\psi(y_{0k}(t)) = 1 + \eta_k^{(1)} + i\phi_k^{(1)} + ig\overline{n}^{-1/2}t\left(1 + \frac{\eta_k^{(1)}}{2}\right) - \frac{g}{2}\overline{n}^{-1/2}t\phi_k^{(1)} \tag{12.58}$$

By use of eqs.(12.54), (12.56) and (12.57), and noticing $\sin x = x$ $(x \ll 1)$, the above equation yields

$$\mathrm{Re}\overline{n}\psi(y_{0k}(t)) \approx -2\overline{n}\left(\frac{g\tau}{2\sqrt{\overline{n}}}\right)^2\left[1 + \frac{g^2t^2}{4\overline{n}}\right]^{-1} + \overline{n}$$

$$
\begin{aligned}
&= -2\overline{n}\sin^2\left(\frac{g\tau}{2\sqrt{\overline{n}}}\right)\left[1+\frac{g^2t^2}{4\overline{n}}\right]^{-1}+\overline{n} \\
&= -2\overline{n}\sin^2\left(\frac{g(t-kT_R)}{2\sqrt{\overline{n}}}\right)\left[1+\frac{g^2t^2}{4\overline{n}}\right]^{-1}+\overline{n} \\
&= -2\overline{n}\sin^2\left(\frac{gt}{2\sqrt{\overline{n}}}\right)\left[1+\frac{g^2t^2}{4\overline{n}}\right]^{-1}+\overline{n}
\end{aligned}
\tag{12.59}
$$

$$
\begin{aligned}
\mathrm{Im}\overline{n}\psi(y_{0k}(t)) &= g\overline{n}^{1/2}t+\overline{n}g\overline{n}^{-1}\tau \\
&= g\overline{n}^{1/2}t+\overline{n}\sin(g\overline{n}^{-1/2}\tau) \\
&= g\overline{n}^{1/2}t+\overline{n}\sin[g\overline{n}^{-1/2}(t-kT_R)] \\
&= g\overline{n}^{1/2}t+\overline{n}\sin(g\overline{n}^{-1/2}t)
\end{aligned}
\tag{12.60}
$$

And it is easy to calculate $\psi''(y_{0k}(t))$

$$
\psi''(y_{0k}(t)) \approx -1 - ig\overline{n}^{-1/2}t/2 \approx \sqrt{1+\frac{g^2t^2}{4\overline{n}}}\exp[i\arctan(g\overline{n}^{-1/2}t/2)] \tag{12.61}
$$

With the help of the above formulae we can write formula (12.45) for the atomic inversion in the form

$$
\langle S_3(t)\rangle = -\frac{1}{2}\left(1+\frac{g^2t^2}{4\overline{n}}\right)^{-1/2}\exp[-C(t)]\cos\Phi(t) \tag{12.62}
$$

$$
C(t) = 2\overline{n}\sin^2\left(\frac{gt}{2\sqrt{\overline{n}}}\right)\left(1+\frac{g^2t^2}{4\overline{n}}\right)^{-1/4} \tag{12.63}
$$

$$
\Phi(t) = g\overline{n}^{1/2}t+\overline{n}\sin(g\overline{n}^{-1/2}t)-\frac{1}{2}\arctan\left(\frac{gt}{2\sqrt{\overline{n}}}\right) \tag{12.64}
$$

This is the analytical formula of $\langle S_3(t)\rangle$ for $\overline{n} \gg 1$. It is evident that the atomic inversion exhibits the cosine oscillations modulated by the exponential function. When $t \ll 1$, $\langle S_3(t)\rangle$ is reduced to

$$
\langle S_3(t)\rangle = -\frac{1}{2}\exp(-g^2t^2/2)\cos\Phi(t) \tag{12.65}
$$

Explicitly, when $t \sim \tau = g_c^{-1}$, $\langle S_3(t)\rangle \approx 0$, thus the collapse phenomenon occurs. And from (12.62), we see that the revival period is $T_R = 2\pi g^{-1}\overline{n}^{1/2}$. In addition, we find from (12.62) that the amplitude of per revival has a tendency to decrease with the time development. Obviously, these analytical results are in agreement with the numerical results (shown in Fig.(12.1)).

In summary, the above analytical and numerical results show that in the one-photon Jaynes-Cummings model with the initial field in a coherent state, the time evolution of the atomic operator exhibits the periodic collapses and revivals, this is an important quantum property of the atomic population evolution. The collapse time depends on the atom-field coupling constant g only, but the revival period depends on the mean photon number $\overline{n}$ and g, and the amplitude of per revival decreases gradually with the time development.

12.3 Periodic collapses and revivals of the atom in the two-photon J-C model

In this section we investigate the time evolution of the atomic inversion in the two-photon Jaynes-Cummings model. The Hamiltonian describing a two-level atom interacting with a single-mode field via two-photon transition processes is given by

$$H = H_1 + H_2 \tag{12.66}$$

$$H_1 = \omega a^\dagger a + \omega_0 S_3 \tag{12.67}$$

$$H_2 = \varepsilon (a^{\dagger 2} S_- + a^2 S_+) \tag{12.68}$$

In the resonant case, i.e., $\omega_0 = 2\omega$, it is easy to verify that $[H_1, H] = 0$, this means that H_1 is a constant of motion. Therefore, in the Heisenberg picture, the time evolution of the atomic inversion operator S_3 satisfies

$$\begin{aligned} S_3(t) &= \exp(iHt) S_3(0) \exp(-iHt) = \exp(iH_2 t) S_3(0) \exp(-iH_2 t) \\ &= S_3(0) + it[H_2, S_3(0)] + \frac{(it)^2}{2!}[H_2, [H_2, S_3(0)]] + \cdots \end{aligned} \tag{12.69}$$

By means of the commutation relations $[a, a^\dagger] = 1$ and $[S_3, S_\pm] = \pm S_\pm$, $[S_+, S_-] = 2S_3$, we have the following relations

$$\begin{aligned} &[H_2, S_3] = [\varepsilon(a^{\dagger 2} S_- + a^2 S_+), S_3] = \varepsilon(S_- a^{\dagger 2} - S_+ a^2) \\ &[H_2, [H_2, S_3]] = 4\varepsilon(a^\dagger a + 2S_3 + 1)(a^\dagger a + 2S_3) S_3 \\ &[H_2, [H_2, [H_2, S_3]]] = 4\varepsilon^2 (a^\dagger a + 2S_3 + 1)(a^\dagger a + 2S_3)\varepsilon(S_- a^{\dagger 2} - S_+ a^2) \end{aligned} \tag{12.70}$$

With the help of (12.70), we can easily obtain

$$S_3(t) = S_3(0)\cos\Omega t + \frac{i\varepsilon}{\Omega}[S_- a^{\dagger 2}(0) - S_+ a^2(0)]\sin\Omega t \tag{12.71}$$

If at t=0, the atom is in its ground state $|1\rangle$ and the field in a coherent state $|\alpha\rangle$, namely, the initial state of the system is

$$|1,\alpha\rangle = \exp(-|\alpha|^2/2)\sum_n \frac{\alpha^n}{\sqrt{n!}}|1,n\rangle \tag{12.72}$$

then from (12.71), the expectation value of $S_3(t)$ gives

$$\langle S_3(t)\rangle = -\frac{1}{2}\exp(-\overline{n})\sum_{n=0}^{\infty}\frac{\overline{n}^n}{n!}\cos(2\varepsilon\sqrt{n(n-1)}t \tag{12.73}$$

here $\overline{n} = |\alpha|^2$. Eq.(12.73) describes the time evolution of the atomic dynamic behavior in the two-photon Jaynes-Cummings model prepared in (12.72).

Fig.(12.2) show the numerical results of (12.73) for different $\overline{n}$ and ε. It is clear that the time evolution of $\langle S_3(t)\rangle$ exhibits the periodic collapses and revivals. In the decay range, the atom does the Rabi oscillations and the Rabi frequency increases with the increasing of the mean photon number $\overline{n}$ (shown in Figs.(12.2-b-c)). Figs.(12.2-a-c) show that the collapse time becomes short with the increasing of the atom-field coupling constant ε and the mean photon number $\overline{n}$. On the other hand, the revival phenomenon takes place after collapse, and the revival interval depends on ε only and is not related to $\overline{n}$. The revival period is longer than the collapse time.

For intensive coherent field ($\overline{n} \gg 1$), we can make an approximation

$$\sqrt{n(n-1)} \approx n - \frac{1}{2} \tag{12.74}$$

taking summation in (12.73), then (12.73) can be approximated as

$$\langle S_3(t)\rangle = -\frac{1}{2}\exp(-2\overline{n}\sin^2\varepsilon t)\cos(\overline{n}\sin(2\varepsilon t) - \varepsilon t) \tag{12.75}$$

here we have used the following relations

$$\sum_{k=0}^{\infty}\frac{p^k}{k!}\sin(kx) = \exp(p\cos x)\sin(p\sin x)$$
$$\sum_{k=0}^{\infty}\frac{p^k}{k!}\cos(kx) = \exp(p\cos x)\cos(p\sin x) \tag{12.76}$$

Fig.(12.2-d) represents the time evolution of the approximation solution (12.75). Clearly, Fig.(12.2-d) is in agreement with Fig.(12.2-c). So we can analyze the

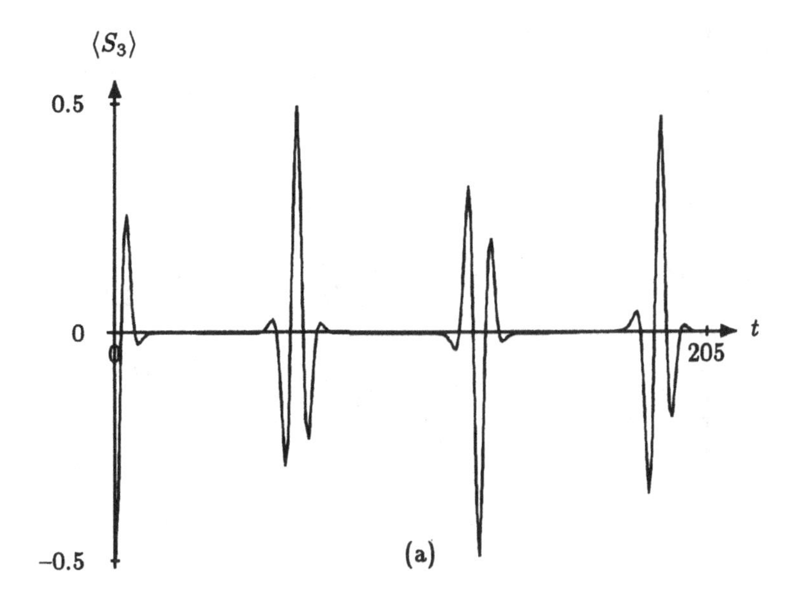

(a)

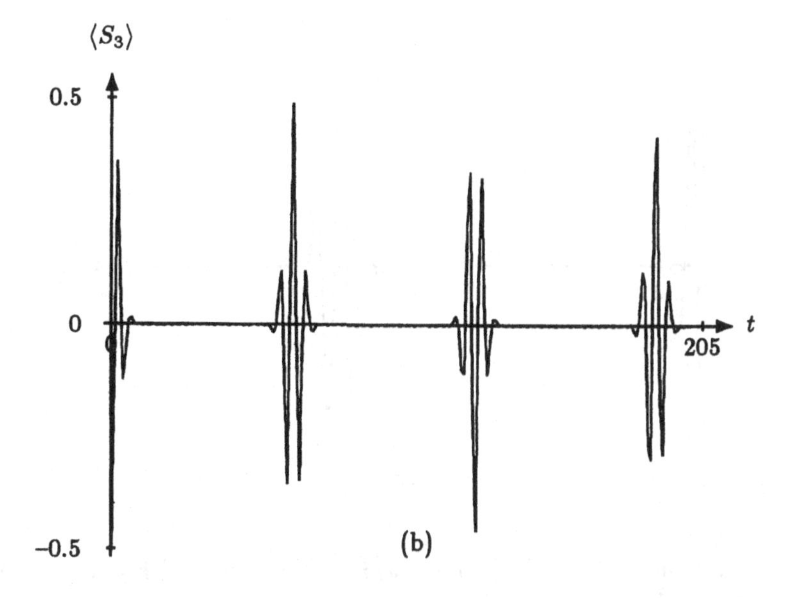

(b)

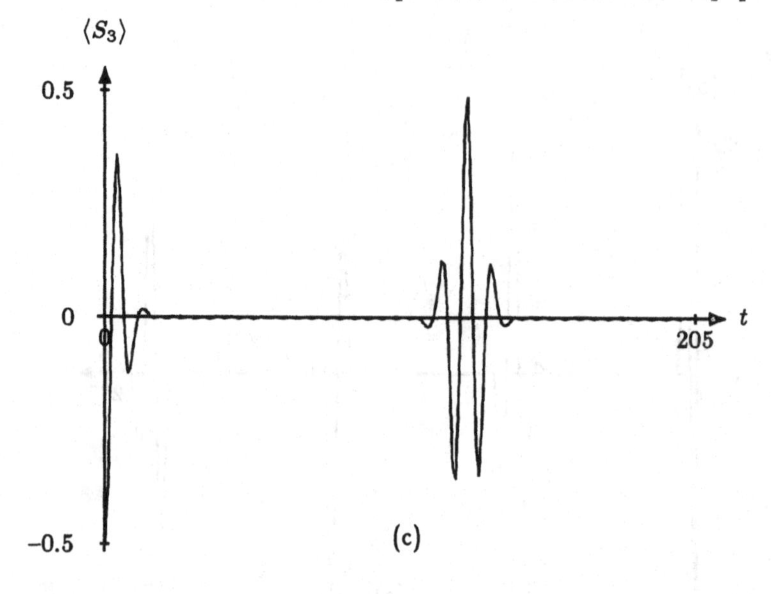

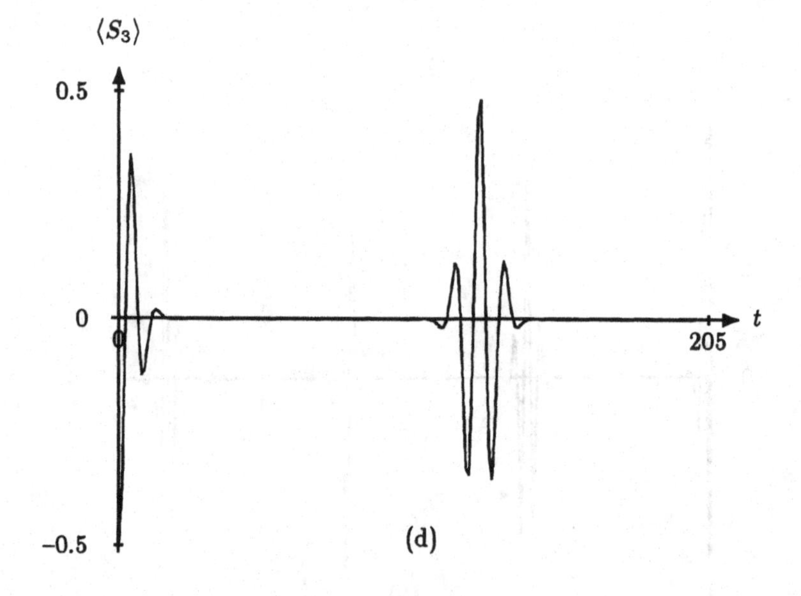

Figure 12.2: Time evolution of $\langle S_3(t) \rangle$, (a) $\overline{n} = 8$, $\varepsilon = 0.05$ (b) $\overline{n} = 15$, $\varepsilon = 0.05$ (c) and (d) $\overline{n} = 15$, $\varepsilon = 0.1$

time evolution of the atomic inversion by use of (12.75).

Equation (12.75) shows that the atomic inversion exhibits fast cosine oscillation and in the meantime these oscillations are suppressed by a Gaussian envelope function $\exp(-2\overline{n}\sin^2(\varepsilon t))$. Thus the atomic dynamic behavior exhibits the periodic collapses and revivals with the time development. From (12.75), the collapse time can be given as

$$2\overline{n}\varepsilon^2 t_c \approx 1$$

i.e.,

$$t_c \approx \frac{1}{\varepsilon\sqrt{2\overline{n}}} \tag{12.77}$$

and the revival period is

$$T_R = \frac{\pi}{\varepsilon} \tag{12.78}$$

Evidently, $\tau_c \ll T_R$ for $\overline{n} \gg 1$. Comparing with the collapse time and revival interval in the one-photon Jaynes-Cummings model, here τ_c is clearly related to $\overline{n}$ but T_R is independent of $\overline{n}$. This means that the features of the periodic collapses and revivals in the two-photon Jaynes-Cummings model are clearly different from those in the one-photon Jaynes-Cummings model.

12.4 Time evolution of the atomic operators for a three-level atom interacting with a single-mode field

Now we study the time behavior of atomic operators for a cascade three-level atom interacting with a single-mode field and discuss the important role of the one-photon detuning in the atomic dynamic behavior.

12.4.1 Time evolution of the state vector of the system

The cascade three-level atomic system, as shown in Fig.(12.3), consists of two dipole-allowed transitions $|3\rangle \to |2\rangle$ and $|2\rangle \to |1\rangle$. Each transition induces an one photon emission with frequency ω. In the rotating-wave approximation, the Hamiltonian of the atom-field coupling system is described by

$$H = H_0 + V \tag{12.79}$$

with

$$H_0 = \sum_{\alpha=1}^{3} \omega_\alpha C_\alpha^\dagger C_\alpha + \omega a^\dagger a \qquad (\hbar = 1) \tag{12.80}$$

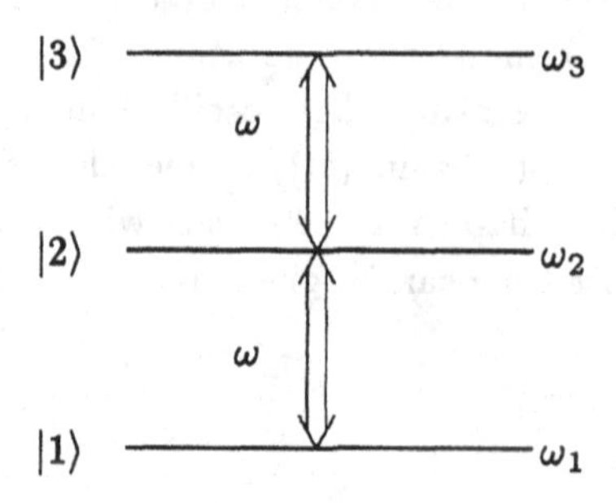

Figure 12.3: Diagram of a cascade three-level atom coupled to a laser field

$$V = g_1(aC_3^\dagger C_2 + a^\dagger C_2^\dagger C_3) + g_2(aC_2^\dagger C_1 + a^\dagger C_1^\dagger C_2) \tag{12.81}$$

here ω_α represents the α-th level energy of the atom, g_1 and g_2 are the atom-field coupling constants. $C_\alpha^\dagger$ and C_α $(\alpha = 1, 2, 3)$ correspond to the Fermi creation and annihilation operators for the α-th level of the atom, respectively, they obey

$$C_\alpha C_\beta^\dagger = 0 \quad (\alpha \neq \beta), \quad C_\alpha C_\alpha^\dagger C_\beta = C_\beta$$

For simplicity to calculate, we only consider the two-photon resonant case, i.e., $\omega_3 - \omega_1 = 2\omega$. In the interaction picture, the interaction Hamiltonian of the system is given as

$$\begin{aligned} V^I(t) &= g_1(aC_3^\dagger C_2 e^{i\Delta t} + a^\dagger C_2^\dagger C_3 e^{-i\Delta t}) \\ &\quad + g_2(aC_2^\dagger C_1 e^{-i\Delta t} + a^\dagger C_1^\dagger C_2 e^{i\Delta t}) \end{aligned} \tag{12.82}$$

$$C_i^\dagger C_i^I(t) = C_i^\dagger C_i^S \qquad (i = 1, 2, 3) \tag{12.83}$$

$$a^\dagger a^I(t) = a^\dagger a^S \tag{12.84}$$

with

$$\Delta = \omega - (\omega_2 - \omega_1) = (\omega_3 - \omega_2) - \omega \quad (\Delta \ll \omega) \tag{12.85}$$

As for the initial condition, here we only concentrate on the case of the atom initially in its ground state $|1\rangle$ and the field in a coherent state. With the time development, the state vector of the system at time t evolves to

$$|\Psi_{AF}^I(t)\rangle = \sum_n a_n(t)|3,n\rangle + b_n(t)|2,n\rangle + c_n(t)|1,n\rangle \tag{12.86}$$

By use of the Schrödinger equation in the interaction picture and the orthogonal relations, we obtain the probability amplitudes as follows

$$a_n(t) = BV_1(n)\left\{\frac{\exp[i(\Delta/2-\gamma(n))t]-1}{\Delta/2-\gamma(n)} - \frac{\exp[i(\Delta/2+\gamma(n))t]-1}{\Delta/2+\gamma(n)}\right\} \tag{12.87}$$

$$b_{n+1}(t) = B\{\exp[-i(\Delta/2-\gamma_2(n))t] - \exp[-i(\Delta/2+\gamma(n))t\} \tag{12.88}$$

$$c_{n+2}(t) = BV_2(n)\left\{\frac{\exp[i(\Delta/2-\gamma(n))t]-1}{\Delta/2-\gamma(n)} - \frac{\exp[i(\Delta/2+\gamma(n))t]-1}{\Delta/2+\gamma(n)}\right\} + F_{n+2} \tag{12.89}$$

where

$$V_1(n) = g_1\sqrt{n+1}, \quad V_2(n) = g_2\sqrt{n+2}$$

$$\gamma_1(n) = \left[\frac{\Delta^2}{4} + V_1^2(n) + V_2^2(n)\right]^{1/2} \tag{12.90}$$

$$B = -\frac{V_2(n)}{2\gamma(n)}F_{n+2}(0) \tag{12.91}$$

$\gamma(n)$ is the parameter related to the atomic Rabi oscillating frequency. So the matrix density of the system at time t is determined by

$$\begin{aligned}\rho^I(t) &= |\Psi_{AF}^I(t)\rangle\langle\Psi_{AF}^I(t)| \\ &= \sum_{n,k}\begin{pmatrix} a_k a_n^*|k\rangle\langle n| & a_k b_{n+1}^*|k\rangle\langle n+1| & a_k c_{n+2}^*|k\rangle\langle n+2| \\ b_{k+1}a_n^*|k+1\rangle\langle n| & b_{k+1}b_{n+1}^*|k+1\rangle\langle n+1| & b_{k+1}c_{n+2}^*|k+1\rangle\langle n+2| \\ c_{k+2}a_n^*|k+2\rangle\langle n| & c_{k+2}b_{n+1}^*|k+2\rangle\langle n+1| & c_{k+2}c_{n+2}^*|k+2\rangle\langle n+2| \end{pmatrix}\end{aligned} \tag{12.92}$$

Starting from (12.92), we can easily evaluate the time evolution of the expectation values of the atomic populations.

12.4.2 Periodic collapses and revivals of the atomic populations

By use of (12.88)-(12.92), we obtain the expressions of the atomic population operators as follows

$$\langle C_3^\dagger C_3(t)\rangle = \mathrm{Tr}(C_3^\dagger C_3(t)\rho^I(t)) = \sum_n |a_n(t)|^2$$

$$= g_1^2 g_2^2 e^{-\bar{n}} \sum_n \frac{\bar{n}^{n+2}}{2\gamma^2 n!}\left[\frac{1}{(\gamma(n)-\Delta/2)^2} + \frac{1}{(\gamma(n)+\Delta/2)^2} + \frac{1}{V_1^2(n)+V_2^2(n)}\right]$$

$$-g_1^2 g_2^2 e^{-\bar{n}} \sum_n \frac{\bar{n}^{n+2}}{2\gamma^2 n!}\left\{\cos(\Delta/2-\gamma(n))t\left[\frac{1}{(\gamma(n)-\Delta/2)^2} + \frac{1}{V_1^2(n)+V_2^2(n)}\right]\right.$$

$$+\cos(\Delta/2+\gamma(n))t\left[\frac{1}{(\gamma(n)+\Delta/2)^2} + \frac{1}{V_1^2(n)+V_2^2(n)}\right]$$

$$\left.-\frac{\cos(2\gamma(n)t)}{V_1^2(n)+V_2^2(n)}\right\} \tag{12.93}$$

$$\langle C_2^\dagger C_2(t)\rangle = \sum_n |b_{n+1}|^2 = g_2^2 e^{-\bar{n}} \sum_n \frac{\bar{n}^{n+2}}{2(n+1)!\gamma^2}(1-\cos(2\gamma)t) \tag{12.94}$$

$$\langle C_1^\dagger C_1(t)\rangle = \sum_n |c_{n+2}(t)|^2 = e^{-\bar{n}} \sum_n \frac{\bar{n}^{n+2}}{(n+2)!}\left[1 - \frac{2V_2^2(n)}{V_1^2(n)+V_2^2(n)}\right.$$

$$\left.+\frac{V_2^4(n)}{2\gamma^2(n)}\frac{3\gamma^2(n)+\Delta^2/4}{(V_1^2(n)+V_2^2(n))^2}\right]$$

$$+e^{-\bar{n}} \sum_n \frac{\bar{n}^{n+2}}{(n+2)!}\left\{\cos(\Delta/2-\gamma(n))t\frac{V_2^2(n)}{\gamma(n)}\left[\frac{1}{\gamma-\Delta/2} - \frac{V_2^2(n)}{2\gamma(\gamma-\Delta/2)^2}\right.\right.$$

$$\left.-\frac{V_2^2(n)}{2\gamma(n)(V_1^2(n)+V_2^2(n))}\right] - \cos(\Delta/2+\gamma(n))t\frac{V_2^2(n)}{\gamma(n)}\left[\frac{1}{\gamma(n)+\Delta/2}\right.$$

$$\left.+\frac{V_2^2(n)}{2\gamma(n)(\gamma(n)+\Delta/2)^2} - \frac{V_2^2(n)}{2\gamma(n)(V_1^2(n)+V_2^2(n))}\right]$$

$$\left.+\frac{V_2^4(n)\cos(2\gamma(n)t)}{2\gamma^2(n)(V_1^2(n)+V_2^2(n))}\right\} \tag{12.95}$$

Eqs.(12.93)-(12.95) describe the time evolution of the atomic populations. In order to exhibit the physical properties of eqs.(12.93)-(12.95) explicitly, we give an analytical analyse under the condition which $\Delta \gg g_1^2(n+1)+g_2^2(n+2)$ holds, and does not hold, respectively.

When the condition $\Delta^2 \gg g_1^2(n+1)+g_2^2(n+2)$ holds, then (12.90) may be simplified as

$$\gamma(n) \approx \frac{\Delta}{2} + \frac{\gamma}{\Delta}[g_1^2(n+1) + g_2^2(n+2)] \tag{12.96}$$

$$\gamma(n) + \frac{\Delta}{2} \approx \Delta \tag{12.97}$$

Noticing the photon number distribution is Poissonian, so for $\bar{n} \gg 1$, eqs. (12.93)-(12.95) are simplified as

$$\langle C_3^\dagger C_3(t)\rangle \approx \frac{1}{2} - \frac{1}{2}\exp\left(-2\bar{n}\sin^2\frac{g^2t}{\Delta}\right)\cos\left[\bar{n}\sin\frac{2g^2t}{\Delta} - \frac{g^2t}{\Delta}\right]$$
$$-\frac{2g^2}{\Delta^2}\bar{n}\cos(\Delta t) \tag{12.98}$$

$$\langle C_2^\dagger C_2(t)\rangle \approx \frac{2g\bar{n}}{\Delta^2}(1 - \cos\Delta t) \tag{12.99}$$

$$\langle C_1^\dagger C_1(t)\rangle \approx \frac{1}{2} + \frac{1}{2}\exp\left(-2\bar{n}\sin^2\frac{g^2t}{\Delta}\right)\cos\left[\bar{n}\sin\frac{2g^2t}{\Delta} - \frac{g^2t}{\Delta}\right]$$
$$+\frac{2g^2}{\Delta^2}\bar{n}\cos(\Delta t) \tag{12.100}$$

Here we have assumed $g_1 = g_2 = g$. From eqs.(12.98)-(12.100) we see that the atomic populations in the atomic states $|1\rangle$ and $|3\rangle$ exhibit the periodic collapses and revivals, the revival interval and the collapse time are

$$T_R = \frac{\pi\Delta}{g^2}, \quad T_D = \frac{\Delta}{g\sqrt{2\bar{n}}} \tag{12.101}$$

Evidently, the revival interval T_R depends on the coupling constant g and the detuning Δ, and when $\bar{n} \gg 1$, the collapse time $T_D \ll T_R$.

By use of (12.98) and (12.100), the atomic inversion between the upper and ground states gives

$$\langle C_3^\dagger C_3(t)\rangle - \langle C_1^\dagger C_1(t)\rangle = -\frac{4g^2\bar{n}}{\Delta^2}\cos\Delta t$$
$$-\exp\left(-2\bar{n}\sin^2\frac{g^2t}{\Delta}\right)\cos\left[\bar{n}\sin 2g^2t\Delta - \frac{g^2t}{\Delta}\right] \tag{12.102}$$

Eq.(12.102) shows that the time behavior of $\langle C_3^\dagger C_3(t)\rangle - \langle C_1^\dagger C_1(t)\rangle$ resembles to that of $\langle C_1^\dagger C_1(t)\rangle$ and $\langle C_3^\dagger C_3(t)\rangle$. Noticing the time behavior of the atomic inversion in the two-photon Jaynes-Cummings model varies according to (12.75), we find when $\varepsilon = \frac{g^2}{\Delta}$, (12.102) is similar to (12.75). This means that for the case of $g_1 = g_2$ and $\Delta^2 \gg g_1^2(n+1) + g_2^2(n+2)$, the time evolution of the atomic inversion in the system of a cascade three-level atom interacting with a single-mode field is in agreement with that in the two-photon Jaynes-Cummings model, and the coupling constants in the two systems obey the relations

$$\varepsilon = \frac{g^2}{\Delta} \tag{12.103}$$

In addition, from equation (12.99) we see that the atomic population in the middle state exhibits fast cosine oscillations. The lifetime for the atom in the middle level is $T_2 = 2\pi/\Delta \ll T_c$. That is to say, for the time evolution of the system, the atom in the middle level $|2\rangle$ is very short. Because $\langle C_2^\dagger C_2\rangle$ represents the contributions of the one-photon transitions $|3\rangle \leftrightarrow |2\rangle$ and $|2\rangle \leftrightarrow |1\rangle$, and the amplitude of $\langle C_2^\dagger C_2\rangle$ is $g^2\overline{n}/\Delta^2 \ll 1$, so the probability of the one-photon transitions related to the the middle level $|2\rangle$ is shorter enough than that of the two-photon transitions between $|3\rangle$ and $|1\rangle$.

In the above, we have verified that for the conditions $g_1 = g_2$ and $\Delta^2 \gg g_1^2(n+1)+g_2^2(n+2)$, the atomic dynamic properties of the system consisted of a cascade three-level atom coupled to a single-mode field can be described by the two-photon Jaynes-Cummings model (characterized by (12.66)). However, if the condition $\Delta^2 \gg g_1^2(n+1)+g_2^2(n+2)$ does not hold, what would happen to the atomic dynamic behavior of the system ? In the following we discuss this question.

Considering the properties of the coherent field for $\overline{n} \gg 1$, we can expand $\gamma(n)$ (12.90) as a Taylor series located at $n = \overline{n}$ as

$$\gamma(n) = \gamma(\overline{n})\left[1 + \frac{n-\overline{n}}{2\gamma^2(\overline{n})}(g_1^2+g_2^2) + \cdots\right] \tag{12.104}$$

here

$$\gamma(\overline{n}) = \sqrt{\Delta^2/4 + V_1^2(\overline{n}) + V_2^2(\overline{n})}$$

Since $|n-\overline{n}| < \sqrt{\overline{n}}$, i.e., $|n-\overline{n}|/\overline{n} < \frac{1}{\sqrt{\overline{n}}} \ll 1$, we can only retain to the second term in (12.104). Substituting (12.104) into (12.93)-(12.95) then have

$$\langle C_3^\dagger C_3(t)\rangle \approx \frac{V_1^2(\overline{n})V_2^2(\overline{n})}{2\gamma^2(\overline{n})}\left[\frac{1}{(\gamma(\overline{n})-\Delta/2)^2} + \frac{1}{(\gamma(\overline{n})+\Delta/2)^2}\right.$$
$$\left. + \frac{1}{V_1^2(\overline{n})+V_2^2(\overline{n})}\right] - \frac{V_1^2(\overline{n})V_2^2(\overline{n})}{2\gamma^2(\overline{n})}\left\{\cos(\gamma(\overline{n}-\Delta/2)t\left[\frac{1}{(\gamma(\overline{n})-\Delta/2)^2}\right.\right.$$

$$+\frac{1}{V_1^2(\bar{n})+V_2^2(\bar{n})}\Bigg] + \cos(\gamma(\bar{n})+\Delta/2)t\left[\frac{1}{\gamma(\bar{n})+\Delta/2)^2}\right.$$
$$\left.+\frac{1}{V_1^2(\bar{n})+V_2^2(\bar{n})}\right]\Bigg\} e^{-2\bar{n}\sin^2\frac{\lambda t}{4}}\cos\left(\bar{n}\sin\frac{\lambda t}{2}-\frac{\lambda\bar{n}t}{2}\right)$$
$$-\frac{V_1^2(\bar{n})V_2^2(\bar{n})}{2\gamma^2(\bar{n})(V_1^2(\bar{n})+V_2^2(\bar{n}))}\cos 2\gamma(\bar{n})t$$
$$\times e^{-2\bar{n}\sin^2\frac{\lambda t}{2}}\cos(\bar{n}\sin\lambda t-\bar{n}\lambda t) \tag{12.105}$$

$$\langle C_2^\dagger C_2(t)\rangle \approx \frac{g^2\bar{n}}{2\gamma^2(\bar{n})}[1-\cos 2\gamma(\bar{n})te^{-2\bar{n}\sin^2\frac{\lambda t}{2}}$$
$$\times\cos(\bar{n}\sin\lambda t-(\bar{n}+1)\lambda t)] \tag{12.106}$$

$$\langle C_1^\dagger C_1(t)\rangle \approx \left[1-\frac{2V_2^2(\bar{n})}{V_1^2(\bar{n})+V_2^2(\bar{n})}+\frac{V_2^4(\bar{n})(3\gamma^2(\bar{n})+\Delta^2/2)}{2\gamma^2(\bar{n})(V_1^2(\bar{n})+V_2^2(\bar{n}))^2}\right]$$
$$+\frac{V_2^2(\bar{n})}{\gamma(\bar{n})}\left\{\left[\frac{1}{\gamma(\bar{n})-\Delta/2}-\frac{V_1^2(\bar{n})}{2\gamma(\bar{n})(\gamma(\bar{n})-\Delta/2)^2}\right.\right.$$
$$\left.-\frac{V_2^2(\bar{n})}{2\gamma(\bar{n})(V_1^2(\bar{n})+V_2^2(\bar{n}))}\right]\cos(\gamma(\bar{n})-\Delta/2)t+\left[\frac{1}{\gamma(\bar{n})+\Delta/2}\right.$$
$$\left.-\frac{V_2^2(\bar{n})}{2\gamma(\bar{n})(\gamma(\bar{n})+\Delta/2)^2}-\frac{V_2^2(\bar{n})}{2\gamma(\bar{n})(V_1^2(\bar{n})+V_2^2(\bar{n}))}\right]$$
$$\times\cos(\gamma(\bar{n})+\Delta/2)t\Big\}\exp\left(-2\bar{n}\sin^2\frac{\lambda t}{4}\right)\cos\left(\bar{n}\sin\frac{\lambda t}{2}-\frac{\bar{n}\lambda t}{2}\right)$$
$$+\frac{V_2^2(\bar{n})}{2\gamma^2(\bar{n})(V_1^2(\bar{n})+V_2^2(\bar{n}))}\cos 2\gamma(\bar{n})t$$
$$\exp\left(-2\bar{n}\sin^2\frac{\lambda t}{2}\right)\cos(\bar{n}\sin\lambda t-\bar{n}\lambda t) \tag{12.107}$$

where

$$\lambda=\frac{g_1^2+g_2^2}{\gamma(\bar{n})} \tag{12.108}$$

It is evident that each of $\langle C_i^\dagger C_i(t)\rangle (i=1,2,3)$ exhibits the periodic collapses and revivals. $\langle C_2^\dagger C_2\rangle$ describes the contributions of one-photon transitions. The revival interval and collapse time of $\langle C_2^\dagger C_2\rangle$ are, respectively,

$$T_{D2}=\sqrt{\frac{2}{\bar{n}}}\frac{\gamma(\bar{n})}{g_1^2+g_2^2},\quad T_{R2}=\frac{2\pi\gamma(\bar{n})}{g_1^2+g_2^2} \tag{12.109}$$

In the one-photon resonant case, T_{R2} and T_{D2} reduce to

$$T_{D2} = \sqrt{\frac{2}{g_1^2 + g_2^2}}, \quad T_{R2} = 2\pi\sqrt{\frac{\overline{n}}{g_1^2 + g_2^2}} \tag{12.110}$$

Evidently, the properties of T_{R2} and T_{D2} are similar to those in the one-photon Jaynes-Cummings model. But in $\langle C_1^\dagger C_1 \rangle$ and $\langle C_3^\dagger C_3 \rangle$, except having the collapse and revival term whose properties resemble to that in $\langle C_2^\dagger C_2 \rangle$, there contain the additional collapse and revival term whose collapse time and revival interval are as follows

$$T_R = 2T_{R2}, \quad T_D = 2T_{D2} \tag{12.111}$$

These terms represent the contributions of the two-photon transitions between $|3\rangle \to |1\rangle$. So in the time range centred at $t = kT_R$ $(k = 0, 1, 2, \cdots)$, the periodic collapses and revivals of the atomic populations $\langle C_1^\dagger C_1 \rangle$ and $\langle C_3^\dagger C_3 \rangle$ result from the one- and two-photon transitions. In this time, the envelope functions of $\langle C_1^\dagger C_1 \rangle$ and $\langle C_3^\dagger C_3 \rangle$ are not the Gaussian-type and are modulated by two Gaussian functions such as $\exp(-2\overline{n}\sin^2\frac{\lambda t}{4})$ and $\exp(-2\overline{n}\sin^2\frac{\lambda t}{2})$. This phenomenon differs from the result in the one-photon Jaynes-Cummings model. In particular, when $\Delta = 0$, $g_1 = g_2 = g$, the ratio of representing the one-photon transitions to the two-photon transitions in $\langle C_1^\dagger C_1 \rangle$ and $\langle C_3^\dagger C_3 \rangle$ is 1:4, so in comparison with the two-photon transitions, the contributions of the one-photon transitions can not be neglected. Since the amplitude of $\langle C_2^\dagger C_2 \rangle$ is equal to 1/4, this means that the atomic lifetime in the middle level can be compared to those in the upper and ground levels. Thus, when the condition $\Delta^2 \gg g_1^2(n+1) + g_2^2(n+2)$ does not hold, the system of a cascade three-level atom interacting with a single-mode field can not be described by the two-photon Jaynes-Cummings model.

The above results show that the cascade three-level atom under the interaction of a single-mode coherent field, exhibits different properties of the periodic collapses and revivals for the different one-photon detunings. That is to say, the one-photon detuning plays an important role in the atomic dynamic behavior of a cascade three-level atom.

References

1. L.Allen, J.H.Eberly, *Optical resonance and two-level atoms* (Wiley, New York, 1975).

2. P.L.Knight and P.W.Milonni, *Phys.Rep.* **66** (1980) 21.

3. J.H.Eberly, N.N.Narozhny, J.J.Sanchez-Mondragon, *Phys.Rev.Lett.* **44** (1980) 1323.

4. N.B.Narozhny, J.J.Sanchez-Mondragron, and J.H.Eberly, *Phys. Rev.* **A23** (1981) 236.

5. H.I.Yoo, J.Sanchez-Mondragon and J.H.Eberly, *J.Phys.* **A14** (1981) 1383.

6. P.L.Knight and P.M.Radmore, *Phys.Rev.* **A26** (1982) 676.

7. G.Milbrun, *Opt.Acta* **31** (1984) 671.

8. H.I.Yoo and J.H.Eberly, *Phys.Rep.* **118** (1985) 239.

9. X.S.Li, D.L.Lin, and C.D.Gong, *Phys.Rev.* **A36** (1987) 5209.

10. P.Alsing and M.S.Zubairy, *J.Opt.Soc.Am.* **B4** (1987) 177.

11. A.Bandilla and H.Ritze, *IEEE.J.Quant.Electronics* **24** (1988) 1388.

12. G.S.Agarwal and R.R.Puri, *J.Opt.Soc.Am.* **B5** (1988) 1669.

13. Jin-sheng Peng and X.Y.Huang, *ACTA Opt.Sin.* **8** (1988) 766.

14. A.Joshi and R.R.Puri, *Phys.Rev.* **A42** (1990) 4336.

15. D.A.Cardimona, V.Kovanis, M.P.Sharma, and A.Gavrielides, *Phys.Rev.* **A43** (1991) 3710.

16. T.Quang, P.L.Knight, and V.Buzek, *Phys.Rev.* **A44** (1991) 6092.

17. W.K.Lai, V.Buzek, and P.L.Knight, *Phys.Rev.* **A44** (1991) 2003.

18. Jin-sheng Peng, Gao-xiang Li and Peng Zhou, *ACTA Phys.Sin.* **40** (1991) 1042.

19. P.Zhou, Z.L.Hu, and Jin-sheng Peng, *J.Mod.Opt.* **39** (1992) 49.

20. A.H.Toor and M.S.Zubairy, *Phys.Rev.* **A45** (1992) 4951.

21. P.Zhou and Jin-sheng Peng, *Physica.x s* **A193** (1993) 114.

22. Jin-sheng Peng and Gao-xiang Li, *Phys.Rev.* **A47** (1993) 4212.

23. Jin-sheng Peng and Gao-xiang Li, *Phys.Lett.* **A176** (1993) 230.

24. B.W.Shore and P.L.Knight, *J.Mod.Opt.* **40** (1993) 1195.

25. Gao-xiang Li, Jin-sheng Peng, and Pneg Zhou, *ACTA Opt.Sin.* **13** (1993) 902.

26. H.Lu and Jin-sheng Peng, *ACTA Phys.Sin.* **43** (1994) 1796.

27. H.Lu, Jin-sheng Peng, and Gao-xiang Li, *ACTA Phys.Sin.* **44** (1995) 708.

CHAPTER 13

SQUEEZING EFFECTS OF THE ATOMIC OPERATORS

As shown in Chapter 7, the squeezing of the radiation field can occur in the atom-field coupling system such as in the Jaynes-Cummings model, the question arisen here is whether the fluctuations of atomic operators can also be squeezed in the atom-field coupling system ? Here we first we introduce the definition of the atomic operator squeezing, then discuss the properties and applications of the atomic squeezed state. As an example, in Section 13.2 we analyze the conditions which induce the squeezing of the atomic operators. The relation between the evolution of atomic inversion and the squeezing of atomic operators in the two-photon Jaynes-Cummings model are also discussed. The last section is devoted to studying the squeezing of atomic operators in a resonance fluorescence system.

13.1 Definition of the atomic operator squeezing

In Section 7.2 we introduced the definitions of the normal squeezing, the higher-order squeezing and the amplitude square squeezing of the radiation field. These definitions can be outlined as the unified formalism as follows. For two arbitrary operators A and B of the field, there exists the commutation relation

$$[A, B] = iC \tag{13.1}$$

correspondingly, the fluctuations of the variables A and B satisfy the Heisenberg uncertainty relation

$$(\Delta A)^2(\Delta B)^2 \geq \frac{1}{4}|\langle C\rangle|^2 \tag{13.2}$$

If the variance $(\Delta A)^2$ or $(\Delta B)^2$ obeys the condition

$$(\Delta A)^2 < \frac{1}{2}|\langle C\rangle|, \quad or \quad (\Delta B)^2 < \frac{1}{2}|\langle C\rangle| \tag{13.3}$$

then the fluctuations of the operator A or B are said to be squeezed, meanwhile, the field is in a squeezed state. Similarly, this definition can be generalized to the atomic operators. For a two-level atom characterized by the pseudo-spin operators $S_\pm$ and S_3 , we can define two Hermitian quadrature operators

$$S_1 = \frac{1}{2}(S_+ + S_-) = \begin{pmatrix} 0 & \frac{1}{2} \\ \frac{1}{2} & 0 \end{pmatrix} \tag{13.4}$$

$$S_2 = \frac{1}{2i}(S_+ - S_-) = \begin{pmatrix} 0 & -\frac{i}{2} \\ \frac{i}{2} & 0 \end{pmatrix} \tag{13.5}$$

Evidently, they satisfy the commutation relation

$$[S_1, S_2] = iS_3 \tag{13.6}$$

and the Heisenberg uncertainty relation

$$(\Delta S_1)^2(\Delta S_2)^2 \geq \frac{1}{4}|\langle S_3\rangle|^2 \tag{13.7}$$

where $(\Delta S_i)^2 = \langle S_i^2\rangle - \langle S_i\rangle^2 (i = 1, 2)$ is the variance of the atomic operator S_i . If there exists a state $|\Psi\rangle$ in which the variance of S_i obeys

$$(\Delta S_i)^2 < \frac{1}{2}|\langle S_3\rangle| \quad (i = 1 \quad or \quad 2) \tag{13.8}$$

or

$$F_i = (\Delta S_i)^2 - \frac{1}{2}|\langle S_3\rangle| < 0 \quad (i = 1 \quad or \quad 2) \tag{13.9}$$

then the fluctuations of the atomic operator are said to be squeezed and the state $|\Psi\rangle$ is called the atomic squeezed state. In fact, from Chapter 2 we know that the operators S_1 and S_2 represent the two quadrature components of the atomic dipole moment, so we sometimes name the atomic operator squeezing as the atomic dipole squeezing.

We now discuss what kind of state $|\Psi\rangle$ can display the atomic operator squeezing. Suppose that a two-level atom is in a coherent superposition of its excited state $|+\rangle$ and its ground state $|-\rangle$, i.e.,

$$|\Psi\rangle = \cos(\theta/2)|+\rangle + \sin(\theta/2)\exp(-i\psi)|-\rangle \tag{13.10}$$

we then can easily obtain the expectation values as

$$\langle S_1\rangle = \left\langle \Psi \left| \frac{S_1 + S_2}{2} \right| \Psi \right\rangle = \frac{1}{2}\sin\theta\cos\psi$$

$$\langle S_2\rangle = \frac{1}{2}\sin\theta\sin\psi$$
$$\langle S_3\rangle = \frac{1}{2}\cos\theta \tag{13.11}$$
$$\langle S_1^2\rangle = \langle S_2^2\rangle = \frac{1}{4}$$

and the variances of atomic operators S_1 and S_2 as

$$(\Delta S_1)^2 = \langle S_1^2\rangle - \langle S_1\rangle^2 = \frac{1}{4}(1-\sin^2\theta\cos^2\psi) \tag{13.12}$$
$$(\Delta S_2)^2 = \langle S_2^2\rangle - \langle S_2\rangle^2 = \frac{1}{4}(1-\sin^2\theta\sin^2\psi) \tag{13.13}$$

They obey the Heisenberg uncertainty relation (13.7). In particular, when $\psi = 0,\ \pi/2,\ 3\pi/2,\ \pi$, (13.7) reduces to

$$(\Delta S_1)^2(\Delta S_2)^2 = \frac{1}{4}|\langle S_3\rangle|^2 \tag{13.14}$$

This means that the atomic state (eq.(13.10)) is the minimum uncertainty state of the atomic operators when $\psi = k\pi/2(k = 0, 1, 2, \cdots)$.

Substituting eqs.(13.11c), (13.12) and (13.13) into (13.9), we obtain

$$F_1 = \frac{1}{4}(1-\sin^2\theta\cos^2\psi) - \frac{1}{4}|\cos\theta| \tag{13.15}$$
$$F_2 = \frac{1}{4}(1-\sin^2\theta\sin^2\psi) - \frac{1}{4}|\cos\theta| \tag{13.16}$$

When $\psi = 0$ or $\pi/2$, $F_2 \geq 0$, which corresponds to that the fluctuations in the component S_2 of the atomic dipole moment can not be squeezed, meanwhile F_1 can be less than zero. On the contrary, when $\psi = \pi/2$ or $3\pi/2$, $F_1 \geq 0$ which means that the fluctuations in the component S_1 do not exhibit the squeezing phenomenon, but F_2 can be less than zero. For the above two cases, the amplitudes of F_1 and F_2 are determined by θ, which satisfy

$$F_1(\psi = 0, \pi) = F_2(\psi = \pi/2, 3\pi/2) = \frac{1}{4}(\cos^2\theta - |\cos\theta|) \tag{13.17}$$

Evidently, for $\psi = 0,\ \pi/2$ and $\theta \neq 0,\ \pi/2,\ \pi,\ 3\pi/2$, $F_1 < 0$ which indicates that the fluctuation of the atomic operator S_1 is squeezed, and we call this atomic state $|\Psi\rangle$ as the atomic squeezed state. Fig.(13.1) displays the relationship between F_1 and $\cos^2(\theta/2)$. It is clear that the optimal squeezing in the

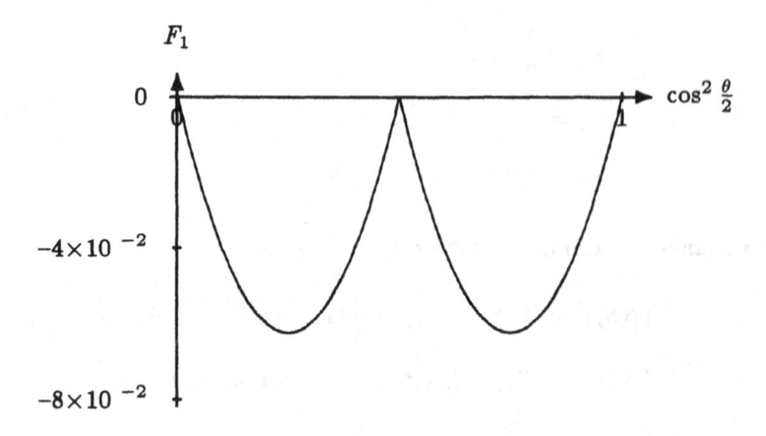

Figure 13.1: F_1 versus $\cos^2(\frac{\theta}{2})$.

component S_1 occurs when $\cos^2(\theta/2) = \frac{1}{4}$ or $\frac{3}{4}$ (i.e., $\theta = \pi/6,\ \pi/3,\ 2\pi/3,\ 5\pi/6$). But when $\cos^2(\theta/2) = 0, 1/2, 1$(i.e., $\theta = 0,\ \pi/2,\ \pi,\ 3\pi/2$) , $F_1 = 0$. That is to say, for such special θ when the two-level atom is initially in its ground state, its excited state or the superposition state

$$|\Psi\rangle = \frac{1}{\sqrt{2}}(|+\rangle + \exp(-i\phi)|-\rangle) \tag{13.18}$$

there displays no squeezing in the component S_1 of the atomic dipole moments. Obviously, for $\psi = \pi/2,\ 3\pi/2$, the similar situation of the squeezing in the component S_2 will appear.

It is worthy mentioning that we have shown in Section 7.2, the two-level atom initially in the state $|\Psi\rangle$ (13.10) interacting with the vacuum field $|0\rangle$ can radiate the squeezed field at time $t = \pi/g$, namely,

$$|\Psi(t = \pi/2g)\rangle = -i[\cos(\theta/2)|-\rangle + \sin(\theta/2)\exp[i(\pi/2 - \psi)]|+\rangle] \otimes |0\rangle \tag{13.19}$$

Here the above result reveals that the initial atomic state $|\Psi\rangle$ (13.10) is just the atomic squeezed state. Thus we may find that the atom in the atomic squeezed state can radiate the squeezed light. So investigating the squeezing phenomenon of atomic operators is very significant not only for restraining the atomic noise, but also for generating the squeezed light.

13.2 Squeezing of atomic operators in the two-photon Jaynes-Cummings model

In the above section, we have shown that the atomic dipole squeezing depends on the initial atomic state. To make a clear explanation of the mechanism of the atomic dipole squeezing, we need to investigate the relations between the fluctuations in the component $S_i (i = 1, 2)$ of atomic dipole moment and the initial atomic state as well as the initial field state in the two-photon Jaynes-Cummings model.

The Hamiltonian of the system in which a single-mode field interacting with a two-level atom involving the emission or absorption of two photons in each transition is written in the rotating-wave approximation as

$$H = H_1 + H_2 \tag{13.20}$$

$$H_1 = \omega(a^\dagger a + 2S_3) \qquad (\hbar = 1) \tag{13.21}$$

$$H_2 = (\omega_0 - 2\omega)S_3 + \varepsilon(S_+ a^2 + a^{\dagger 2} S_-) \tag{13.22}$$

Since H_1 and H_2 satisfy the commutation relation

$$[H_1, H_2] = 0 \tag{13.23}$$

the time-dependent unitary transformation operator may be written as

$$U(t) = \exp(-iHt) = U_1(t)U_2(t) \tag{13.24}$$

where

$$U_1(t) = \exp(-iH_1 t) \tag{13.25}$$

$$U_2(t) = \exp(-iH_2 t) \tag{13.26}$$

In the representation of the atomic eigenstates $|+\rangle$ and $|-\rangle$, we have

$$U_1(t) = \begin{pmatrix} \exp[-i(a^\dagger a + 1)\omega t] & 0 \\ 0 & \exp[-i(a^\dagger a - 1)\omega t] \end{pmatrix} \tag{13.27}$$

$$U_2(t) = \begin{pmatrix} \cos At - i\frac{\Delta}{2A}\sin At & -i\varepsilon a^2 \frac{\sin Bt}{B} \\ -i\varepsilon a^{\dagger 2}\frac{\sin At}{A} & \cos Bt + i\frac{\Delta \sin Bt}{2B} \end{pmatrix} \tag{13.28}$$

where

$$A = [(\Delta/2)^2 + \varepsilon^2 a^2 a^{\dagger 2}]^{1/2}$$

$$B = [(\Delta/2)^2 + \varepsilon^2 a^{\dagger 2} a^2]^{1/2}$$

$$\Delta = \omega_0 - 2\omega$$

If the atom is initially in the state $|\Psi_A\rangle$ (13.10), i.e.,

$$|\Psi_A\rangle = \cos(\theta/2)|+\rangle + \sin(\theta/2)e^{-i\psi}|-\rangle = \begin{pmatrix} \cos(\theta/2) \\ \sin(\theta/2)e^{-i\psi} \end{pmatrix} \tag{13.29}$$

and the field initially in the superposition of the photon number states

$$|\Psi_F(0)\rangle = \sum_{n=0}^{\infty} f_n|n\rangle \tag{13.30}$$

then the initial state of the atom-field coupling system is

$$|\Psi(0)\rangle = \sum_n f_n \begin{pmatrix} \cos(\theta/2)|n\rangle \\ \sin(\theta/2)e^{-i\psi}|n\rangle \end{pmatrix} \tag{13.31}$$

With the time development, the general time-dependent state of the system can be given as

$$\begin{aligned} |\Psi(t)\rangle =& e^{iHt}|\Psi(0)\rangle \\ =& U_1(t)U_2(t) \begin{pmatrix} \sum_n f_n \cos\frac{\theta}{2}|n\rangle \\ \sum_n f_n \sin\frac{\theta}{2}e^{-i\psi}|n\rangle \end{pmatrix} = \begin{pmatrix} \psi_1 \\ \psi_2 \end{pmatrix} \end{aligned} \tag{13.32}$$

after straightforward calculating, we obtain

$$\begin{aligned} \psi_1 =& \cos(\theta/2)\sum_{n=0}^{\infty} f_n \exp[-i(n+1)\omega t]\left[\cos\Omega_n^+ t - i\frac{\Delta}{2}\frac{\sin\Omega_n^+ t}{\Omega_n^+}\right]|n\rangle \\ &-i\sin\frac{\theta}{2}e^{-i\psi}\sum_{n=2}^{\infty} f_n e^{-i(n-1)\omega t}\varepsilon\sqrt{(n-2)(n-1)}\frac{\sin\Omega_n^- t}{\omega_n^-}|n-2\rangle \end{aligned} \tag{13.33}$$

$$\begin{aligned} \psi_2 =& \sin\frac{\theta}{2}e^{-i\psi}\sum_{n=0}^{\infty} f_n \exp[-i(n-1)\omega t]\left[\cos\Omega_n^+ t + i\frac{\Delta}{2}\frac{\sin\Omega_n^- t}{\Omega_n^-}\right]|n\rangle \\ &-i\cos\frac{\theta}{2}\sum_{n=0}^{\infty} f_n e^{-i(n+1)\omega t}\varepsilon\sqrt{(n+2)(n+1)}\frac{\sin\Omega_n^+ t}{\Omega_n^+}|n+2\rangle \end{aligned} \tag{13.34}$$

$$\begin{aligned} \Omega_n^+ &= [(\Delta/2)^2 + \varepsilon^2(n+2)(n+1)]^{1/2} \\ \Omega_n^- &= [(\Delta/2)^2 + \varepsilon^2(n-1)n]^{1/2} \end{aligned} \tag{13.35}$$

From the above equations we can discuss the influences of different initial parameters θ, ψ and f_n on the atomic dipole squeezing.

13.2.1 Squeezing of atomic operators in the vacuum field

In practical detections, the phase-sensitive detectors do not respond to the rapid oscillations at atomic frequency ω_0 but to the envelope of the evolution of atomic variables. In order to suit the needs of detections, we define two slowly varying Hermitian quadrature operators

$$\begin{aligned} S_1 &= \frac{1}{2}[S_+ \exp(-i\omega_0 t) + S_- \exp(i\omega_0 t)] \\ S_2 &= \frac{1}{2i}[S_+ \exp(-i\omega_0 t) - S_- \exp(i\omega_0 t)] \end{aligned} \tag{13.36}$$

They obey the commutation relation

$$[S_1, S_2] = iS_3$$

Correspondingly, the Heisenberg uncertainty relation is

$$(\Delta S_1)^2(\Delta S_2)^2 \geq \frac{1}{4}|\langle S_3\rangle|^2$$

If there exists the squeezing of atomic operators, the fluctuations in the component $S_i (i = 1 or 2)$ must obey

$$(\Delta S_i)^2 < \frac{1}{2}|\langle S_3\rangle| \quad (i = 1 \text{ or } 2) \tag{13.37}$$

or

$$F_i = (\Delta S_i)^2 - \frac{1}{2}|\langle S_3\rangle| < 0 \qquad (i = 1 \text{ or } 2) \tag{13.38}$$

Supposing that the field is initially in the vacuum state $|0\rangle$ and the atom initially in the superposition of its excited and ground states which is described by (13.29), namely, the system is initially in

$$|\Psi(0)\rangle = \begin{pmatrix} \cos(\theta/2)|0\rangle \\ \sin(\theta/2)e^{-i\psi}|0\rangle \end{pmatrix} \tag{13.39}$$

The state of the system at time t can be obtained from (13.32), it gives

$$|\Psi(t)\rangle = \begin{pmatrix} \cos(\theta/2)\exp(-i\omega t)\cos(\sqrt{2}\varepsilon t)|0\rangle \\ \sin(\theta/2)e^{-i\psi}e^{i\omega t}|0\rangle - i\cos(\theta/2)\sin(\sqrt{2}\varepsilon t)e^{-i\omega t}|2\rangle \end{pmatrix} \tag{13.40}$$

here we have assumed $\Delta = 0$, i.e., $\omega_0 = 2\omega$. The expectation values of the atomic operators $S_i (i = 1, 2, \text{ and } 3)$ in the state $|\Psi(t)\rangle$ are easily given by

$$\langle S_1\rangle = \left\langle \Psi(t) \left| \frac{S_+ \exp(-2i\omega t) + S_- \exp(2i\omega t)}{2} \right| \Psi(t) \right\rangle$$
$$= \frac{1}{2} \sin\theta \cos\psi \cos(\sqrt{2}\varepsilon t) \tag{13.41}$$
$$\langle S_2\rangle = \frac{1}{2} \sin\theta \sin\psi \cos(\sqrt{2}\varepsilon t) \tag{13.42}$$
$$\langle S_3\rangle = \cos^2(\theta/2) \cos^2(\sqrt{2}\varepsilon t) - \frac{1}{2} \tag{13.43}$$
$$\langle S_i^2\rangle = \frac{1}{4} \qquad (i = 1, 2) \tag{13.44}$$

Substituting (13.41) into (13.38), we obtain

$$F_1 = \frac{1}{4} - \frac{1}{4} \sin^2(\theta/2) \cos^2\psi \cos^2(\sqrt{2}\varepsilon t) - \frac{1}{2} \left| \cos^2(\theta/2) \cos^2(\sqrt{2}\varepsilon t) - \frac{1}{2} \right| \tag{13.45}$$
$$F_2 = \frac{1}{4} - \frac{1}{4} \sin^2(\theta/2) \sin^2\psi \cos^2(\sqrt{2}\varepsilon t) - \frac{1}{2} \left| \cos^2(\theta/2) \cos^2(\sqrt{2}\varepsilon t) - \frac{1}{2} \right| \tag{13.46}$$

In analogy with the discussion of Section 13.1, we choose $\psi = 0$. In this case, $F_2 \geq 0$, which means that the fluctuations in S_2 cannot be squeezed, however, F_1 can be negative, we therefore discuss the varying of F_1 with θ and t. When $\psi = 0$, (13.45) reduces to

$$F_1 = \frac{1}{4} - \frac{1}{4} \sin^2\theta \cos^2(\sqrt{2}\varepsilon t) - \frac{1}{2} \left| \cos^2(\theta/2) \cos^2(\sqrt{2}\varepsilon t) - \frac{1}{2} \right| \tag{13.47}$$

If $\cos^2(\theta/2) < 1/2$, then

$$F_1 = -\left(\frac{1}{2} - \cos^2(\theta/2)\right) \cos^2(\theta/2) \cos^2(\sqrt{2}\varepsilon t) \leq 0 \tag{13.48}$$

It is clearly shown from (13.48) that the fluctuations in S_1 will be squeezed at all time except $t_k = (k + \frac{1}{2})\frac{\pi}{\sqrt{2\varepsilon}} (k = 0, 1, 2, \cdots)$, and the reduced amplitude of

F_1 oscillates with the Rabi frequency $2\sqrt{2}\varepsilon$ with respect to t. In addition, the maximum squeezing gives

$$F_1|_{\cos^2(\theta/2)=1/4} = -\frac{1}{16}\cos^2(\sqrt{2}\varepsilon t) \tag{13.49}$$

and the optimal reduction of the quantum fluctuations in the component S_1 takes place at time to be

$$t_m = \frac{k\pi}{\sqrt{2}\varepsilon} \quad (k = 0, 1, 2, \cdots) \tag{13.50}$$

When $\cos^2\frac{\theta}{2} > 1/2$, the squeezing in S_1 displays only when $\cos^2\frac{\theta}{2}\cos^2\sqrt{2}\varepsilon t > 1/2$. In this case

$$F_1 = \frac{1}{2} - \cos^2(\theta/2)\left[\frac{3}{2} - \cos^2(\theta/2)\right]\cos^2(\sqrt{2}\varepsilon t) \tag{13.51}$$

Evidently, when

$$0 \le t < \frac{t_c}{\sqrt{2}\varepsilon}$$

or

$$\frac{k\pi - t_c}{\sqrt{2}\varepsilon} < t < \frac{k\pi + t_c}{\sqrt{2}\varepsilon} \quad (k = 0, 1, 2, \cdots)$$

$$t_c = \arccos\frac{1}{[\cos^2(\theta/2)(3/2 - \cos^2(\theta/2))]^{1/2}}$$

we have $F_1 < 0$. This means that for $\cos^2(\theta/2) > 1/2$, the fluctuations in the atomic dipole variable S_1 are periodically squeezed, and the period is $T_R = \frac{\pi}{\sqrt{2}\varepsilon}$. When $\cos^2(\theta/2) = 3/4$, (13.51) gives the time evolution of the maximum dipole squeezing in S_1 as

$$F_1|_{\cos^2(\theta/2)=3/4} = \frac{1}{2} - \frac{9}{16}\cos^2(\sqrt{2}\varepsilon t) \tag{13.52}$$

$$0 \le t < \frac{1}{\sqrt{2}\varepsilon}\arccos\left(\sqrt{\frac{8}{9}}\right)$$

or

$$k\pi - \arccos\left(\sqrt{\frac{8}{9}}\right) < \sqrt{2}\varepsilon t < k\pi + \arccos\left(\sqrt{\frac{8}{9}}\right) \quad (k = 1, 2, \cdots)$$

and the optimal squeezing occurs at

$$t_m = \frac{k\pi}{\sqrt{2}\varepsilon} \qquad (k = 0, 1, 2, \cdots)$$

Fig.(13.2) displays the time evolution of F_1 for $\psi = 0$.

Fig.(13.3) shows the evolution of F_1 for various values of ψ when $\cos^2(\theta/2) =$

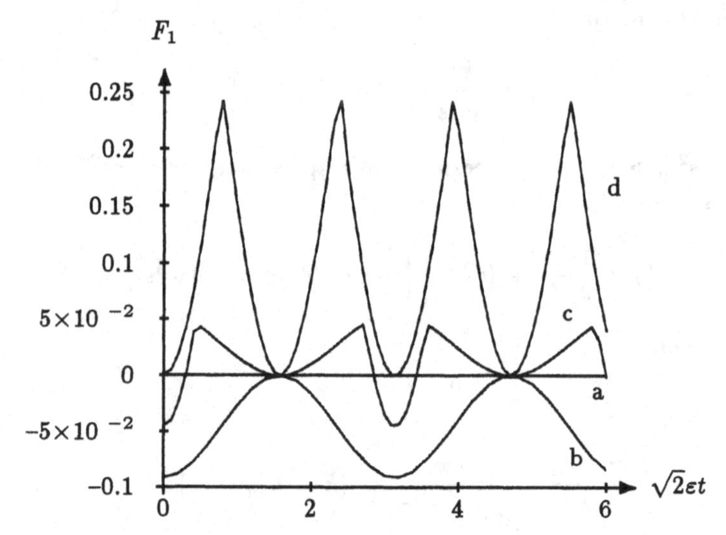

Figure 13.2: Time evolution of F_1, $\phi = 0$ (a) $\cos^2\frac{\theta}{2} = 0$; (b) $\cos^2\frac{\theta}{2} = 0.25$; (c) $\cos^2\frac{\theta}{2} = 0.6$; (d) $\cos^2\frac{\theta}{2} = 1$

1/4. We can see for $\psi = 0$ and $\pi/6$, $F_1 < 0$, the fluctuations of the atomic dipole variable S_1 is squeezed. But for $\psi = \pi/3$, the squeezing in S_1 can nearly not occur, and for $\psi = \pi/2$ there is no squeezing in S_1 . This means that when the atom is initially in the superposition state the squeezing of atomic operators in the system of a two-level atom interacting with a vacuum field is not only dependent on θ but also on the phase angle ψ.

It is worthwhile to point out that when $\cos(\theta/2) = 0$ or $\sin(\theta/2) = 0$, we can see from (12.45) and (12.46) that $F_1 \geq 0$ and $F_2 \geq 0$. This implies that when the atom is initially in the excited state ($\sin(\theta/2) = 0$) or ground state ($\cos(\theta/2) = 0$), there is no squeezing appeared due to the interaction of vacuum field. Inasmuch as the atomic spontaneous radiation results from the atom being a excited state interacting with the vacuum field, the atomic spon-

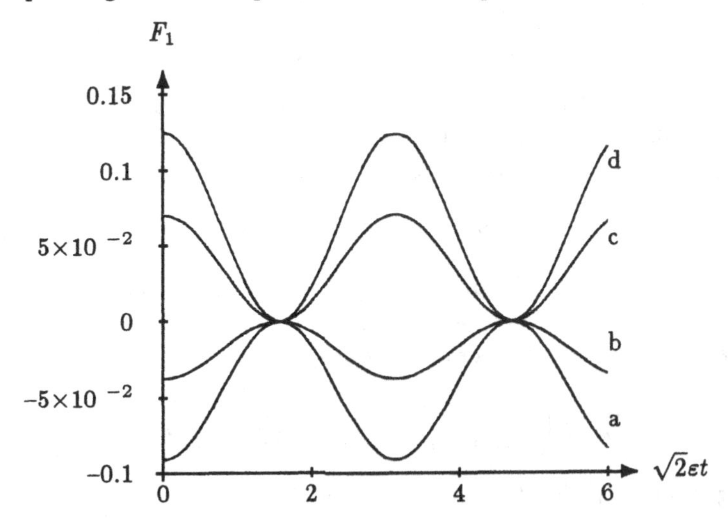

Figure 13.3: Time evolution of F_1 for $\cos^2(\theta/2) = 0.25$, (a) $\phi = 0$; (b) $\phi = \pi/6$; (c) $\phi = \pi/3$; (d) $\phi = \pi/2$

taneous emission can not radiate a squeezing light, and also can not induce the squeezing of the atomic dipole moments.

In addition, from (13.40) we know that the state $|\Psi(t)\rangle$ of the system at time $t_0 = (k + 1/2)\frac{\pi}{\sqrt{2\varepsilon}}$ $(k = 0, 1, 2, \cdots)$ evolves into

$$
\begin{aligned}
|\Psi(t)\rangle &= \begin{pmatrix} 0 \\ \sin(\theta/2)e^{-i\psi}e^{i\omega t_0}|0\rangle - i\cos(\theta/2)e^{-i\omega t_0}|2\rangle \end{pmatrix} \\
&= -ie^{-i\omega t_0}[\cos(\theta/2)|2\rangle + \sin(\theta/2)e^{2i\omega t_0}e^{i(\pi/2-\psi)}|0\rangle] \otimes |-\rangle
\end{aligned}
$$

Employing the similar method in Section 7.2, it is easy to verify that the field in the state

$$
|\Psi_F\rangle = \cos(\theta/2)|2\rangle + \sin(\theta/2)\exp(i\psi')|0\rangle \tag{13.53}
$$

is a squeezed one. That is to say, in resemblance with the one-photon Jaynes-Cummings model, the atom initially in the atomic squeezed state under the interaction of vacuum field in the two-photon Jaynes-Cummings model, can spontaneously decay into its ground state, and meanwhile, the atom radiates the squeezed light.

13.2.2 Squeezing of atomic operators in the superposition state field

We have verified that the atom initially in its ground state interacting with the vacuum field can not exhibit the squeezing of atomic operators. However, if the field is initially in the superposition state

$$|\Psi_F\rangle = \cos(\eta/2)|0\rangle + \sin(\eta/2)\exp(-i\xi)|2\rangle \tag{13.54}$$

can the fluctuations in the component S_i (i=1 or 2) of the atomic dipole moment be squeezed ? Now we discuss this question.

In this condition, the initial state of the system is

$$|\Psi(0)\rangle = \begin{pmatrix} 0 \\ \cos(\eta/2)|0\rangle + \sin(\eta/2)e^{-i\xi}|2\rangle \end{pmatrix} \tag{13.55}$$

According to (13.32), the state of the system at time t evolves to

$$|\Psi(t)\rangle = \begin{pmatrix} -i\sin(\eta/2)e^{-i\xi}e^{-i\omega t}\sin(\sqrt{2}\varepsilon t)|0\rangle \\ \cos(\eta/2)e^{i\omega t}|0\rangle + \sin(\eta/2)e^{-i\xi}e^{-i\omega t}\cos(\sqrt{2}\varepsilon t)|2\rangle \end{pmatrix} \tag{13.56}$$

Here we have supposed $\Delta = 0$. By use of (13.56), the expectation values of the atomic operators $S_i (i = 1, 2, 3)$ are, respectively,

$$\begin{aligned} \langle S_1\rangle &= \frac{1}{2}\sin\eta\sin(\sqrt{2}\varepsilon t)\sin\xi \\ \langle S_2\rangle &= \frac{1}{2}\sin\eta\sin(\sqrt{2}\varepsilon t)\cos\xi \\ \langle S_3\rangle &= \sin^2(\eta/2)\sin(\sqrt{2}\varepsilon t) - 1/2 \end{aligned} \tag{13.57}$$

Substituting (13.57) into (13.38), F_i may be expressed as

$$\begin{aligned} F_1 &= \frac{1}{4} - \frac{1}{4}\sin^2\eta\sin^2(\sqrt{2}\varepsilon t)\sin^2\xi \\ &\quad - \frac{1}{2}\left|\sin^2(\eta/2)\sin(\sqrt{2}\varepsilon t) - \frac{1}{2}\right| \\ F_2 &= \frac{1}{4} - \frac{1}{4}\sin^2\eta\sin^2(\sqrt{2}\varepsilon t)\cos^2\xi \\ &\quad - \frac{1}{2}\left|\sin^2\eta\sin(\sqrt{2}\varepsilon t) - \frac{1}{2}\right| \end{aligned} \tag{13.58}$$

It is clear that for $\xi = 0$ and π, $F_1 \geq 0$, thus the fluctuations in the component S_1 of the atomic dipole moment can not be squeezed. And for $\xi = \pi/2$ and

$3\pi/2$, $F_2 \geq 0$, which means that there is no squeezing in the component S_2. For simplicity, we only discuss the case of $\xi = 0$. Correspondingly, F_2 is simplified as

$$F_2 = \frac{1}{4} - \frac{1}{4}\sin^2\eta \sin^2(\sqrt{2}\varepsilon t) - \frac{1}{2}\left|\sin^2(\eta/2)\sin^2\sqrt{2}\varepsilon t - \frac{1}{2}\right| \tag{13.59}$$

Equation (13.59) shows that no squeezing in S_1 component, but the fluctuations in S_2 will be squeezed at all time except $t = \frac{k\pi}{\sqrt{2}\varepsilon}(k = 0, 1, 2, \cdots)$ if $\sin^2(\eta/2) < 1/2$. When $\sin^2(\eta/2) = 1/4$, the maximum squeezing goes to

$$F_2 = -\frac{1}{16}\sin^2(\sqrt{2}\varepsilon t) \tag{13.60}$$

the time while the optimal squeezing occurs is $t_m = \frac{k+1/2}{\sqrt{2}\pi}(k = 0, 1, 2, \cdots)$, as shown in Fig.(13.4-b-c). However, if $\sin^2(\eta/2) > 1/2$, the squeezing in S_2 can be exhibited only while

$$k\pi + t_s < \sqrt{2}\varepsilon t < (k+1)\pi - t_s \quad (k = 0, 1, 2, \cdots) \tag{13.61}$$

where $t_s = \arcsin\frac{1}{[(\frac{3}{2} - \sin^2(\eta/2))\sin^2(\eta/2)]^{1/2}}$, and the maximum squeezing occurs for $\sin^2(\eta/2) = 3/4$. It is

$$F_2 = \frac{1}{2} - \frac{9}{16}\sin^2(\sqrt{2}\varepsilon t) \tag{13.62}$$

The time evolution of F_2 corresponding to the squeezing of S_2 is shown in Fig.(13.4) for different values $\sin(\eta/2)$. Moreover, (13.58) also shows that the dipole squeezing depends upon the phase angle ξ. This phase sensitivity can also be found from Fig.(13.5).

On the other hand, when $\sin(\eta/2) = 0$ or $\cos(\eta/2) = 0$, i.e., the field is initially in the number state $|2\rangle$ or the vacuum state $|0\rangle$, the dipole squeezing does not occur.

In conclusion, through discussing the atomic dipole squeezing in the system of a two-level atom initially in a coherent superposition of its excited and ground states interacting with a vacuum field, and in the system of an atom initially in its ground state coupled to a field initially in the coherent superposition of number states $|0\rangle$ and $|2\rangle$, we showed that the atomic dipole squeezing can appear only in the atom-field coupling system initially in the coherent superposition of $\{|+, n\rangle, |-, n\rangle\}$. If the system is initially in the pure state $|+, n\rangle$

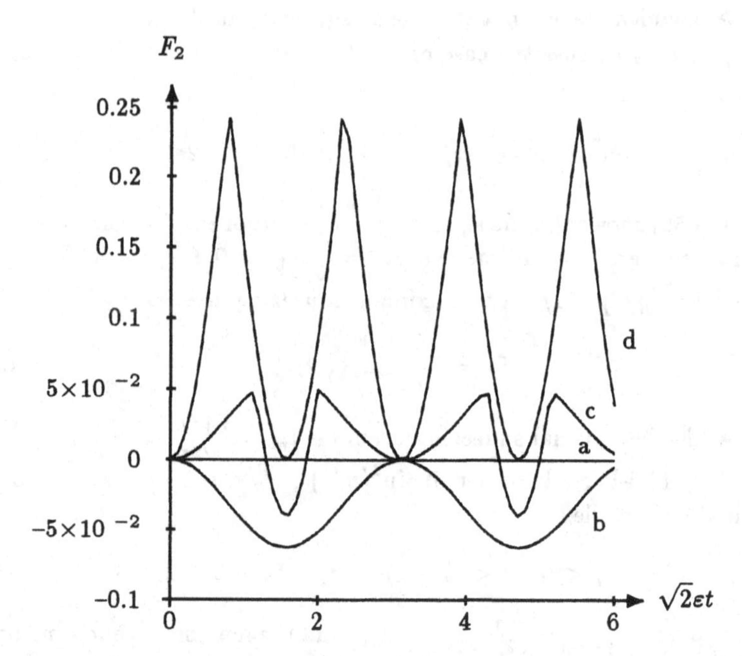

Figure 13.4: Time evolution of F_2 for $\phi = 0$, (a) $\sin^2\frac{\eta}{2} = 0$; (b) $\sin^2\frac{\eta}{2} = 0.25$; (c) $\sin^2\frac{\eta}{2} = 0.6$; (d) $\sin^2\frac{\eta}{2} = 1$

or $|-, n\rangle$, there is no squeezing of the atomic operators.

From (13.56) we know that after a half Rabi period, the state of the system evolves into

$$\left|\Psi\left(t_0 = \frac{\pi}{2\sqrt{2}\varepsilon}\right)\right\rangle = \begin{pmatrix} -i\sin(\eta/2)e^{-i\xi}e^{-i\omega t_0}|0\rangle \\ \cos(\eta/2)e^{i\omega t_0}|0\rangle \end{pmatrix}$$
$$= -ie^{-i(\omega t_0+\xi)}\left[\sin\frac{\eta}{2}|+\rangle + \cos\frac{\eta}{2}\exp[i(2\omega t_0 + \xi - \pi/2)]|-\rangle\right] \qquad (13.63)$$

Clearly, in the meantime, the field evolves into the vacuum state and the atom evolves into the coherent superposition state

$$\sin\frac{\eta}{2}|+\rangle + \cos\frac{\eta}{2}\exp[i(2\omega t_0 + \xi - \pi/2)]|-\rangle \qquad (13.64)$$

This state is also an atomic squeezed state as shown in (13.10). Since the initial state (13.53) is a squeezed state of the field, we can see that in the atom-field

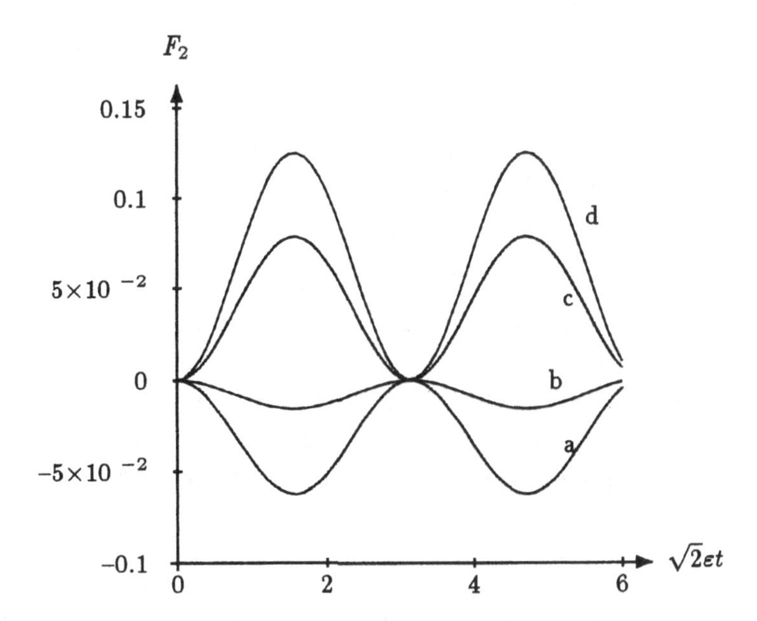

Figure 13.5: Time evolution of F_2 for $\sin^2\frac{\eta}{2}=0$, (a) $\xi=0$; (b) $\xi=\pi/6$; (c) $\xi=\pi/3$; (d) $\xi=\pi/2$

coupling system, not only the atom which is in the atomic squeezed state can radiate the squeezed light, but also the squeezed field can induce the atom to evolve into the atomic squeezed state.

13.2.3 Squeezing of atomic operators in the coherent state field

The above results showed that under the interaction of a squeezed field (described by (13.55)), the atom initial in its ground state can exhibit the squeezing of atomic operators. However, if the initial field is in a coherent state, what happens to the fluctuations of the atomic operators ?

Now, we consider the case of the atom initially in its ground state and the field initially in a coherent state, in addition, we assume $\omega_0=2\omega$. Under these conditions, the state of the system at t is (from (13.32))

$$|\Psi(t)\rangle=\begin{pmatrix} -ie^{-|\alpha|^2/2}\sum\limits_{n=2}^{\infty}\frac{\alpha^n}{\sqrt{n!}}e^{-i(n-1)\omega t}\sin\varepsilon\sqrt{n(n-1)}t|n-2\rangle \\ e^{-|\alpha|^2/2}\sum\limits_{n=0}^{\infty}\frac{\alpha^n}{\sqrt{n!}}e^{-i(n-1)\omega t}\cos\varepsilon\sqrt{n(n-1)}t|n\rangle \end{pmatrix} \tag{13.65}$$

By use of (13.65), the expectation values of the atomic operators are

$$\begin{aligned}
\langle S_1\rangle &= -\exp(-\bar{n})\sum_{n=0}^{\infty}\frac{\bar{n}^{n+1}}{n!\sqrt{(n+1)(n+2)}} \\
&\quad\times\sin\varepsilon\sqrt{(n+1)(n+2)}t\cos\varepsilon\sqrt{n(n-1)}t\sin\phi \\
\langle S_2\rangle &= -\exp(-\bar{n})\sum_{n=0}^{\infty}\frac{\bar{n}^{n+1}}{n!\sqrt{(n+1)(n+2)}} \\
&\quad\times\sin\varepsilon\sqrt{(n+1)(n+2)}t\cos\varepsilon\sqrt{n(n-1)}t\cos\phi \\
\langle S_3\rangle &= -\frac{1}{2}\exp(-\bar{n})\sum_{n=0}^{\infty}\frac{\bar{n}^{n}}{n!}\cos(2\varepsilon\sqrt{n(n-1)}t)
\end{aligned} \tag{13.66}$$

here $\alpha=\sqrt{\bar{n}}\exp(i\psi)$. Substituting (13.66) into (13.45), it gives

$$F_i=\frac{1}{4}-\langle S_i\rangle^2-\frac{1}{2}|\langle S_3\rangle| \quad (i=1,2) \tag{13.67}$$

Starting from (13.67) we can discuss the fluctuations of atomic operators. Since the formalism of F_i is very complicated, we adopt the numerical method to reveal the time evolution of F_i as shown in Fig.(13.6). For simplicity, we choose $\psi=0$, in this case, $F_1\geq 0$. This means that the fluctuations in S_1 can not be squeezed. But from Fig.(13.6) we know that in some certain time ranges, F_2 may be negative. That is to say, the fluctuations in S_2 can be squeezed. Furthermore, if the mean photon number $\bar{n}$ is small, the squeezing of S_2 is not periodic (shown in Fig.13.6-a). But when $\bar{n}$ is large, the squeezing phenomena appear periodically as shown in Fig.(13.6-b,c) and occur in the collapse ranges of the atomic inversion $\langle S_3\rangle$. The physical origin is that for large $\bar{n}$, $\langle S_3\rangle$ exhibits the periodic revival-and-collapse phenomenon. In the time ranges of $\langle S_3\rangle=0$, the Rabi oscillations with different frequencies lose the phase correlation, the fluctuations of atomic dipole moments become large, so in these time ranges, there is no squeezing of the atomic operators. But in the neighborhood of time points $t=kT_R(k=0,1,2,\cdots)$, the Rabi oscillations with different frequencies have strong phase correlation and there is strong interference among these Rabi oscillations, which lead to the fluctuations of atomic operators to be reduced significantly and the squeezing of atomic operators displays. On the other hand, since the collapse time of $\langle S_3\rangle$ is $t_c=\frac{1}{\sqrt{2\bar{n}}\varepsilon}$, the phase correlation time of the Rabi oscillations decreases with the increasing of $\bar{n}$. So the time in

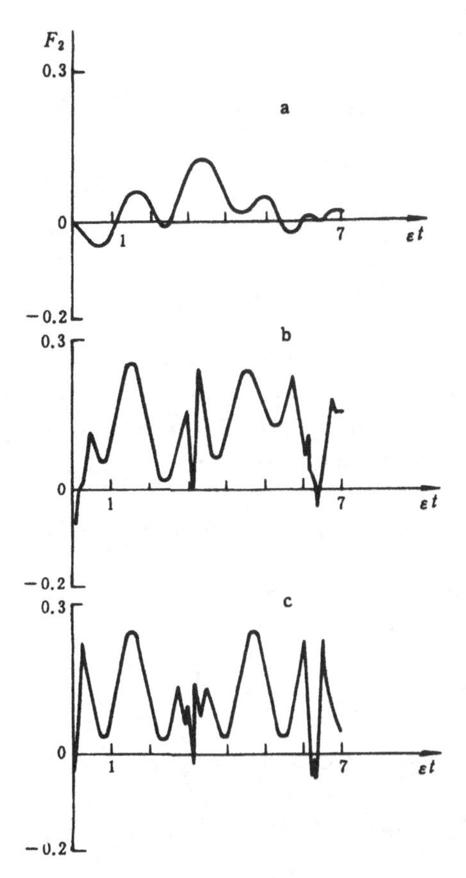

Figure 13.6: Time evolution of F_2, (a) $\overline{n} = 1$, (b) $\overline{n} = 5$, (c) $\overline{n} = 10$

which the first squeezing appears becomes short with the increasing of $\overline{n}$.

13.3 Squeezing of atomic operators in the resonance fluorescence system

The two-level atomic resonance fluorescence system is another typical atom-field coupling system. Can the fluctuations of atomic operators be squeezed in this system ? In order to make use of the results shown in Section 8.3, here we first transform the atomic operators $S_{\pm}$ and S_3 into the dressed atomic representation. That is

$$S_+ = |+\rangle\langle -| = \frac{1}{2}(|2\rangle + |1\rangle)(\langle 2| - \langle 1|)$$
$$= \frac{1}{2}(|2\rangle\langle 2| - |1\rangle\langle 1| - |2\rangle\langle 1| + |1\rangle\langle 2|) \tag{13.68}$$
$$S_- = \frac{1}{2}(|2\rangle\langle 2| - |1\rangle\langle 1| + |2\rangle\langle 1| - |1\rangle\langle 2|) \tag{13.69}$$
$$S_3 = \frac{1}{2}(|2\rangle\langle 1| + |1\rangle\langle 2|) \tag{13.70}$$

here $|2\rangle$ and $|1\rangle$ are the atomic dressed states (described by (8.166)), $|+\rangle$ and $|-\rangle$ represent the bare atomic states. Since the density matrix in section 8.3 is obtained in the interaction picture, the atomic operators $S_i (i = 1, 2, 3)$ (defined by (13.36)) must be transformed into the interaction picture. This means that

$$S_1^I = \frac{1}{2}(S_+ + S_-), \quad S_2^I = \frac{1}{2i}(S_+ - S_-), \quad S_3^I = S_3 \tag{13.71}$$

By use of the density operator as given in (8.183), the expectation values of atomic operators $S_i (i = 1, 2, 3)$ and $S_i^2 (i = 1, 2)$ are, respectively, written as

$$\langle S_1 \rangle = \mathrm{Tr}\rho S_1^I = \frac{1}{2}\mathrm{Tr}\rho(S_+ + S_-) = \frac{1}{2}(\rho_{22} - \rho_{11}) = \tilde{\rho}_2(t) \tag{13.72}$$
$$\langle S_2 \rangle = \mathrm{Tr}\rho S_2^I = \frac{1}{2i}\mathrm{Tr}\rho(S_+ - S_-) = \frac{1}{2i}(\rho_{21} - \rho_{12}) = -i\tilde{\rho}_4(t) \tag{13.73}$$
$$\langle S_3 \rangle = \mathrm{Tr}\rho S_3^I = \frac{1}{2}(\rho_{22} - \rho_{11}) = \tilde{\rho}_3(t) \tag{13.74}$$
$$\langle S_1^2 \rangle = \frac{1}{4}\mathrm{Tr}\rho(|2\rangle\langle 2| + |1\rangle\langle 1|) = \frac{1}{4} \tag{13.75}$$
$$\langle S_2^2 \rangle = \frac{1}{4} \tag{13.76}$$

Noticing equation (8.185), we can discuss the evolution of the functions F_i against the variations of the parameters Ω, Γ and t. Here Ω and Γ are defined in Chapter 8.

We usually focus on the steady-state characteristics in the resonance fluorescence system, in the following we consider the properties of F_i for $t \to \infty$. According to eq.(8.185) we know

$$\tilde{\rho}_2(\infty) = 0, \ \tilde{\rho}_3(\infty) = -\frac{1}{4}\frac{\Gamma^2}{4\Omega^2 + \Gamma^2/2}, \ \tilde{\rho}_4(\infty) = \frac{i\Omega\Gamma}{4\Omega^2 + \Gamma^2/2} \tag{13.77}$$

Substituting (13.77) into (13.72)-(13.74), we may obtain the steady-state expectation values of the atomic operators S_i. Thus

$$F_1 = \frac{1}{4} - \langle S_1\rangle^2 - \frac{1}{2}|\langle S_3\rangle| = \frac{1}{4} - \frac{\Gamma^2/2}{4\Omega^2+\Gamma^2/2} \tag{13.78}$$

$$\begin{aligned} F_2 &= \frac{1}{4} - \langle S_2\rangle^2 - \frac{1}{2}|\langle S_3\rangle| \\ &= \frac{1}{4} - \frac{\Omega^2\Gamma^2}{4\Omega^2+\Gamma^2/2} - \frac{1}{8}\frac{\Gamma^2}{4\Omega^2+\Gamma^2/2} \end{aligned} \tag{13.79}$$

From (13.78) we know that for $t \to \infty$, $F_1 \geq 0$, so the fluctuations in the component S_1 of the atomic dipole moment can not be squeezed. Simplifying equation (13.79) we get

$$F_2 = \frac{8\Omega^2 - \Omega^2\Gamma^2}{2(4\Omega^2+\Gamma^2/2)^2} \tag{13.80}$$

It is evident that $F_2 < 0$ if $\Omega^2 < \Gamma^2/8$ ($\Omega = \varepsilon E$). That is to say, if $\Omega^2 < \Gamma^2/8$ the component S_2 of the atomic dipole moment can display steady-state squeezing. Inasmuch as Γ is the atomic spontaneous decay rate, this means that the steady-state squeezing of the atomic operators displays only for the weak driven laser field. From (13.80) we know that the maximum squeezing occurs when $\Omega^2 = \Gamma^2/16$, in this moment, $F_2 = -1/36$. As shown in the previous section that the atom in the atomic squeezed state can radiate the squeezed field, thus in this resonance fluorescence system, the steady-state fluorescent field is also a squeezed field when $\Omega^2 < \Gamma^2/8$.

References

1. D.F.Walls and P.Zoller, *Phys.Rev.Lett.* **47** (1981) 709.

2. Z.Ficek, R.Tanas, and S.Kielich, *Opt.Commun.* **46** (1983) 23.

3. Z.Ficek, R.Tanas, and S.Kielich, *J.Opt.Soc.Am.* **B1** (1984) 882.

4. Z.Ficek, R.Tanas, and S.Kielich, *Phys.Rev.* **A29** (1984) 2004.

5. S.M.Barnett, *Opt.Commun.* **61** (1987) 432.

6. K.Wodkiewicz, P.L.Knight, P.L.Buckle and S.M.Barnett, *Phys.Rev.* **A35** (1987) 2667.

7. R.Loudon and P.L.Knight, *J.Mod.Opt.* **34** (1987) 709.

8. S.M.Barnett and M.A.Dupetuis, *J.Opt.Soc.Am.* **B4** (1987) 505.

9. X.S.Li, D.L.Lin, T.F.George and Z.D.Liu, *Phys.Rev.* **A40** (1989) 228.

10. Peng Zhou and Jin-sheng Peng, *ACTA Phys.Sin.* **38** (1989) 2004.

11. Peng Zhou and Jin-sheng Peng, *ACTA Opt.Sin.* **10** (1990) 837.

12. Peng Zhou and Jin-sheng Peng, *Phys.Rev.* **A44** (1991) 3331.

13. M.H.Mahran, *Phys.Rev.* **A45** (1992) 5113.

14. M.Ashraf and M.S.K.Razmi, *Phys.Rev.* **A45** (1992) 8121.

15. D.Cohen, A.Mann and Y.Ben-Aryeh, *Opt.Commun.* **99** (1993) 123.

16. Gao-xiang Li and Jin-sheng Peng, *ACTA Phys.Sin.* **45** (1995) 1670.

CHAPTER 14

COHERENT TRAPPING OF THE ATOMIC POPULATION

In Chapter 12, we have verified that the atomic populations of a two-level or three-level atom interacting with a coherent field exhibit the periodic collapse-and-revival phenomenon. And from Section 11.2, we know that the coherence of the fields can be destroyed due to the atom-field coupling. However, these results are obtained under the situation of the atom initially in its ground state or its excited state. But if the atom is initially in its coherent superposition of its lower and upper states, how about the atomic populations and the coherence of the field evolve with the time development ? Here as an example, we investigate the population evolutions in the system of a V-type three-level atom driven by a bimodal field. In Section 14.1, we discuss the time evolution of the atomic populations of a V-configuration three-level atom interacting with a bimodal uncorrelated coherent field, in which the atom initially in the coherent superposition of its two upper states. We first give the state vector of the system at time t, then discuss the time evolution of the phase operator of the field, furthermore, the influences of atomic initial coherence on the atomic populations and the coherence of field are analyzed. The results show that with an appropriate initial atomic preparation, not only the initial coherence of the field but also the atomic populations can not vary with time. This phenomenon is interpreted as the atomic coherent population trapping. In the last of Section 14.1, we give an explanation for this phenomenon. Section 14.2, is devoted to revealing the effects of the bath field and the relative direction between two atomic dipole moments of a V-type three-level atom driven by a laser field on the atomic population coherent trapping. We find when the two atomic dipole moments are equal, the atomic population coherent trapping can take place in the two upper levels even for the damping of bath field, and in the steady-state limit, the atom can evolve into its coherent population trapping state.

14.1 Atomic population coherent trapping and phase properties in the system of a V-configuration three-level atom interacting with a bimodal field

The scheme of the V-type three-level atom system, as shown in Fig.(14.1), consists of two allowed transitions $|a\rangle \leftrightarrow |c\rangle$ and $|b\rangle \leftrightarrow |c\rangle$ due to the interaction

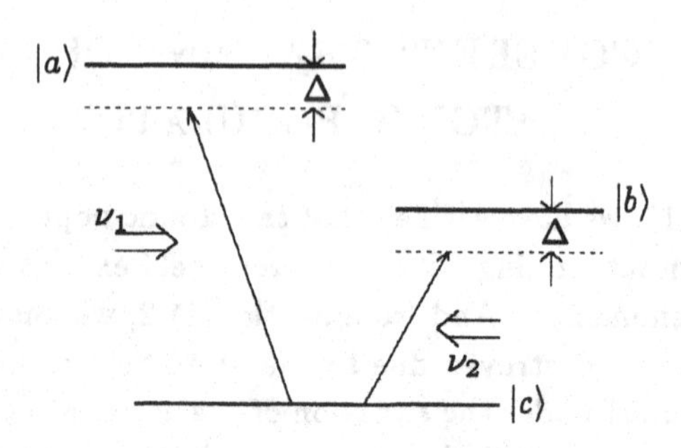

Figure 14.1: Diagram of a V-configuration three-level atom coupled to a two-mode field

of the bimodal field. Each field mode has different frequency. In the rotating-wave approximation, the Hamiltonian of the system is described by

$$H = H_0 + V \tag{14.1}$$

where

$$H_0 = \sum_{i=a,b,c} \omega_i |i\rangle\langle i| + \sum_{j=1}^{2} \nu_j a_j^\dagger a_j \quad (\hbar = 1) \tag{14.2}$$

$$V = g_1 a_1 |a\rangle\langle c| + g_1 a_1^\dagger |c\rangle\langle a| + g_2 a_2 |b\rangle\langle c| + g_2 a_2^\dagger |c\rangle\langle b| \tag{14.3}$$

Here $a_j^\dagger$ and $a_j (j = 1, 2)$ are, respectively, the creation and annihilation operators for the field of frequency ν_j. $|i\rangle (i = a, b, c)$ is the eigenstate of the atom with eigenfrequency ω_i, and $g_i (i = 1, 2)$ is the corresponding coupling constant. In order to discuss the atomic coherent population trapping phenomenon, one must find the state vector of the system at time t.

14.1.1 Time evolution of the state vector of the system

In the interaction picture, the state vector of this atom-field coupling system at time t can be described in general by

$$|\Psi^I(t)\rangle = \sum_{n_1,n_2} (C_{a,n_1,n_2}|a,n_1,n_2\rangle + C_{b,n_1,n_2}|b,n_1,n_2\rangle + C_{c,n_1,n_2}|c,n_1,n_2\rangle) \tag{14.4}$$

Substituting (14.4) into the Schrödinger equation in the interaction picture, we obtain

$$\begin{aligned}
\frac{d}{dt}C_{a,n_1-1,n_2} &= -ig_1\sqrt{n_1}\exp(i\Delta t)C_{c,n_1,n_2} \\
\frac{d}{dt}C_{b,n_1,n_2-1} &= -ig_2\sqrt{n_2}\exp(i\Delta t)C_{c,n_1,n_2} \\
\frac{d}{dt}C_{c,n_1,n_2} &= -i(g_1\sqrt{n_1}C_{a,n_1-1,n_2} + g_2\sqrt{n_2}C_{b,n_1,n_2-1})\exp(-i\Delta t)
\end{aligned} \tag{14.5}$$

where we have taken the system satisfying the two-photon resonant condition and the one-photon detuning $\Delta = \omega_a - \omega_c - \nu_1 = \omega_b - \omega_c - \nu_2$.

If the atom is initially in the state $|\Psi_A(0)\rangle$

$$|\Psi_A(0)\rangle = \cos\frac{\theta}{2}|a\rangle + \sin\frac{\theta}{2}e^{-i\phi}|b\rangle \tag{14.6}$$

which means that the atom is in the coherent superposition state of its eigenkets $|a\rangle$ and $|b\rangle$, meanwhile we also assume that the field is in the superposition of the photon number states at time t=0

$$|\Psi_F(0)\rangle = \sum_{n_1,n_2} F_{n_1,n_2}|n_1,n_2\rangle \tag{14.7}$$

where F_{n_1,n_2} is the probability amplitude which corresponds to the bimodal field in the number state $|n_1,n_2\rangle$, then the state vector of the total system at t=0 can be written by

$$|\Psi(0)\rangle = \sum_{n_1,n_2}\left[\cos\frac{\theta}{2}F_{n_1-1,n_2}|a,n_1-1,n_2\rangle + \sin\frac{\theta}{2}e^{-i\phi}F_{n_1,n_2-1}|b,n_1,n_2-1\rangle\right] \tag{14.8}$$

Using the initial condition (14.8), we obtain the solution of (14.5) to be

$$
\begin{aligned}
C_a(t) &= -A_1 g_1\sqrt{n_1}\left[\frac{\exp[i(\frac{\Delta}{2}+\beta)t]-1}{\frac{\Delta}{2}+\beta}-\frac{\exp[i(\frac{\Delta}{2}-\beta)t]-1}{\frac{\Delta}{2}-\beta}\right] \\
&\quad +\cos\frac{\theta}{2}F_{n_1-1,n_2} \\
C_b(t) &= -A_1 g_2\sqrt{n_2}\left[\frac{\exp[i(\frac{\Delta}{2}+\beta)t]-1}{\frac{\Delta}{2}+\beta}-\frac{\exp[i(\frac{\Delta}{2}-\beta)t]-1}{\frac{\Delta}{2}-\beta}\right] \\
&\quad +\sin\frac{\theta}{2}e^{-i\phi}F_{n_1,n_2-1} \\
C_c(t) &= -A_1\left\{\exp\left[-i\left(\frac{\Delta}{2}-\beta\right)t\right]-\exp\left[-i\left(\frac{\Delta}{2}+\beta\right)t\right]\right\}
\end{aligned}
\tag{14.9}
$$

Here we have used abbreviate symbols $C_a(t)$, $C_b(t)$, and $C_c(t)$ to replace the symbols $C_{a,n_1-1,n_2}(t)$, $C_{b,n_1,n_2-1}(t)$, and $C_{c,n_1,n_2}(t)$, respectively, and A_1 and β obey

$$
\begin{aligned}
A_1 &= -\frac{1}{2\beta}\left[g_1\sqrt{n_1}\cos\frac{\theta}{2}F_{n_1-1,n_2}+g_2\sqrt{n_2}\sin\frac{\theta}{2}e^{-i\phi}F_{n_1,n_2-1}\right] \\
\beta &= \sqrt{\frac{\Delta^2}{4}+g_1^2 n_1+g_2^2 n_2}
\end{aligned}
$$

Clearly, β is associated with the frequency of the atomic Rabi oscillation. Substituting equation (14.9) into (14.4), we may obtain the state vector of the system at time t in the interaction picture. Then transferring it to the Schrödinger picture, we have

$$
\begin{aligned}
|\Psi(t)\rangle = \sum_{n_1,n_2}&\{C_a\exp\{-i[(n_1-1)\nu_1+n_2\nu_2+\omega_a)t]\}|a,n_1-1,n_2\rangle \\
&+C_b\exp\{-i[n_1\nu_1+(n_2-1)\nu_2+\omega_b)t]\}|b,n_1,n_2-1\rangle \\
&+C_c\exp\{-i[n_1\nu_1+n_2\nu_2+\omega_c)t]\}|c,n_1,n_2\rangle\}
\end{aligned}
\tag{14.10}
$$

Starting from (14.10), we can discuss the time evolution of phase operators of the bimodal field, and analyze the influence of the different values of θ and ϕ on the properties of the atomic population coherent trapping of the system.

14.1.2 Time evolution of the phase operator in the atom-field coupling system

Here we also adopt the phase theory introduced by Pegg and Barnett (shown in Section 3.3) to discuss the time evolution of the phase operator in the atom-field coupling system.

For a two-mode field, the phase states are defined by

$$|\theta_{m1},\theta_{m2}\rangle = [(s_1+1)(s_2+1)]^{-1/2}\sum_{m_1}^{s_1}\sum_{m_2}^{s_2}\exp[i(n_1\theta_{m_1}+n_2\theta_{m_2})]|n_1,n_2\rangle \tag{14.11}$$

where

$$\theta_{m_i} = \theta_{0_i} + \frac{2\pi m_i}{s_i+1}, \quad m_i = 0,1,2\cdots,s_i;\ i=1,2$$

here θ_{0_i} is the reference phase, $(s_1+1)(s_2+1)$ represents the dimension of the Hilbert space spanned by the two-mode phase states. Since the values of s_1 and s_2 will be made to tend to infinity after all necessary expectation values have been calculated, here we assume $s_1 = s_2 = s$ without loss generality. So the phase Hermitian operator of the i-th mode in the two-mode field can be described as

$$\Phi_i = \sum_{m_1,m_2=0}^{s} \theta_{m_i}|\theta_{m_1},\theta_{m_2}\rangle\langle\theta_{m_1},\theta_{m_2}| \quad (i=1,2) \tag{14.12}$$

Explicitly, Φ_i is the i-th mode phase operator of the two-mode field, θ_{m_i} is the eigenvalue of the eigenket $|\theta_{m_i}\rangle$ with frequency ν_i. Therefore, the state vector $|\Psi(t)\rangle$ of the atom-field coupling system can be spanned by the phase eigenstates as

$$\begin{aligned}|\Psi(t)\rangle = \sum_{m_1,m_2} \{&\langle a,\theta_{m_1},\theta_{m_2}|\Psi(t)\rangle|a,\theta_{m_1},\theta_{m_2}\rangle \\ &+\langle b,\theta_{m_1},\theta_{m_2}|\Psi(t)\rangle|b,\theta_{m_1},\theta_{m_2}\rangle \\ &+\langle c,\theta_{m_1},\theta_{m_2}|\Psi(t)\rangle|c,\theta_{m_1},\theta_{m_2}\rangle\}\end{aligned} \tag{14.13}$$

and

$$\begin{aligned}P(\theta_{m_1},\theta_{m_2},t) = &|\langle a,\theta_{m_1},\theta_{m_2}|\Psi(t)\rangle|^2 + |\langle b,\theta_{m_1},\theta_{m_2}|\Psi(t)\rangle|^2 \\ &+|\langle c,\theta_{m_1},\theta_{m_2}|\Psi(t)\rangle|^2\end{aligned} \tag{14.14}$$

represents the phase probability distribution function. Thus the expectation value of the phase operator gives

$$\langle \Phi_1^n \Phi_2^k \rangle = \sum_{m_1, m_2 = 0}^{s} \theta_{m_1}^n \theta_{m_2}^k P(\theta_{m_1}, \theta_{m_2}, t) \quad (n, k = 0, 1, 2) \tag{14.15}$$

If the radiation field is initially in a two-mode uncorrelated coherent state, i.e.,

$$F_{n_1, n_2} = \exp[-(\overline{n_1} + \overline{n_2})/2] \frac{\alpha_1^{n_1} \alpha_2^{n_2}}{\sqrt{n_1! n_2!}} \tag{14.16}$$

where

$$\alpha_i = \sqrt{\overline{n}_i} \exp(i\xi_i) \qquad (i = 1, 2)$$

and $\overline{n}_i$ is the mean photon number of the i-th mode, and ξ_i is the phase angle of α_i. For $\overline{n}_1$, $\overline{n}_2 \gg 1$, employing the same approximation of (3.173), F_{n_1, n_2} can be well approximated as

$$\begin{aligned} F_{n_1, n_2} &\approx (4\pi^2 \overline{n}_1 \overline{n}_2)^{-1/4} \exp[i(n_1 \xi_1 + n_2 \xi_2)] \\ &\times \exp\left[-\frac{(n_1 - \overline{n}_1)^2}{4\overline{n}_1} - \frac{(n_2 - \overline{n}_2)^2}{4\overline{n}_2} \right] \end{aligned} \tag{14.17}$$

Considering the feature of the photon number distribution function, the approximations

$$F_{n_1, n_2} \approx F_{n_1, n_2 - 1} \approx F_{n_1 - 1, n_2} \tag{14.18}$$

and

$$\beta(n_1, n_2) = \overline{\beta} + \frac{g_1^2}{2\overline{\beta}}(n_1 - \overline{n}_1) + \frac{g_2^2}{2\overline{\beta}}(n_2 - \overline{n}_2) \tag{14.19}$$

are reasonable when we sum with respect to n_1 and n_2. Here $\overline{\beta} = \overline{\beta}(\overline{n}_1, \overline{n}_2)$. So the phase probability distribution $P(\theta_{m1}, \theta_{m2}, t)$ can be approximated as

$$\begin{aligned} P(\theta_{m1}, \theta_{m2}, t) = &\left(\frac{2\pi}{s+1}\right)^2 \left(\frac{4\overline{n}_1 \overline{n}_2}{\pi^2}\right)^{1/2} \{\exp[-2(\overline{n}_1 z_1 + \overline{n}_2 z_2)] \\ &\times \frac{g_1^2 \overline{n}_1 \sin^2 \frac{\theta}{2} + g_2^2 \overline{n}_2 \cos^2 \frac{\theta}{2} - g_1 g_2 \sqrt{\overline{n}_1 \overline{n}_2} \sin\theta \cos\phi}{g_1^2 \overline{n}_1 + g_2^2 \overline{n}_2} \\ &+ \frac{1}{2\overline{\beta}} \left[\frac{\exp[-2(\overline{n}_1 x_1 + \overline{n}_2 x_2)]}{\overline{\beta} + \Delta/2} + \frac{\exp[-2(\overline{n}_1 y_1 + \overline{n}_2 y_2)]}{\overline{\beta} - \Delta/2} \right] \\ &\times \left[g_1^2 \overline{n}_1 \cos^2 \frac{\theta}{2} + g_2^2 \overline{n}_2 \sin^2 \frac{\theta}{2} + g_1 g_2 \sqrt{\overline{n}_1 \overline{n}_2} \sin\theta \cos\phi \right] \Bigg\} \end{aligned} \tag{14.20}$$

where

$$z_i = (\xi_i - \theta_{m_i} - \nu_i t)^2 \quad , x_i = \left(\xi_i - \theta_{m_i} - \nu_i t + \frac{g_i^2 t}{2\overline{\beta}}\right)^2$$

$$y_i = \left(\xi_i - \theta_{m_i} - \nu_i t - \frac{g_i^2 t}{2\overline{\beta}}\right)^2$$

In the continued limit, i.e., $s \to \infty$, θ_{m_1} and θ_{m_2} are two continued variables. The phase probability density is normalized according to

$$\int_{\theta_{0_1}}^{\theta_{0_1}+2\pi} d\theta_1 \int_{\theta_{0_2}}^{\theta_{0_2}+2\pi} d\theta_2 P(\theta_1, \theta_2, t) \left(\frac{s+1}{2\pi}\right)^2$$

$$= \int \int_{-\infty}^{\infty} d\theta_1 d\theta_2 P(\theta_1, \theta_2, t) \left(\frac{s+1}{2\pi}\right)^2 = 1$$

Since $P(\theta_1, \theta_2, t)$ exhibits a Gaussian distribution, the integral range can be expanded to $-\infty \to \infty$. $(\frac{s+1}{2\pi})^2$ is the density of phase states.

Using (14.15) and (14.20), we obtain the time evolution of expectation values of phase operators to be

$$\langle \Phi_1 \rangle = \xi_1 - \nu_1 t - A_2 g_1^2 t \Delta \tag{14.21}$$

$$\langle \Phi_2 \rangle = \xi_2 - \nu_2 t - A_2 g_2^2 t \Delta \tag{14.22}$$

$$\langle \Phi_1^2 \rangle = \frac{1}{4\overline{n}_1} + (\xi_1 - \nu_1 t)^2 + A_2 g_1^4 t^2 - 2A_2 g_1^2 t \Delta (\xi_1 - \nu_1 t) t \tag{14.23}$$

$$\langle \Phi_2^2 \rangle = \frac{1}{4\overline{n}_2} + (\xi_2 - \nu_2 t)^2 + A_2 g_2^4 t^2 - 2A_2 g_2^2 t \Delta (\xi_2 - \nu_2 t) t \tag{14.24}$$

So the phase fluctuations can be expressed as

$$(\Delta\Phi_1)^2 = \langle \Phi_1^2 \rangle - \langle \Phi_1 \rangle^2 = \frac{1}{4\overline{n}_1} + g_1^4 t^2 A_2 (1 - A_2 \Delta^2) \tag{14.25}$$

$$(\Delta\Phi_2)^2 = \langle \Phi_2^2 \rangle - \langle \Phi_2 \rangle^2 = \frac{1}{4\overline{n}_2} + g_2^4 t^2 A_2 (1 - A_2 \Delta^2) \tag{14.26}$$

with

$$A_2 = \frac{g_1^2 \overline{n}_1 \cos^2 \frac{\theta}{2} + g_2^2 \overline{n}_2 \sin^2 \frac{\theta}{2} + g_1 g_2 \sqrt{\overline{n}_1 \overline{n}_2} \sin\theta \cos\phi}{4\overline{\beta}^2 (g_1^2 \overline{n}_1 + g_2^2 \overline{n}_2)}$$

From (14.21)-(14.26) we can see that the expectation values and the fluctuations of the phase operators ((14.21), (14.23), and (14.25)) for frequency ν_1 contain the constants $\overline{n}_2$ and g_2 related to the field mode with frequency ν_2, this is because both modes of the field are interacting with the atom and become to be correlated. The case for the mode with frequency ν_2 is the same. In addition, from eqs.(14.25) and (14.26) we find that the phase fluctuations of the both modes depend on the parameters θ and ϕ, which describe the atomic initial preparation. In the following, we discuss the influences of the atomic initial state parameters θ and ϕ on the atomic populations and the coherence of the field.

14.1.3 Coherent trapping of the atomic population

When $\theta = 0$, corresponding to the atom initially in its upper state $|a\rangle$, the phase fluctuations of the both modes are reduced to be

$$(\Delta\Phi_1)^2 = \frac{1}{4\overline{n}_1} + \frac{g_1^6 \overline{n}_1 t^2}{4\overline{\beta}^2 (g_1^2\overline{n}_1 + g_2^2\overline{n}_2)} \left[1 - \frac{\Delta^2 g_1^2 \overline{n}_1}{4\overline{\beta}^2 (g_1^2\overline{n}_1 + g_2^2\overline{n}_2)}\right] \tag{14.27}$$

$$(\Delta\Phi_2)^2 = \frac{1}{4\overline{n}_2} + \frac{g_1^2 g_2^4 \overline{n}_1 t^2}{4\overline{\beta}^2 (g_1^2\overline{n}_1 + g_2^2\overline{n}_2)} \left[1 - \frac{\Delta^2 g_1^2 \overline{n}_1}{4\overline{\beta}^2 (g_1^2\overline{n}_1 + g_2^2\overline{n}_2)}\right] \tag{14.28}$$

If the atom is initially in the state $|b\rangle$, i.e., $\theta = \pi$, then (14.25) and (14.26) are simplified as

$$(\Delta\Phi_1)^2 = \frac{1}{4\overline{n}_1} + \frac{g_1^4 g_2^2 \overline{n}_2 t^2}{4\overline{\beta}^2 (g_1^2\overline{n}_1 + g_2^2\overline{n}_2)} \left[1 - \frac{\Delta^2 g_2^2 \overline{n}_2}{4\overline{\beta}^2 (g_1^2\overline{n}_1 + g_2^2\overline{n}_2)}\right] \tag{14.29}$$

$$(\Delta\Phi_2)^2 = \frac{1}{4\overline{n}_2} + \frac{g_2^6 \overline{n}_2 t^2}{4\overline{\beta}^2 (g_1^2\overline{n}_1 + \overline{n}_2)} \left[1 - \frac{\Delta^2 g_2^2 \overline{n}_2}{4\overline{\beta}^2 (g_1^2\overline{n}_1 + g_2^2\overline{n}_2)}\right] \tag{14.30}$$

Evidently, the phase fluctuations of the both modes are enhanced with the time development. (14.27)-(14.30) show that when the atom is initially in the non-coherent state such as $|a\rangle$ or $|b\rangle$, the initial coherence of the bimodal field will be destroyed due to the atom-field coupling. In the meantime, the atomic populations (described by the equations resemblance to eqs.(12.98) and (12.100)) exhibit the periodic collapses and revivals.

However, if the atom is initially in the coherent superposition $|\Psi(0)\rangle$ in which $\theta \neq 0$ or π, then from eqs.(14.25) and (14.26) we see when θ and ϕ

satisfy

$$g_1^2\overline{n}_1 \cos^2\frac{\theta}{2} + g_2^2\overline{n}_2 \sin^2\frac{\theta}{2} + g_1 g_2\sqrt{\overline{n}_1\overline{n}_2}\sin\theta\cos\phi = 0$$

namely,

$$\phi = \pi, \qquad \theta = 2rctan\left(\frac{g_1}{g_2}\sqrt{\frac{\overline{n}_1}{\overline{n}_2}}\right) \tag{14.31}$$

the phase fluctuations become

$$(\Delta\Phi_1)^2 = \frac{1}{4\overline{n}_1} \tag{14.32}$$

$$(\Delta\Phi_2)^2 = \frac{1}{4\overline{n}_2} \tag{14.33}$$

This means that under this atomic preparation, the phase fluctuations of the both modes do not vary with the time development, and retain each initial value respectively. Since for $\overline{n}_1$, $\overline{n}_2 \gg 1$, the photon number distributions of the both modes can be regarded as Poissonian function with the time development, i.e., $(\Delta N_1)^2 = \overline{n}_1$, $(\Delta N_2)^2 = \overline{n}_2$, so that the number-phase uncertainty products of the two-mode field give

$$(\Delta\Phi_1)^2(\Delta N_1)^2 = (\Delta\Phi_2)^2(\Delta N_2)^2 = \frac{1}{4} \tag{14.34}$$

It shows that for the appropriate atomic preparation (described by (14.31)), each mode of the two-mode field, which is initially in a coherent state, retains its coherence with the time development even though the existence of atom-field coupling described by (14.3).

Now, we examine the variation of atomic populations with the time development when the atom is initially in the state in which the field can retain its coherence. It is clear that the initial populations of the atom are given from (14.9) and (14.31) as

$$P_a(0) = \cos^2(\theta/2) = \frac{g_2^2\overline{n}_2}{g_1^2\overline{n}_1 + g_2^2\overline{n}_2} \tag{14.35}$$

$$P_b(0) = \sin^2(\theta/2) = \frac{g_1^2\overline{n}_1}{g_1^2\overline{n}_1 + g_2^2\overline{n}_2} \tag{14.36}$$

$$P_c(0) = 0 \tag{14.37}$$

With the time development, in the Hilbert space spanned by the eigenkets of the atom and the field, the populations of the atom become

$$P_a(t) = \int\int_{-\infty}^{\infty} \left(\frac{s+1}{2\pi}\right)^2 |\langle a, \theta_1, \theta_2 | \Psi(t)\rangle|^2 d\theta_1 d\theta_2$$

$$= \left(\frac{4\overline{n}_1\overline{n}_2}{\pi^2}\right)^{1/2} \int\int_{-\infty}^{\infty} \frac{g_2^2\overline{n}_2}{g_1^2\overline{n}_1 + g_2^2\overline{n}_2} \exp[-2(\overline{n}_1 z_1 + \overline{n}_2 z_2)] d\theta_1 d\theta_2$$

$$= \frac{g_2^2\overline{n}_2}{g_1^2\overline{n}_1 + g_2^2\overline{n}_2} \tag{14.38}$$

$$P_b(t) = \frac{g_1^2\overline{n}_1}{g_1^2\overline{n}_1 + g_2^2\overline{n}_2} \tag{14.39}$$

$$P_c(t) = 0 \tag{14.40}$$

This means that the dynamic behavior of the atom does not exhibit the periodic collapses and revivals, and $P_a(t)$, $P_b(t)$ and $P_c(t)$ are the same as these for the initial preparations. So in the two-photon resonant case, under the interaction of the uncorrelated two-mode coherent field, the populations of the atom initially in the coherent superposition state $\frac{1}{\sqrt{g_1^2\overline{n}_1+g_2^2\overline{n}_2}}(g_1^2\sqrt{\overline{n}_2}|a\rangle - g_1^2\sqrt{\overline{n}_1}|b\rangle)$ reach to their steady values, we interpret this phenomenon as atomic population coherent trapping. In this case, both the properties of atomic populations and the field coherent property are the same as that of their initial preparations. A possible explanation for such behavior can be understood as follows.

In the system shown in Fig.(14.1), there exists two different one-photon transition processes such as $|a\rangle \to |c\rangle$ and $|b\rangle \to |c\rangle$, and these two processes interfere each other. Since the atom is initially in $\frac{1}{\sqrt{g_1^2\overline{n}_1+g_2^2\overline{n}_2}}(g_1^2\sqrt{\overline{n}_2}|a\rangle - g_1^2\sqrt{\overline{n}_1}|b\rangle)$, this interference between the transition processes $|a\rangle \to |c\rangle$ and $|b\rangle \to |c\rangle$ makes counter balance, which induces the atom-field coupling system to be decoupled. Thus the atomic population in $|c\rangle$ is equal to zero. Meanwhile, the atomic populations in the upper states retain their initial values, and the coherence of the field does not change, so the atomic population coherent trapping takes place. We call the state leading to the atomic population coherent trapping as the atomic population coherent trapping state. The above results show that the atomic initial state plays an important role in the time evolution of the atom and the field in the atom-field coupling system. What are the features of the atomic population trapping state ? We now turn to discuss this question.

As the atomic population coherent trapping state $\frac{1}{\sqrt{g_1^2\bar{n}_1+g_2^2\bar{n}_2}}(g_2\sqrt{\bar{n}_2}|a\rangle - g_1\sqrt{\bar{n}_1}|b\rangle)$ is not related to the one-photon detuning Δ, for simplicity to calculate and without loss of the generality, we can assume $\Delta = 0$ in the following. First, let us solve the eigenkets (dressed states) of the atom-field coupling system. The dressed states can be written as

$$\begin{aligned} |i\rangle &= \alpha_i|a, n_1 - 1, n_2\rangle + \beta_i|b, n_1, n_2 - 1\rangle + \gamma_i|c, n_1, n_2\rangle \\ &= \begin{pmatrix} \alpha_i \\ \beta_i \\ \gamma_i \end{pmatrix} \qquad (i = 1, 2, 3) \end{aligned} \tag{14.41}$$

correspondingly, the matrix representation of the Hamiltonian (14.1) is

$$\begin{aligned} H &= (n_1\nu_1 + n_2\nu_2 + \omega_c) \begin{pmatrix} 1 & 0 & 0 \\ 0 & 1 & 0 \\ 0 & 0 & 1 \end{pmatrix} \\ &+ \begin{pmatrix} 0 & 0 & g_1\sqrt{\bar{n}_1} \\ 0 & 0 & g_2\sqrt{\bar{n}_2} \\ g_1\sqrt{\bar{n}_1} & g_2\sqrt{\bar{n}_2} & 0 \end{pmatrix} \end{aligned} \tag{14.42}$$

and the eigenvalue of the Hamiltonian may be assumed as

$$E_i = n_1\nu_1 + n_2\nu_2 + \omega_c + \lambda_i \tag{14.43}$$

Then the eigenvalue and its corresponding eigenket can be solved by

$$\begin{pmatrix} 0 & 0 & g_1\sqrt{\bar{n}_1} \\ 0 & 0 & g_2\sqrt{\bar{n}_2} \\ g_1\sqrt{\bar{n}_1} & g_2\sqrt{\bar{n}_2} & 0 \end{pmatrix} \begin{pmatrix} \alpha_i \\ \beta_i \\ \gamma_i \end{pmatrix} = \lambda_i \begin{pmatrix} \alpha_i \\ \beta_i \\ \gamma_i \end{pmatrix} \tag{14.44}$$

The solutions of the above equation are

$$\begin{aligned} E_1 &= n_1\nu_1 + n_2\nu_2 + \omega_c, \\ |1\rangle &= (g_1^2 n_1 + g_2^2 n_2)^{-1/2}(g_2\sqrt{n_2}|a, n_1 - 1, n_2\rangle \\ &\quad - g_1\sqrt{n_1}|b, n_1, n_2 - 1\rangle) \end{aligned} \tag{14.45}$$

$$\begin{aligned} E_2 &= n_1\nu_1 + n_2\nu_2 + \omega_c + \sqrt{g_1^2 n_1 + g_2^2 n_2}, \\ |2\rangle &= 2^{-1/2}[|c, n_1, n_2\rangle + (g_1^2 n_1 + g_2^2 n_2)^{-1/2} \end{aligned}$$

$$\times\left(g_2\sqrt{n_2}|a,n_1-1,n_2\rangle+g_1\sqrt{n_1}|b,n_1,n_2-1\rangle\right) \tag{14.46}$$

$$E_3=n_1\nu_1+n_2\nu_2+\omega_c-\sqrt{g_1^2n_1+g_2^2n_2},$$

$$|3\rangle=2^{-1/2}[|c,n_1,n_2\rangle-(g_1^2n_1+g_2^2n_2)^{-1/2}$$
$$\times\left(g_2\sqrt{n_2}|a,n_1-1,n_2\rangle+g_1\sqrt{n_1}|b,n_1,n_2-1\rangle\right) \tag{14.47}$$

Eq.(14.45) shows that the eigenvalue corresponding to the dressed state $|1\rangle$ is the free energy of the atom and the field, so $|1\rangle$ is the decoupled state of the system. Noticing the two-mode field is intensive, namely, its photon number distribution $|F_{n_1,n_2}|^2$ satisfies (14.18), so the neighborhood of $\overline{n}_1$ and $\overline{n}_2$ gives the dominant contribution with respect to the summation of n_1 and n_2. Thus the initial state of the atom-field coupling system (described by (14.8) and (14.31)) can be expressed as

$$|\Psi(0)\rangle=\sum_{n_1,n_2}F_{n_1,n_2}|1\rangle \tag{14.48}$$

The state vector at time t becomes

$$|\Psi(t)\rangle=\exp(-iHt)|\Psi(0)\rangle=\sum_{n_1,n_2}F_{n_1,n_2}|1\rangle\exp[-i(n_1\nu_1+n_2\nu_2+\omega_c)t] \tag{14.49}$$

This means that the probability of the system in $|1\rangle$ is independent of time t. So the atomic populations retain their initial values with the time evolution. That is to say, the initial atomic state $(g_1^2\overline{n}_1+g_2^2\overline{n}_2)^{-1/2}(g_2\overline{n}_2^{1/2}|a\rangle-g_1\overline{n}_1^{1/2}|b\rangle)$ leading to the atomic population coherent trapping is just the dressed state of the atom-field system. Because the dressed state is a stationary state of the system, the atomic populations and the coherence of field do not vary with the time evolution.

14.2 Coherent trapping of the atomic population for a V-configuration three-level atom driven by a classical field in a heat bath

As we know, the atom in its excited state must be damped to its ground state due to the fluctuations of the vacuum field. But the atomic population coherent trapping discussed in the above section occurs in a linear superposition state of the atomic upper states. The question arisen here is whether this coherent trapping of atomic populations is steady. In other words, can the atom in $|\Psi(0)\rangle$ (described by (14.31)) be damped due to the fluctuations of the vacuum field ? Now we focus our attention on this question.

14.2.1 Time evolution of the reduced density matrix ρ of the atom

For simplicity, we consider a V-type three-level atom shown in Fig.(14.2). The atom interacts with both a single-mode cw classical laser field of frequency ω_L and a quantized multi-mode radiation field. The laser drives the atomic transitions which take place from $|2\rangle$ to $|1\rangle$, and $|3\rangle$ to $|1\rangle$. The atomic dipole moment can be described by

$$\mathbf{D}^{\dagger} = \mathbf{d}_{21} A_{21} + \mathbf{d}_{31} A_{31}, \quad \mathbf{D}^{-} = (\mathbf{D}^{\dagger})^{\dagger} \tag{14.50}$$

Here $\mathbf{d}_{i1} = \langle i|\mathbf{d}|1\rangle$ represents the atomic dipole element for the $|i\rangle \to |1\rangle$ transition (i=2 or 3), $A_{ij} = |i\rangle\langle j|$ is the atomic operator (i, j=1,2,3). The laser field is given by

$$\mathbf{E} = E_0\{\varepsilon \exp[-i(\omega t + \phi_L)] + \text{c.c.}\} \tag{14.51}$$

where ε is the polarization unit vector, ϕ_L is the phase angle of the laser field and E_0 is the real parameter associated with the laser intensity. In the dipole approximation and the rotating-wave approximation, the interaction between the atom and the laser field can be characterized in the form

$$V_{A-L} = \exp[i(\omega t + \phi_L)](\Omega_{12} A_{12} + \Omega_{13} A_{13}) + \text{h.c.} \tag{14.52}$$

here we have assumed $\Omega_{1i} = E_0 \varepsilon \cdot \mathbf{d}_{1i} (i = 2, 3)$. Meanwhile, the atom is damped by the heat bath. In the rotating-wave approximation, the interaction between the bath and the atom is characterized by the Hamiltonian

$$V_{A-B} = i \sum_{kj} \left(\frac{2\pi\omega_k}{V} \right)^{1/2} [\mathbf{e}_{kj} \cdot \mathbf{d}_{21} A_{21} + \mathbf{e}_{kj} \cdot \mathbf{d}_{31} A_{31}] a_{kj} + \text{h.c.} \tag{14.53}$$

where a_{kj} ($a_{kj}^{\dagger}$) is the annihilation (creation) operator for the photon with the frequency ω_k and the polarization unit vector $\mathbf{e}_{kj}$. The characteristic of the bath is defined by

$$\begin{aligned}
&\langle a_{kj}(t)\rangle = \langle a_{kj}^{\dagger}(t)\rangle = 0 \\
&\langle a_{kj}^{\dagger}(t) a_{k'j'}(t')\rangle = \overline{n}_0 \delta(\omega_k - \omega_{k'}) \delta_{j,j'} \delta(t - t') \\
&\langle a_{kj}(t) a_{k'j'}^{\dagger}(t')\rangle = (\overline{n}_0 + 1) \delta(\omega_k - \omega_{k'}) \delta_{j,j'} \delta(t - t')
\end{aligned} \tag{14.54}$$

here $\overline{n}_0$ (defined by (5.88)) is the mean photon number of the bath. The Hamiltonian of the total system is therefore described by

$$H = \sum_{i=1}^{3} \omega_i A_{ii} + \sum_{kj} \omega_k a_{kj}^{\dagger} a_{kj} + V_{A-L} + V_{A-B} \tag{14.55}$$

Following the same method introduced in Section 5.2, we can derive the master equation of the reduced density matrix ρ for the atom as follows

$$\begin{aligned}\frac{\partial}{\partial t}\rho = & -i\left[\sum_{i=1}^{3} \omega_i A_{ii}, \rho\right] - i[V_{A-L}, \rho] + (\overline{n}_0 + 1)(2\gamma_2 A_{12}\rho A_{21} \\ & + 2\gamma_{1321} A_{13}\rho A_{21} + 2\gamma_{1231} A_{12}\rho A_{31} + 2\gamma_3 A_{13}\rho A_{31} - \gamma_2 A_{22}\rho \\ & - \gamma_{3112} A_{23}\rho - \gamma_3 A_{33}\rho - \gamma_2 \rho A_{22} - \gamma_{3112}\rho A_{32} - \gamma_{2113}\rho A_{23} - \gamma_3 \rho A_{33}) \\ & + \overline{n}_0 (2\gamma_2 A_{21}\rho A_{12} + 2\gamma_{3112} A_{31}\rho A_{12} + 2\gamma_{2113} A_{21}\rho A_{13} \\ & + 2\gamma_3 A_{31}\rho A_{13} - \gamma_2 A_{11}\rho - \gamma_3 A_{11}\rho - \gamma_2 \rho A_{11} - \gamma_3 \rho A_{11})\end{aligned} \tag{14.56}$$

where

$$\gamma_{\ell\sigma mn} = \frac{2\pi^2}{V} \sum_{k} \sum_{j=1}^{2} (\mathbf{d}_{\ell\sigma} \cdot e_{kj})(\mathbf{d}_{mn} \cdot e_{kj}) \omega_k \delta(\omega_k - |\omega_{mn}|) \tag{14.57}$$

and we have used the abbreviation $\gamma_i = \gamma_{\ell ii\ell} (i = 2, 3)$. In order to show the effect of the relative angle between the atomic dipole moments $\mathbf{d}_{21}$ and $\mathbf{d}_{31}$ clearly, we simplify (14.57) in the following.

For simplicity to calculate, we choose the polarization unit vectors $\mathbf{e}_{k1}$ and $\mathbf{e}_{k2}$, and the propagation unit vector $\mathbf{k}$ to be

$$\begin{aligned}\mathbf{k} = & \sin\theta\cos\phi\mathbf{e}_1 + \sin\theta\sin\phi\mathbf{e}_2 + \cos\theta\mathbf{e}_3 \\ \mathbf{e}_{k1} = & -\cos\theta\cos\phi\mathbf{e}_1 - \cos\theta\sin\phi\mathbf{e}_2 + \sin\theta\mathbf{e}_3 \\ \mathbf{e}_{k2} = & \sin\phi\mathbf{e}_1 - \cos\phi\mathbf{e}_2\end{aligned} \tag{14.58}$$

here $\mathbf{e}_p (p = 1, 2, 3)$ are the three mutually perpendicular unit vectors. Employing the summation rule of polarization unit vectors of the field

$$\sum_{j=1}^{2} (\mathbf{e}_{kj})_p (\mathbf{e}_{kj})_q = \delta_{p,q} - \mathbf{k}_p \mathbf{k}_q \tag{14.59}$$

where $(\mathbf{e}_{kj})_p$ and $(\mathbf{k})_p$ are the projections of $\mathbf{e}_{kj}$ and $\mathbf{k}$ in the direction $\mathbf{e}_p$. Thus, (14.57) may be reduced to

$$\begin{aligned}\gamma_{\ell\sigma mn} &= \frac{2\pi^2}{V}\sum_k \sum_{p,q=1}^{3} \sum_{j=1}^{2} (\mathbf{d}_{\ell\sigma})_p (\mathbf{d}_{mn})_q (\mathbf{e}_{kj})_p (\mathbf{e}_{kj})_q \omega_k \delta(\omega_k - |\omega_{mn}|) \\ &= \frac{2\pi^2}{V}\sum_k \sum_{p,q=1}^{3} (\mathbf{d}_{\ell\sigma})_p (\mathbf{d}_{mn})_q \omega_\delta (\omega_k - |\omega_{mn}|)(\delta_{p,q} - \mathbf{k}_p \mathbf{k}_q) \qquad (14.60)\end{aligned}$$

In the continued limit, $\frac{1}{V}\sum_k$ can be replaced by $\int \frac{d^3\mathbf{k}}{(2\pi)^3} = \frac{1}{(2\pi c)^3}\int_0^\infty \omega_k^2 d\omega_k \times \int_0^\pi \sin\theta d\theta \int_0^{2\pi} d\phi$, so (14.60) becomes

$$\begin{aligned}\gamma_{\ell\sigma mn} = \frac{1}{4\pi c^3} \sum_{p,q=1}^{3} (\mathbf{d}_{\ell\sigma})_p (\mathbf{d}_{mn})_q \int_0^\infty \omega_k^3 \delta(\omega_k - |\omega_{mn}|) d\omega_k \\ \times \int_0^\pi d\theta \sin\theta \int_0^{2\pi} d\phi (\delta_{p,q} - \mathbf{k}_p \mathbf{k}_q) \qquad (14.61)\end{aligned}$$

Substituting (14.58) into the above equation, we obtain

$$\gamma_{\ell\sigma mn} = \frac{2}{3c^3} \mathbf{d}_{\ell\sigma} \cdot \mathbf{d}_{mn} |\omega_{mn}|^3 \qquad (14.62)$$

It is evident to see in (14.56) that γ_i is the linewidth of the transition $|i\rangle \rightarrow |1\rangle$ and $\gamma_{\ell\sigma mn}$ represents the decay rate of the interference between the dipole moments $\mathbf{d}_{\ell\sigma}$ and $\mathbf{d}_{mn}$. The amplitude of $\gamma_{\ell\sigma mn}$ is related to the relative directions between $\mathbf{d}_{\ell\sigma}$ and $\mathbf{d}_{mn}$. If $\mathbf{d}_{\ell\sigma} \cdot \mathbf{d}_{mn} = 0$ which corresponds to no interference between $\mathbf{d}_{\ell\sigma}$ and $\mathbf{d}_{mn}$, then $\gamma_{\ell\sigma mn} = 0$. On the contrary, when $\mathbf{d}_{\ell\sigma}$ is parallel to $\mathbf{d}_{mn}$, $\gamma_{\ell\sigma mn}$ is a real constant and indicates that there is strong interference between the two atomic dipole moments.

If we assume

$$\rho_{21} = \tilde{\rho}_{21} \exp[-i(\omega_L t + \phi)], \quad \rho_{31} = \tilde{\rho}_{31} \exp[-i(\omega_L t + \phi)] \qquad (14.63)$$

then starting from (14.56), the equations for the time evolution of atomic density elements can be easily obtained as follows

$$\frac{d}{dt}\tilde{\rho}_{21} = i\Delta\tilde{\rho}_{21} + i\Omega_{12}^*(\rho_{22} - \rho_{11}) + i\Omega_{13}^*\rho_{23}$$

$$
\begin{aligned}
&-[\gamma_3\bar{n}_0+\gamma_2(2\bar{n}_0+1)]\tilde{\rho}_{21}-\gamma_{2113}(\bar{n}_0+1)\tilde{\rho}_{31}\\
\frac{d}{dt}\tilde{\rho}_{31}=&\,i(\Delta-\omega_{32})\tilde{\rho}_{31}+i\Omega_{13}^*(\rho_{33}-\rho_{11})+i\Omega_{13}^*\rho_{23}\\
&-[\gamma_2\bar{n}_0+\gamma_3(2\bar{n}_0+1)]\tilde{\rho}_{31}-\gamma_{3112}(\bar{n}_0+1)\tilde{\rho}_{21}]\\
\frac{d}{dt}\rho_{32}=&-i\omega\rho_{32}+i\Omega_{12}\tilde{\rho}_{31}+i\Omega_{13}^*\tilde{\rho}_{12}-(\gamma_3+\gamma_3)(\bar{n}_0+1)\tilde{\rho}_{32}\\
&-\gamma_{3112}(\bar{n}_0+1)(\rho_{33}+\rho_{22})+2\gamma_{3112}\bar{n}_0\rho_{11}\\
\frac{d}{dt}\rho_{33}=&-i(\Omega_{13}^*\tilde{\rho}_{13}-\Omega_{13}\tilde{\rho}_{31})-2\gamma_3(\bar{n}_0+1)\rho_{33}\\
&-(\bar{n}_0+1)(\gamma_{3112}\rho_{32}+\gamma_{2113}\rho_{23})\\
\frac{d}{dt}\rho_{22}=&-i(\Omega_{12}^*\tilde{\rho}_{12}-\Omega_{12}\tilde{\rho}_{21})-2\gamma_3(\bar{n}_0+1)\rho_{22}\\
&-(\bar{n}_0+1)(\gamma_{2113}\rho_{32}+\gamma_{3113}\rho_{23})\\
&\rho_{33}+\rho_{22}+\rho_{11}=1
\end{aligned}
\tag{14.64}
$$

where $\Delta=\omega_L-(\omega_2-\omega_1)$, $\omega_{32}=\omega_3-\omega_2$. Hence, we can discuss the influence of relative directions between the atomic dipole moments $\mathbf{d}_{\ell\sigma}$ and $\mathbf{d}_{mn}$ on the steady-state behavior of the atom, and realize whether the coherent trapping of the atomic populations occurs in the two upper atomic states.

14.2.2 Steady-state behavior and the coherent trapping of the atomic populations

For simplicity, we only consider the resonant case ($\Delta=0$, $\omega_{32}=0$), and restrict ourselves to the cases in which both atomic dipole $\mathbf{d}_{21}$ and $\mathbf{d}_{31}$ are perpendicular or parallel, and the moduli of $\mathbf{d}_{21}$ and $\mathbf{d}_{31}$ are equal. For the case $\mathbf{d}_{21}\cdot\mathbf{d}_{31}=0$, we choose them as follows

$$
\mathbf{d}_{31}=|\mathbf{d}_{31}|\mathbf{e}_x,\quad \mathbf{d}_{21}=|\mathbf{d}_{21}|\mathbf{e}_y \tag{14.65}
$$

When $\mathbf{d}_{31}$ is parallel to $\mathbf{d}_{21}$, we may write them in the form

$$
\mathbf{d}_{31}=\mathbf{d}_{21}=|\mathbf{d}_{31}|\mathbf{e}_x \tag{14.66}
$$

For these two cases, $\gamma_{\ell\sigma mn}$ may be simplified as

$$
\gamma_2=\gamma_3=\gamma_{2112}=\gamma_{3113}=\gamma,\ \gamma_{3112}=\gamma_{2113}=\gamma_{2131}=\gamma_{3121}=\chi \tag{14.67}
$$

Here γ and χ are the real constants. The polarization unit vector of the laser field is chosen as $\varepsilon=\frac{1}{\sqrt{2}}(\mathbf{e}_x+\mathbf{e}_y)$. So the parameters Ω_{1i} (i=2,3) relating to the atomic Rabi oscillating frequency satisfy

$$\Omega_{12} = \Omega_{13} = \Omega_{13}^* = \Omega_{12}^* = \Omega \tag{14.68}$$

In this case, the steady-state equations for the atomic density elements are

$$\begin{aligned}
&i\Omega(\rho_{22} - \rho_{11}) + i\Omega\rho_{23} - \gamma(3\overline{n}_0 + 1)\tilde{\rho}_{21} - \chi(\overline{n}_0 + 1)\tilde{\rho}_{31} = 0 \\
&i\Omega(\rho_{33} - \rho_{11}) + i\Omega\rho_{32} - \gamma(3\overline{n}_0 + 1)\tilde{\rho}_{31} - \chi(\overline{n}_0 + 1)\tilde{\rho}_{21} = 0 \\
&i\Omega\tilde{\rho}_{31} - i\Omega\tilde{\rho}_{12} - 2\gamma(\overline{n}_0 + 1)\rho_{23} - \chi(\overline{n}_0 + 1)(\rho_{22} + \rho_{33}) \\
&+2\chi\overline{n}_0\rho_{11} = 0 \\
&-i\Omega(\tilde{\rho}_{13} - \tilde{\rho}_{31}) - 2\gamma(\overline{n}_0 + 1)\rho_{33} - \chi(\overline{n}_0 + 1)(\rho_{32} + \rho_{23}) = 0 \\
&-i\Omega(\tilde{\rho}_{12} - \tilde{\rho}_{21}) - 2\gamma(\overline{n}_0 + 1)\rho_{22} - \chi(\overline{n}_0 + 1)(\rho_{32} + \rho_{23}) = 0 \\
&\rho_{33} + \rho_{22} + \rho_{11} = 1
\end{aligned} \tag{14.69}$$

(1). When $\mathbf{d}_{31} \cdot \mathbf{d}_{21} = 0$

First we study the steady-state populations $\rho_{ii}^{st}(i = 1, 2, 3)$ when the atomic dipole moments $\mathbf{d}_{21}$ is perpendicular to $\mathbf{d}_{31}$. In this case, $\chi = 0$. Starting from eq.(14.69) and after lengthy but straightforward calculations, we may obtain the steady-state populations ρ_{ii}^{st} as

$$\rho_{22}^{st} = \rho_{33}^{st} = \frac{\Omega^2}{4\Omega^2 + \gamma^2(3\overline{n}_0 + 1)(\overline{n}_0 + 1)} \tag{14.70}$$

$$\rho_{11}^{st} = \frac{2\Omega^2 + \gamma^2(\overline{n}_0 + 1)(3\overline{n}_0 + 1)}{4\Omega^2 + \gamma(3\overline{n}_0 + 1)(\overline{n}_0 + 1)} \tag{14.71}$$

Evidently, $\rho_{ii}^{st}(i = 1, 2, 3)$ are obviously dependent on the parameters $\overline{n}_0$ which represents the feature of the heat bath. If the bath is a normal vacuum reservoir, i.e., $\overline{n}_0 = 0$, then

$$\rho_{22}^{st} = \rho_{33}^{st} = \frac{\Omega^2}{4\Omega^2 + \gamma^2}, \quad \rho_{11}^{st} = \frac{2\Omega^2 + \gamma^2}{4\Omega^2 + \gamma^2} \tag{14.72}$$

and the atomic inversion gives

$$\rho_{22}^{st} + \rho_{33}^{st} - \rho_{11}^{st} = -\frac{\gamma^2}{4\Omega^2 + \gamma^2} \tag{14.73}$$

It is easily seen that (14.73) is identified with the steady-state inversion (described by (8.103)) of a two-level atom driven by a laser field in a vacuum bath.

That is to say, when the atomic dipole moment $\mathbf{d}_{21}$ is perpendicular to $\mathbf{d}_{31}$, the steady-state behavior of the V-type three-level atom driven by a classical laser field in a heat bath is similar to that of a two-level atom system. In this case, the coherent trapping of atomic populations in the two upper atomic states can not occur.

(2) For the case of $\mathbf{d}_{21} = \mathbf{d}_{31}$

Second, we discuss the case for which the atomic dipole moments $\mathbf{d}_{21} = \mathbf{d}_{31}$. In this condition $\chi = \gamma$, which means that there exists strong interference between $\mathbf{d}_{21}$ and $\mathbf{d}_{31}$. From (14.69), the steady-state populations ρ_{ii}^{st} are written in the form

$$\rho_{33}^{st} = \rho_{22}^{st} = \frac{1}{2}, \qquad \rho_{11}^{st} = 0 \tag{14.74}$$

The above equations show that the V-configuration three-level atom driven by a laser field can exhibit the atomic coherent population trapping in its two upper states even under the damping of the heat bath. And this atomic population coherent trapping phenomenon is not related to the characteristics of the bath because $\overline{n}_0$ does not appear in (14.74). This means that for a fixed polarization of the driven laser field, the relative direction of the atomic dipole moments $\mathbf{d}_{21}$ and $\mathbf{d}_{31}$ plays an important role in the steady-state population distribution of the atom. So the strong interference between $\mathbf{d}_{31}$ and $\mathbf{d}_{21}$ brings about this atomic population coherent trapping phenomenon.

In order to realize this phenomenon in depth, we inspect the steady-state off-diagonal elements of the atomic density matrix. Substituting (14.74) into (14.69), we have

$$\rho_{21}^{st} = \rho_{31}^{st} = 0, \quad \rho_{23}^{st} = \rho_{32}^{st} = -\frac{1}{2} \tag{14.75}$$

From (14.74) and (14.75), we can see that in the steady-state limit, the atom driven by a laser field in the heat bath evolves into the pure state

$$|\Psi_A(\infty)\rangle = \exp(i\xi)\frac{(|3\rangle - |2\rangle)}{\sqrt{2}} \tag{14.76}$$

It is evident that $|\Psi(\infty)\rangle$ is just the population coherent trapping state (shown in (14.6) and (14.31)) of the V-configuration three-level atom. In this state, there is a strong cancellation between the atomic transitions $|2\rangle \rightarrow |1\rangle$ and $|3\rangle \rightarrow |1\rangle$. This cancellation induces that the interactions between the atom and the radiation fields (laser field + bath field) are decoupled, and the atom is trapped in its two upper states.

So we find when the atomic dipole moments $\mathbf{d}_{21} = \mathbf{d}_{31}$, the atomic population coherent trapping of the V-type three-level atom in its upper state can occur even if there exists the damping of the heat bath. And it is just the damping of the heat bath that induces the atom to evolve into its population coherent trapping state in the long time limit.

References

1. P.M.Radmore and P.L.Knight, *J.Phys.* **B15** (1982) 561.
2. S.Swain, *J.Phys.* **B15** (1982) 3045.
3. D.A.Cardimona, M.G.Raymer and C.R.Stroud Jr., *J.Phys.* **B 15** (1982) 55.
4. H.I.Yoo and J.H.Eberly, *Phys.Rep.* **118** (1985) 232.
5. B.J.Dalton, R.McDuff and P.L.Knight, *Opt.Acta* **32** (1985) 61.
6. D.A.Cardimona, *Phys.Rev.* **A41** (1990) 5016.
7. Jin-sheng Peng, Gao-xiang Li and Peng Zhou, *Phys.Rev.* **A46** (1992) 5016.
8. Gao-xiang Li, Jin-sheng Peng and Peng Zhou, *Chin.J.Laser* **1** (1992) 221.
9. G.C.Hegerfeldt and M.B.Plenio, *Phys.Rev.* **A47** (1992) 2186.
10. B.Deb, G.Gangopadhyay and D.S.Ray, *Phys.Rev.* **A48** (1993) 1400.
11. Gao-xiang Li and Jin-sheng Peng, *Phys.Lett.* **A189** (1994) 449.
12. Gao-xiang Li and Jin-sheng Peng, *ACTA Phys.Sin.* **44** (1995) 700.
13. Gao-xiang Li and Jin-sheng Peng, *Phys.Rev.* **A52** (1995) 465.
14. Gao-xiang Li and Jin-sheng Peng, *ACTA Phys.Sin.* **46** (1996) 37.
15. Gao-xiang Li and Jin-sheng Peng, *Opt.Commun.* **123** (1996) 94.
16. Gao-xiang Li and Jin-sheng Peng, *Phys.Lett.* **A218** (1996) 41.
17. Gao-xiang Li and Jin-sheng Peng, *Z.Phys.* **B102** (1997) 223.
18. Gap-xiang Li and Jin-sheng Peng, *Opt. Commun.* **138** (1997) 59.

CHAPTER 15

QUANTUM CHARACTERISTICS OF A TWO-ATOM SYSTEM UNDER THE INTERACTION OF THE RADIATION FIELD

Up to now, we have treated many problems in the system of a single atom interacting with the radiation field. In treating the superfluorescence system which including a number of atoms, we also assumed that the relative distance between any two atoms is so far that the dipole-dipole interaction between atoms can be ignored. However, if the relative distance between atoms in some dense-atom systems is not large, then the dipole-dipole interaction between atoms can not be neglected. The question arisen here is what are the influences of the dipole-dipole interaction among the atoms on the atomic behaviors in the atom-field coupling system. The present chapter is devoted to discussing this subject. Since the characteristics in a two-atom system under the interaction of a radiation field can be generalized into the multi-atom system, here we focus our attention on the two-atom system. In Section 15.1 we deduce the Hamiltonian of the system of two two-level atoms with the dipole-dipole interaction under the interaction of a radiation field. Then in Section 15.2, we treat the atomic inversion and the atomic dipole squeezing in a two-atom system under the interaction of a weak field. And Section 15.3 is devoted to revealing the effects of the dipole-dipole interaction among the atoms on the collapse-and-revival of atomic populations and the atomic population coherent trapping in the system of two atoms interacting with a strong coherent field.

15.1 Hamiltonian of a two-atom system with the dipole-dipole interaction

In order to simplify our calculations, we suppose that both atoms have respective two levels. In general sense, the dipole-dipole interaction between the atoms results from two parts, one is the electric dipole-dipole interaction and the other is the dipole-dipole interaction induced by the fluctuations of the vacuum field. We first discuss the electric dipole-dipole interaction.

15.1.1 Hamiltonian of the electric dipole-dipole interaction between two atoms

As we know in the classical electrodynamics, two one-electron atoms or hydrogen-like atoms can be regarded as two electric dipoles as shown in Figure (15.1). The kernels of the two atoms are assumed to remain motionless at

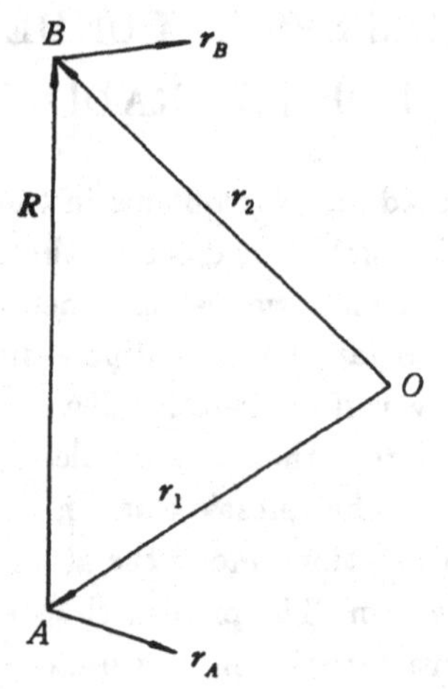

Figure 15.1: Schematic diagram of two atoms system including dipole-dipole interaction

points A and B, respectively. $\mathbf{r}_A$ is the position vector of the electron attached to atom (A) with respect to point A, and $\mathbf{r}_B$ the position vector of the electron attached to atom (B) with respect to B. $\mathbf{R} = \mathbf{r}_2 - \mathbf{r}_1$ is the relative distance vector. Correspondingly, the electric dipole moments of the two atoms are, respectively, as: $\mathbf{d}_1 = e\mathbf{r}_1$, $\mathbf{d}_2 = e\mathbf{r}_B$. For the case of $|\mathbf{d}_1| = |\mathbf{d}_2| = d$, the electrostatic interaction energy between the two atoms is written in the familiar form

$$V_1 = \mathbf{d}_1 \cdot \mathbf{d}_2 \frac{(1 - 3\cos^2\theta)}{R^3} \tag{15.1}$$

where θ is the relative angle between $\mathbf{d}_1$ (or $\mathbf{d}_2$) and $\mathbf{R}$. By means of the result obtained in Chapter 2, the electric dipole moment $\mathbf{d}_i$ of a two-level atom can be expressed by the pseudo-spin operators $S_+^{(i)}$ and $S_-^{(i)}$ as follows

$$\mathbf{d}_i = d(S_+^{(i)} + S_-^{(i)})\mathbf{e}_d \tag{15.2}$$

here $\mathbf{e}_d$ is the unit vector of the electric dipole moment $\mathbf{d}$. Substituting eq.(15.2) into (15.1), V_1 becomes

$$V_1 = \frac{d^2}{R^3}(1 - 3\cos^2\theta)(S_+^{(1)}S_+^{2)} + S_-^{(1)}S_-^{(2)} + S_+^{(1)}S_-^{(2)} + S_-^{(1)}S_+^{(2)}) \tag{15.3}$$

Here $S_+^{(1)}S_-^{(2)}$ corresponds to the excitation of the atom A from its ground state $|-\rangle$ to its excited state $|+\rangle$ and in the meantime the deexcitation of the atom B from its excited state $|+\rangle$ to its ground state $|-\rangle$. Conversely, $S_-^{(1)}S_+^{(2)}$ describes the deexcitation of the atom A from $|+\rangle$ to $|-\rangle$ and the excitation of the atom B from $|+\rangle$ to $|-\rangle$ simultaneously. Clearly, the varying values of the energy in these two transition processes are very small. However, $S_+^{(1)}S_+^{(2)}$ ($S_-^{(1)}S_-^{(2)}$) represents the transition processes of both atoms excited (deexcited) from $|-\rangle$ ($|+\rangle$) to $|+\rangle$ ($|-\rangle$), and in these two processes, the varying values of the energy are large. In the analogy with the rotating-wave-approximation, the terms $S_+^{(1)}S_+^{(2)}$ and $S_-^{(1)}S_-^{(2)}$ can be dropped in the interaction Hamiltonian of the electric dipole-dipole interaction between the two two-level atoms. Consequently, (15.3) takes the form

$$V_1 = \Omega_1(S_+^{(1)}S_-^{(2)} + S_-^{(1)}S_+^{(2)}) \tag{15.4}$$

here

$$\Omega_1 = \frac{d^2}{R^3}(1 - 3\cos^2\theta) \tag{15.5}$$

Eq.(15.4) is exactly the Hamiltonian of the electric dipole-dipole interaction between the two two-level atoms.

15.1.2 Hamiltonian of a two-atom system with the dipole-dipole interaction induced by the fluctuations of the vacuum field

In Chapter 11 we have shown that under the interaction of the vacuum field, a single atom in its ground state can continuously radiate and reabsorb the virtual photons. For the many-atom system under the interaction of the vacuum field, there exist the exchanges of virtual photons among the atoms, which induce the dipole-dipole interaction different from the electric dipole-diploe interaction.

Here we also discuss the two two-level atoms system and assume these two atoms have equal transition frequency ω. Under the interaction of the vacuum field, the Hamiltonian of the two-atom system including the electric dipole-dipole interaction is

$$\begin{aligned}H = &\omega_0 \sum_{\ell=1}^{2} S_3^{(\ell)} + \sum_{kj} \omega_{kj} a_{kj}^{\dagger} a_{kj} + \Omega_1 (S_+^{(1)} S_-^{(2)} + S_-^{(1)} S_+^{(2)}) \\ &+ \sum_{\ell=1}^{2} \sum_{kj} \varepsilon_{kj} [a_{kj} \exp(i\mathbf{k}\cdot\mathbf{r})(S_+^{(\ell)} + S_-^{(\ell)}) \\ &+ a_{kj}^{\dagger} \exp(-i\mathbf{k}\cdot\mathbf{r})(S_+^{(\ell)} + S_-^{(\ell)})] \quad (\hbar = 1)\end{aligned} \tag{15.6}$$

where $\sum_{kj} \omega_{kj} a_{kj}^{\dagger} a_{kj}$ represents the Hamiltonian of the free field, a_{kj} is the annihilation operator for the photons with wave vector $\mathbf{k}$ and unit polarization vector $\mathbf{e}_{kj}$. ε_{kj} is the atom-field coupling constant and from Chapter 4 we know it is defined as

$$\varepsilon_{kj} = \left(\frac{2\pi\omega_0^2}{V\omega_{kj}} \right)^{1/2} d\mathbf{e}_{kj} \cdot \mathbf{e}_d \tag{15.7}$$

Employing the Heisenberg equation, we obtain

$$i\frac{d}{dt} a_{kj} = [a_{kj}, H] = \omega_{kj} a_{kj} + \sum_{\ell=1}^{2} \varepsilon_{kj} \exp(-i\mathbf{k}\cdot\mathbf{r}_\ell)(S_+^{(\ell)} + S_-^{(\ell)}) \tag{15.8}$$

$$\begin{aligned}i\frac{d}{dt} S_-^{(\ell)} = &\omega_0 S_-^{(\ell)} - 2\Omega_1 S_3^{(\ell)} \sum_{\ell'} S_-^{(\ell')} (1 - \delta_{\ell,\ell'}) \\ &- 2\sum_{kj} \varepsilon_{kj} S_3^{(\ell)} [a_{kj} \exp(i\mathbf{k}\cdot\mathbf{r}_\ell) + a_{kj}^{\dagger} \exp(-i\mathbf{k}\cdot\mathbf{r}_\ell)]\end{aligned} \tag{15.9}$$

The second term in (15.9) corresponds to the effect of the electric dipole-dipole interaction on the atomic operator $S_-^{(\ell)}$ for the ℓ-th atom. The third term represents the influence of the vacuum field on $S_-^{(\ell)}$. In order to discuss this influence, we must solve $a_{kj}(t)$ and its conjugation operator $a_{kj}^{\dagger}(t)$. Integrating eq.(15.8) formally, we have

$$a_{kj}(t) = a_{kj}(0) \exp(-i\omega_{kj} t)$$

$$-i\sum_{kj}\varepsilon_{kj}e^{-i\mathbf{k}\cdot\mathbf{r}}\int_0^t \exp[-i\omega_{kj}(t-t')][S_-^{(\ell)}(t')+S_+^{(\ell)}(t')]dt' \tag{15.10}$$

Inasmuch as the dipole-dipole interaction energy is sufficiently smaller than ω_0, $S_-^{(\ell)}(t')$ and $S_+^{(\ell)}(t')$ in (15.10) can be written approximately as

$$S_-^{(\ell)}(t') = S_-^{(\ell)}(t)\exp[i\omega_0(t-t')], \quad S_+^{(\ell)}(t') = (S_+^{(\ell)}(t'))^\dagger \tag{15.11}$$

Inserting (15.11) into (15.10), we have

$$\begin{aligned}
a_{kj}(t) &= a_{kj}(0)\exp(-i\omega_{kj}t) \\
&-i\sum_{\ell=1}^{2}\varepsilon_{kj}\exp(-i\mathbf{k}\cdot\mathbf{r}_\ell)\left\{S_-^{(\ell)}(t)\int_0^t \exp[i(\omega_0-\omega_{kj})(t-t')]\right. \\
&\left.+S_+^{(\ell)}(t)\int_0^t \exp[-i(\omega_0+\omega_k)(t-t')]\right\}dt' \\
&= a_{kj}(0)\exp(-i\omega_{kj}t) + \sum_{\ell=1}^{2}\varepsilon_{kj}\exp(-i\mathbf{k}\cdot\mathbf{r}_\ell) \\
&\times\left\{S_-^{(\ell)}(t)\frac{1}{\omega_0-\omega_{kj}} - S_+^{(\ell)}(t)\frac{1}{\omega_0+\omega_{kj}}\right\}
\end{aligned} \tag{15.12}$$

By the substitution of the above equation and its adjoint into (15.9), we find

$$\begin{aligned}
i\frac{d}{dt}S_-^{(\ell)} &= \omega_0 S_-^{(\ell)} - 2\Omega_1 S_3^{(\ell)}\sum_{\ell'}S_-^{(\ell')}(1-\delta_{\ell,\ell'}) \\
&-2\sum_{kj}\varepsilon_{kj}^2 S_3^{(\ell)}[a_{kj}(0)\exp(i\mathbf{k}\cdot\mathbf{r}_\ell)\exp(-i\omega_{kj}t)+\text{h.c.}] \\
&-2\sum_{\ell'}\varepsilon_{kj}^2 S_3^{\ell}\left\{\exp(i\mathbf{k}\cdot\mathbf{R}_{\ell\ell'})\left[S_-^{(\ell')}(t)\frac{1}{\omega_0-\omega_{kj}}\right.\right. \\
&\left.\left.-S_+^{(\ell')}(t)\frac{1}{\omega_0+\omega_{kj}}\right]+\text{h.c.}\right\}
\end{aligned} \tag{15.13}$$

The first term on the right-hand side of (15.13) describes the free oscillation of $S_-^{(\ell)}$ in the absence of the vacuum field interaction. The second term represents the effect of the ℓ'-th atom ($\ell' \neq \ell$) on the operator $S_-^{(\ell)}$ by the electric dipole-dipole interaction. The third term corresponds to the contribution of the vacuum field to $S_-^{(\ell)}$. The term $\ell \neq \ell'$ in the last term reflects the reaction interaction of the ℓ-th atom itself by emitting and reabsorbing of the virtual photons due to the effect of the vacuum field, which is a source of the Lamb shift of the ℓ-th atomic ground level. The term $S_3^{(\ell)} S_+^{(\ell')}$ $(\ell \neq \ell')$ in the last term is a fast varying one corresponding to the two-photon transition processes, so this term can be neglected. And the formalism of the term $S_3^{(\ell)} S_-^{(\ell')}$ $(\ell \neq \ell')$ is similar to the second term which describes the effect of the electric dipole-dipole interaction, so it represents the interaction effect resulting from the exchanges of virtual photons between the ℓ'-th atom and the ℓ-th atom. In analogy to V_1, this interaction Hamiltonian can be written as

$$V_2 = \Omega_2 (S_+^{(1)} S_-^{(2)} + S_-^{(1)} S_+^{(2)}) \tag{15.14}$$

where

$$\Omega_2 = \sum_{kj} \varepsilon_{kj}^2 \left\{ \exp(i\mathbf{k} \cdot \mathbf{R}_{\ell\ell'}) \frac{1}{\omega_0 - \omega_{kj}} - \exp(-i\mathbf{k} \cdot \mathbf{R}_{\ell\ell'}) \frac{1}{\omega_0 + \omega_{kj}} \right\} \tag{15.15}$$

As yet we have found that V_2 describes the dipole-dipole interaction in the two two-level atoms system due to the effects of the vacuum field, this dipole-dipole interaction results from the exchange of virtual photons between two atoms. Consequently, we also call this interaction between two atoms as the radiated dipole-dipole interaction.

The above results show that the dipole-dipole interaction between two atoms includes two parts, one is the electric dipole-dipole interaction, the other is the radiated dipole-dipole interaction. So the total dipole-dipole interaction can be characterized by

$$V = V_1 + V_2 = (\Omega_1 + \Omega_2)(S_+^{(1)} S_-^{(2)} + S_-^{(1)} S_+^{(+)}) \tag{15.16}$$

What is the relation between the relative intensities of these two kinds of the dipole-dipole interactions ? In other words, for Ω_1 and Ω_2 which one is larger ? We now turn to discuss it in the following.

First we simplify Ω_2 (15.15). Substituting (15.7) into (15.15), we have

$$\Omega_2 = \sum_k \frac{2\pi\omega_0^2}{V\omega_k} d^2 \left\{ \exp(i\mathbf{k} \cdot \mathbf{R}_{\ell\ell'}) \frac{1}{\omega_0 - \omega_{kj}} - \exp(-i\mathbf{k} \cdot \mathbf{R}_{\ell\ell'}) \frac{1}{\omega_0 + \omega_{kj}} \right\} \sum_{j=1}^{2} (\mathbf{e}_{kj} \cdot \mathbf{e}_d)^2 \tag{15.17}$$

Employing the summation rule of the unit polarization vectors of photons

$$\sum_{j=1}(\mathbf{e}_{kj}\cdot\mathbf{e}_d)^2=(\mathbf{e}_{kd}^{(1)})^2+(\mathbf{e}_{kd}^{(2)})^2=1-(\mathbf{k}_d)^2 \tag{15.18}$$

where $\mathbf{k}_d$ is the projection of the unit wave vector $\mathbf{k}$ on the direction of $\mathbf{e}_{kj}$. Noticing in the continuum limit, the summation $\frac{1}{V}\sum_k$ in (15.17) can be replaced by the integral $(2\pi)^{-3}\int d^3\mathbf{k}$, i.e.,

$$\frac{1}{V}\sum_k\Longrightarrow\frac{1}{(2\pi)^3}\int d^3\mathbf{k} \tag{15.19}$$

After substituting (15.18) and (15.19) into (15.17), Ω_2 becomes

$$\begin{aligned}\Omega_2=&\int d^3\mathbf{k}\frac{d^2\omega_0^2}{4\pi^2\omega_k}[1-(\mathbf{k}_d)^2]\frac{1}{\omega_0^2-\omega_k^2}\{\omega_0[\exp(i\mathbf{k}\cdot\mathbf{R}_{\ell\ell'})\\&-\exp(-i\mathbf{k}\cdot\mathbf{R}_{\ell\ell'})]+\omega_k[\exp(i\mathbf{k}\cdot\mathbf{R}_{\ell\ell'})+\exp(-i\mathbf{k}\cdot\mathbf{R}_{\ell\ell'})]\}\\=&\int dk\frac{d^2k_0^2k}{4\pi^2(k_0^2-k^2)}\int d\Omega[1-(\mathbf{k}_d)^2]\{k_0[\exp(i\mathbf{k}\cdot\mathbf{R}_{\ell\ell'})\\&-\exp(-i\mathbf{k}\cdot\mathbf{R}_{\ell\ell'})]+k[\exp(i\mathbf{k}\cdot\mathbf{R}_{\ell\ell'})+\exp(-i\mathbf{k}\cdot\mathbf{R}_{\ell\ell'})]\}\end{aligned} \tag{15.20}$$

Since

$$\begin{aligned}\frac{1}{4\pi}\int\exp(\pm i\mathbf{k}\cdot\mathbf{R}_{\ell\ell'})d\Omega=&\frac{1}{4\pi}\int_0^{2\pi}d\phi\exp(\pm ikR\cos\gamma)d(\cos\gamma)\\=&\frac{\sin(kR)}{kR}\end{aligned} \tag{15.21}$$

and

$$\frac{1}{k}\nabla_d\int\exp(\pm i\mathbf{k}\cdot\mathbf{R}_{\ell\ell'})d\Omega=\pm i\int\mathbf{k}_d\exp(\pm i\mathbf{k}\cdot\mathbf{R}_{\ell\ell})d\Omega \tag{15.22}$$

$$\begin{aligned}\frac{1}{4\pi}\int(\mathbf{k})_d^2\exp(\pm i\mathbf{k}\cdot\mathbf{R}_{\ell\ell'})d\Omega=&-\frac{1}{4\pi k^2}\nabla_d^2\int\exp(\pm i\mathbf{k}\cdot\mathbf{R}_{\ell\ell'})d\Omega\\=&-\frac{1}{k^2}\nabla_d^2\left[\frac{\sin(kR)}{kR}\right]\\=&-(1-3\cos^2\theta)\left(\frac{\cos(kR)}{k^2R^2}-\frac{\sin(kR)}{k^3R^3}\right)\\&+\cos^2\theta\frac{\sin(kR)}{kR}\end{aligned} \tag{15.23}$$

we see that

$$\frac{1}{4\pi}\int[[1-(\mathbf{k}_d)^2]\exp(\pm i\mathbf{k}\cdot\mathbf{R}_{\ell\ell'})d\Omega$$
$$=(1-\cos^2\theta)\frac{\sin(kR)}{kR}+(1-3\cos^2\theta)\left(\frac{\cos(kR)}{k^2R^2}-\frac{\sin(kR)}{k^3R^3}\right)\quad(15.24)$$

Inserting (15.24) into (15.20), Ω_2 is reduced as

$$\Omega_2=\frac{2d^2k_0^2}{\pi}\int_0^\infty dk\frac{k^2}{k_0^2-k^2}\left[(1-\cos^2\theta)\frac{\sin(kR)}{kR}\right.$$
$$\left.+(1-3\cos^2\theta)\left(\frac{\cos(kR)}{k^2R^2}-\frac{\sin(kR)}{k^3R^3}\right)\right]\quad(15.25)$$

The integral $\int_0^\infty \frac{k^2}{k_0^2-k^2}f(kR)dk$ in the above equation can be simplified by the residue theorem. For example, the integral

$$I=\int_0^\infty dk\frac{k\sin(kR)}{(k_0^2-k^2)R}=\frac{1}{2i}\int_{-\infty}^{\infty}\frac{ke^{ikR}}{(k_0^2-k^2)R}dk$$

can be calculated by an appropriate contour shown in Fig.(15.2), here C is a half circle with the radius $R\to\infty$. By the residue theorem, the integral

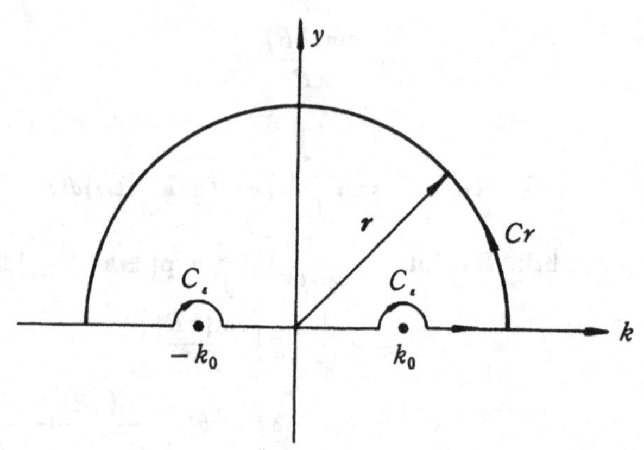

Figure 15.2: Diagram of the integration contour of I

becomes

$$I = \frac{\pi i}{2i}\frac{1}{R}\left[-\frac{1}{2}\exp(ik_0R) - \frac{1}{2}\exp(ik_0R)\right] = -\frac{\pi}{2R}\cos(k_0R)$$

Following the same method to calculation the last two integrals in equation (15.25), then

$$\Omega_2 = -d^2k_0^3\left[(1-\cos^2\theta)\frac{\cos(k_0R)}{k_0R}\right.$$
$$\left.-(1-3\cos^2\theta)\left(\frac{\sin(k_0R)}{k_0^2R^2}+\frac{\cos(k_0R)-1}{k_0^3R^3}\right)\right] \quad (15.26)$$

Therefore, the intensity of the dipole-dipole interaction between the two two-level atoms is

$$\Omega = \Omega_1 + \Omega_2 = d^2k_0^3\left\{(1-3\cos^2\theta)\left[\frac{\sin(k_0R)}{k_0^2R^2}+\frac{\cos(k_0R)}{k_0^3R^3}\right]\right.$$
$$\left.-(1-\cos^2\theta)\frac{\cos(k_0R)}{k_0^3R^3}\right\} \quad (15.27)$$

Comparing (15.5) with (15.26), we find if the relative distance between the two two-level atoms is so far that $k_0R \gg 1$, then $\Omega_1 = \Omega_2 = 0$ corresponding to the dipole-dipole interaction can be neglected, in this case, the two atoms are isolated. If the distance between the two atoms are not so far such as $k_0R = 3$, (15.27) is dominant by the terms including $(k_0R)^{-1}$, the ratio $\Omega_2/\Omega_1 \sim (k_0R)^2 = 9$, so the dominant contribution to the dipole-dipole interaction between the two atoms is the one induced by the vacuum field. When the two atoms close nearly such as $k_0R = 1$, $\Omega_2/\Omega_1 \sim 1$, the dipole-dipole interaction is determined by the summation of V_1 and V_2 . However, if the distance R is so small that $k_0R \ll 1$ (but the electron clouds of the two atoms do not overlap), thus $\Omega_2 \propto (k_0R)^{-2}$ and $\Omega_1 \propto (k_0R)^{-3}$, then the dipole-dipole interaction is governed by the electric one.

15.2 Quantum characteristics of the two-atom coupling system under the interaction of a weak field

Now we deal with the time evolution of the atomic inversion and the atomic dipole squeezing in the system of two two-level atoms interacting with a weak field, so as to analyze the effects of the dipole-dipole interaction between the atoms on the quantum characteristics of the system.

The Hamiltonian of the system of two two-level atoms interacting with a single-mode field is

$$H = \omega_0 \sum_{i=1}^{2} S_3^{(i)} + \omega a^\dagger a + \sum_{i=1}^{2} g(a^\dagger S_-^{(i)} + a S_+^{(i)}) + \Omega(S_+^{(1)} S_-^{(2)} + S_-^{(1)} S_+^{(2)}) \quad (15.28)$$

To simplify the calculations, we only discuss the resonant case, i.e., $\omega_0 = \omega$.

Supposing the two atoms initially in the superposition state of the excited state $|+,+\rangle$ and the ground state $|-,-\rangle$, i.e.,

$$|\Psi_A(0)\rangle = \cos(\theta/2)|-,-\rangle - \sin(\theta/2)\exp(i\phi)|+,+\rangle$$
$$(0 \leq \theta \leq \pi, \ -\pi \leq \phi < \pi) \quad (15.29)$$

and the single-mode field initially in the vacuum state $|0\rangle$, that is to say, the system is initially in the state as

$$|\Psi(0)\rangle = \cos(\theta/2)|-,-,0\rangle - \sin(\theta/2)\exp(i\phi)|+,+,0\rangle \quad (15.30)$$

With the time development, the state vector of the system in the interaction picture evolves to

$$\begin{aligned} |\Psi^I(t)\rangle = & \cos(\theta/2)|-,-,0\rangle - \sin(\theta/2)\exp(i\phi)[C_0(t)|+,+,0\rangle \\ & + C_1(t)|+,-,1\rangle + C_2(t)|-,+,1\rangle + C_3(t)|-,-,2\rangle] \end{aligned} \quad (15.31)$$

By means of the Schrödinger equation, we obtain

$$\begin{aligned} \frac{d}{dt}C_0(t) &= -ig(C_1(t) + C_2(t)) \\ \frac{d}{dt}C_1(t) &= -ig(\sqrt{2}C_3(t) + C_0(t)) - i\Omega C_2(t) \\ \frac{d}{dt}C_2(t) &= -ig(\sqrt{2}C_3(t) + C_0(t)) - i\Omega C_1(t) \\ \frac{d}{dt}C_3(t) &= -ig\sqrt{2}(C_1(t) + C_2(t)) \end{aligned} \quad (15.32)$$

Considering the initial condition (15.30), the solution of the above equation gives

$$C_0(t) = \frac{2g^2}{\sqrt{\Omega^2 + 24g^2}} \left[\frac{\exp(iat)}{a} - \frac{\exp(ibt)}{b} \right] + \frac{2}{3}$$

$$C_1(t)=C_2(t) = -\frac{g}{\sqrt{\Omega^2+24g^2}}[\exp(iat)-\exp(ibt)] \tag{15.33}$$
$$C_3(t)=\frac{2\sqrt{2}g^2}{\sqrt{\Omega^2+24g^2}}\left[\frac{\exp(iat)}{a}-\frac{\exp(ibt)}{b}\right]-\frac{\sqrt{2}}{3}$$

where

$$a=[-\Omega+\sqrt{\Omega^2+24g^2}]/2, \quad b=[-\Omega-\sqrt{\Omega^2+24g^2}]/2 \tag{15.34}$$

Starting from (15.31) and (15.33), we may discuss the atomic characteristics in the two-atom system including the dipole-dipole interaction as follows.

15.2.1 Time evolution of the atomic population inversion of a two-atom system

If both of the atoms are initially in its excited state $|+\rangle$, i.e., $\theta=\pi$, then the atomic population inversion of the two-atom system between the excited state $|+,+\rangle$ and $|-,-\rangle$ yields from eq.(15.31)

$$\begin{aligned}\langle S_3\rangle &= \langle\Psi^I(t)|S_3^{(1)}+S_3^{(2)}|\Psi^I(t)\rangle\\ &= \frac{1}{9}+\frac{8g^2}{3(\Omega^2+24g^2)}\sin^2\left(\frac{\sqrt{\Omega^2+24g^2}}{2}t\right)\\ &\quad -\frac{8}{9\sqrt{\Omega^2+24g^2}}[b\cos(at)-a\cos(bt)]\end{aligned} \tag{15.35}$$

The above equation shows that the atomic population inversion in a two-atom system is more complex than that in a single-atom system. When the relative distance of the two atoms is so far that the dipole-dipole interaction between the two atoms can be ignored (i.e., $\Omega=0$), $\langle S_3\rangle$ reduces to

$$\langle S_3\rangle = \frac{1}{9}+\frac{1}{9}\sin^2(\sqrt{6}gt)+\frac{8}{9}\cos^2(\sqrt{6}gt) \tag{15.36}$$

We find that in the system of two two-level atoms interacting with a single-mode field, $\langle S_3\rangle$ is composed of two series of cosine oscillations with different frequencies due to the atomic cooperative effect. We see from Fig.(15.3a) that $\langle S_3\rangle$ varies from 1 to -7/9 with the period time $\frac{2\pi}{\sqrt{6}g}$. This means that under the interaction of the vacuum field, the two atoms initially in the excited state $|+,+\rangle$ can not completely evolve into the ground state $|-,-\rangle$ due to the atomic cooperative effect. That is to say, the two atoms in their excited states can not

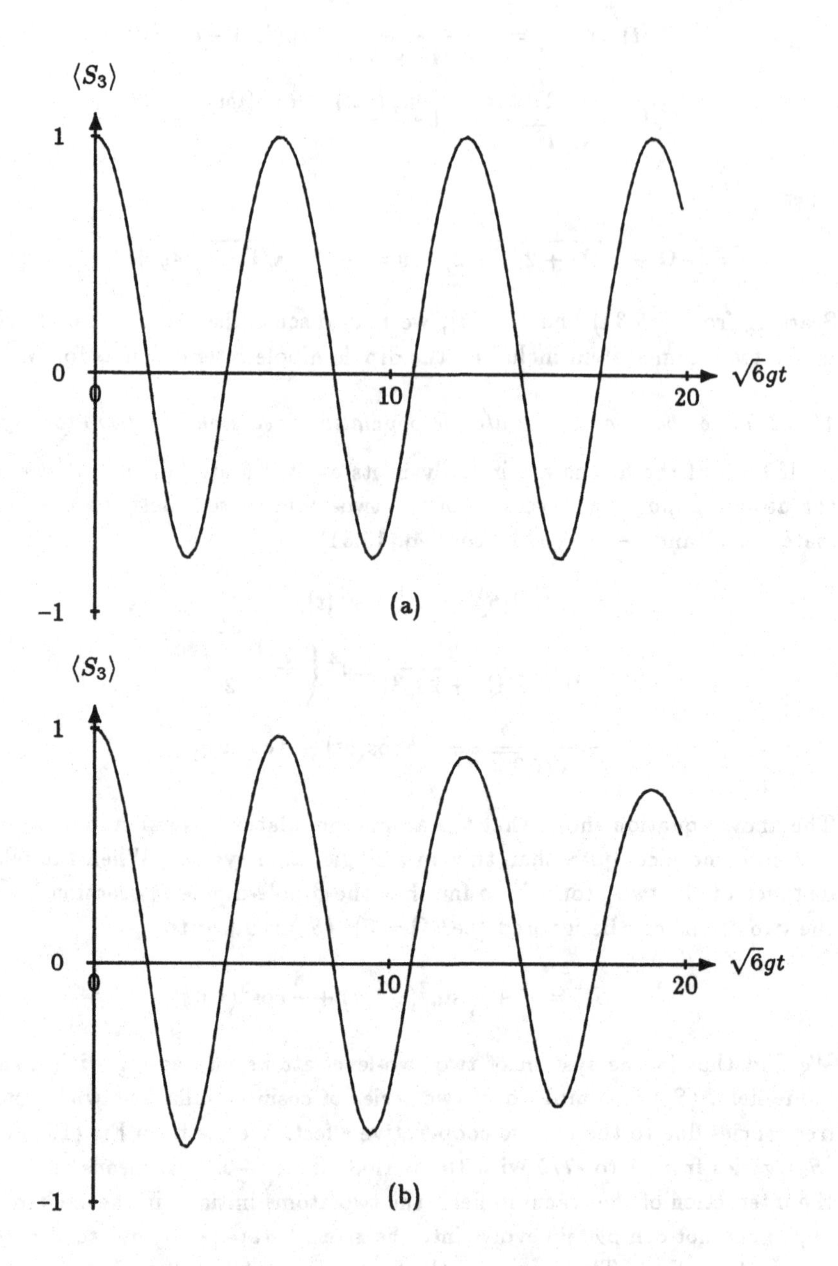

⟨S₃⟩
1
0
−1
0
10
20
√6gt
(a)
⟨S₃⟩
1
0
−1
0
10
20
√6gt
(b)

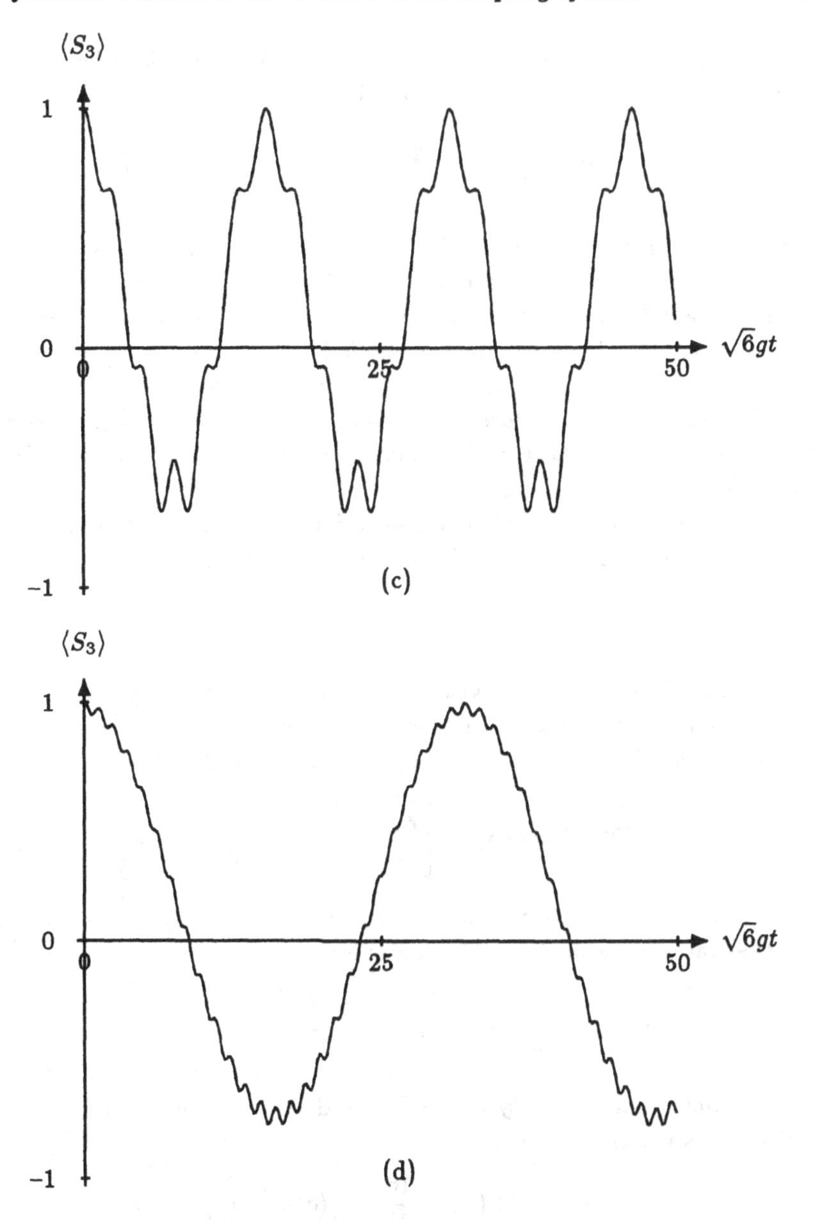

Figure 15.3: Time-dependence of $\langle S_3 \rangle$ for different Ω, here $\theta = \pi$, (a) $\Omega = 0$; (b) $\Omega = 0.2g$; (c) $\Omega = 5g$; (d) $\Omega = 12g$

simultaneously decay into their ground states. For example, when $t' = \frac{\pi}{\sqrt{6}g}$, according to (15.31) and (15.33) we obtain the state vector as

$$|\Psi^I(t')\rangle = \exp(i\phi)\left[\frac{2\sqrt{2}}{3}|-,-,2\rangle - \frac{1}{3}|+,+,0\rangle\right] \tag{15.37}$$

This result is different from that in Section 12.1. As we know when the time develops to the half Rabi oscillating period, a two-level atom initially in its excited state will decay into its ground state and simultaneously radiate one photon due to the influence of the vacuum field. But here, (15.37) indicates that with the time development, the two-atom system can partially remain in its excited state $|+,+\rangle$. Therefore, in the two-atom system under the interaction of the vacuum field, the atomic cooperative effect will partially inhibit the atomic decay from its excited state.

In order to understand this phenomenon in detail, we inspect the intensity of the field at time $t = t'$. From (15.37), the intensity of the field gives

$$\langle a^\dagger a\rangle^{(1)}(t') = \frac{16}{9} \tag{15.38}$$

We assume when the time develops to t', the photons radiated by the two atoms fly out of the cavity, furthermore the time t' set to be a new starting time point, that is to say, the system is in a new initial state

$$|\Psi(0)\rangle^{(1)} = \exp(i\phi)\left[\frac{2\sqrt{2}}{3}|-,-,0\rangle - \frac{1}{3}|+,+,0\rangle\right] \tag{15.39}$$

From (15.31) and (15.33), we know when the system evolves again to time t', then the state vector of the system becomes

$$|\Psi^I(t')\rangle^{(2)} = \exp(i\phi)\left[\frac{2\sqrt{2}}{3}|-,-,0\rangle + \frac{1}{3}(\frac{2\sqrt{2}}{3}|-,-,2\rangle - \frac{1}{3}|+,+,0\rangle)\right] \tag{15.40}$$

In this moment, the intensity of the field radiated by the atoms in the cavity reaches to its maximum value

$$\langle a^\dagger a\rangle^{(2)}(t') = \frac{16}{81} = \frac{1}{9}\langle a^\dagger a\rangle^{(1)}(t') \tag{15.41}$$

Eq.(15.41) shows that the two-atom system can repeatedly radiate photons due to the influence of vacuum fluctuations, and the intensity of the field is sufficiently smaller than that in the first radiation. This means the atomic cooperative effect inhibits the decay of the excited states, the system therefore exhibits many times damping radiations. This is also one physical cause of the ring phenomenon existed in superfluorescence.

Fig.(15.3.b) displays the case of the weak dipole-dipole interaction ($\Omega = 0.2g$). In comparison Fig.(15.3.b) with (15.3.a), we see that the oscillating frequencies of $\langle S_3 \rangle$ change slightly, but the oscillating amplitudes decrease gradually with the time development. This means the dipole-dipole interaction increases the inhibition ability for the decay of the two-atom excited energy. When $\Omega = 5g$ (shown in Fig.(15.3.c)), the oscillating frequencies of $\langle S_3 \rangle$ becomes obviously small due to the relative strong dipole-dipole interaction, and $\langle S_3 \rangle$ is difficult to evolve to the negative maximum value -7/9.

Fig.(15.3.d) describes the case of very strong dipole-dipole interaction ($\Omega = 12g$), i.e., $\Omega^2 \gg 24g^2$. Explicitly, $\langle S_3 \rangle$ exhibits the regular cosine oscillations with the amplitude varying from -7/9 to 1. In order to understand the time evolution of $\langle S_3 \rangle$ in this case, we give an analytical explanation. When $\Omega^2 \gg 24g^2$, a and b are approximated to be

$$a \approx \frac{6g^2}{\Omega}, \quad b \approx -\Omega \tag{15.42}$$

Inserting equation (15.42) into (15.31) and (15.35), the state vector of the system and the atomic population inversion are expressed as

$$\begin{aligned} |\Psi^I(t)\rangle = & -\frac{\exp(i\phi)}{3} \left\{ 2\sqrt{2}i \sin\left(\frac{3g^2}{\Omega}t\right) \exp\left(\frac{i3g^2}{\Omega}t\right) |-,-,2\rangle \right. \\ & \left. + \left[2\exp\left(\frac{i3g^2}{\Omega}t\right) \cos\left(\frac{3g^2}{\Omega}t\right) + 1 \right] |+,+,0\rangle \right\} \end{aligned} \tag{15.43}$$

$$\langle S_3 \rangle = \frac{1}{9} + \frac{8}{9} \cos^2\left(\frac{6g^2}{\Omega}t\right) \tag{15.44}$$

It is clear that (15.44) is coincident with Fig.(15.3.d). The cause leading to this phenomenon can be explained as follows. The two-atom system we discuss here can be regarded as a system with three symmetric states

$$\begin{aligned} E_3 &= \omega_0, & |3\rangle &= |+,+\rangle \\ E_2 &= \omega_0/2, & |2\rangle &= \frac{1}{\sqrt{2}}(|+,-\rangle + |-,+\rangle) \\ E_1 &= 0, & |1\rangle &= |-,-\rangle \end{aligned} \tag{15.45}$$

and one asymmetric state

$$E_0 = 0, \quad |0\rangle = \frac{1}{\sqrt{2}}(|+,-\rangle - |-,+\rangle)$$

here we have assumed that the eigenenergy of the ground state $|-,-\rangle$ is zero. Inasmuch as the state $|0\rangle$ is decoupled from the field, the two-atom system can be identified as a three-level atom. Because there displays the level shift of the state $|2\rangle$ resulting from the dipole-dipole interaction

$$\Omega(S_+^{(1)}S_-^{(2)} + S_-^{(1)}S_+^{(2)})|2\rangle = \Omega|2\rangle \tag{15.46}$$

the one-photon detuning is induced in the one-photon transition processes such as $|1\rangle \leftrightarrow |2\rangle$ and $|2\rangle \leftrightarrow |3\rangle$ in the atom-field coupling system. Therefore, when $\Omega^2 \gg 24g^2$, the two-photon processes play dominant role which leads to that the probability amplitude of the one-photon term $|+,-,1\rangle + |-,+,1\rangle$ is nearly equal to zero. Consequently, $\langle S_3\rangle$ exhibits the cosine oscillations with one frequency. Since the atomic cooperative effect and the atomic dipole-dipole interaction inhibit the decay of the excited state energy, then $\langle S_3\rangle$ can not decay to -1.

15.2.2 Influence of the dipole-dipole interaction on the squeezing of atomic operators

In order to discuss the influence of the dipole-dipole interaction on the squeezing of collective atomic operators in multi-atom system, we first introduce the definition of the atomic operator squeezing for the two-atom system.

A. Definition of the atomic operators squeezing for the two-atom system

As the squeezing definition of a single-mode field can be extended to the two-mode field in Section 7.2, here we generalize the squeezing definition of atomic operators for one-atom system to two-atom system in a similar way. Defining two Hermitian quadrature operators

$$S_1 = \frac{1}{2}(D_+ + D_-), \qquad S_2 = \frac{1}{2i}(D_+ - D_-) \tag{15.47}$$

where D_+ and D_- are the collective pseudo-spin operators for the two-atom system

$$D_+ = S_+^{(1)} + S_+^{(2)}, \quad D_- = S_-^{(1)} + S_-^{(2)}, \quad S_3 = S_3^{(1)} + S_3^{(2)} \tag{15.48}$$

S_1 and S_2 represent the two quadrature components of the atomic dipole moment of the two-atom system, they satisfy the commutation relation

$$[S_1, S_2] = iS_3 \tag{15.49}$$

Correspondingly, the Heisenberg uncertainty relation is

$$(\Delta S_1)^2(\Delta S_2)^2 \geq \frac{1}{4}\langle S_3\rangle^2 \tag{15.50}$$

If there exists a state $|\Psi\rangle$ in which the variance of $S_i (i{=}1$ or $2)$ satisfies

$$(\Delta S_i)^2 < \frac{1}{2}|\langle S_3\rangle| \qquad (i = 1 \text{ or } 2)$$

or

$$F_i = (\Delta S_i)^2 - \frac{1}{2}|\langle S_3\rangle| < 0 \qquad (i = 1 \text{ or } 2) \tag{15.51}$$

then the fluctuations of the operators S_i in the two-atom system are said to be squeezed. Meanwhile, the state $|\Psi\rangle$ is called the two-atom dipole squeezed state.

For the initial state (15.29) of the two-atom system, it is easy to obtain

$$\begin{aligned}
&\langle S_i\rangle = 0 \qquad (i = 1,\ 2) \\
&\langle S_3\rangle = \sin^2(\theta/2) - \cos^2(\theta/2) = -\cos(\theta) \\
&\langle S_1^2\rangle = \left\langle \Psi_A(0) \left| \frac{1}{2}(S_+^{(1)} + S_+^{(2)} + S_-^{(1)} + S_-^{(2)})^2 \right| \Psi_A(0) \right\rangle \\
&\qquad = \frac{1}{2}(1 - \sin\theta\cos\phi) \\
&\langle S_2^2\rangle = -\frac{1}{2}(1 + \sin\theta\cos\phi)
\end{aligned} \tag{15.52}$$

and the variances of the atomic operators S_1 and S_2 are, respectively,

$$(\Delta S_1)^2 = \langle S_1^2\rangle - \langle S_1\rangle^2 = \frac{1}{2}(1 - \sin\theta\cos\phi) \tag{15.53}$$

$$(\Delta S_2)^2 = \langle S_2^2\rangle - \langle S_2\rangle^2 = \frac{1}{2}(1 + \sin\theta\cos\phi) \tag{15.54}$$

From eqs.(15.52b), (15.53) and (15.54) we know when $\psi = -\pi$ or 0, the fluctuations of S_i obey

$$(\Delta S_1)^2(\Delta S_2)^2 = \frac{1}{4}\langle S_3\rangle^2 \tag{15.55}$$

which means when $\phi = -\pi$ or 0, the atomic state vector (15.29) is the minimum uncertainty state of the two-atom system.

Now we examine whether or not the two-atom system in the state $|\Psi_A(0)\rangle$ (eq. (15.29)) can exhibit the atomic dipole squeezing. By use of eqs.(15.52b), (15.53) and (15.54), we have

$$F_1 = \frac{1}{2}(1 - \sin\theta\cos\phi - |\cos\theta|) \tag{15.56}$$

$$F_2 = \frac{1}{2}(1 + \sin\theta\cos\phi - |\cos\theta|) \tag{15.57}$$

For $\phi = 0$, if $\theta \neq 0, \pi/2$ and π, then $F_1 < 0$ and $F_2 > 0$, which means that the fluctuations of the component S_1 can be squeezed in the two-atom system. But for $\phi = -\pi$ and $\theta \neq 0, \pi/2$ and π , $F_2 < 0$ and $F_1 > 0$, which corresponds to that the fluctuations of the component S_2 can be squeezed. From (15.56) and (15.57) we also find when $\theta = \pi/4$, $\phi = 0$ or $-\pi$, the fluctuations of the component S_1 or S_2 have the maximum reduction value $\frac{1}{4}(1 - \sqrt{2})$. In comparison with the value in one-atom system, the maximum reduction value here is larger than that in one-atom system (in one-atom system, the maximum reduction value of the fluctuations of the component S_i is -1/16). The reason leading to this difference is that the mechanisms resulting in the squeezing are different in these two systems. In one-atom system, $\langle S_i^2\rangle = 1/4$, the reduction value of the fluctuations of the component S_i depends on $\langle S_+ \pm S_-\rangle$ corresponding to one-atom transition processes, but in two-atom system, $\langle S_i\rangle = 0$, the reduction value depends on $\langle S_+^{(1)} S_-^{(2)} + S_-^{(1)} S_+^{(2)}\rangle$ corresponding to two-atom transition processes. So in these two systems, the different mechanisms leading to the dipole squeezing cause the different maximum reduction values.

B. Influence of the dipole-dipole interaction on the squeezing of atomic operators

Now we discuss the influence of the dipole-dipole interaction on the squeezing of atomic operators in the system of two two-level atoms interacting with the vacuum field. As in Chapter 13, we are concerned with the slowly varying amplitude of the fluctuations of atomic operators. So we define two slowly varying Hermitian quadrature operators in the two-atom system as

$$S_1 = \frac{1}{2}\sum_{i=1}^{2}(S_+^{(i)}\exp(-i\omega_0 t) + S_-^{(i)}\exp(i\omega_0 t))$$

$$S_2 = \frac{1}{2i}\sum_{i=1}^{2}(S_+^{(i)}\exp(-i\omega_0 t) - S_-^{(i)}\exp(i\omega_0 t)) \tag{15.58}$$

Evidently, they obey the commutation relation

$$[S_1, S_2] = i\sum_{i=1}^{2} S_3^{(i)} = iS_3$$

If

$$(\Delta S_j)^2 < \frac{1}{2}|\langle S_3\rangle| \qquad (j = 1 \; or \; 2)$$

or

$$Q_j = (\Delta S_j)^2 - \frac{1}{2}|\langle S_3\rangle| < 0 \qquad (j = 1 \; or \; 2) \tag{15.59}$$

then the fluctuations of the component S_i of the atomic operators are squeezed. By use of eqs.(15.31), (15.33) and (15.58) we have

$$\begin{aligned} Q_{\substack{1\\2}} &= \frac{1}{2} \mp \frac{1}{2}\sin\theta\left\{\frac{2}{3}\cos\phi - \frac{1}{3\sqrt{\Omega^2+24g^2}}[b\cos(at+\phi) - a\cos(bt+\phi)]\right\} \\ &+4\sin^2\frac{\theta}{2}\frac{g^2}{\Omega^2+24g^2}\sin^2\left(\frac{\sqrt{\Omega^2+24g^2 t}}{2}\right) - \frac{1}{2}\left|\sin^2\frac{\theta}{2}\left[\frac{1}{9}\right.\right. \\ &\left.+\frac{8g^2\sin^2\frac{\sqrt{\Omega^2+24g^2 t}}{2}}{3\Omega^2+24g^2} - \frac{8(b\cos(at) - a\cos(bt))}{9\sqrt{\Omega^2+24g^2}}\right] \\ &\left. - \cos^2\left(\frac{\theta}{2}\right)\right| \end{aligned} \tag{15.60}$$

If $\phi = 0$, then $Q_2 \geq 0$, which corresponding to that the fluctuations of the component S_2 can not be squeezed. And for $\phi = \pi$, $Q_1 \geq 0$, which means the squeezing of the fluctuations of S_1 can not appear. In order to understand whether the squeezing can exhibit in the component S_1 or not, we discuss the time evolution of Q_1 for $\phi = 0$ in the following.

If the interatomic coupling is ignored, i.e., $\Omega = 0$, then

$$\begin{aligned} Q_1 &= \frac{1}{2} - \frac{1}{2}\sin\theta\left(\frac{2}{3} + \frac{1}{3}\cos(\sqrt{6}gt) + \frac{1}{6}\sin^2(\theta/2)\right) + \frac{1}{6}\sin^2(\theta/2)\sin^2(\sqrt{6}gt) \\ &-\frac{1}{2}\left|\sin^2(\theta/2)\left(\frac{1}{9} + \frac{1}{9}\sin^2(\sqrt{6}gt) + \frac{8}{9}\cos(\sqrt{6}gt)\right) - \cos^2(\theta/2)\right| \end{aligned} \tag{15.61}$$

Fig.(15.4.a) is the diagrammatic sketch of (15.61). It shows that the value of Q_1 at $t_k = \frac{2k\pi}{\sqrt{6}g}(k = 1, 2, \cdots)$ is equal to that at $t = 0$. So if the two-atom system is initially in its optimal atomic squeezing state $|\Psi_A(0)\rangle$ ($\theta = \pi/4$, $\phi = 0$), then at $t = t_k$ the maximum negative value of Q_1 is also equal to $Q_{1M} = (1-\sqrt{2})/4$. That is to say, in these certain time points, the fluctuations of the dipole moment in the two-atom system can be reduced optimally and the time evolution of the reduction value is periodic. Fig.(15.4a) also indicates if the two atoms initially in the squeezed state, then they remain in a squeezing state with the time development. This is different from that in the one-atom system, in which the squeezing disappears at some time points. The cause leading to this difference is that the two-atom system under the interaction of the vacuum field can never evolve into the incoherent state such as $|-,-\rangle$ and $|+,+\rangle$. With the time development the system is always in the coherent superposition state, and the squeezing may appear stably.

When the dipole-dipole interaction is included, i.e., $\Omega \neq 0$, the time evolution of Q_1 is displayed as shown in Fig.(15.4b-d). We found that if Ω is not very large ($\Omega = 0.2g$ and 5g), with the time evolution the reduction value of the fluctuations in the component S_1 is smaller than that at t=0, which means that the dipole-dipole interaction decreases the squeezing degree when the dipole-dipole interaction is not intense. But when the dipole-dipole interaction is intense ($\Omega = 12g$), the maximum squeezing at time t=0 displays periodic. All these results indicate that the dipole-dipole interaction have a nonlinear influence on the squeezing of the atomic operators in the two-atom system.

15.3 Periodic collapses and revivals and the coherent population trapping in the two-atom system under the interaction of a coherent field

For the two-atom system described by the Hamiltonian (15.28), if we suppose the i-th atom initially in the coherent superposition of its excited state $|e\rangle$ and its ground state $|g\rangle$, i.e.,

$$|\Psi_A^{(i)}(0)\rangle = \cos\theta_i|+\rangle + \sin\theta_i \exp(-i\phi_i)|-\rangle \quad (i = 1, 2) \tag{15.62}$$

then the two two-level atoms are initially in the state

$$|\Psi_A(0)\rangle = |\Psi_{(1)}(0)\rangle \otimes |\Psi_{(2)}(0)\rangle = \cos\theta_1 \cos\theta_2|+,+\rangle$$

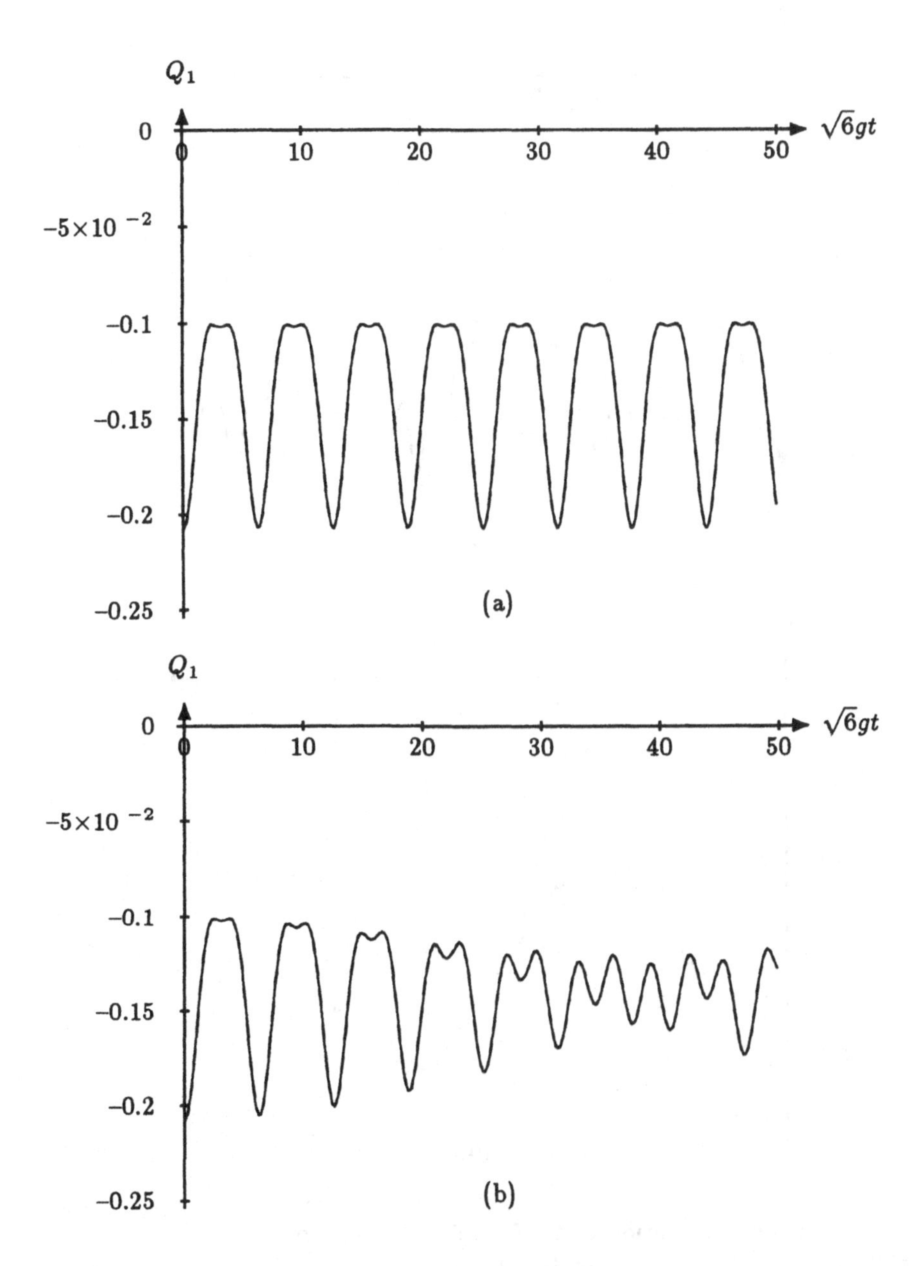

Q_1
0
-5×10^{-2}
−0.1
−0.15
−0.2
−0.25
0
10
20
30
40
50
$\sqrt{6}gt$
(a)
Q_1
0
-5×10^{-2}
−0.1
−0.15
−0.2
−0.25
0
10
20
30
40
50
$\sqrt{6}gt$
(b)

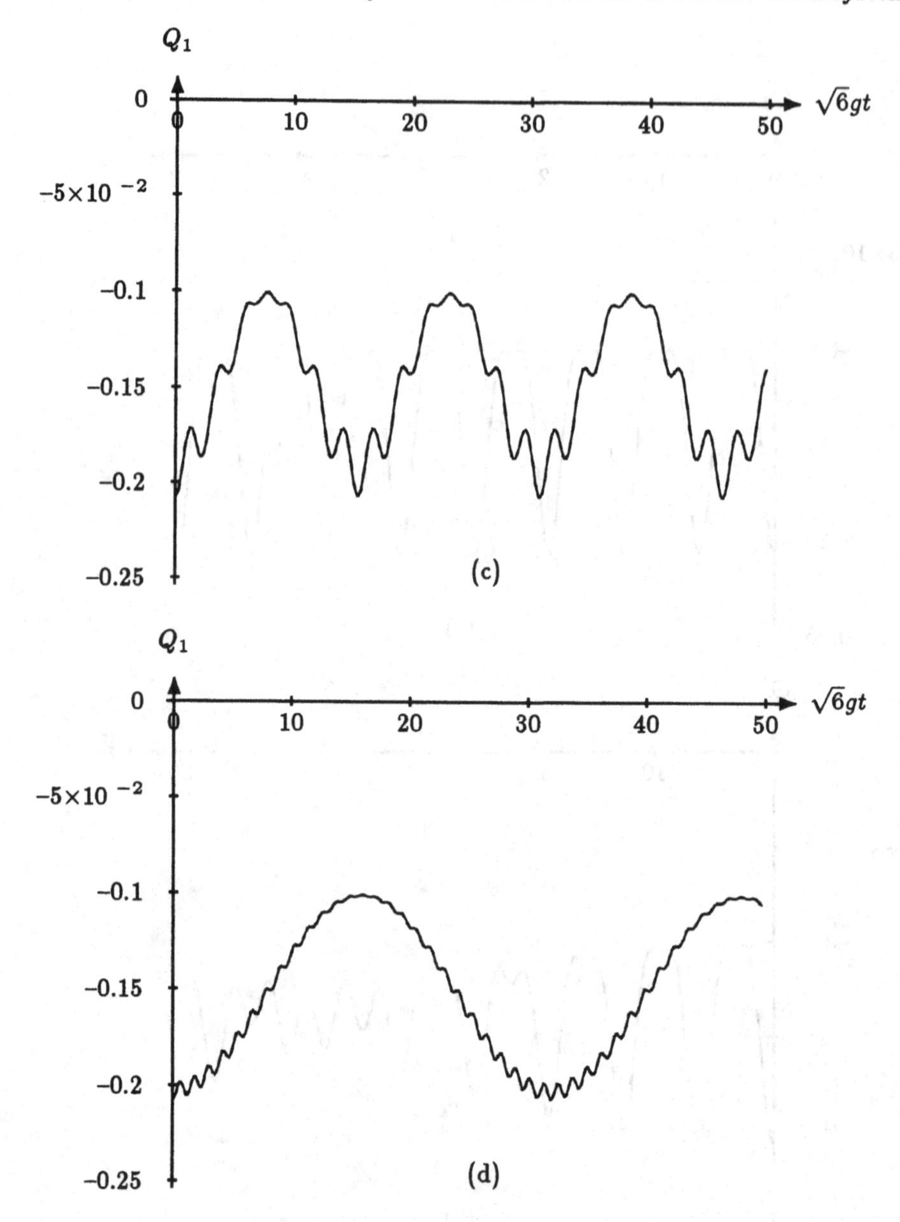

Figure 15.4: Time evolution of Q_1 for different Ω, here $\theta = \pi/4$, $\phi = 0$ (a) $\Omega = 0$; (b) $\Omega = 0.2g$; (c) $\Omega = 5g$; (d) $\Omega = 12g$

$$+\sin\theta_1\cos\theta_2\exp(-i\phi_1)|-,+\rangle+\sin\theta_2\cos\theta_1\exp(-i\phi_2)|+,-\rangle$$
$$+\sin\theta_1\sin\theta_2\exp[-i(\phi_1+\phi_2)]|-,-\rangle \tag{15.63}$$

furthermore, we also assume the initial field in a coherent state

$$|\Psi_F(0)\rangle=\sum_n F_n|n\rangle,\qquad F_n=\exp(-\overline{n}/2)\overline{n}^{n/2}\exp(in\xi)/\sqrt{n!} \tag{15.64}$$

here $\alpha=\sqrt{\overline{n}}\exp(i\xi)$. For simplicity, we let $\xi=0$. Correspondingly, the state vector of the total system at t=0 can be defined as

$$|\Psi_{AF}(0)\rangle=|\Psi_A(0)\rangle\otimes|\Psi_F(0)\rangle \tag{15.65}$$

Therefore, the state vector of this atom-field coupling system at time t can be described in the interaction picture by

$$|\Psi^I_{AF}(t)\rangle=\sum_n[C_1^n(t)|-,-,n\rangle+C_2^{n-1}(t)|+,-,n-1\rangle$$
$$+C_3^{n-1}(t)|-,+,n-1\rangle+C_4^{n-2}(t)|+,+,n-2\rangle] \tag{15.66}$$

Substituting (15.66) into the Schrödinger equation in the interaction picture, we obtain

$$\begin{aligned}
i\frac{d}{dt}C_1^n(t)&=DC_2^{n-1}(t)+DC_3^{n-1}(t)\\
i\frac{d}{dt}C_2^{n-1}(t)&=DC_1^n(t)+D'C_4^{n-2}(t)+\Omega C_3^{n-1}(t)\\
i\frac{d}{dt}C_3^{n-1}(t)&=DC_1^n(t)+D'C_4^{n-2}(t)+\Omega C_2^{n-1}(t)\\
i\frac{d}{dt}C_4^{n-2}(t)&=D'C_2^{n-1}(t)+D'C_3^{n-1}(t)
\end{aligned} \tag{15.67}$$

with

$$D=g\sqrt{n},\qquad D'=g\sqrt{n-1}$$

Considering the initial condition of the system as described by (15.65), the solution of (15.67) is

$$C_1^n(t)=B_1\exp(iat)+B_2\exp(ibt)-\frac{ADD'}{2(D^2+D'^2)} \tag{15.68}$$

$$C_2^{n-1}(t) = -\frac{1}{2D}[aB_1 \exp(iat) + bB_2 \exp(ibt)] + \frac{D_1}{2} \exp(i\Omega t) \tag{15.69}$$

$$C_3^{n-1}(t) = -\frac{1}{2D}[aB_2 \exp(iat) + bB_2 \exp(ibt)] - \frac{D_1}{2} \exp(i\Omega t) \tag{15.70}$$

$$C_4^{n-2}(t) = -\frac{D'}{D}[B_1 \exp(iat) + B_2 \exp(ibt)] + \frac{AD^2}{2(D^2 + D'^2)} \tag{15.71}$$

where

$$A = 2\left\{F_{n-2} \cos\theta_1 \cos\theta_2 - \frac{D'}{D} F_n \sin\theta_1 \sin\theta_2 \exp[-i(\phi_1 + \phi_2)]\right\} \tag{15.72}$$

$$\begin{aligned} B_1 = \frac{1}{b-a}\Big\{ & \frac{bD^2}{D^2 + D'^2} F_n \sin\theta_1 \sin\theta_2 \exp[-i(\phi_1 + \phi_2)] \\ & + DF_{n-1}[\sin\theta_1 \cos\theta_2 \exp(-i\phi_1) + \sin\theta_2 \cos\theta_1 \exp(-i\phi_2)] \\ & + \frac{aDD'}{D^2 + D'^2} F_{n-2} \cos\theta_1 \cos\theta_2 \Big\} \end{aligned} \tag{15.73}$$

$$\begin{aligned} B_2 = \frac{1}{a-b}\Big\{ & \frac{bD^2}{D^2 + D'^2} F_n \sin\theta_1 \sin\theta_2 \exp[-i(\phi_1 + \phi_2) \\ & + DF_{n-1}[\sin\theta_1 \cos\theta_2 \exp(-i\phi_1) + \sin\theta_2 \cos\theta_1 \exp(-i\phi_2)] \\ & + \frac{bDD'}{D^2 + D'^2} F_{n-2} \cos\theta_1 \cos\theta_2 \Big\} \end{aligned} \tag{15.74}$$

$$\begin{aligned} a &= \frac{1}{2}[-\Omega + \sqrt{\Omega^2 + 8(D^2 + D'^2)}] \\ b &= \frac{1}{2}[-\Omega - \sqrt{\Omega^2 + 8(D^2 + D'^2)}] \end{aligned} \tag{15.75}$$

$$D_1 = F_{n-1}[\sin\theta_1 \cos\theta_2 \exp(-i\phi_1) - \sin\theta_2 \cos\theta_1 \exp(-i\phi_2)] \tag{15.76}$$

Inserting eqs.(15.72)-(15.76) into (15.66) we may get the state vector of the system at time t. By using the state vector, then we can discuss the effect of the dipole-dipole interaction on the time evolution of atomic populations and the atomic population coherent trapping in the two-atom system under the interaction of a coherent field.

15.3.1 Periodic collapses and revivals of atomic populations in the two-atom system

By use of eqs.(15.68)-(15.76), we can obtain the atomic populations in the states $|+,+\rangle$, $|-,-\rangle$, $|+,-\rangle$ and $|-,+\rangle$, respectively. Here we only consider the case of the initial intense field, i.e., $\overline{n} \gg 1$. Considering the Poissonian distribution of the photon number is well located around $\overline{n}$ with $\Delta n = \sqrt{\overline{n}}$, the approximation

$$F_n \approx F_{n-1} \approx F_{n-2}, \qquad D \approx D'$$

is reasonable while we sum with respect to n. Then the population for both of the atoms to be in their ground states $|-\rangle$ at time t is given by

$$\begin{aligned}
P_1(t) &= \sum_n |C_1^n(t)|^2 \\
&= \sum_n F_n^2 \Bigg\{ \frac{1}{4}\left[\cos^2\theta_1\cos^2\theta_2 + \sin^2\theta_1\sin^2\theta_2 - \frac{1}{2}\sin 2\theta_1 \sin 2\theta_2 \cos(\phi_1+\phi_2)\right] \\
&+ \frac{1}{4(b-a)^2}\Bigg\{(b^2+a^2)\left[\frac{1}{2}\sin 2\theta_1\sin 2\theta_2\cos(\phi_1+\phi_2) + \cos^2\theta_1\cos^2\theta_2\right. \\
&\left. + \sin^2\theta_1\sin^2\theta_2\right] - 2\Omega D(\sin 2\theta_1\cos\phi_1 + \sin 2\theta_2\cos\phi_2) \\
&+ 8D^2\left[\cos^2\theta_1\sin^2\theta_2 + \cos^2\theta_2\sin^2\theta_1 + \frac{1}{2}\sin 2\theta_1\sin 2\theta_2\cos(\phi_1+\phi_2)\right]\Bigg\} \\
&- \frac{2}{(a-b)^2}\Bigg\{\cos(a-b)t\left[-\frac{\Omega D}{4}(\sin 2\theta_1\cos\phi_1 + \sin 2\theta_2\cos\phi_2)\right. \\
&\left. + D^2(\sin 2\theta_1\sin 2\theta_2\sin\phi_1\sin\phi_2 - \cos 2\theta_1\cos 2\theta_2)\right] \\
&+ \frac{D\sqrt{\Omega^2+16D^2}}{4}\sin(a-b)t[\sin\phi_1\cos 2\theta_2\sin 2\theta_1 + \sin 2\theta_2\cos 2\theta_1\sin\phi_2]\Bigg\} \\
&+ \frac{1}{2(a-b)}\Big\{\cos(at)[b(\cos^2\theta_1\cos^2\theta_2 - \sin^2\theta_1\sin^2\theta_2)] \\
&+ D[\sin 2\theta_1\cos 2\theta_2\cos\phi_1 + \sin 2\theta_2\cos 2\theta_1\cos\phi_2] \\
&+ \sin(at)\left[\frac{b}{2}\sin 2\theta_1\sin 2\theta_2\sin(\phi_1+\phi_2) + D(\sin 2\theta_1\sin\phi_1 + \sin 2\theta_2\sin\phi_2)\right] \\
&- \cos(bt)[a(\cos^2\theta_1\cos^2\theta_2 - \sin^2\theta_1\sin^2\theta_2) \\
&+ D(\sin 2\theta_1\cos 2\theta_2\cos\phi_1 + \sin 2\theta_2\cos 2\theta_1\cos\phi_2)] \\
&+ \sin(bt)[D(\sin 2\theta_1\sin\phi_1 + \sin 2\theta_2\sin\phi_2) \\
&+ \frac{a}{2}\sin 2\theta_1\sin 2\theta_2\sin(\phi_1+\phi_2)]\Big\}\Bigg\}
\end{aligned} \tag{15.77}$$

and the atomic population inversion between in the states $|+,+\rangle$ and $|-,-\rangle$ is

$$\langle S_3\rangle = \langle\Psi^I(t)|S_3^{(1)} + S_3^{(2)}|\Psi^I(t)\rangle = \sum_n |C_4^{n-2}(t)|^2 - \sum_n |C_1^n(t)|^2$$

$$= P_4(t) - P_1(t) = \sum_n F_n^2 \frac{1}{b-a} \{\cos(at)[b(\cos^2\theta_1\cos^2\theta_2$$
$$-\sin^2\theta_1\sin^2\theta_2)] + D(\sin 2\theta_1 \cos 2\theta_2 \cos\phi_2 + \sin 2\theta_2 \cos 2\theta_1 \cos\phi_1)]$$
$$+\sin(at)\left[\frac{b}{2}\sin 2\theta_1 \sin 2\theta_2 \sin(\phi_1+\phi_2) + D(\sin 2\theta_1 \sin\phi_1 + \sin 2\theta_2 \sin\phi_2)\right]$$
$$-\cos(bt)[a(\cos^2\theta_1\cos^2\theta_2 - \sin^2\theta_1\sin^2\theta_2) + D(\sin 2\theta_1 \cos 2\theta_2 \cos\phi_1$$
$$+\sin 2\theta_2 \cos 2\theta_1 \cos\phi_2)] + \sin(bt)\left[\frac{a}{2}\sin 2\theta_1 \sin 2\theta_2 \sin(\phi_1+\phi_2)\right.$$
$$\left. + D(\sin 2\theta_1 \sin\phi_1 + \sin 2\theta_2 \sin\phi_2)\right]\} \tag{15.78}$$

Similarly, we can obtain the expressions of the atomic populations in the states $|+,-\rangle$ and $|-,+\rangle$.

If we choose $\theta_1 = \theta_2 = \pi/2$ corresponding to both atoms in their ground states $|-\rangle$ initially, then eq.(15.76) may be simplified as

$$P_1(t) = \sum_n F_n^2 \left\{ -\frac{ab}{2(a-b)^2}\cos(a-b)t + \frac{1}{4} + \frac{a^2+b^2}{4(a-b)^2} \right.$$
$$\left. + \frac{\Omega}{2(a-b)}\sin\frac{\Omega t}{2}\sin\frac{a-b}{2}t + \frac{1}{2}\cos\frac{\Omega t}{2}\cos\frac{a-b}{2}t \right\} \tag{15.79}$$

Evidently, the time evolution of $P_1(t)$ depends on the strength Ω of the dipole-dipole interaction. If the distance between the two atoms is so far that the effect of the dipole-dipole interaction can be neglected, i.e., $\Omega = 0$, then (15.79) becomes

$$P_1(t) = \sum_n F_n^2 \left\{ \frac{3}{8} + \frac{1}{8}\cos(2\sqrt{4n-2}gt) + \frac{1}{2}\cos(\sqrt{4n-2}gt) \right\} \tag{15.80}$$

In this case, the atomic population $P_1(t)$ exhibits two series of collapses and revivals. The reason leading to this phenomenon is that in the system of two two-level atoms interacting with the radiation field, there exist not only one-photon transition processes but also two-photon transition processes. These two kinds of transition processes induce that $P_1(t)$ is composed of two cosine oscillating functions with different frequencies and exhibits two series of collapses and revivals as shown in Fig.(15.5a). In Fig.(15.5a), the second revival results from the superposition of the two different cosine oscillations with the ratio of amplitudes to be 1/4, so the corrseponding collapse after the second revival is no longer Gaussian form.

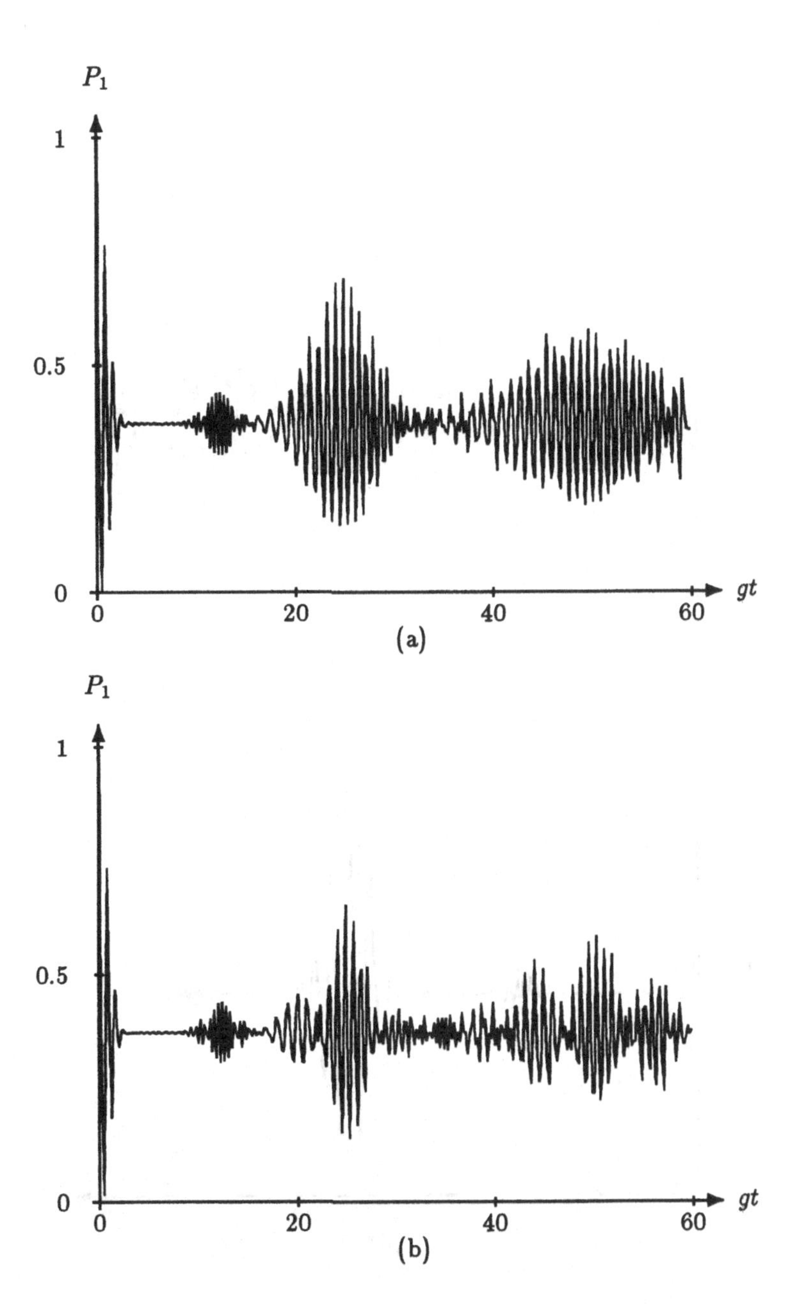

P_1
1
0.5
0
0
20
40
60
gt
(a)
P_1
1
0.5
0
0
20
40
60
gt
(b)

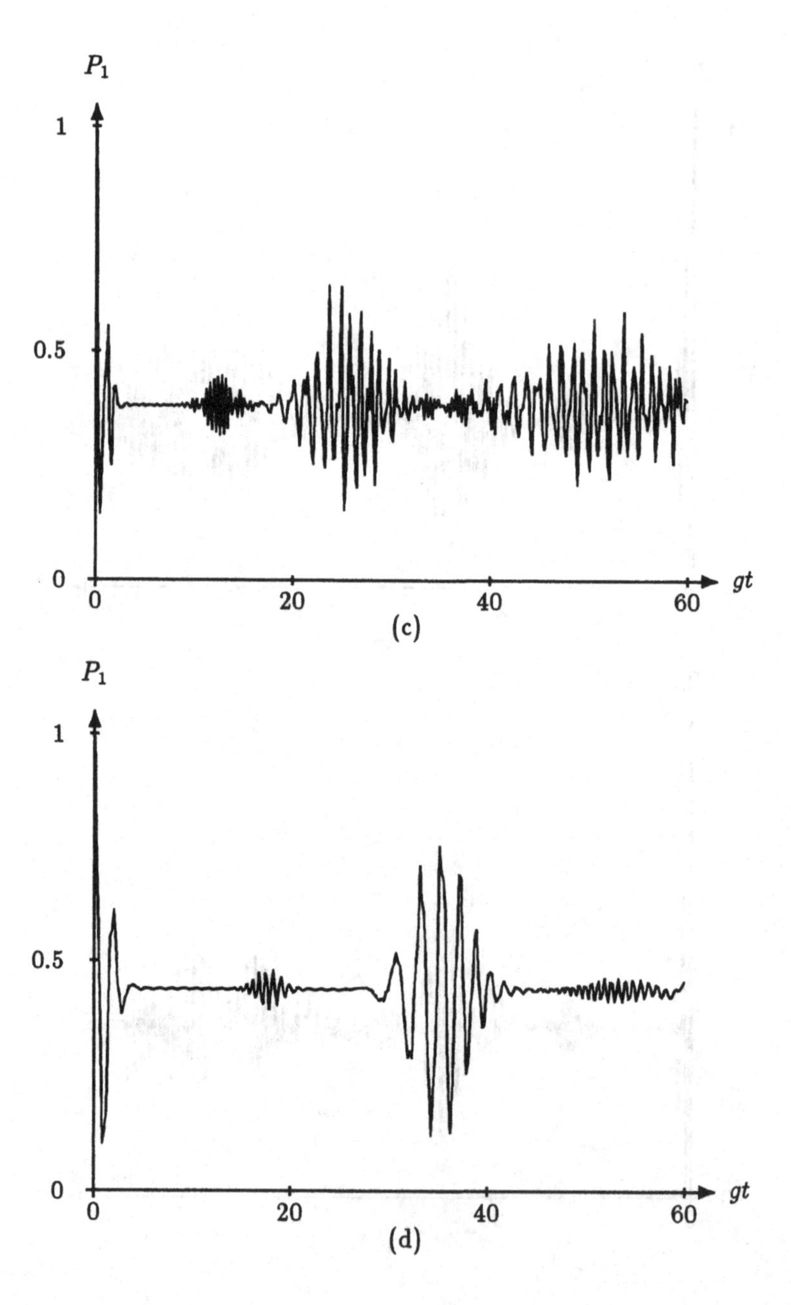
P_1
1
0.5
0
0
20
40
60
gt
(c)
P_1
1
0.5
0
0
20
40
60
gt
(d)

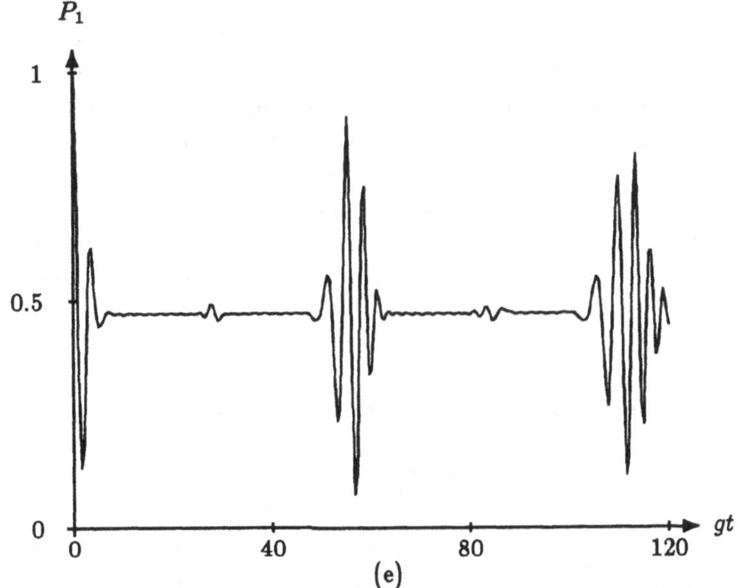

Figure 15.5: Time-dependence of P_1 for different Ω, here $\theta = \pi$, (a) $\Omega = 0$; (b) $\Omega = g$; (c) $\Omega = 4g$; (d) $\Omega = 32g$

If the strength of the dipole-dipole interaction between two atoms is so weak that $g^2\bar{n} \gg \Omega^2$, then $P_1(t)$ can be approximated as

$$P_1(t) = \sum_n F_n^2 \left\{ \frac{1}{8}\cos(a-b)t + \frac{1}{8} + \frac{\Omega}{2(a-b)} \sin\frac{\Omega t}{2} \sin\frac{a-b}{2}t + \frac{1}{2}\cos\frac{\Omega t}{2}\cos\frac{a-b}{2}t \right\} \tag{15.81}$$

Here we plotted $P_1(t)$ in Fig.(15.5b). It shows that the time evolution of $P_1(t)$ is different from the case of $\Omega = 0$ after the first revival. Clearly, this is because of the influence of the dipole-dipole interaction.

Figs.(15.5c-d) display $P_1(t)$ for $\Omega = 4g, 16g$, respectively. With the time development, the amplitude of first revival of $P_1(t)$ decreases evidently, and the oscillations of the second revival is regular ($\Omega = 16g$). From Fig.(15.5a-d), we also find that the revival time of $P_1(t)$ becomes large as Ω increasing .

When the two atoms are close, the strength of the dipole-dipole interaction becomes so large that $\Omega^2 \gg 16g^2\bar{n}$. The time-dependence of the population $P_1(t)$ is shown in Fig.(15.5e). It shows that $P_1(t)$ exhibits regular collapses and revivals and the amplitude of the first revival is nearly equal to zero. Comparing with Fig.(12.2) in Chapter 12, here the laws of $P_1(t)$ is akin to that in the effective two-photon Jaynes-Cummings model. That is to say, for $\Omega^2 \gg 16g^2\bar{n}$, the time evolution of populations in the two-atom system is identified as that in the effective two-photon Jaynes-Cummings model. The reason may be understood as follows. When the dipole-dipole interaction is so strong that Ω satisfies $\Omega^2 \gg 16g^2\bar{n}$, then we can make an approximation for a and b while considering the characteristics of the Poissonian of the coherent field as

$$a \approx \frac{2g^2}{\Omega}(2n-1), \quad b \approx -\Omega \tag{15.82}$$

After substituting (15.82) into (15.79), the population $P_1(t)$ may be written as

$$\begin{aligned} P_1(t) &\approx \sum_n F_n^2 \left\{\frac{1}{2} + \frac{1}{2}\cos[2g^2(2n-1)t/\Omega]\right\} \\ &= \frac{1}{2} + \frac{1}{2}\exp\left[-2\bar{n}\sin^2\left(\frac{2g^2t}{\Omega}\right)\right]\cos\left[\bar{n}\sin\left(\frac{4g^2t}{\Omega}\right) - \frac{2g^2t}{\Omega}\right] \end{aligned} \tag{15.83}$$

If we let $\varepsilon = 2g^2/\Omega$, then (15.83) is coincident with (12.76) which describing the population inversion in the effective two-photon Jaynes-Cummings model. This means that the dynamic features of two two-level atoms are similar to that of the effective two-photon Jaynes-Cummings model due to the strong dipole-dipole interaction. In this moment, the two-photon transition processes dominate the properties of the two-atom system. In order to understand this phenomenon in detail, we inspect the state vector of the system at time t. When $\theta_1 = \theta_2 = \pi/2$, eqs.(15.72)-(15.76) may be taken in the form

$$\begin{aligned} A &= -2F_n, \qquad B_1 = \frac{bD^2}{(b-a)(D^2+D'^2)}, \\ B_2 &= \frac{aD^2}{(a-b)(D^2+D'^2)}, \qquad D_1 = 0 \end{aligned} \tag{15.84}$$

Considering $\Omega^2 \gg 16g^2\bar{n}$ and inserting (15.84) into (15.68) and (15.71), we obtain

$$\begin{aligned} C_2^{n-1}(t) &= C_3^{n-1}(t) \approx 0 \\ C_1^n(t) &\approx \frac{1}{2}F_n \exp\left[i\frac{2g^2}{\Omega}(2n-1)t\right] - \frac{1}{2}F_n \end{aligned} \tag{15.85}$$

$$= iF_n \exp\left[i\frac{g^2t}{\Omega}(2n-1)\right]\sin\left[\frac{g^2t}{\Omega}(2n-1)\right] \quad (15.86)$$

$$C_4^{n-2}(t) \approx F_n \exp\left[i\frac{g^2}{\Omega}(2n-1)t\right]\cos\left[\frac{g^2t}{\Omega}(2n-1)t\right] \quad (15.87)$$

Since the coherent field considering here is intensive, we have

$$n - 1/2 \approx \sqrt{n(n-1)} \quad (15.88)$$

Inserting (15.88) into (15.86) and (15.87), then

$$C_1^n(t) = iF_n \exp[i\varepsilon\sqrt{n(n-1)}t]\sin[\varepsilon\sqrt{n(n-1)}t] \quad (15.89)$$

$$C_4^{n-2}(t) = F_n \exp[i\varepsilon n(n-1)t]\cos[\varepsilon\sqrt{n(n-1)}t] \quad (15.90)$$

here $\varepsilon = 2g^2/\Omega$. If we neglect the phase factor $\exp(i\varepsilon\sqrt{n(n-1)}t)$ in (15.89) and (15.90), then the state vector of the system of two two-level atoms interacting with a single-mode field is coincident with the state vector of the effective two-photon Jaynes-Cummings model. Therefore, when the coherent field is intensive, this two-atom-field coupling system can be characterized by the effective two-photon Jaynes-Cummings model.

Under the interaction of an intensive coherent field, that the evolution of populations in the two-atom system with strong interatomic coupling exhibits the two-photon collapses and revivals can be explained as follows. The two-atom system with strong interatomic coupling can be equivalent to a cascade three-level atom. The dipole-dipole coupling between two atoms induces the shift of the middle level. This shift causes the one-photon transitions to be detuned. If the dipole-dipole interaction is intensive, then the probabilities of one-photon transitions are very small and the dynamic features of the system is mainly governed by two-photon transitions. So the atomic dynamic behavior of the system resembles to the dynamic properties of the effective two-photon J-C model.

From the above, we can see that the strength of the dipole-dipole coupling plays an important role in the collapses and revivals of populations in the two-atom system.

15.3.2 Atomic population coherent trapping in the two-atom coupling system

In Chapter 14 we have verified that when the atom initially in an appropriate state , the atomic populations can not vary with the time development and the atomic population coherent trapping takes place. Does the system of two two-level atoms interacting with a strong coherent field can exhibit the atomic population coherent trapping ?

From (15.77) we know when θ and ϕ satisfy

$$D(\sin 2\theta_1 \sin 2\theta_2 \sin\phi_1 \sin\phi_2 - \cos 2\theta_1 \cos 2\theta_2)$$
$$-\frac{\Omega}{4}(cos\phi_1 \sin 2\theta_1 + \cos\phi_2 \sin 2\theta_2) = 0 \tag{15.91}$$
$$\sin 2\theta_1 \cos 2\theta_2 \sin\phi_1 + \sin\phi_2 \cos 2\theta_1 \sin 2\theta_2 = 0 \tag{15.92}$$
$$b(\cos^2\theta_1 \cos^2\theta_2 - \sin^2\theta_1 \sin^2\theta_2)$$
$$+D(\cos\phi_1 \sin 2\theta_1 \cos 2\theta_2 + \cos\phi_2 \cos 2\theta_1 \sin 2\theta_2) = 0 \tag{15.93}$$
$$a(\cos^2\theta_1 \cos^2\theta_2 - \sin^2\theta_1 \sin^2\theta_2)$$
$$+D(\cos\phi_1 \sin 2\theta_1 \cos 2\theta_2 + \cos\phi_2 \cos 2\theta_1 \sin 2\theta_2) = 0 \tag{15.94}$$
$$\frac{b}{2}\sin 2\theta_1 \sin 2\theta_2 \sin(\phi_1 + \phi_2)$$
$$+D(\sin\phi_1 \sin 2\theta_1 + \sin\phi_2 \sin 2\theta_2) = 0 \tag{15.95}$$
$$\frac{a}{2}\sin 2\theta_1 \sin 2\theta_2 \sin(\phi_1 + \phi_2)$$
$$+D(\sin\phi_1 \sin 2\theta_1 + \sin\phi_2 \sin 2\theta_2) = 0 \tag{15.96}$$

$P_1(t)$=constant and $P_4(t) - P_1(t)$=constant, which correspond to the case of the atomic population coherent trapping. Solving eqs.(15.91)-(15.96) we obtain the atomic population coherent trapping conditions as

$$I.\ \phi_1 = \phi_2 = 0, \quad \theta_1 + \theta_2 = \frac{\pi}{2}, \quad \theta_1 = \frac{1}{2}\arcsin(a/2D) \tag{15.97}$$
$$II.\ \phi_1 = -\phi_2 = \arccos(-a/2D), \quad \theta_1 = \theta_2 = \frac{\pi}{4} \tag{15.98}$$
$$III.\ \phi_1 = \phi_2 = 0, \quad \theta = \frac{\pi}{4}, \quad \theta_2 = -\frac{\pi}{4} \tag{15.99}$$

This means that the atomic population coherent trapping can occur when the two-atom system initial in one of the above three states. And we see that (15.97) and (15.98) contain the parameter a, so the conditions inducing the atomic population coherent trapping depend on the strength Ω of the dipole-dipole interaction between the atoms. But the initial state described by (15.99) is independent of Ω.

When $\Omega = 0$, (15.97) and (15.98) reduce to

$$\theta_1 = \theta_2 = \frac{\pi}{4}, \quad \phi_1 = \phi_2 = 0 \tag{15.100}$$

in this case, the populations of the two-atom system can be deduced from (15.77) as

$$P_i(t) = P_i(0) = \frac{1}{4} \qquad (i = 1, 2, 3, 4) \tag{15.101}$$

Similarly, by use of (15.99) and (15.31) we also find that the populations of the two-atom system are consistent with (15.101). So the atomic initial conditions leading to the atomic population coherent trapping are eqs.(15.99) and (15.100) when the dipole-dipole interaction between the atoms is omitted.

When $\Omega \neq 0$, after substituting (15.98) into (15.77), the population in state $|-,-\rangle$ is obtained as

$$P_1(t) = P_1(0) = \frac{1}{4} \tag{15.102}$$

It is clear that $P_1(t)$ is not related to the parameter Ω although the initial condition depends on Ω. But for the condition I, by using the photon number distribution properties of the coherent field, we obtain that $P_1(t)$ satisfies

$$P_1(t) \approx \frac{(-\Omega + \sqrt{\Omega^2 + 16g^2\overline{n}})^2}{64g^2\overline{n}} = P_1(0) \tag{15.103}$$

Eq.(15.103) (plotted in Fig.15.6) shows that $P_1(t)$ is a nonlinear function of Ω. In Fig.(15.6) we find that $P_1(t)$ varies nonlinearly form 1/4 to 0 when Ω increases. The reason is that the energy levels corresponding to the states $|+,-\rangle$ and $|-,+\rangle$ are shifted by the effect of the dipole-dipole interaction. Hence, the probability of one-photon transitions $|+,+\rangle \leftrightarrow |-,+\rangle \leftrightarrow |-,-\rangle$ and $|+,+\rangle \leftrightarrow |+,-\rangle \leftrightarrow |-,-\rangle$ decreases, while the populations of the states $|+,-\rangle$ and $|-,+\rangle$ increase with increasing Ω, and the populations in the state $|-,-\rangle$ decreases. If $\Omega^2 \gg 16g^2\overline{n}$, which means that the interatomic coupling is very strong. In this case, the two atoms in condition I initially in its excited state $|+,-\rangle$ or $|-,+\rangle$ only oscillate between the states $|+,-\rangle$ and $|-,+\rangle$ and the probability representing one-photon transition is nearly equal to zero due to the extremely coupling between the two atoms. Then the two-atom system does not radiate photons. The probability of the two atoms in the state $|-,-\rangle$ (or $|+,+\rangle$) is therefore equal to zero.

Furthermore, for the condition III, we can find that

$$P_i(t) = P_i(0) = \frac{1}{4} \quad (i = 1, 2, 3, 4) \tag{15.104}$$

This means that under the interaction of an intensive coherent field, if the two atoms are initially in the state III which is independent of Ω, then the

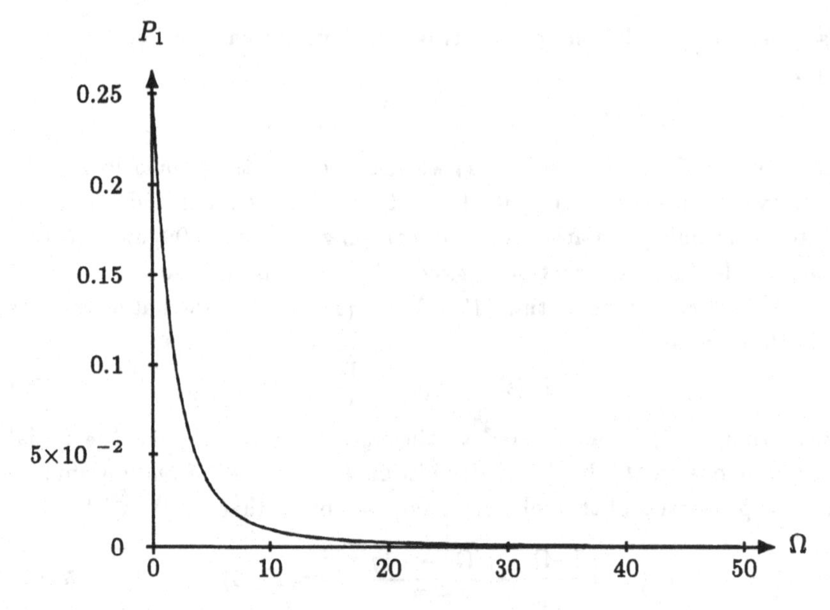

Figure 15.6: Ω-dependence of $P_I(t)$

time evolution of the populations is also not related to Ω. That is to say, the atomic dynamic characteristics of the two-atom system are not related to the dipole-dipole interaction between the atoms, which is the same as the result for $\Omega = 0$. Because both the initial states of the atoms and the time evolution of $P_i(t)(i = 1, 2, 3, 4)$ are equal to the results when $\Omega = 0$, the effect of the dipole-dipole interaction between the two atoms is seen to be screened by the atom-field strong coupling.

In order to discover the features of the three initial states which lead to the atomic population coherent trapping, we employ the dressed states of the atom-field coupling system as introduced in Section 14.1. Generally, the dressed states of this two-atom system can be described by

$$|n(j)\rangle = C_{1jn}|-,-,n\rangle + C_{2jn}|-,+,n-1\rangle + C_{3jn}|+,-,n-1\rangle + C_{4jn}|+,+,n-2\rangle \qquad (j = 1, 2, 3, 4) \tag{15.105}$$

In the interaction picture, solving the eigenvalue equation

$$V^I|n(j)\rangle = \omega_{jn}|n(j)\rangle \tag{15.106}$$

with

$$V^I = \sum_{j=1}^{2} g(a^\dagger S_-^{(i)} + aS_+^{(i)}) + \Omega(S_+^{(1)}S_-^{(2)} + S_-^{(1)}S_+^{(2)}) \tag{15.107}$$

we obtain the eigenvalues ω_{jn} and their corresponding eigenstates as follows

$$\omega_{1n} = 0, \quad |n(1)\rangle = -|-,-,n\rangle + |+,+,n-2\rangle \tag{15.108}$$

$$\omega_{2n} = -\Omega, \quad |n(2)\rangle = |-,+,n-1\rangle - |+,-,n-1\rangle \tag{15.109}$$

$$\omega_{3n} = -a, \quad |n(3)\rangle = |-,-,n\rangle - \frac{a}{2D}(|-,+,n-1\rangle + |+,-,n-1\rangle) + |+,+,n-2\rangle \tag{15.110}$$

$$\omega_{4n} = -b, \quad |n(4)\rangle = |-,-,n\rangle - \frac{b}{2D}(|-,+,n-1\rangle + |+,-,n-1\rangle) + |+,+,n-2\rangle \tag{15.111}$$

Thus, we can express the particular state by use of the dressed states as

$$I: \ |\Psi_{AF}^I(0)\rangle = \sum_n F_n \left[\frac{1}{2}\sin 2\theta_1 |n(4)\rangle - \cos 2\theta_1 |n(2)\rangle\right] \tag{15.112}$$

$$II: \ |\Psi_{AF}^{II}(0)\rangle = \sum_n F_n \left[\frac{1}{2}|n(3)\rangle - \frac{i}{2}\sin\phi_1 |n(2)\rangle\right] \tag{15.113}$$

$$III: \ |\Psi_{AF}^{III}(0)\rangle = \sum_n \frac{1}{2} F_n [|n(1)\rangle - |n(2)\rangle] \tag{15.114}$$

here we have considered the intensive field approximation

$$F_n \approx F_{n-1} \approx F_{n-2} \tag{15.115}$$

From (15.112)-(15.114) we can see that the initial states in which the atomic population coherent trapping occurs are the mixed states of the dressed states $|n(2)\rangle$ and $|n(j)\rangle (j = 1,3,4)$, respectively. This result is in disagreement with that presented in Section 14.1 which shows that the atomic population coherent trapping in single-atom system can occur only in one of its dressed states. To analyze this difference, we need to find the time-dependent state vector $|\Psi_{AF}^i(t)\rangle (i = I, II, III)$ for the three different initial conditions in the interaction picture. According to (15.71), $|\Psi_{AF}^i(t)\rangle$ can be described as follows in the dressed-state representation,

$$|\Psi^{I}_{AF}(t)\rangle = \sum_n F_n \left[\frac{1}{2}\sin(2\theta_1)e^{ibt}|n(4)\rangle - \frac{1}{2}\cos(2\theta_1)e^{i\Omega t}|n(2)\rangle\right] \quad (15.116)$$

$$|\Psi^{II}_{AF}(t)\rangle = \sum_n F_n \left[\frac{1}{2}e^{iat}|n(3)\rangle - \frac{i}{2}\sin\phi_1 e^{i\Omega t}|n(2)\rangle\right] \quad (15.117)$$

$$|\Psi^{III}_{AF}(t)\rangle = \sum_n \frac{1}{2}F_n[|n(1)\rangle - e^{i\Omega t}|n(2)\rangle] \quad (15.118)$$

We can see that the probability of the state vector $|\Psi^{i}_{AF}(t)\rangle$ $(i = I, II, III)$ in dressed states $|n(j)\rangle (j = 1,2,3,4)$ at time t is identified as that at time t=0. Becaues the dressed state $|n(2)\rangle$ is independent of the bare atomic states $|-,-\rangle$ and $|+,+\rangle$, the atomic populations in the states $|+,+\rangle$ and $|-,-\rangle$ at time t are only related to the probability amplitude of one of the dressed states $|n(j)\rangle (j$=1,3,4) for the different initial conditions, respectively. Since the dressed states $|n(j)\rangle$ are the stationary states, the atomic population coherent in the states $|+,+\rangle$ and $|-,-\rangle$ are the same as those at time t=0. So the atomic inversion is independent of time t and does not exhibit collapses and revivals. And the atomic population coherent trapping takes place in the above three cases as eqs.(15.97)-(15.99), respectively. In view of the interaction between the two atoms and the field, the asymmertric state $|n(2)\rangle$ with respect to an exchange of the two atoms is independent of the atom-field coupling and is only related to the dipole-dipole interaction. So the population initially in the dressed state $|n(2)\rangle$ can never contribute to the probability amplitude of $|n(j)\rangle$ (j=1,3,4) at time t. Contributions to the oscillations of $P_1(t)$ and $P_4(t)$ can eventually be expected only from the symmetric states $|n(j)\rangle$ with j=1,3, and 4.

In conclusion, under the interaction of a strong coherent field, the two-atom system including atom-atom coupling can exhibit the atomic population coherent trapping. In disagreement with the one-atom system, the atomic population coherent trapping takes place in the superposition states of the asymmetric dressed state and one of the three symmetric dressed states.

References

1. C.Cohen-Tannoudji, B.Diu, and F.Laloe, *Quatnum Mechanics* (John Wiley & Sons, New York, 1977) p.1120.
2. D.P.Craig, *Molecular Quantum Electrodynamics–An Introduction to Rad-iation-Molecular Interaction* (Acamedic Press, 1984) p.142.
3. R.H.Lehmerg, *Phys.Rev.* **A2** (1970) 889.
4. R.Friederg, S.R.Hartmann, and J.T.Manssan, *Phys.Rep.* **7** (1973) 101.
5. P.Carbonaro and R.Persico, *Phys.Lett.* **A76** (1980) 37.
6. M.Gross and S.Haroche, *Phys.Rep.* **93** (1982) 301.
7. R.D.Griffin and S.E.Harris, *Phys.Rev.* **A25** (1982) 1528.
8. S.M.Barnett and M.D.A.Dupertuis, *J.Opt.Soc.Am.* **B4** (1987) 505.
9. K.Yamada and P.R.Berman, *Phys.Rev.* **A41** (1990) 453.
10. F.Seminara and C.Leonardi, *Phys.Rev.* **A43** (1991) 2599.
11. Z.Ficek, *Phys.Rev.* **A44** (1991) 7759.
12. A.Joshi, R.R.Puri, and S.V.Lawande, *Phys.Rev.* **A44** (1991) 2135.
13. Jin-sheng Peng, Gao-xiang Li, and Peng Zhou, *Chin.J.Lasers* **1** (1992) 221.
14. G.V.Varada and G.S.Agarwal, *Phys.Rev.* **A45** (1992) 6721.
15. Jin-sheng Peng and Gao-xiang Li, *Phys.Rev.* **A47** (1993) 4212.
16. Jin-sheng Peng and Gao-xiang Li, *Phys.Lett.* **A176** (1993) 230.
17. P.A.Brau, V.N.Ostrovsky, and N.V.Prudpv, *Phys.Rev.* **A48** (1993) 941.
18. Gerog Lenz and P.Meystre, *Phys.Rev.* **A48** (1993) 3365.

CHAPTER 16

AUTOIONIZATION OF THE ATOM IN A LASER FIELD

As we know, when an atom in its ground state absorbs certain energy, its electron will be excited to its excited levels. If the excited energy is so high to exceed its ionization threshold energy, the electron can escape from the atom and become a free one, in this case, the atom is referred to being ionized. Because the energy of the free electron is continuous, the atomic system will jump into the continuum states from its bound state when the atom is ionized. For a multi-electron atom, the ionization threshold of its outer-shell electrons is lower than that of its inner-shell electrons, so the outer-shell electrons are easier to be ionized than the inner-shell ones.

Under the interaction of a strong laser field, the atom can exhibit the quantum characteristics such as the periodic collapse-revival effects, the squeezing of atomic operators and the atomic population coherent trapping due to the atomic transitions among the atomic discrete levels. However, if the energy of laser photons is very high, then the outer-shell electron in bound state can escape from the atom by absorbing a high-energy photon to jump into the atomic continuum state and becomes a free one, this is just a kind of photoionization processes. On the other hand, the inner-shell electron in bound state can probably absorbs a high energy photon and jumps to a new discrete state. The electron in the new discrete state will be associated with the electrons in continuum states through Coulomb interaction, and causes the system to be mixed in the configuration state. In addition, the ionization threshold of exciting two outer-shell electrons simultaneously is higher than that of exciting one inner-shell electron. Thus when the two outer-shell electrons are excited, the atom may evolve into the configuration state in which one electron is in the continuum state and the other in the discrete state. Because this kind of atom being in the configuration state may be autoionized, we interpret these states as atomic autoionizing states. That the atom in a strong laser field may be autoionized is an important quantum characteristic of the multi-electron atom under the interaction of the radiation field. This quantum characteristic has been most investigated recently in atomic and molecular physics.

This chapter is devoted to discussing the principal theory of the atomic autoionization. We first introduce the theoretical treatment introduced by Fano to deal with the system of a discrete state interacting with the continuum states, and discuss the autoionization of the atom in a week laser field. Secondly, we discuss the autoionization of the atom in a strong laser field, and analyze the properties of the photoelectron spectrum. Finally, the above threshold ionization is also investigated in detail.

16.1 Autoionization of the atom in a weak laser field

Under the interaction of a laser field, a multi-electron atom can be in the autoionizing state, i.e., the atom can be in the configuration of one electron in a bound state and the other one in a continuum, the level-diagram is illustrated in Fig.(16.1). When the atom initially in the bound state $|a\rangle$ absorbs a photon whose energy is higher than the ionization threshold, it jumps from the bound

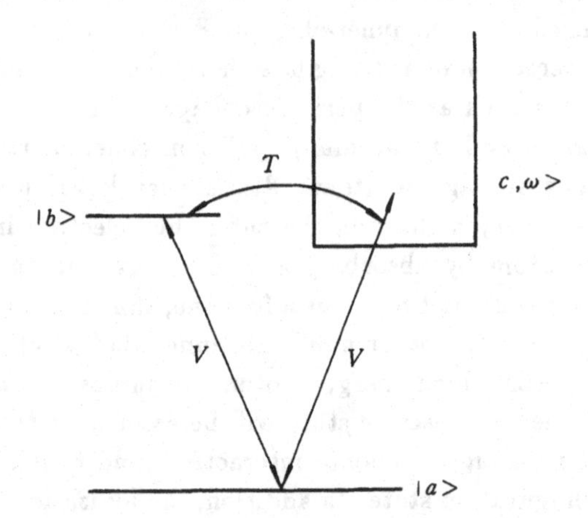

Figure 16.1: Diagram of autoionizing transitions of the atom under the interaction of a laser field

state $|a\rangle$ to the continuum $|c,\omega\rangle$. The electron in the continuum state has positive energy, it will escape from the atom and becomes a free one. This process is called the photoionization process. On the other hand, the inner-shell electron can also probably jump to a new bound state $|b\rangle$, which is embedded in the continuum $|c,\omega\rangle$. Because the Coulomb interaction between electrons enables the bound state to be mixed with the continuum, then forms an autoionizing state. This Coulomb interaction between electrons is called the configuration interaction. The electron in the bound state $|b\rangle$ can jump to $|c,\omega\rangle$ due to the effects of the configuration interaction and become a free one, this autoionizing process is due to the configuration interaction. Inasmuch as there does not exist the absorption and emission of photons, we interpret the bound state $|b\rangle$ embedded in continuum as the autoionizing state, and call this ionization resulting from the configuration interaction as the autoionization. Therefore, we find that there exist two ionization channels for the atom in bound state $|a\rangle$ under the interaction of the laser field, one channel is $|a\rangle \leftrightarrow |c,\omega\rangle$ and the other $|a\rangle \leftrightarrow |b\rangle \leftrightarrow |c,\omega\rangle$, so we must take into account these two effects. Since there are involved two transition channels, the theoretical treatment is certainly more complex, so it is necessary to simplify the model illustrated in Fig.(16.1).

As we see, the atomic configuration interaction between the autoionizing state $|b\rangle$ and the continuum state $\{|c,\omega\rangle\}$ is isolated from the bound state $|a\rangle$, then we can treat $|b\rangle$ and $\{|c,\omega\rangle\}$ as a subspace of the atomic states. In this subspace, the atomic Hamiltonian is written as

$$H_A = H_{A0} + T \tag{16.1}$$

here H_{A0} represents the atomic Hamiltonian in the subspace consisted of $|b\rangle$ and $\{|c,\omega\rangle\}$ in the absence of the configuration interaction, i.e.,

$$H_{A0} = E_b|b\rangle\langle b| + \int_0^\infty \omega|c,\omega\rangle\langle c,\omega|d\omega \qquad (\hbar = 1) \tag{16.2}$$

T stands for the configuration interaction between $|b\rangle$ and $\{|c,\omega\rangle\}$, which can be characterized as

$$T = \int_0^\infty d\omega T_{b\omega}|b\rangle\langle c,\omega| + \text{h.c.} \tag{16.3}$$

where $T_{b\omega}$ are the matrix elements of the configuration interaction

$$T_{b\omega} = \langle b|T|c,\omega\rangle \tag{16.4}$$

If one can find a new set of continuum $\{|F,\omega\rangle\}$ which obey

$$H_A|F,\omega\rangle = \omega|F,\omega\rangle \tag{16.5}$$

that is to say, $\{|F,\omega\rangle\}$ is a set of eigenvectors of the atomic Hamiltonian H_A including the configuration interaction, then the model illustrated in Fig.(16.2) is equivalent to the one shown in Fig.(16.1). In this case, the interaction between

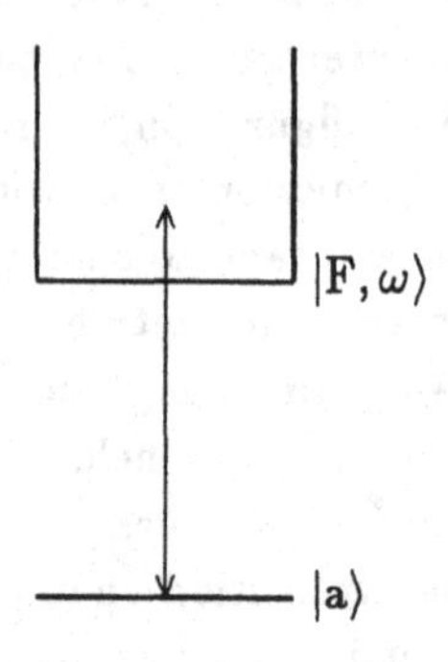

Figure 16.2: Diagram of Fano eigenvectors $\{|F,\omega\rangle\}$ including the configuration interaction

$|b\rangle$ and $|c,\omega\rangle$ is incorporated into $|F,\omega\rangle$. This leads to the photoionization including the autoionizing process becomes simple. Next we turn to solve the eigenstates $|F,\omega\rangle$ of H_A, which are usually termed as the Fano eigenstates.

Evidently, the Fano eigenstates can be expanded in terms of the state $|b\rangle$ and $\{|c,\omega\rangle\}$ as

$$|F,\omega\rangle = B(\omega)|b\rangle + \int_0^\infty d\omega' C(\omega,\omega')|c,\omega'\rangle \tag{16.6}$$

Substituting the above equation into equation (16.5) and multiplying it from the left by $\langle b|$, then we have

$$\langle b|H_{A0} + T|\left[B(\omega)|b\rangle + \int_0^\infty d\omega' C(\omega,\omega')|c,\omega'\rangle\right]$$

$$= \omega\langle b|\left[B(\omega)|b\rangle + \int_0^\infty d\omega' C(\omega,\omega')|c,\omega'\rangle\right] \tag{16.7}$$

Since $|b\rangle$ and $|c,\omega\rangle$ obey

$$\begin{aligned} \langle b|b\rangle &= 1 \\ \langle c,\omega'|c,\omega\rangle &= \delta(\omega-\omega') \\ \langle b|c,\omega\rangle &= 0 \end{aligned} \tag{16.8}$$

(16.7) becomes

$$(\omega - E_b)B(\omega) = \int_0^\infty d\omega' C(\omega,\omega')T_{b\omega'} \tag{16.9}$$

Similarly, multiplying (16.5) by $\langle c,\omega'|$ gives

$$(\omega-\omega')C(\omega,\omega') = T_{\omega' b}B(\omega) \tag{16.10}$$

Substitute (16.10) into (16.9), one can obtain an equation for $B(\omega)$. However, there is a singularity at $\omega = \omega'$ so that $(\omega-\omega')^{-1}$ must be divided into two parts, one is its principal and the other is a delta function as

$$C(\omega,\omega') = B(\omega)T_{\omega' b}\left[\frac{P}{\omega-\omega'} + Z(\omega)\delta(\omega-\omega')\right] \tag{16.11}$$

where $z(\omega)$ is a function of ω . Substituting equation (16.11) into (16.9) we obtain

$$(\omega - E_b)B(\omega) = B(\omega)\int_0^\infty d\omega'|T_{b\omega'}|^2\left[\frac{P}{\omega-\omega'} + Z(\omega)\delta(\omega-\omega')\right] \tag{16.12}$$

In general $B(\omega) \neq 0$, so that $Z(\omega)$ is given by

$$Z(\omega) = \frac{\omega - E_b'(\omega)}{|T_{b\omega'}|^2} \tag{16.13}$$

where

$$E_b'(\omega) = E_b + \int_0^\infty d\omega'|T_{b\omega'}|^2\frac{P}{\omega-\omega'} \tag{16.14}$$

The above equation shows that the eigenenergy of $|b\rangle$ is shifted by the configuration interaction. Substituting eqs.(16.11) and (16.13) into (16.6) yields

$$|F,\omega\rangle = B(\omega)|b\rangle + \int_0^\infty d\omega' B(\omega)T_{\omega' b}\frac{P}{\omega-\omega'}|c,\omega'\rangle$$

$$
\begin{aligned}
&+\int_0^\infty d\omega' B(\omega)T_{\omega' b}Z(\omega)\delta(\omega-\omega')|c,\omega'\rangle \\
&= B(\omega)\left(|b\rangle + \frac{\omega - E_b'(\omega)}{T_{b\omega}}|c,\omega\rangle\right. \\
&\left. + \int_0^\infty d\omega' T_{\omega' b}\frac{P}{\omega-\omega'}|c,\omega'\rangle\right)
\end{aligned}
\tag{16.15}
$$

here $B(\omega)$ can be determined by the normalization condition of $|F,\omega\rangle$. It is

$$
\begin{aligned}
\langle F,\omega'|F,\omega\rangle = \delta(\omega-\omega') &= B^*(\omega')B(\omega)\left\{1+\int_0^\infty d\omega''|T_{b\omega''}|^2\left[\frac{P}{\omega'-\omega''}\right.\right. \\
&\left.\left. +Z(\omega')\delta(\omega'-\omega)\right]\left[\frac{P}{\omega-\omega''}+Z(\omega)\delta(\omega-\omega'')\right]\right\} \\
&= B^*(\omega')B(\omega)\left\{1+\int_0^\infty d\omega''|T_{b\omega''}|^2\left[\frac{P}{\omega'-\omega''}\cdot\frac{P}{\omega-\omega''}\right.\right. \\
&+\frac{P}{\omega'-\omega''}Z(\omega)\delta(\omega-\omega'')+Z(\omega')Z(\omega)\delta(\omega'-\omega'')\delta(\omega-\omega'') \\
&\left.\left.+\frac{P}{\omega-\omega''}Z(\omega')\delta(\omega'-\omega'')\right]\right\}
\end{aligned}
\tag{16.16}
$$

In order to calculate the integration in the above equation, we introduce the expression of the Fourier transformation of $(\omega'-\omega'')^{-1}$ as

$$
\frac{P}{\omega'-\omega''} = -i\pi\int_{-\infty}^{\infty} dk\frac{|k|}{k}\exp[2\pi ik(\omega'-\omega'')]
\tag{16.17}
$$

therefore

$$
\frac{P}{\omega'-\omega''}\cdot\frac{P}{\omega-\omega''} = -\pi^2\int_{-\infty}^{\infty}dk\int_{-\infty}^{\infty}dk'\frac{kk'}{|kk'|}\exp\{2\pi i[k(\omega'-\omega'')+k'(\omega-\omega'')]\}
\tag{16.18}
$$

Let

$$
u = k+k', \qquad v = \frac{k-k'}{2}
\tag{16.19}
$$

then

$$
\frac{kk'}{|kk'|} = \frac{u^2-4v^2}{|u^2-4v^2|} = -1+2S_t(u^2-4v^2)
\tag{16.20}
$$

where $S_t(u^2 - 4v^2)$ is a step function which defined as

$$S_t(u^2 - 4v^2) = \begin{cases} 0, & (|u| < 2|v|), \\ 1, & (|u| > 2|v|) \end{cases} \tag{16.21}$$

Thus (16.18) becomes

$$\begin{aligned} \frac{P}{\omega' - \omega''} \cdot \frac{P}{\omega - \omega''} &= -\pi^2 \int_{-\infty}^{\infty} du \exp\left[2\pi i u \left(\frac{\omega + \omega'}{2} - \omega''\right)\right] \\ &\times \left[\int_{-\infty}^{\infty} dv - 2 \int_{-|v|/2}^{|v|/2} dv\right] \exp[2\pi i v(\omega' - \omega)] \\ &= \pi^2 \int_{-\infty}^{\infty} du \exp\left[2\pi i u \left(\frac{\omega + \omega'}{2} - \omega''\right)\right] \left[\delta(\omega' - \omega) - \frac{2 \sin \pi |u|(\omega' - \omega)}{\pi(\omega' - \omega)}\right] \\ &= \pi^2 \delta\left(\omega'' - \frac{\omega + \omega'}{2}\right) \delta(\omega - \omega') + \frac{\pi i}{\omega' - \omega} \int_{-\infty}^{\infty} \{\exp[2\pi i u(\omega' - \omega'')] \\ &- \exp[2\pi i u(\omega - \omega'')]\} \frac{u}{|u|} du \end{aligned} \tag{16.22}$$

Noticing (16.17), then

$$\begin{aligned} \frac{P}{\omega' - \omega''} \cdot \frac{P}{\omega - \omega''} = \pi^2 \delta\left(\omega'' - \frac{\omega + \omega'}{2}\right) \delta(\omega - \omega') \\ + \frac{1}{\omega' - \omega} \left[\frac{P}{\omega - \omega''} - \frac{P}{\omega' - \omega''}\right] \end{aligned} \tag{16.23}$$

Substituting equation (16.23) and $\delta(\omega - \omega'')\delta(\omega' - \omega'') = \delta(\omega'' - \frac{\omega + \omega'}{2})\delta(\omega - \omega')$ into (16.16), we have

$$\begin{aligned} \delta(\omega - \omega') = B^*(\omega')B(\omega) \{ 1 + [|T_{\omega b}|^2 \pi^2 + Z(\omega)Z(\omega')]\delta(\omega - \omega') \\ + \int_0^{\infty} d\omega'' |T_{b\omega''}|^2 \left[\frac{1}{\omega' - \omega} \left(\frac{P}{\omega - \omega''} - \frac{P}{\omega' - \omega''}\right)\right. \\ \left.\left. + \frac{P}{\omega' - \omega''} Z(\omega)\delta(\omega - \omega'') + \frac{P}{\omega - \omega''} Z(\omega')\delta(\omega' - \omega'')\right]\right\} \end{aligned} \tag{16.24}$$

Recalling (16.13) and (16.14), the above equation reduces to

$$\delta(\omega-\omega') = B^*(\omega')B(\omega)\delta(\omega-\omega')[\pi^2 + Z(\omega)Z(\omega')]|T_{\omega b}|^2 \tag{16.25}$$

Therefore, we have

$$|B(\omega)|^2 = \frac{|T_{\omega b}|^2}{\pi^2|T_{\omega b}|^4 + [\omega - E_b'(\omega)]^2} \tag{16.26}$$

If we choose the phase factor appropriately, then $|F,\omega\rangle$ can be expressed as

$$\begin{aligned}|F,\omega\rangle = &\frac{1}{\sqrt{(\pi|T_{\omega b}|^2)^2 + [\omega - E_b'(\omega)]^2}}\{T_{\omega b}|b\rangle + [\omega - E_b'(\omega)]|c,\omega\rangle \\ &+T_{b\omega}\int_0^\infty d\omega' T_{\omega' b}\frac{P}{\omega-\omega'}|c,\omega'\rangle\Bigg\}\end{aligned} \tag{16.27}$$

Now we have obtained the Fano eigenstates of the system including the configuration interaction, they are a new set of continuum. The above equation can also be rewritten as

$$\begin{aligned}|F,\omega\rangle = &\{(\pi|T_{b\omega}|^2)^2 + [\omega - E_b'(\omega)]^2\}^{-1/2}\{T_{b\omega}|\phi(\omega)\rangle \\ &+[\omega - E_b'(\omega)]|c,\omega\rangle\}\end{aligned} \tag{16.28}$$

where

$$|\phi(\omega)\rangle = |b\rangle + \int_0^\infty d\omega' T_{\omega' b}\frac{P}{\omega-\omega'}|c,\omega'\rangle \tag{16.29}$$

represents the new discrete state resulting from the interaction between the autoionizing state $|b\rangle$ and the set of continuum $\{|c,\omega\rangle\}$. Eq.(16.28) shows that there exists a singularity point $\omega = E_b'(\omega)$ in the continuum $|F,\omega\rangle$. If assuming the matrix element $T_{\omega b}$ of the configuration interaction is independent of ω, then the old continuum $|c,\omega\rangle$ is modified by a dispersive factor at point $\omega = E_b'(\omega)$ and the modified new bound state $|\phi_(\omega)\rangle$ has a square-root Lorentzian factor centered at $\omega = E_b'(\omega)$.

From the Fano eigenstate $|F,\omega\rangle$ including the configuration interaction, we may discuss the photoionization properties of the atom under the interaction of a weak field in terms of the Fano eigenvectors $\{|F,\omega\rangle\}$. For the model illustrated in Fig.(16.2), if the laser field is very weak, then in the Hamiltonian

$H = H_A + V$, the atom-field coupling Hamiltonian V is comparatively small and can be treated as a perturbation term. According to the Fermi golden rule, the probability finding the atom jumping from its initial bound state $|a\rangle$ to the Fano eigenvector $|F, \omega\rangle$ is

$$\Gamma_{aF} = \frac{\pi}{2} \int_0^{\pi} \delta(\omega - \omega_i) |\langle a|V|F, \omega\rangle|^2 \tag{16.30}$$

with

$$\omega_i = \omega_a + \omega_L$$

here ω_L is the frequency of laser field, ω_a is the eigen frequency of the atom being the bound level $|a\rangle$, i.e., $H_{A0} = \omega_a|\omega\rangle$. The transition matrix element $\langle a|V|F, \omega\rangle$ in (16.30) is expressed as

$$\begin{aligned}\langle a|V|F, \omega\rangle = &\{(\pi|T_{b\omega}|^2)^2 + [\omega - E_b'(\omega)]^2\}^{-1/2}[T_{b\omega}\langle a|V|\phi(\omega)\rangle \\ &+(\omega - E_b'(\omega))\langle a|V|c, \omega\rangle]\end{aligned} \tag{16.31}$$

Let

$$\gamma = \pi|T_{b\omega}|^2 \tag{16.32}$$

$$\varepsilon = \frac{\omega - E_b'(\omega)}{\gamma} \tag{16.33}$$

$$q(\omega) = \frac{\langle a|V|\phi(\omega)\rangle}{\langle a|V|c, \omega\rangle \pi T_{\omega b}} \tag{16.34}$$

here γ is the decay rate from $|b\rangle$ to the continuum $|c, \omega\rangle$, which represents the autoionizing strength due to the configuration interaction between $|b\rangle$ and $|c, \omega\rangle$, ε is called the Fano reduced energy, $q(\omega)$ is termed the Fano parameter which can be chosen to be real by adopting an appropriate phase factor, evidently it stands for the relative transition strength between $|a\rangle \leftrightarrow |b\rangle$ and $|a\rangle \leftrightarrow |c, \omega\rangle$. If the coupling strength of the transition $|a\rangle \leftrightarrow |c, \omega\rangle$ is very weak, then the value of $q(\omega)$ is large. In general sense, the dependence of the matrix elements $\langle a|V|\phi(\omega)\rangle$ and $T_{\omega b}$ on ω is not explicit, so for the sake of calculating simply, we assume that q is independent of ω. Combining eqs.(16.32)-(16.34) with (16.31) gives

$$\langle a|V|F, \omega\rangle = (1 + \varepsilon^2)^{-1/2}\langle a|V|c, \omega\rangle(\varepsilon + q)$$

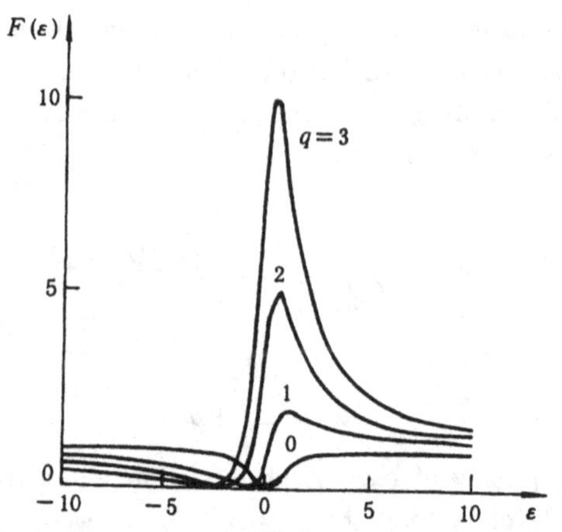

Figure 16.3: The curve of $F(\varepsilon)$ versus ε for different q

$$
\begin{aligned}
&= (\varepsilon - i)^{-i} \langle a|V|c,\omega\rangle e^{i\psi}(\varepsilon + q) \\
&= \langle a|V|c,\omega\rangle (q+i) e^{i\psi} \left[\frac{1}{\varepsilon - i} + \frac{1}{q+i} \right]
\end{aligned} \tag{16.35}
$$

here ψ is a phase factor. Substituting (16.35) into (16.30), we obtain the probability of the atom jumping from $|a\rangle$ to $|F,\omega\rangle$, i.e., the ionization rate of the atom under the interaction of a weak field as

$$
\Gamma_{aF} = \Gamma_{a\omega} \frac{(\varepsilon + q)^2}{1 + \varepsilon^2} \tag{16.36}
$$

with

$$
\Gamma_{a\omega} = \frac{\pi}{2} |\langle a|V|c,\omega\rangle|^2 = \frac{\pi}{2} |V_{a\omega}|^2 \tag{16.37}
$$

where $\Gamma_{a\omega}$ represents the probability of the atom jumping from $|a\rangle$ to $|c,\omega\rangle$. $\Gamma_{aF}/\Gamma_{a\omega}$ stands for the ratio of the decay from state $|a\rangle$ into the Fano continuum $|F,\omega\rangle$ to the decay from state $|a\rangle$ into the old one $|c,\omega\rangle$. Because q is assumed to be independent of ω, $\Gamma_{aF}/\Gamma_{a\omega}$ is the function of ε, i.e.,

$$
F(\varepsilon) = \frac{\Gamma_{aF}}{\Gamma_{a\omega}} = \frac{(\varepsilon + q)^2}{1 + \varepsilon^2} \tag{16.38}
$$

In Fig.(16.3) we plot the $F(\varepsilon)$ versus ε for different q. For different q, i.e., the ratios of the coupling strengths $|a\rangle \leftrightarrow |b\rangle$ and $|a\rangle \leftrightarrow |c,\omega\rangle$ are different, then the values of $F(\varepsilon)$ are different. When q=0, corresponding to that the transition

$|a\rangle \leftrightarrow |b\rangle$ is forbidden, $F(\varepsilon)$ is symmetrically distributed in the two sides of $\varepsilon = 0$. For $\varepsilon = 0$, $F(0)$=0, which means that there exhibits a range for the atom not to be ionized at $\omega = E'_b(\omega)$. This is because that the level $|c, E'_b(\omega)\rangle$ with eigenenergy $E'_b(\omega)$ is shifted by the configuration interaction between $|b\rangle$ and $|c,\omega\rangle$, which leads to that the level $|c, E'_b(\omega)\rangle$ does not exist in the Fano eigenvectors $|F,\omega\rangle$ (16.29). So the atomic transition from $|a\rangle$ to $|c, E'_b(\omega)\rangle$ can not happen. Therefore, for q=0, $F(0)$=0. When $\varepsilon \neq 0$, $F(\varepsilon)$ tends to be 1 with the increasing of $|\varepsilon|$. Inasmuch as the atomic transition from $|a\rangle$ to $|b\rangle$ is forbidden, the dominant contribution to Γ_{aF} is the transition process from $|a\rangle$ to $\{|c,\omega\rangle\}$.

With the increasing of q, which means that the coupling strength of $|a\rangle \leftrightarrow |c,\omega\rangle$ decreases, from Fig.(16.3) (q=1,2, 3) we see that the lineshapes of $F(\varepsilon)$ against ε are no longer symmetric. And the zero point of $F(\varepsilon)$ appears at $\varepsilon = -q$. In the range of $\varepsilon < -q$, the values of $F(\varepsilon)$ are smaller than those for q=0. But in the range of $\varepsilon > 0$, the values of $F(\varepsilon)$ are larger than those for q=0. The maximum value of $F(\varepsilon)$ is located at $\varepsilon = 1/q$. These results can be explained as follows. It can be found from (16.31) and (16.35) that the atomic transition process from the bound state $|a\rangle$ to the Fano eigenvector $|F,\omega\rangle$ are the combination of the transition processes from $|a\rangle$ to the new discrete state $|\phi(\omega)\rangle$ and from $|a\rangle$ to the continuum $|c,\omega\rangle$, these two kinds of transitions interfere each other. When $\varepsilon < -q$, i.e., $\omega - E'_b(\omega) < -\gamma q$, the difference of the phase factors in the two matrix elements in (16.31) is π, there is a destructive interference between these two transition processes, so that the values of $F(\varepsilon)$ for $\varepsilon < -q$ are smaller than the corresponding values of $F(\varepsilon)$ for q=0. As we find when $\varepsilon = -q$, this destructive interference is strongest, therefore $F(-q)$=0. We usually call the point $\varepsilon = -q$ as the Fano zero point. When $\varepsilon > 0$, i.e., $\omega - E'_b(\omega) > 0$, the two transition processes have the same phase, there occurs the enhancing interference, and at $\varepsilon = 1/q$, this enhancing interference is strongest. In addition, when $q \neq 0$, the probability for the direct photoionization transition from $|a\rangle$ to $|c,\omega\rangle$ decreases. Consequently for $\varepsilon > 0$, the values of $F(\varepsilon)$ are larger than those for q=0. For the narrowest neighbour of $\varepsilon \to 0^-$, when $q \neq 0$, the interference between the two transition processes $|a\rangle \leftrightarrow |\phi(\omega)\rangle$ and $|a\rangle \leftrightarrow |c,\omega\rangle$ is a destructive one, but at q=0 the atom can not be ionized in the narrowest range of $\varepsilon = 0^-$, so at $q \neq 0$ $F(\varepsilon)$ is still larger than that at q=0 in this narrowest range.

Fig.(16.4) illustrates another limited case, where $F^{1/10}(\varepsilon)$ varies with ε

when q is very large. In order to compare with the result for q being not very

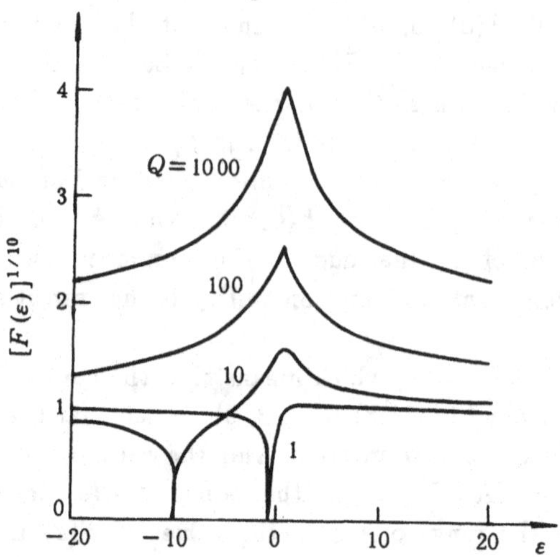

Figure 16.4: The curve of $[F(\varepsilon)]^{1/10}$ versus ε for large q

large, we also plot the curve of $F^{1/10}(\varepsilon)$ for q=1. As wee see, although the direction transition from $|a\rangle$ to $|c,\omega\rangle$ is very weak, the atomic autoionization can also happen through the channel $|a\rangle \leftrightarrow |b\rangle \leftrightarrow |c,\omega\rangle$. Evidently, $\varepsilon = -q$ is a zero point of $F(\varepsilon)$ and $F(\varepsilon)$ reaches its maximum value at $\varepsilon = 1/q$.

It can be found from the above discussions that the atomic autoionization process is an important one in the photoionization processes. So when we discuss the atomic photoionization, the effects of the autoionization process must be included. Up to now we have studied the atomic autoionization in the weak field, however, if the laser is intensive, what is about the autoionization of atom ? Next we discuss the atomic autoionization under the interaction of a strong laser field.

16.2 Autoionization of the atom under the interaction of a strong laser field

Here we also study the model shown in Fig.(16.2), but assume the field is a strong one so that the interaction Hamiltonian V is not small and the perturbation theory is unsuitable to treat this problem. So we must adopt a new treatment to re-examine it.

Assuming that the laser field in Fig.(16.2) is so strong that it can be characterized by a classical electromagnetic field, then in the rotating-wave -approximation, the Hamiltonian of the system is written as

$$H = E_a|a\rangle\langle a| + \int d\omega\omega|\omega\rangle\langle\omega| + \left[\int d\omega\Omega_\omega|a\rangle\langle\omega| + \text{h.c.}\right] \tag{16.39}$$

For simplicity to write, here the symbol $|F,\omega\rangle$ is replaced by $|\omega\rangle$, Ω_ω represents the matrix element for the transition $|a\rangle \rightarrow |c,\omega\rangle$, i.e.,

$$\Omega_\omega = (q+i)e^{i\psi}\langle a|V|c,\omega\rangle[(\varepsilon-i)^{-1} + (q+i)^{-1}] \tag{16.40}$$

If defining the parameter Ω_0 associated with the coupling strengths as

$$\Omega_0 = (4\pi\gamma)^{1/2}V_{a\omega}\exp(i\psi) \tag{16.41}$$

where

$$V_{a\omega} = \langle a|V|c,\omega\rangle \tag{16.42}$$

and choosing an appropriate phase factor to make Ω_0 to be real, then eq.(16.40) becomes

$$\Omega_\omega = [(\omega-i)^{-1} + (q+i)^{-1}]\Omega_0(4\pi\gamma)^{-1/2} \tag{16.43}$$

If the atom is initially in a bound state $|a\rangle$, then with the time development, the state vector of the system evolves into

$$|\Psi(t)\rangle = \alpha(t)|a\rangle + \int \beta_\omega(t)|\omega\rangle d\omega \tag{16.44}$$

In order to obtain the expression of $|\Psi(t)\rangle$, we must solve $\alpha(t)$ and $\beta_\omega(t)$. By means of the Schrödinger equation, we have

$$\frac{d}{dt}\alpha(t) = -i\int \Omega_\omega\beta_\omega(t)d\omega \tag{16.45}$$

$$\frac{d}{dt}\beta_\omega(t) = -i(\omega-\omega_\ell)\beta_\omega(t) - i\Omega_\omega^*\alpha(t) \tag{16.46}$$

here E_a is assumed to be zero, ω_L is the laser frequency. Making the Laplace transformation for (16.45) and (16.46), i.e., defining

$$\tilde{\beta}_\omega(z) = \int_0^\infty \beta_\omega(\tau)e^{-z\tau}d\tau, \qquad \tilde{\alpha}(z) = \int_0^\infty \alpha(\tau)e^{-z\tau}d\tau \tag{16.47}$$

then (16.45) and (16.46) become

$$z\tilde{\alpha} - 1 = -i\int d\omega \Omega_\omega \tilde{\beta}_\omega \tag{16.48}$$

$$z\tilde{\beta}_\omega = -i\delta\tilde{\beta}_\omega - i\Omega^*_\omega \tilde{\alpha} \tag{16.49}$$

here $\delta = \omega - \omega_L$. It is obtained from (16.49) that

$$\tilde{\beta}_\omega = -i\frac{\Omega^*_\omega \tilde{\alpha}}{(z + i\delta)} \tag{16.50}$$

Inserting (16.51) into (16.48) we have

$$\tilde{\alpha} = 1/(z + N) \tag{16.51}$$

with

$$N = \int d\omega \Omega^*_\omega \Omega_\omega/(z + i\delta) \tag{16.52}$$

Substituting (16.43) into (16.52) gives

$$N = [(q^2+1)4\pi\gamma]^{-1}\Omega_0^2 \int d\omega \frac{(\varepsilon + q)^2}{(z + i\delta)(1 + \varepsilon^2)} \tag{16.53}$$

Because the frequency shift due to the configuration interaction is smaller than E_b in (16.14), so in the following discussion, we may let $E'_b(\omega) = E_b$. And assuming that $\omega_L = E_b$, which corresponds to that the frequency of the laser field is resonant to the frequency of the atomic transition between $|a\rangle$ and $|b\rangle$. Then inserting (16.33) into (16.53) and using the residues theorem to calculate the integration, we obtain

$$N = \frac{\Omega_0^2(q^2 + z/\gamma - 2iq)}{4(q^2+1)(z+\gamma)} \tag{16.54}$$

Inserting eqs.(16.51) and (16.54) into eq.(16.50), the expression of $\tilde{\beta}_\omega(z)$ is given as

$$\tilde{\beta}_\omega(z) = -\frac{i\Omega^*_\omega}{(z + i\delta)(z + N)} \tag{16.55}$$

Making the Laplace inverse transformation for $\tilde{\alpha}(z)$ and $\tilde{\beta}_\omega(z)$, then obtain the expressions of $\alpha(t)$ and $\beta_\omega(t)$. Furthermore, the probabilities finding the atom in $|a\rangle$ and $|\omega\rangle$ at time t can be given, and the properties of the time evolution of photoelectrons can also be found.

However, the measurement results in experiments are mainly the steady-state photoelectron spectrum, so next we only discuss the steady-state photoelectron spectrum. The result of the Fourier transformation of $\beta_\omega(t)$ is just the spectrum of photoelectrons. So the steady-state photoelectron spectrum is the Fourier transformation of $\beta_\omega(t)$ in the long time limit, i.e., $t \to \infty$. Assuming

$$x = \frac{\Omega_0^2}{4(1+q^2)}$$

then (16.55) becomes

$$\begin{aligned}\tilde{\beta}_\omega(z) &= -i\Omega_\omega^* \frac{z+\gamma}{(z+i\delta)(z-z_1)(z-z_2)} \\ &= -i\Omega_\omega^* \left[\frac{z_1+\gamma}{z_1+i\delta}\left(\frac{1}{z-z_1} - \frac{1}{z+i\delta}\right)\right. \\ &\quad \left. - \frac{z_2+\gamma}{z_2+i\delta}\left(\frac{1}{z-z_2} - \frac{1}{z+i\delta}\right)\right] \end{aligned} \tag{16.56}$$

with

$$z_{1,2} = \frac{1}{2}\left[-\left(\gamma+\frac{x}{\gamma}\right) \pm \sqrt{\left(\gamma+\frac{x}{\gamma}\right)^2 - 4(xq^2 - 2iqx)}\right]$$

Making the Laplace inverse transformation for (16.56) and using the relation

$$\int_0^\infty e^{-zt} e^{at}\, dt = \frac{1}{z-a} \qquad (\mathrm{Re}(z) > \mathrm{Re}(a))$$

then

$$\begin{aligned}\beta_\omega(t) &= -i\frac{\Omega_\omega^*}{z_1 - z_2}\left\{\frac{z_1+\gamma}{z_1+i\delta}[\exp(z_1 t) - \exp(-i\delta t)]\right. \\ &\quad \left. - \frac{z_2+\gamma}{z_2+i\delta}[\exp(z_2 t) - \exp(-i\delta t)]\right\} \end{aligned} \tag{16.57}$$

When $t \to \infty$, the above equation reduces to

$$\beta_\omega(t)|_{t\to\infty} = -i\Omega_\omega^* \frac{\gamma - i\delta}{(z_1+i\delta)(z_2+i\delta)} \exp(-i\delta t) \tag{16.58}$$

Evidently,

$$\pi|\beta_\omega(t)|^2_{t\to\infty} = \frac{\Omega_0^2}{4\gamma(1+q^2)} \frac{(\delta+q\gamma)^2}{\{[\delta^2 - \frac{\Omega_0^2 q^2}{4(1+q^2)}]^2 + [\delta\gamma + \frac{\Omega_0^2}{4(1+q^2)\gamma}(\delta+2q\gamma)]^2\}} \tag{16.59}$$

which corresponds to the spectrum of the photoelectron. Noticing eqs.(16.56), (16.58), and (16.59), the steady-state photoelectron spectrum can be rewritten as

$$\pi W(\omega) = \lim_{z\to -i\delta} \pi|\tilde{\beta}_\omega(z)(z+i\delta)|^2 \tag{16.60}$$

From (16.59), we can discuss the influences of the Fano factor q and the parameter Ω_0 associated with the laser intensity on the steady-state photoelectron spectrum. First, we study the characteristics of the steady-state photoelectron spectrum when $q \to \infty$, in which the probability of the atom jumping from $|a\rangle$ to $|c,\omega\rangle$ is very small but the probability of the atom jumping from $|a\rangle$ to $|b\rangle$ is comparatively large.

1. When $q \to \infty$, (16.59) reduces to

$$\pi W(\omega) = \frac{\gamma\Omega_0^2/4}{(\delta^2 - \Omega^2/4)^2 + \gamma^2\delta^2} \tag{16.61}$$

Fig.(16.5) gives the distribution of the photoelectron spectrum according to (16.61) for different values of Ω_0^2. If the laser field is weak, i.e., $\Omega_0^2 \leq 2\gamma^2$, the

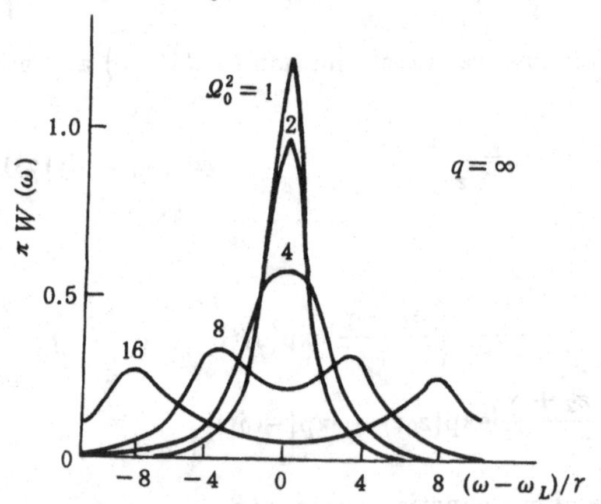

Figure 16.5: Photoelectron spectrum for different Ω_0^2 when $q \to 0$

photoelectron spectrum has a maximum value $\frac{4\gamma}{\Omega_0^2}$ at $\omega = \omega_L$. This means that under the interaction of a weak field, the probability of the transition $|a\rangle \rightarrow |b\rangle \rightarrow |c,\omega\rangle$ is maximum, so the probability of producing the photoelectron with frequency ω_L is maximum. This result is in agreement with that obtained by the perturbation theory previously. Meanwhile, Fig.(16.5) also indicates that with the increasing of the intensity of laser field ($\Omega_0^2 = 1$, $\Omega_0^2 = 2$), the peak value of $\pi W(\omega)$ at $\omega = \omega_L$ decreases and the linewidth of the photoelectron spectrum becomes wide. Since with the increasing of the laser field, the atomic oscillating frequency between $|a\rangle$ and $|b\rangle$ increases, the lifetime of the atom in $|b\rangle$ becomes short. This leads to that the linewidth of the photoelectron spectrum becomes wide.

If $\Omega_0^2 > 2\gamma^2$, i.e., the intensity of the laser field is comparatively strong, then $\pi W(\omega)$ reaches its maximum value $\frac{\Omega_0^2}{\gamma(\Omega_0^2-\gamma^2)}$ at two points $\omega = \omega_L \pm \sqrt{\frac{\Omega_0^2}{4} - \frac{\gamma^2}{2}}$ as shown in Fig.(16.5). This is because , under the resonant interaction of the strong laser field, the autoionization level is splitted into two levels with frequencies $\omega_L \pm \sqrt{\frac{\Omega_0^2}{4} - \frac{\gamma^2}{2}}$, the atom initially in $|a\rangle$ has the same probability to jump into these two levels, then under the influences of the configuration interaction, the atom in these two levels will jump into the continuum $|c, \omega_L \pm \sqrt{\frac{\Omega_0^2}{4} - \frac{\gamma^2}{2}}\rangle$. So the peak point at $\omega = \omega_L$ under the interaction of the weak laser is splitted into two ones at $\omega = \omega_L \pm \sqrt{\frac{\Omega_0^2}{4} - \frac{\gamma^2}{2}}$ due to the interaction of strong field. Fig.(16.5) also shows that with the increasing of the laser field intensity ($\Omega_0^2 = 4$, $\Omega_0^2 = 16$), the peak values of $\pi W(\omega)$ decrease and the linewidth become wide, which is similar to the results of the weak field case ($\Omega_0^2 = 1$ and $\Omega_0^2 = 2$).

2. If the Fano factor q is chosen to be a definite value, then the direct transition from $|a\rangle$ to $|c,\omega\rangle$ can happen under the interaction of laser field. From (16.59) we find that $\delta = -\gamma q$ is always the zero point of the steady-state photoelectron spectrum. That is to say, whether the field is strong or weak, there exists the destructive interference between the two ionization channels, so the Fano zero point always appears in the photoelectron spectrum.

Fig.(16.6) shows the curves of $\pi W(\omega)$ versus δ for q = -1 and different Ω_0. It can be seen from Fig.(16.6) that the lineshape of the photoelectron spectrum for $\Omega_0 = 3$ is very different from that for $\Omega_0 = 1$. For $\Omega_0 = 3$, there displays a sharp peak whose peak value is very high and linewidth is very narrow in the neighbourhood of $\delta = 0.85\gamma$, which implies that the atomic autoionization characteristics change evidently. In order to understand the physical meaning

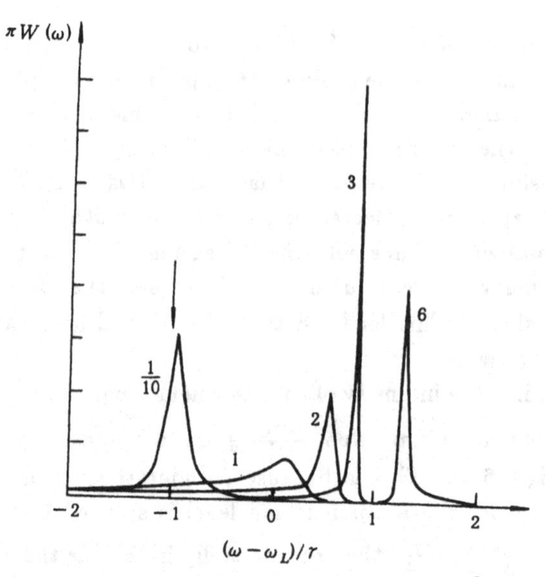

Figure 16.6: Photoelectron spectrum for different Ω_0^2 when $q = -1$

of this change, we rewrite (16.59) as

$$\pi W(\omega) = W_q \left| \frac{\delta + q\gamma}{z_+ z_-} \right|^2 \tag{16.62}$$

with

$$\begin{aligned} W_q &= \frac{\Omega_0^2}{4\gamma(1+q^2)} \\ z_\pm &= \left(\delta \pm R\cos\frac{\theta}{2}\right) - i\frac{\gamma + W_q \pm R\sin\theta}{2} \\ Re^{i\theta} &= \sqrt{-(\gamma - W_q)^2 + 4\gamma W_q (q - i)^2} \end{aligned} \tag{16.63}$$

Eq.(16.62) shows that $\pi W(\omega)$ is the product of two Lorentz functions $\frac{1}{|z_\pm|^2}$, the centers of these two Lorentz functions are $\delta = \pm R\cos(\theta/2)$ and the widthes are $(\gamma + W_q \pm R\sin\theta)/2$, respectively. Clearly, if z_+ or z_- contains the factor $\delta + \gamma q$, then there will display a very sharp peak. This demands

$$\left(\delta \pm R\cos\frac{\theta}{2}\right) - i\frac{\gamma + W_q \pm R\sin\theta}{2} = \delta + \gamma q$$

The solution of above equation gives

$$W_q = \gamma$$

Correspondingly, when the intensity of the laser field obeys

$$\Omega_0^2 = \Omega_c^2 = 4\gamma^2(1+q^2) \tag{16.64}$$

a sharp peak can appear in the photoelectron spectrum $\pi W(\omega)$. For the case of q = -1, $\Omega_c = 2\sqrt{2}\gamma$ and $\delta = \gamma$, which is in agreement with $\Omega_0 = 3\gamma$ and $\delta = 0.85\gamma$.

Now we study the total probability of photoelectron for $W_q = \gamma$. From (16.62) and (16.63) we obtain

$$P_e(\infty) = \int W(\omega)d\omega = \int d\omega \frac{\gamma}{(\omega - \omega_L + \gamma q)^2 + 4\gamma^2} = \frac{1}{2} \tag{16.65}$$

then the unionization probability of the atom initially in the bound state $|a\rangle$ gives

$$P_a(\infty) = 1 - P_e(\infty) = \frac{1}{2} \tag{16.66}$$

The above equation shows when the intensity of laser field obeys (16.64), the atom initially in its bound state $|a\rangle$ can be partially trapped in $|a\rangle$ and not to be completely ionized. This phenomenon is called the confluence of coherence. The reason resulting in this phenomenon is that there are two ionization channels $|a\rangle \leftrightarrow |c,\omega\rangle$ and $|a\rangle \leftrightarrow |b\rangle \leftrightarrow |c,\omega\rangle$ of the atom under the interaction of laser field, these two channels interfere each other. When $W_q = \gamma$, these two channels interference destructively, so the atom initially in $|a\rangle$ can not be completely ionized.

16.3. Above threshold ionization of the atom in a strong laser field

As we know, the atom initially in the bound state $|a\rangle$ can absorb a photon with frequency nearly equal to the above ionization threshold and jump to the continuum $|\omega_1\rangle$, then the atom can release a photoelectron, this is just the Einstein photoelectronic effect including the configuration interaction. However, under the interaction of a strong laser field, if there exist the continuum $|\omega_i\rangle (i = 2, 3, \cdots)$ whose eigenfrequencies are $n\hbar\omega_L$ higher than $|\omega_1\rangle (n = 1, 2, \cdots)$ (as shown in Fig.16.7), the atom in the continuum $|\omega_1\rangle$

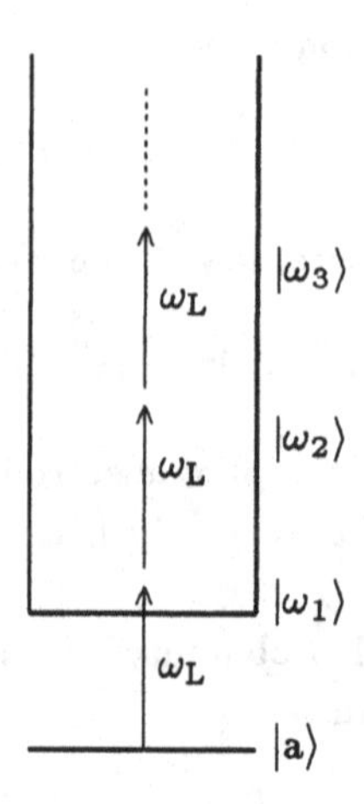

Figure 16.7: Level-diagram of above threshold ionization of atom

can reabsorb photons to jump into the continuum $|\omega_i\rangle$. The atom in $|\omega_i\rangle$ can release a electron with high energy, i.e., a fast electron. We interpret this photoionization as the above threshold ionization. With the development of high power laser technology, the above threshold ionization phenomena are verified by the recent experimental results. The experimental results show that there are multi-peak in the photoelectron spectrum, and the energy difference between two neighbour peaks is the laser frequency ω_L. If the laser field is not very strong, the intensity of the first peak in photoelectron spectrum is stronger than those of other peaks (as shown in Fig.16.8a). With the increasing of the laser power, the first peak reduces, but the second and third peaks increases evidently, furthermore they may exceed the first one (Fig.16.8b). If the laser power is very strong, the first peak, even the first several peaks disappear, and the high-energy peaks increase apparently as shown in Fig.(16.8c). This phenomenon is termed as "peak switching". Evidently, the first peak shown in Fig.(16.8) is the intensity distribution of the photoelectron emitted by the atom in continuum $|\omega_1\rangle$ when the atom is jumped from its initial bound state to the continuum $|\omega_1\rangle$ by absorbing a laser photon. The rest peaks show that the distributions of the photoelectrons emitted by the atom in other new continuum states $|\omega_i\rangle$ $(i \neq 1)$ [as shown in Fig.(16.7)] transited from the con-

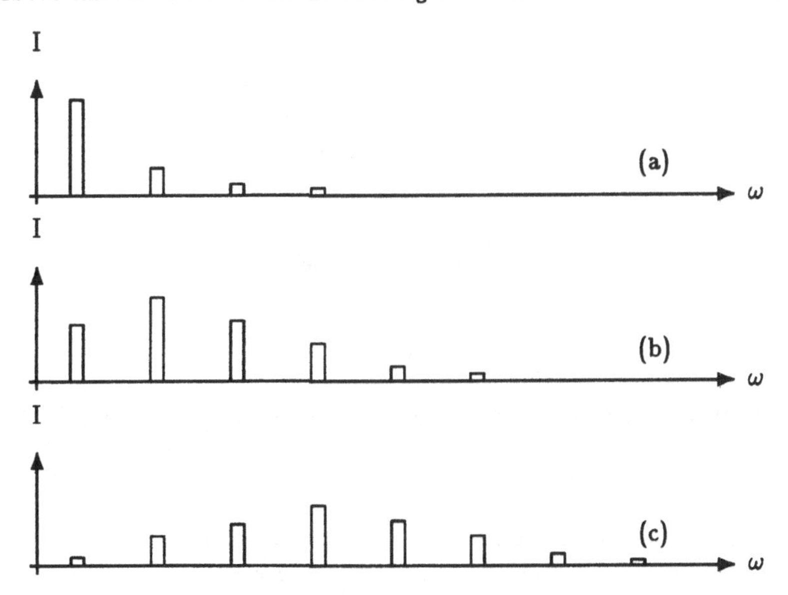

Figure 16.8: Schematic diagram of photoelectron spectra of the above threshold ionization, the gap betwen two neighbouring peaks is $\hbar\omega_L$, (a) weak field case, (b) middle field case, (c) strong field case

tinuum $|\omega_1\rangle$ by absorbing continuously the laser photons. The photoelectron intensities corresponding to the rest peaks are higher than that of first peak, i.e., the above threshold ionization happens in the atom due to the interaction of the strong laser field. Next, we turn to discuss the second-order ionization processes, that is the atom in the continuum $|\omega_1\rangle$ transits to continuum $|\omega_2\rangle$ by absorbing another laser photon, and follows an ionization process, which is called the second-order ionization process. We first investigate the influence of the second-order ionization on the photoelectron spectrum, and further analyze the properties of the above threshold ionization.

The model to be considered here is sketched in Fig.(16.9). The atomic states are composed of a bound state $|a\rangle$, an autoionizing state $|b\rangle$, a set of continuum $|c_1\rangle$ which is coupled to $|b\rangle$ by the configuration interaction, and a second set of continuum $|c_2\rangle$ whose frequencies are higher than those of $|c_1\rangle$. Evidently, under the interaction of the laser field, the transitions $|b\rangle \leftrightarrow |c_2\rangle$ and $|c_1\rangle \leftrightarrow |c_2\rangle$ are the second-order ionization processes. First we use the Fano

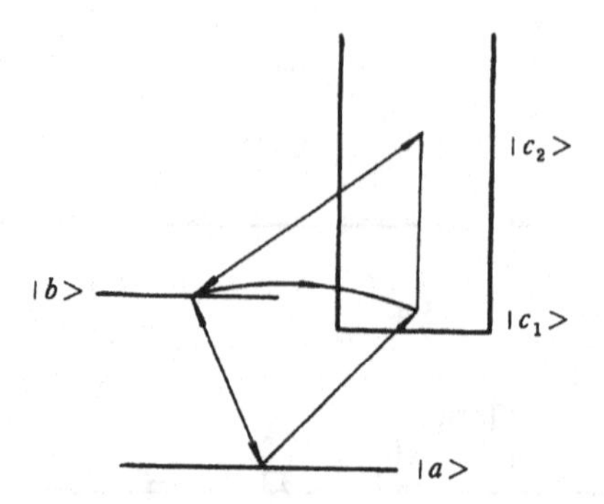

Figure 16.9: Diagram of autoionization model including second-order ionization processes

diagonalization method to replace the autoionization state $|b\rangle$ and the continuum $\{|c_1\rangle\}$ by the Fano states $\{|\omega\rangle\}$, then the model illustrated in Fig.(16.9) is reduced to the cascade two-photon transition model as shown in Fig.(16.10). Therefore, the Hamiltonian of the atom-field coupling system is written as

$$\begin{aligned} H = & E_a|a\rangle\langle a| + \int d\omega\omega|\omega\rangle\langle\omega| + \int dE_{c_2}E_{c_2}|c_2\rangle\langle c_2| \\ & + \left[\int\int dE_{c_2}d\omega\Omega^*_{1\omega}|\omega\rangle\langle c_2| + \int d\omega\Omega_{0\omega}|a\rangle\langle\omega| + \text{h.c.}\right] \end{aligned} \tag{16.67}$$

Similar to (16.40), the matrix elements $\Omega_{0\omega}$ and $\Omega^*_{1\omega}$ are respectively expressed as

$$\begin{aligned} \Omega_{0\omega} &= (q_0+i)\exp(i\psi_0)\langle a|V|c_1\rangle[(\varepsilon-i)^{-1}+(q_0+i)^{-1}] \\ \Omega^*_{1\omega} &= (q_1+i)\exp(i\psi_1)\langle c_1|V_1|c_2\rangle[(\varepsilon-i)^{-1}+(q_1+i)^{-1}] \end{aligned} \tag{16.68}$$

here q_0 is the Fano factor related to the bound state $|a\rangle$ and q_1 is the Fano factor associated with the continuum $|c_2\rangle$, it is defined by

$$q_1 = \frac{\langle b|V_1|c_2\rangle}{\pi\langle b|T|c_1\rangle\langle c_1|V_1|c_2\rangle} \tag{16.69}$$

V and V_1 are the atom-field coupling Hamiltonians. In addition, defining

$$\Omega_0 = \sqrt{4\pi\gamma}(q_0+i)\exp(i\psi_0)\langle a|V|c_1\rangle$$

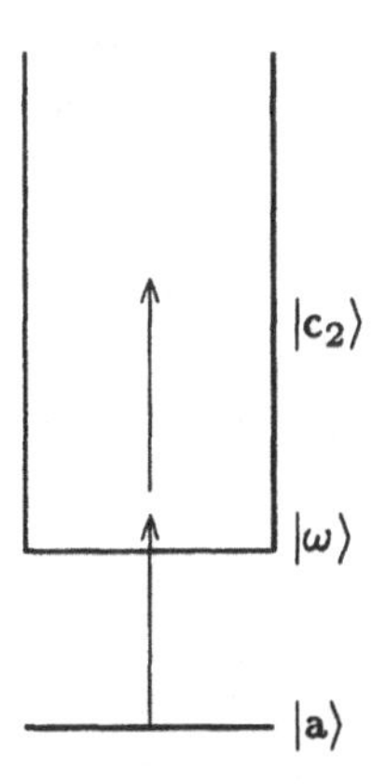

Figure 16.10: Atomic level-diagram after Fano-diagonalization in the presence of second-order ionization processes

$$\Omega_1 = \sqrt{4\pi\gamma}(q_1 + i)\exp(i\psi_1)\langle c_1|V_1|c_2\rangle$$

Since ψ_0 and ψ_1 are the phase factors, we can choose appropriate ψ_0 and ψ_1 so that Ω_0 and Ω_1 may be real. In the following discussion, we assume that Ω_0 and Ω_1 are real. In this case, (16.68) becomes

$$\Omega_{0\omega} = (4\pi\gamma)^{-1/2}\Omega_0[(\varepsilon - i)^{-1} + (q_0 + i)^{-1}] \tag{16.70}$$

$$\Omega^*_{1\omega} = (4\pi\gamma)^{-1/2}\Omega_1[(\varepsilon - i)^{-1} + (q_1 + i)^{-1}] \tag{16.71}$$

Supposing that the atom is initially in the bound state $|a\rangle$, then with the time development, the state vector of the system at time t can be expanded as

$$|\Psi(t)\rangle = C_a(t)|a\rangle + \int d\omega C_\omega(t)|\omega\rangle + \int dE_{c_2} C_{c_2}(t)|c_2\rangle \tag{16.72}$$

Inserting (16.67) and (16.72) into the Schrödinger equation, we have

$$\begin{aligned}
\frac{d}{dt}C_a &= -i\int d\omega\Omega_{0\omega}C_\omega \\
\frac{d}{dt}C_\omega &= -i\delta_{\omega 1}C_\omega - i\Omega^*_{0\omega}C_a - i\int dE_{c_2}\Omega^*_{1\omega}C_{c_2} \\
\frac{d}{dt}C_{c_2} &= -i\delta_{\omega 2}C_{c_2} - i\int d\omega\Omega_{1\omega}C_\omega
\end{aligned} \tag{16.73}$$

where we have assumed that $E_a = 0$, $E_b = \omega_L$, $\delta_{\omega 1} = \omega - \omega_L$, and $\delta_{\omega 2} = E_{c_2} - 2\omega_L$. By the aid of the Laplace transformation defined by (16.47), the above equations become

$$z\tilde{C}_a - 1 = -i\int d\omega \Omega_{0\omega}\tilde{C}_\omega \tag{16.74}$$

$$z\tilde{C}_\omega = -i\delta_{\omega 1}\tilde{C}_\omega - i\Omega^*_{0\omega}\tilde{C}_a - i\int dE_{c_2}\Omega^*_{1\omega}\tilde{C}_{c_2} \tag{16.75}$$

$$z\tilde{C}_{c_2} = -i\delta_{\omega 2}\tilde{C}_{c_2} - i\int d\omega \Omega_{1\omega}\tilde{C}_\omega \tag{16.76}$$

Rewriting (16.76) as

$$\tilde{C}_{c_2} = \frac{-i}{z + i\delta_{\omega_2}}\int d\omega \Omega_{1\omega}\tilde{C}_\omega \tag{16.77}$$

and inserting the above equation into (16.75), we have

$$\tilde{C}_\omega = -i\frac{\Omega^*_{0\omega}}{z + i\delta_{0\omega_1}}\tilde{C}_a - \frac{\pi\Omega^*_{1\omega}}{z + i\delta_{\omega_1}}\int d\omega'\Omega_{1\omega'}\tilde{C}_{\omega'} \tag{16.78}$$

Multiplying the above equation by $\Omega_{1\omega}$ and integrating with respect to ω we obtain

$$\int \Omega_{1\omega}\tilde{C}_\omega d\omega = -i\tilde{C}_a\int \frac{\Omega_{1\omega}\Omega^*_{0\omega}}{z + i\delta_{\omega 1}}d\omega - \pi\int d\omega\frac{\Omega_{1\omega}\Omega^*_{1\omega}}{z + i\delta_{\omega 1}}\int \Omega_{1\omega'}\tilde{C}_{\omega'}d\omega' \tag{16.79}$$

Assuming

$$N_{ij} = \int d\omega\frac{\Omega_{i\omega}\Omega^*_{j\omega}}{z + i\delta_{\omega 1}}(i, j = 0, 1) \tag{16.80}$$

$$K = \int \Omega_{1\omega}\tilde{C}_\omega d\omega \tag{16.81}$$

then

$$K = -\frac{iN_{10}}{1 + \pi N_{11}}\tilde{C}_a \tag{16.82}$$

Inserting (16.68) into (16.80) and using the residue theorem in complex function, N_{ij} becomes

$$N_{ij} = \frac{1}{z + \gamma}\frac{\Omega_i\Omega_j}{4(1 + q_i^2)^{1/2}(1 + q_j^2)^{1/2}}\left[q_iq_j - i(q_i + q_j) + \frac{z}{\gamma}\right]\ (i, j = 0, 1) \tag{16.83}$$

Substituting (16.82) into (16.78) gives

$$\tilde{C}_\omega = -i\frac{\tilde{C}_a}{z+i\delta_{\omega 1}}\left[\Omega^*_{0\omega} - \frac{\pi N_{10}\Omega^*_{1\omega}}{1+\pi N_{11}}\right] \tag{16.84}$$

Combining (16.74) with (16.84) we have

$$\begin{aligned} z\tilde{C}_a &= 1 - \int d\omega \Omega_{0\omega}\left[\frac{\tilde{C}_a}{z+i\delta_{\omega 1}}\left(\Omega^*_{0\omega} - \frac{\pi N_{10}\Omega^*_{1\omega}}{1+\pi N_{11}}\right)\right] \\ &= 1 - \tilde{C}_a\left(N_{00} - \frac{\pi N_{10}N_{01}}{1+\pi N_{11}}\right) \end{aligned}$$

i.e.,

$$\tilde{C}_a = \left[z + N_{00} - \frac{\pi N_{10}N_{01}}{1+\pi N_{11}}\right]^{-1} \tag{16.85}$$

Solving eqs.(16.71), (16.84), and (16.85) yield

$$\tilde{C}_\omega = -\frac{i}{z+i\delta_{\omega 1}}\frac{\Omega^*_{0\omega} - \frac{\pi N_{10}\Omega^*_{1\omega}}{1+\pi N_{11}}}{z+N_{00} - \frac{\pi N_{10}N_{01}}{1+\pi N_{11}}} \tag{16.86}$$

$$\tilde{C}_{c_2} = -\frac{N_{10}}{(z+i\delta_{\omega 2})(1+\pi N_{11})(z+N_{00}-\frac{\pi N_{10}N_{01}}{1+\pi N_{11}})} \tag{16.87}$$

where

$$1+\pi N_{11} = \frac{1+\pi^2 V^2_{c_1c_2}}{1+z'}\left[1+z'+\frac{\gamma_\alpha}{A}\left(1-\frac{i}{q_1}\right)^2\right] \tag{16.88}$$

$$(z+N_{00})(1+\pi N_{11}) - \pi N_{10}N_{01} = \frac{1+\pi^2 V^2_{c_1c_2}}{1+z'}S(z') \tag{16.89}$$

$$\begin{aligned} S(z') &= z'^2 + \frac{z'}{A}\left(1+I+\gamma_\alpha - 2i\frac{\gamma_\alpha}{q_1}\right) \\ &+ \frac{I}{A}\left[q_0^2 + \gamma_\alpha\left(1-\frac{q_0}{q_1}\right)^2\right] - 2iq_0 I/A \end{aligned} \tag{16.90}$$

$$\gamma_\alpha = \frac{V^2_{bc_2}}{T^2_{bc_1}}, \qquad I = \frac{V^2_{ac_1}}{T^2_{bc_1}}, \qquad A = 1+\frac{\gamma_\alpha}{q_1^2} \tag{16.91}$$

$$z' = z/\gamma \tag{16.92}$$

Here we also only concern on the steady-state photoelectron spectrum. In the long time limit, i.e., $t \rightarrow \infty$, from eqs.(18.60) and (16.86) -(16.92) we obtain the steady-state photoelectron spectrum as

$$W_1(\varepsilon) = \lim_{z \rightarrow -i\delta_{\omega 1}} |\tilde{C}_\omega(z)(z + i\delta_{\omega 1})|^2 = \frac{I(\varepsilon + q_0)^2 + \gamma_\alpha^2(1 - q_0/q_1)^2}{A^2|S(\varepsilon)|^2} \quad (16.93)$$

$$W_2(\varepsilon') = \lim_{z \rightarrow -i\delta_{\omega 2}} |\tilde{C}_{c_2}(z)(z + i\delta_{\omega 2})|^2 = \frac{\gamma_\alpha I(\varepsilon' + q_0 + q_1)^2}{(q_1^2 + 1)A^2|S(\varepsilon')|^2} \quad (16.94)$$

where

$$\varepsilon = \frac{\omega - \omega_L}{\gamma}, \qquad \varepsilon' = \frac{E_{c_2} - 2\omega_L}{\gamma} \quad (16.95)$$

(16.93) and (16.94) are just the analytical solutions of the steady-state photoelectron spectrum including the effects of two second-order ionization processes $|b\rangle \leftrightarrow |c_2\rangle$ and $|c_1\rangle \leftrightarrow |c_2\rangle$. Similar to (16.56), $W_1(\varepsilon)$ describes the distribution of photoelectrons whose energy are nearly E_b. From (16.94) we see that there exists another kind of photoelectrons whose energy are nearly $E_b + \omega_L$, the velocity of these photoelectrons is faster than that of the photoelectrons belonging to $W_1(\varepsilon)$. These photoelectrons result from the second-order ionization processes. In order to distinguish $W_1(\varepsilon)$, we interpret $W_2(\varepsilon')$ as the high-energy photoelectron spectrum and $W_1(\varepsilon)$ as the low-energy photoelectron spectrum. Now we discuss the effects of the second-order ionization processes on the photoelectron spectrum from (16.93) and (16.94).

16.8.1 Influences of the second-order ionization processes on the low-energy photoelectron spectrum

A. Influences of the transition $|b\rangle \leftrightarrow |c_2\rangle$ on the low-energy photoelectron spectrum

Eq.(16.93) shows that the influence of the transition $|b\rangle \leftrightarrow |c_2\rangle$ on $W(\varepsilon)$ is determined by the ionization rate $\gamma_\alpha = \pi V_{bc_2}^2/\gamma$ of the autoionizing state $|b\rangle$. Inasmuch as the second-order ionization process $|b\rangle \leftrightarrow |c_2\rangle$ and the first-order ionization process $|a\rangle \leftrightarrow |c_1\rangle$ are induced by the same laser field, the rates $\mu_b = \pi V_{bc_2}^2$ and $\mu_a = \pi V_{ac_1}^2$ have the same magnitude, i.e., $\alpha = \mu_b/\mu_a \approx 1$. And (16.91) indicates that $\gamma_\alpha = \alpha I$, which means that the second-order ionization rate increases with the increasing of the laser power. This result implies that the second-order ionization process plays an important role in the atomic autoionization due to the interaction of a strong laser field.

Fig.(16.11) describes the lower-energy photoelectron spectrum for $q_0 = q_1 = 2$, I=3.2, and different α. The curve for $\alpha = 0$ corresponds to the photoelectron

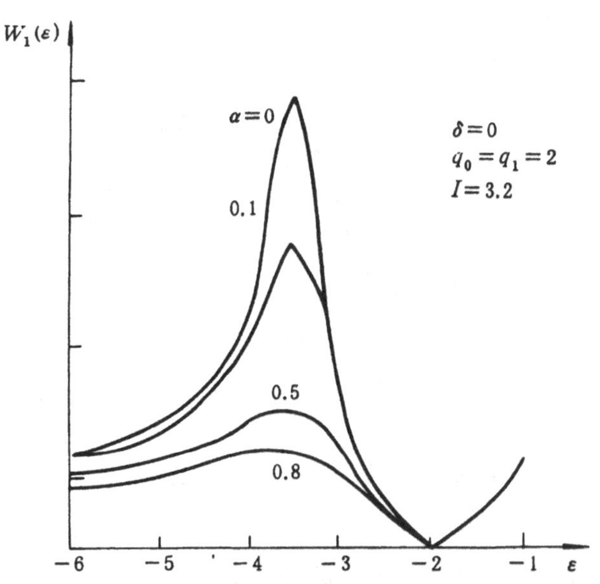

Figure 16.11: Lower-energy photoelectron spectrum for different α when $q_0 = q_1 = 2$ and $I = 3.2$

spectrum neglecting the second-order ionization process $|b\rangle \leftrightarrow |c_2\rangle$, and the curves for $\alpha \neq 0$ are the photoelectron spectrum including the effects of $|b\rangle \leftrightarrow |c_2\rangle$. Here the value of α represents the coupling strength of the transition $|b\rangle \leftrightarrow |c_2\rangle$. It is easily found that the transition $|b\rangle \leftrightarrow |c_2\rangle$ leads to a reduction of the peak of the lower-energy photoelectron spectrum even if a weak coupling α ($\alpha = 0.1$). So the existence of the transition $|b\rangle \leftrightarrow |c_2\rangle$ results in the reduction of the intensity of the low-energy photoelectron spectrum.

B. Influences of the transition $|c_1\rangle \leftrightarrow |c_2\rangle$ on the low-energy photoelectron spectrum

From the definition of q_1 (16.69) we know that q_1 describes the importance role of the transition $|c_1\rangle \leftrightarrow |c_2\rangle$ in comparison $|b\rangle \leftrightarrow |c_2\rangle$. The case for $q_1 \to \infty$ corresponds to neglect the transition $|c_1\rangle \leftrightarrow |c_2\rangle$, and the small q_1 represents the apparent effect of this transition.

Fig.(16.12) gives the lower-energy photoelectron spectrum $W_1(\varepsilon)$ versus ε

for I=1, q_0 = 2, α = 0.8, and different q_1. Fig.(16.12) indicates that the

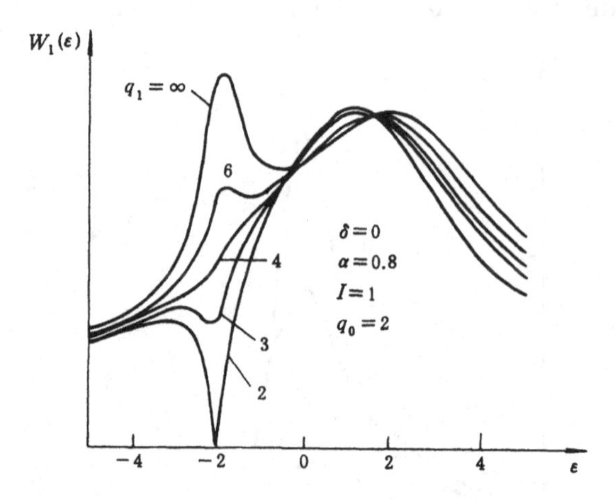

Figure 16.12: Lower-energy photoelectron spectrum for different q_1 when $q_0 = 2$, $I = 1$ and $\alpha = 0.8$

different q_1 leads to that $W_1(\varepsilon)$ displays different lineshapes. Comparing with the lineshape for $q_1 \to \infty$, the transition $|c_1\rangle \leftrightarrow |c_2\rangle$ leads to the previous peak to disappear for $q_1 = 4$, and to become a hollow for $q_1 = 3$, and to close zero point for $q_1 = 2$.

It is valuable to point out that the point $\varepsilon = -\gamma q_0$ is always the Fano zero point of $W_1(\varepsilon)$ when the transition $|c_1\rangle \leftrightarrow |c_2\rangle$ is absent, but (16.93) shows that the Fano zero-point disappears. That is to say, the property of the lower-energy photoelectron spectrum changes evidently due to the influences of the second-order ionization processes.

In conclusion, the main effect of the transition $|b\rangle \leftrightarrow |c_2\rangle$ is to reduce the intensity of $W_1(\varepsilon)$, and the transition $|c_1\rangle \leftrightarrow |c_2\rangle$ changes the lineshape of $W_1(\varepsilon)$.

16.3.2 Higher-energy photoelectron spectrum and the peak switching effect

Since the rate γ_α of the second-order ionization processes is proportional to the laser intensity, the intensity of higher-energy photoelectron spectrum is mainly determined by the intensity of laser field. Fig.(16.13) shows the higher-energy and lower-energy photoelectron spectra for different laser intensities. We can find that for the weak field case (I=0.02), the higher-energy photoelectron spectrum can be neglected, the lower-energy photoelectron plays

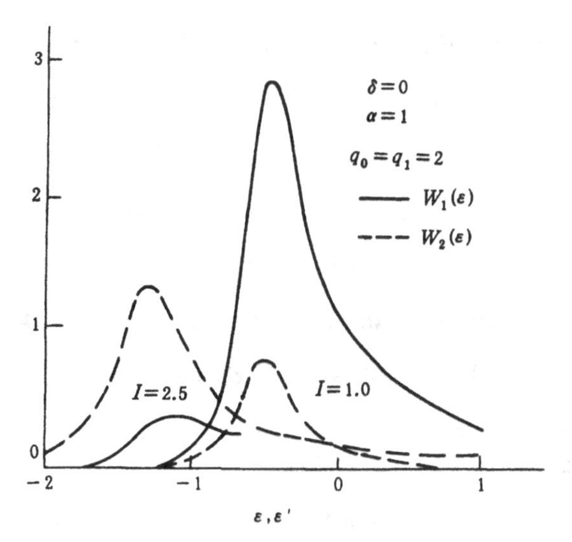

Figure 16.13: The variation of lower-energy and higher-energy photoelectron spectra

a dominant role. But with the increasing of the laser intensity, the intensity of the higher-energy photoelectron spectrum increases evidently and the intensity of the lower-energy photoelectron spectrum decreases. When the intensity of laser field is very high, the higher-energy photoelectron spectrum exceeds the lower-energy one, and the peak-switching effect occurs. So we see that under the interaction of the intensive laser field, the higher-order ionization processes play an important role in the photoionization.

References

1. U.Fano, *Phys.Rev.* **124** (1961) 1866.

2. P.Agostini, G.Petite, and N.K.Rahman, *Phys.Rev.Lett.* **42** (1979) 1127.

3. K.Rzazewski and J.H.Eberly, *Phys.Rev.Lett.* **47** (1981) 408.

4. P.Lambropoulos and P.Zoller, *Phys.Rev.* **A24** (1981) 379.

5. K.Rzazewski and J.H.Eberly, *Phys.Rev.* **A27** (1983) 2026.

6. G.S.Agarwal, S.L.Haan, and J.Cooper, *Phys.Rev.* **A29** (1984) 2552.

7. G.S.Agarwal, S.L.Haan, and J.Cooper, *Phys.Rev.* **A29** (1984) 2565.

8. Z.Deng and J.H.Eberly, *J.Opt.Soc.Am.* **B1** (1984) 102.

9. M.H.R.Hutchinson and K.M.M.Ness, *Phys.Rev.Lett.* **60** (1988) 105.

10. L.Roso-Franco and J.H.Eberly, *J.Opt.Soc.Am.* **B7** (1990) 407.

11. P.L.Knight, M.Lauder, and B.J.Dalton, *Phys.Rep.* **190** (1990) 1.

12. J.H.Eberly, J.Javanien, and K.Rzazewski, *Phys.Rep.* **204** (1991) 331.

13. L.Roso-Franco, Kzazewski, and J.H.Eberly, *J.Mod.Opt.* **38** (1991) 997.

14. Y.L.Shao, D.Charamlambidis, C.Fotakis, J.Zhang, and P.Lambropoulos, *Phys. Rev. Lett.* **67** (1991) 3669.

15. S.Cavalieri, F.S.Pavone, and M.Matera, *Phys. Rev. Lett.* **67** (1991) 3673.

16. O.Faucher, D.Charalambidis, C.Fotakis, J.Zhang, and P.Lambropoulos, *Phys. Rev. Lett.* **70** (1993) 3004.

17. J.Zhang and P.Lambropoulos, *Phys.Rev.* **A45** (1992) 489.

18. G.X.Li and J.S.Peng, *Acta Phys.Sin. (Overseas Edition)* **2** (1993) 569 .

19. R.Grobe and J.H.Eberly, *Phys.Rev.Lett.* **68** (1993) 2905.

20. T.Nakajima and P.Lambropoulos, *Phys.Rev.Lett.* **70** (1993) 1081.

21. T.Nakajima and P.Lambropoulos, *Phys.Rev.* **A50** (1994) 595.

22. R.R.Jones, C.S.Raman, D.W.Schumacher, and P.H.Bucksbaum, *Phys. Rev. Lett.* **71** (1993) 2575.

23. L.D.Noordam, H.Stapelfeldt, D.J.Duncan, and T.F.Gallagher, *Phys. Rev. Lett.* **68** (1992) 1946.

24. A.Wojcik and R.Parzynski, *Phys.Rev.* **A50** (1994) 2475.

25. A.Wojcik and R.Parzynski, *Phys.Rev.* **A51** (1995) 4787.

26. R.R.Jones, *Phys.Rev.Lett.* **74** (1995) 1091; *ibid*, **75** (1995) 1491.

27. L.G.Hanson and P.Lambropoulos, *Phys.Rev.Lett.* **74** (1995) 5009.

28. G.X.Li and J.S.Peng, *Phys.Rev.* **A52** (1995) 465.

29. G.X.Li and J.S.Peng, *Opt.Commun.* **123** (1996) 99.

30. R.B.Vrijen, J.H.Hoogenraad, and L.D.Noordam, *Phys.Rev.* **A52** (1995) 2279.

31. G.X.Li and J.S.Peng, *Phys.Lett.* **A218** (1996) 41.

32. G.X.Li and J.S.Peng, *Z.Phys.* **B** (1997) (in press).

33. G.X.Li and J.S.Peng, *J.Mod.Opt.* **44** (1997) 5.

34. G.X.Li and J.S.Peng, *ACTA Phys.Sinica (O.E.)* (1997) (in press).

CHAPTER 17

MOTION OF THE ATOM IN A LASER FIELD

Up to now we have stressed to study the atomic internal-state properties in the atom-field coupling system, such as the time evolution of atomic populations and the atomic dipole momentum, the atomic population coherent trapping and the squeezing of atomic operators. However, the motion of the atom due to the interaction of the laser field is not considered yet. Since every photon has the momentum, the atomic momentum can be changed in the processes of absorbing and emitting the photons, the dynamic behavior of the atom can be therefore varied by the interaction of a radiation field. As we know, the intensities of the light emitted by usual sources are very weak, the influences of this weak radiation field on the atomic motion can be neglected . But the intensities of laser fields generally are very strong, so the motion of the atom may be evidently changed by the interaction of the laser field. In this chapter we focus our attention on discussing this aspect, especially on the atomic momentum distributions and the force acting on the atom by the field. Usually, the laser field may be classified as the running wave field and the standing wave field. Here we first investigate the atomic diffraction and deflection phenomena under the interaction of a standing wave field when the spontaneous emission effect is neglected. In fact, when the atom passing through a standing wave field, the influence of the spontaneous emission on the atomic diffraction and deflection is very weak, it may be neglected as an approximation. However, when we discuss the force acting on the atom by the radiation field, we will find that the effect of spontaneous emissions plays an important role, it is the physical origin of the radiation pressure or scattering force acting on the atom. In this chapter we will adopt two kinds of approaches to study the force acting on the atom by radiation field. One kind is by use of the quasi-classical theoretical approach, that is, starting from the optical Bloch equations to discuss the force acting on the atom and analyze the properties of the radiation force. The other one is by use of the dressed-state approach, here we investigate the atomic dipole force exerted by the laser field.

17.1 Atomic diffraction and deflection in a standing-wave field

17.1.1 State function of the system of an atom interacting with a standing-wave field

First we discuss the diffraction and deflection of a moving atom passing through a standing-wave laser field. Here the momentum of a two-level atom with mass M before it entering into the laser field is supposed to be $\mathbf{P}_0$ as shown in Fig.(17.1). In the frame of quasiclassical theory, the standing-wave

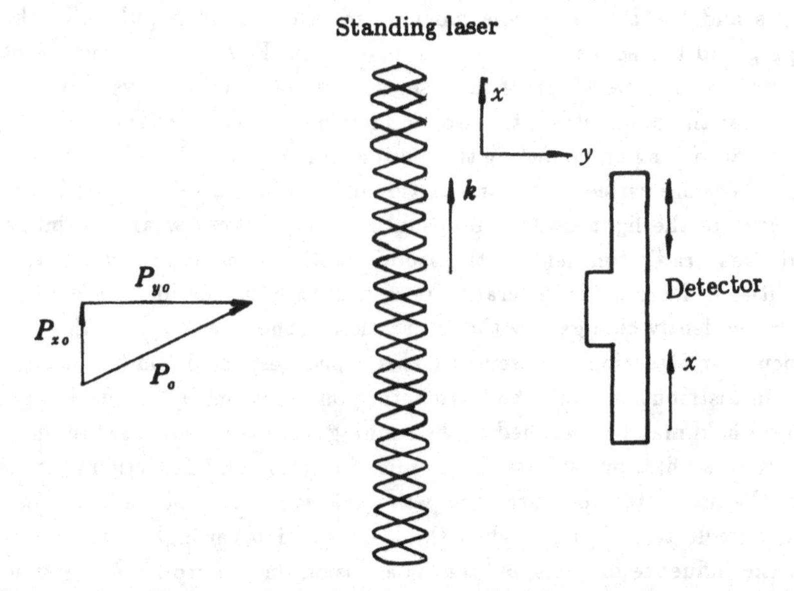

Figure 17.1: Diagram of an atom interacted with a standing-wave field

laser field can be equivalent to the superposition of two running-wave fields with equal intensities but anti-parallel propagating directions, the electric intensity of the standing-wave field can be written in the form

$$\mathbf{E}(x,t) = \varepsilon 2E_0 \cos(kx) = \varepsilon[E_0 \cos(\omega t - kx) + E_0 \cos(\omega t + kx)] \quad (17.1)$$

When a moving atom entering into the standing-wave field, the atom will interacting with the field, so that the Hamiltonian of the system may be described by

$$H = H_A + H_F + V \tag{17.2}$$

Here H_A is the free Hamiltonian of the two-level atom, it must include both the atomic kinetic energy $\mathbf{P}^2/2M$ and the unperturbed internal energy $\hbar\omega_0 S_z$, that is,

$$H_A = \hbar\omega_0 S_z + \frac{\mathbf{P}^2}{2M}$$

Therefore, the ground and excited states of the two-level atom with momentum $\mathbf{P}$ may be denoted by ψ_+ and ψ_-, respectively. Evidently, $\psi_\pm(\mathbf{P})$ obey the eigenvalue equation as

$$H_A\psi_\pm(\mathbf{P}) = \left(\frac{\mathbf{P}^2}{2M} \pm \frac{\hbar\omega_0}{2}\right)\psi_\pm(\mathbf{P}) \tag{17.3}$$

Since a single-mode standing-wave field with frequency ω and wave vector $\mathbf{k}$ can be equivalent to be the sum of two running-wave radiation fields which are in opposite sense with respect to their propagation vectors. Thus the free Hamiltonian of the field in (17.2) is

$$H_F = \sum_{i=1}^{2} \hbar\omega a_i^\dagger a_i \tag{17.4}$$

where $a_i^\dagger$ and $a_i (i{=}1,2)$ stand for the creation and annihilation operators for photons with frequency ω and wave vector $\mathbf{k}_i$, here $\mathbf{k}_i$ satisfy

$$\mathbf{k}_1 = -\mathbf{k}_2 = \mathbf{k} \tag{17.5}$$

In this case, the interaction Hamiltonian V of this atom-field coupling system may be expressed by

$$V = \hbar\sum_{i=1}^{2} \varepsilon_0[S_+a_i \exp(ik_ix) + S_-a_i^\dagger \exp(-ik_ix)] \tag{17.6}$$

here x represents the coordinate of the atom along the wave vector direction. The difference between (17.6) and (4.36) is that (17.6) contains the terms $\exp(\pm ik_ix)$ associated with the wave vector. This is because we only concern on the evolution of the atomic internal state in (4.36), where the value of x is restricted within one atomic radius, so in the dipole approximation there exists $\exp(\pm ik_i x) \approx 1$. But here the subject under consideration is the atomic motion in the laser field, then x corresponds to the coordinate of the atomic center-of-mass in the laser field, the dipole approximation is therefore invalid. Thus the interaction Hamiltonian of the atom-field coupling system is determined by (17.6). Starting from the Hamiltonian (17.2), we may obtain the state vector of the system, then can find the time evolution of the atomic momentum due to the interaction of the laser field. Before solving the state vector, we first analyse the interaction processes between the atom and the field.

It is supposed that the atom is in the ground state $\psi_-(\mathbf{P})$ with momentum $\mathbf{P}$ before entering into the laser field, and both running-wave fields are in the coherent states $|\alpha_i\rangle$, whose mean photon numbers $N_i = |\alpha_i|^2$. For $N_i \gg 1$, the coherent state $|\alpha_i\rangle$ can be approximated as a number state $\phi_i(N_i) = |N_i\rangle$ in the quasi-classical approximation. Therefore, the atom-field coupling system is initially in the state

$$\Phi_0 = \phi_1(N_1)\phi_2(N_2)\psi_-(\mathbf{P}) \tag{17.7}$$

When the atom entering into the laser field, the atom can absorb photons from either of the two running-wave fields and vice versa. For the atom being in the initial state (17.7), it can absorb a photon from the first running-wave field to reach its excited state, in this case the state vector of the first running-wave field evolves from $\phi_1(N_1)$ to $\phi_1(N_1 - 1)$, and simultaneously the atomic momentum is increased by $\hbar\mathbf{k}_1$. That is to say, the state vector of the system develops into

$$\Phi_1 = \phi_1(N_1 - 1)\phi_2(N_2)\psi_+(\mathbf{P} + \hbar\mathbf{k}_1) \tag{17.8}$$

Similarly, the atom initially in the state Φ_0 may absorb a photon from the second running-wave field with wave vector $\mathbf{k}_2$ to reach the excited state Φ_{-1}

$$\Phi_{-1} = \phi_1(N_1)\phi_2(N_2 - 1)\psi_+(\mathbf{P} + \hbar\mathbf{k}_2) \tag{17.9}$$

When the system is in the state Φ_1, the atom can stimulately emit a photon due to the interaction of laser field. Either the atom can emit a photon with the wave vector $\mathbf{k}_1$ and the system returns to the initial state Φ_0, or it emits a photon with the wave vector $\mathbf{k}_2$ and goes back its ground state. If the atom in the state Φ_1 emits a photon with the wave vector $\mathbf{k}_2$, then the state vector of the second running-wave field evolves into $\phi_2(N_2 + 1)$, and simultaneously the atomic momentum is decreased by $\hbar\mathbf{k}_2$, the atom-field coupling system therefore evolves into the state

$$\Phi_2 = \phi_1(N_1 - 1)\phi_2(N_2 + 1)\psi_-(\mathbf{P} + \hbar\mathbf{k}_1 - \hbar\mathbf{k}_2) \tag{17.10}$$

Similarly, that the atom emitting a photon with the momentum $\hbar\mathbf{k}_1$ evolves into the first running-wave field from the state Φ_{-1} generates the state

$$\Phi_{-2} = \phi_1(N_1 + 1)\phi_2(N_2 - 1)\psi_-(\mathbf{P} - \hbar\mathbf{k}_1 + \hbar\mathbf{k}_2) \tag{17.11}$$

In fact, when the atom passing through the standing-wave field, there exist a number of emitting and absorbing processes. If the atom initially in the state Φ_0 undergoing even number times of absorbing and emitting processes, the state of the system evolves to be the form as

$$\Phi_n(\mathbf{P}) = \phi_1\left(N_1 - \frac{n}{2}\right)\phi_2\left(N_2 + \frac{n}{2}\right)\psi_-\left(\mathbf{P} + \frac{n}{2}\hbar\mathbf{k}_1 - \frac{n}{2}\hbar\mathbf{k}_2\right) \tag{17.12}$$

here n is an even integer. When the atom undergoes odd number times of absorbing and emitting processes, the state of the atom-field coupling system may evolves into

$$\Phi_n(\mathbf{P}) = \phi_1\left(N_1 - \frac{n+1}{2}\right)\phi_2\left(N_2 + \frac{n-1}{2}\right)\psi_+\left(\mathbf{P} + \frac{n+1}{2}\hbar\mathbf{k}_1 - \frac{n-1}{2}\hbar\mathbf{k}_2\right) \tag{17.13}$$

Fig.(17.2) illustrates the variations of the states of atom-field coupling system in the transition processes. Evidently, the state vectors $\Phi_n(\mathbf{P})$ consist of the set of complete basis vectors of the atom-field coupling system, by using this set, the state of the system at time t can be expanded as

$$\Psi(t) = \int d\mathbf{P}\sum_n C_n(\mathbf{p}, t)\Phi_n(\mathbf{P})\exp(-i\alpha_n t) \tag{17.14}$$

where α_n is an arbitrary phase at liberty to choose for subsequent mathematical convenience.

Substituting (17.14) and (17.2) into the Schrödinger equation

$$i\hbar\frac{\partial}{\partial t}|\Psi(t)\rangle = H|\Psi(t)\rangle \tag{17.15}$$

and noticing the relation

$$\exp(\pm ik_i x)\Psi(P) = \Psi(P \pm \hbar k_i) \tag{17.16}$$

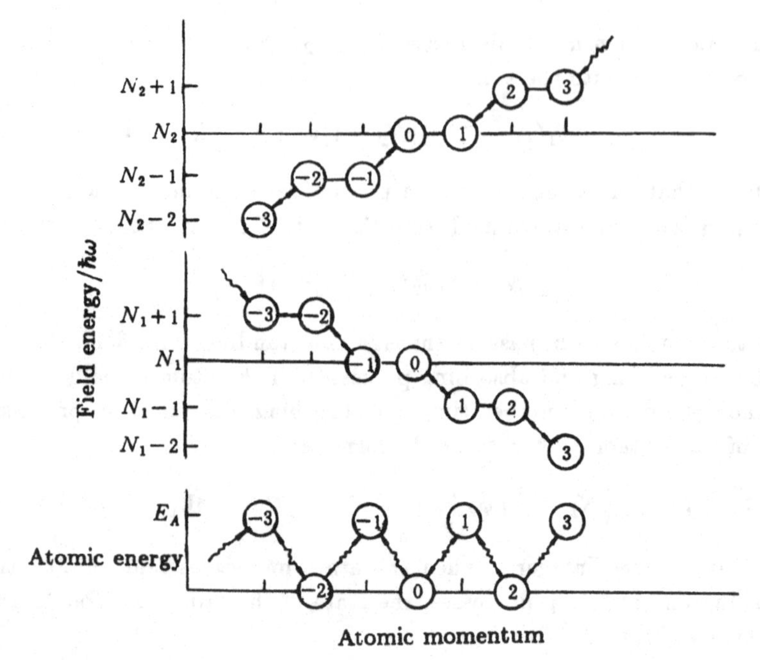

Figure 17.2: Variations of states of the atom-field coupling system in the transition processes

we find the following set of coupled ordinary differential equations for the probability amplitudes $C_n(\mathbf{P}, t)$,
(1). For n is an even integer, we have

$$\begin{aligned} i\hbar\frac{d}{dt}C_n = &\left[\left(N_1 - \frac{n}{2}\right)\hbar\omega + \left(N_2 + \frac{n}{2}\right)\hbar\omega\right. \\ &\left.+\frac{1}{2M}\left(\mathbf{P} + \frac{n}{2}\hbar\mathbf{k}_1 - \frac{n}{2}\hbar\mathbf{k}_2\right)^2 - \frac{1}{2}\hbar\omega_0 - \hbar\alpha_n\right] C_n \\ &+\hbar\varepsilon_0\left(\sqrt{N_1 - \frac{n}{2}}C_{n+1} + \sqrt{N_2 + \frac{n}{2}}C_{n-1}\right) \end{aligned} \tag{17.17}$$

(2). For n is an odd integer, we have

$$i\hbar\frac{d}{dt}C_n = \left[\left(N_1 - \frac{n+1}{2}\right)\hbar\omega + \left(N_2 + \frac{n-1}{2}\right)\hbar\omega\right.$$

$$+\frac{1}{2M}\left(\mathbf{P}+\frac{n+1}{2}\hbar\mathbf{k}_1-\frac{n-1}{2}\hbar\mathbf{k}_2\right)^2+\hbar\omega_0/2-\hbar\alpha_n\bigg]C_n$$

$$+\hbar\varepsilon_0\left(\sqrt{N_1-\frac{n-1}{2}}C_{n-1}+\sqrt{N_2+\frac{n+1}{2}}C_{n+1}\right) \tag{17.18}$$

Inasmuch as the phase factor α_n can be chosen arbitrarily, for making the expressions simple, we make the choice in (17.17) and (17.18) for α_n as follows. For n is even

$$\hbar\alpha_n=(N_1-n/2)\hbar\omega+(N_2+n/2)\hbar\omega+\hbar\omega/2+E_0 \tag{17.19}$$

and for n is odd

$$\hbar\alpha_n=\left(N_1-\frac{n+1}{2}\right)\hbar\omega+\left(N_2+\frac{n-1}{2}\right)\hbar\omega-\frac{\hbar\omega}{2}+E_0 \tag{17.20}$$

Here E_0 is the initial kinetic energy of the atom, i.e.,

$$E_0=\frac{P_x^2+P_y^2}{2M}=\frac{P^2}{2M} \tag{17.21}$$

Substituting (17.19)-(17.21) into (17.17) and (17.18) we obtain
(1). For n is even

$$i\frac{d}{dt}C_n=\frac{\Delta}{2}C_n+\left[\frac{1}{2M}\left(\frac{n}{2}\hbar\mathbf{k}_1-\frac{n}{2}\hbar\mathbf{k}_2\right)^2+\frac{\mathbf{P}}{\hbar M}\cdot\left(\frac{n}{2}\hbar\mathbf{k}_1-\frac{n}{2}\hbar\mathbf{k}_2\right)\right]C_n$$

$$+\varepsilon\left(\sqrt{N_1-\frac{n}{2}}C_{n+1}+\sqrt{N_2+\frac{n}{2}}C_{n-1}\right) \tag{17.22}$$

(2). For n is odd

$$i\frac{d}{dt}C_n=-\frac{\Delta}{2}C_n+\left[\frac{1}{2M}\left(\frac{n+1}{2}\hbar\mathbf{k}_1-\frac{n-1}{2}\hbar\mathbf{k}_2\right)^2\right.$$

$$\left.+\frac{\mathbf{P}}{\hbar M}\cdot\left(\frac{n+1}{2}\hbar\mathbf{k}_1-\frac{n-1}{2}\hbar\mathbf{k}_2\right)\right]C_n$$

$$+\varepsilon_0\left(\sqrt{N_1-\frac{n-1}{2}}C_{n-1}+\sqrt{N_2+\frac{n+1}{2}}C_{n+1}\right) \tag{17.23}$$

here the term $\Delta=\omega-\omega_0$ is the off-resonant detuning.

Considering the case of the strong coherent field, i.e., $N_1, N_2 \gg 1$, and the range of laser field is infinite, so the value of n which represents the number of absorptions and emissions of photons is sufficiently less than N_1 and N_2, the approximation may be therefore as follows

$$\sqrt{N_2 + n/2} \approx \sqrt{N_2 + (n+1)/2} \approx \sqrt{N_2}$$
$$\sqrt{N_1 - n/2} \approx \sqrt{N_1 - \frac{n-1}{2}} \approx \sqrt{N_1} \tag{17.24}$$

Since the laser field is a standing-wave one with amplitude $2E_0$, it can be equivalent to be the sum of two running-wave ones with identical intensity, therefore

$$\sqrt{N_1} = \sqrt{N_2} = \sqrt{N} \tag{17.25}$$

Substituting (17.5) and (17.25) into (17.22) and (17.23) we obtain
(1). For n is even

$$i\frac{d}{dt}C_n = \frac{\Delta C_n}{2} + \left[\hbar\frac{n^2k^2}{2M} + \frac{nP_xk}{M}\right]C_n + \frac{\Omega}{2}(C_{n+1} + C_{n-1}) \tag{17.26}$$

(2). For n is odd

$$i\frac{d}{dt}C_n = -\frac{\Delta C_n}{2} + \left[\hbar\frac{n^2k^2}{2M} + \frac{nP_xk}{M}\right]C_n + \frac{\Omega}{2}(C_{n+1} + C_{n-1}) \tag{17.27}$$

here $\Omega = 2\varepsilon_0\sqrt{N}$ is the Rabi frequency of the atomic oscillation. In (17.26) and (17.27) we introduce the parameters

$$\hbar b = \frac{\hbar^2k^2}{2M}, \qquad q = \frac{P_x}{\hbar b} \tag{17.28}$$

Evidently, $\hbar b$ corresponds to the recoil kinetic energy after the atom absorbs or emits one photon, q is the ratio of the initial atomic momentum along the x direction to the momentum of photons. The term $\frac{nP_xk}{M}$ means the Doppler frequency shifts and $\frac{n^2\hbar^2k^2}{2M}$ represents the recoil kinetic energy. Then (17.26) and (17.27) reduce to

$$i\frac{d}{dt}C_n = \left[\frac{\Delta}{2} + b(n^2 + 2qn)\right]C_n + \frac{\Omega}{2}(C_{n+1} + C_{n-1}) \tag{17.29}$$

and

$$i\frac{d}{dt}C_n = \left[-\frac{\Delta}{2} + b(n^2 + 2qn)\right]C_n + \frac{\Omega}{2}(C_{n+1} + C_{n-1}) \tag{17.30}$$

By solving (17.29) and (17.30) we can obtain the probability amplitude $C_n(p,t)$ at time t, then we can find the time evolution of the atomic momentum under the interaction of a standing-wave laser field. Next we analyze the atomic diffraction and deflection phenomena when the atom passed through the standing-wave laser field.

17.1.2 Diffraction of the atom under the interaction of a laser field

If we assume the initial atomic momentum $\mathbf{P}_0$ is perpendicular to the wave vector $\mathbf{k}$ of the laser field, i.e., $\mathbf{P}_x = 0, q = 0$, and the atom-field coupling system obeys the resonant condition, i.e., $\Delta = 0$. And for the case of the strong laser field, the atomic Rabi frequency Ω is sufficiently larger than the recoil kinetic energy term nb^2, so that the term nb^2 in (17.29) and (17.30) can be ignored, then (17.29) and (17.30) are simplified to

$$i\frac{d}{dt}C_n(t) = \frac{\Omega}{2}(C_{n+1} + C_{n-1}) \tag{17.31}$$

If making a transformation for the above equation as follows

$$C_n(t) = (-i)^n R_n(t) \tag{17.32}$$

then (17.31) changes as

$$2\frac{d}{dt}R_n(t) = \Omega(R_{n-1} - R_{n+1}) \tag{17.33}$$

Introducing the parameter $\xi = \Omega t$, then the above equation becomes

$$2\frac{d}{d\xi}R_n(\xi) = R_{n-1}(\xi) - R_{n+1}(\xi) \tag{17.34}$$

Noticing the recurrence relation of the Bessel function

$$2\frac{d}{dx}J_n(x) = J_{n-1}(x) - J_{n+1}(x) \tag{17.35}$$

we find the solution of (17.34) is to be a Bessel function, that is

$$R_n(t) = AJ_n(\Omega t) \tag{17.36}$$

here A is a constant. The n-order Bessel function $J_n(\Omega t)$ obeys

$$J_n(\Omega t) = J_{-n}(\Omega t) \tag{17.37}$$

Substituting (17.36) and (17.37) into (17.32) and considering the normalization condition

$$\sum_{n=-\infty}^{\infty} |C_n(t)|^2 = 1$$

we obtain

$$|A|^2 \sum_{n=-\infty}^{\infty} J_n^2(\Omega t) = |A|^2 = 1$$

here we have used the relation

$$\sum_{n=-\infty}^{\infty} J_n^2(x) = 1$$

If we choose an appropriate phase factor to lead to A=1, then

$$C_n(t) = (-i)^n J_n(\Omega t)$$

Therefore, the probability to find the atomic momentum $n\hbar k$ in the x-direction is

$$P_n(t) = |C_n(t)|^2 = J_n^2(\Omega t) \tag{17.38}$$

Inasmuch as the n-order Bessel function $J_n(\Omega t)$ and the -n-order Bessel function $J_{-n}(\Omega t)$ obey (17.37), the atomic momentum exhibits symmetric distribution in the positive x-direction and the negative x-direction. Fig.(17.3) shows the distribution of the atomic momentum at different time, these patterns are similar to the pattern of diffraction photons from a thick grating, that is to say, the atom exhibits the diffraction characteristic when it leaves from a standing-wave field. The different momentum and momentum distribution of the atom in the x-direction can be exhibited by the fringes with different intensities detected by the detector at different positions. This diffraction characteristics of the atom passing through the standing-wave laser field reflects the wave property of the atom. Fig.(17.3) shows that with the increasing of the interaction time, in other words, the range of the laser field is widen, the probability finding the atomic momentum in the neighbourhood of n=0 decreases. When $\Omega t = 12.5$, $P_n(t)$ is equal to zero for $|n| > 15$, and when $\Omega t = 25$, $P_n(t)$=0 only for $|n| > 27$. This means that with the increasing of the interaction time, the distribution of the atomic momentum becomes wide, that is to say, the atom absorbs more and more photons from the first running-wave field and transfers these photons into the second running-wave field, and vice versa. From

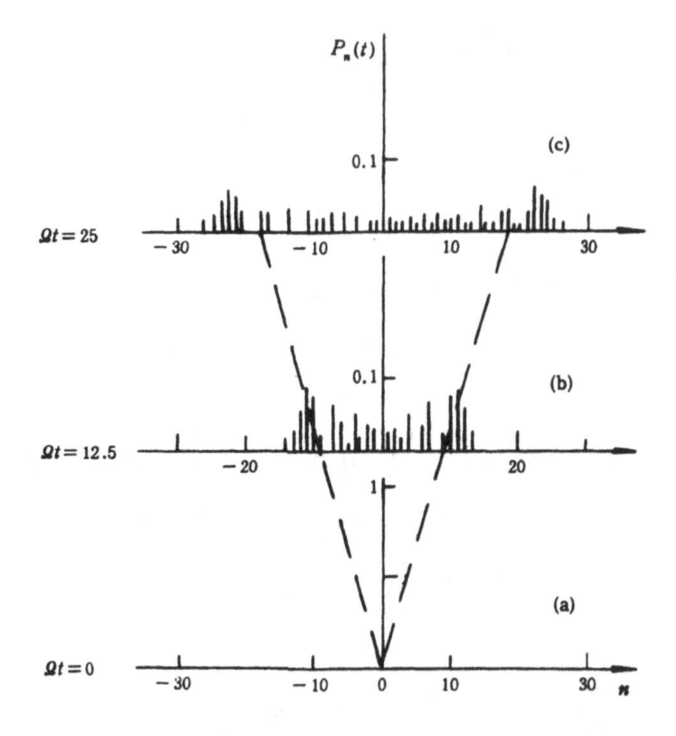

Figure 17.3: Distribution of the atomic momentum at different time when $\Delta = 0$ and $P_x = 0$

Figs.(17.3b) and (17.3c) we see that when $\Omega t = 12.5$ and 25, $P_n(t)$ reaches its maximum value at the neighbourhood of $|n| = 11$ and 22, respectively, which means that the momentum distribution of the atom under the interaction of a standing-wave field exhibits clear two-peak structure along the wave vector. However, for $|n| > 13$ (or 26), $P_n(t)$ decreases to zero rapidly. This result shows that since the atomic momentum distribution function $J_n^2(\Omega t)$ reaches its maximum value at $n \cong \Omega t$, the change of the atomic momentum develops with time at the speed of Ω corresponding to the Rabi frequency.

As yet we have discussed the diffraction phenomenon of the atom with initial momentum $P_x = 0$ in a standing-wave field at the resonant case ($\Delta = 0$). Next we study the case of the atomic motion for $P_x \neq 0$ at initial time. Since $P_x \neq 0$, the term $2bqnC_n$ corresponding to the Doppler effect can not be neglected in (17.29-30). If the atomic Rabi frequency Ω and the Doppler shift

$2bqn$ are remarkably larger than the atomic recoil energy term, i.e.,

$$bn^2 \ll |2bqn|, \quad \Omega/2 \tag{17.39}$$

then (17.29) and (17.30) reduce to

$$i\frac{d}{dt}C_n = 2bqnC_n + \frac{\Omega}{2}(C_{n+1} + C_{n-1}) \tag{17.40}$$

Making the transformation for the above equation as follows

$$C_n(t) = (-i)^n \tilde{C}_n(t) \exp(-ibqnt) \tag{17.41}$$

we obtain

$$\begin{aligned}\frac{d}{dt}\tilde{C}_n(t) &= -ibq\tilde{C}_n(t) + \frac{\Omega}{2}[\tilde{C}_{n-1}\exp(ibqt) - \tilde{C}_{n+1}\exp(-ibqt)] \\ &= -ibqn\tilde{C}_n + \frac{1}{2}\Omega\cos(bqt)(\tilde{C}_{n-1} - \tilde{C}_{n+1}) \\ &\quad + \frac{i}{2}\Omega\sin(bqt)(\tilde{C}_{n-1} + \tilde{C}_{n+1})\end{aligned}$$

that is,

$$\begin{aligned}2\frac{d}{dt}\tilde{C}_n &= -i[2bqn\overline{C}_n - \Omega\sin(bqt)(\tilde{C}_{n+1} + \tilde{C}_{n-1})] \\ &\quad + \Omega\cos(bqt)(\tilde{C}_{n-1} - \tilde{C}_{n+1})\end{aligned} \tag{17.42}$$

If we assume

$$\eta = \frac{\Omega}{bq}\sin(bqt) \tag{17.43}$$

then (17.42) becomes

$$\begin{aligned}2\frac{d}{dt}\tilde{C}_n(\eta) &= -i\frac{bq}{\Omega\cos(bqt)}\{2n\tilde{C}_n(\eta) - \eta[\tilde{C}_{n+1}(\eta) + \tilde{C}_{n-1}(\eta)]\} \\ &\quad + \tilde{C}_{n-1}(\eta) - \tilde{C}_{n+1}(\eta)\end{aligned} \tag{17.44}$$

Noticing (17.35) and the recurrence formula of the Bessel function

$$2nJ_n(x) = x[J_{n-1}(x) + J_{n+1}(x)] \tag{17.45}$$

The solution of (17.44) reads immediately as

$$\tilde{C}_n(t) = BJ_n\left(\frac{\Omega}{bq}\sin bqt\right) \tag{17.46}$$

here B is the normalization constant, $J_n(\frac{\Omega}{bq}\sin bqt)$ is the n-order Bessel function with special parameter $\frac{\Omega}{bq}\sin(bqt)$. By means of $\sum\limits_{n=-\infty}^{\infty} J_n^2(x) = 1$ it is easily found that B=1. Thus,

$$\begin{aligned} C_n(t) &= (-i)^n \exp(-ibqnt) J_n\left(\frac{\Omega}{bq}\sin bqt\right) \\ &= \exp[-in(2bqt+\pi)/2] J_n\left(\frac{\Omega}{bq}\sin bqt\right) \end{aligned} \tag{17.47}$$

Correspondingly, the atomic momentum distribution function is to be

$$P_n = |C_n(t)|^2 = J_n^2\left(\frac{\Omega}{bq}\sin bqt\right) \tag{17.48}$$

To analyze the time evolution of the probability distribution function (17.48), one can find the moving characteristics of the atom. Different from eq.(17.38), here the special parameter $\frac{\Omega}{bq}\sin bqt$ changes nonlinearly with the time evolution, it is a sine function with frequency bq. Fig.(17.4) plots the atomic momentum distribution function $P_n(t)$ versus n for $2bq = 1.63\times 10^8 Hz$ and $\frac{\Omega}{bq} = 20$ at time bqt=0, $\pi/4$, $\pi/2$ and $3\pi/4$. We see that with the time development, the atomic momentum distribution function alters periodically and the period is $\pi/(bq)$. The reason is

$$\begin{aligned} P_n(t+\frac{\pi}{bq}) &= J_n^2\left[\frac{\Omega}{bq}\sin(\pi+bqt)\right] = J_n^2\left(-\frac{\Omega}{bq}\sin bqt\right) \\ &= J_n^2\left(\frac{\Omega}{bq}\sin bqt\right) = P_n(t) \end{aligned}$$

This is very different from Fig.(17.3) which shows that with the time development, the atomic momentum distribution becomes wide, but here only at $bqt = (k+1/2)\pi$ $(k = 0,1,2,\cdots)$ the width of the momentum distribution reaches its maximum value. That is to say, at $bqt = (k+1/2)\pi$, the atom can transfer most photons from the first running-wave field into the second running-wave field and vice versa. Fig.(17.4e) also shows that the atomic momentum distribution at time $t = k\pi/(bq)(k = 1,2,\cdots)$ is the same as that t=0. This means that the atomic diffraction does not occur, which different from the case neglecting the Doppler shift. In addition, when $t \neq 0,\pi$, the atomic momentum distribution exhibits distinctive two-peak structure. For the different time points, the maximum value of $P_n(t)$ appears at the neighbourhood of

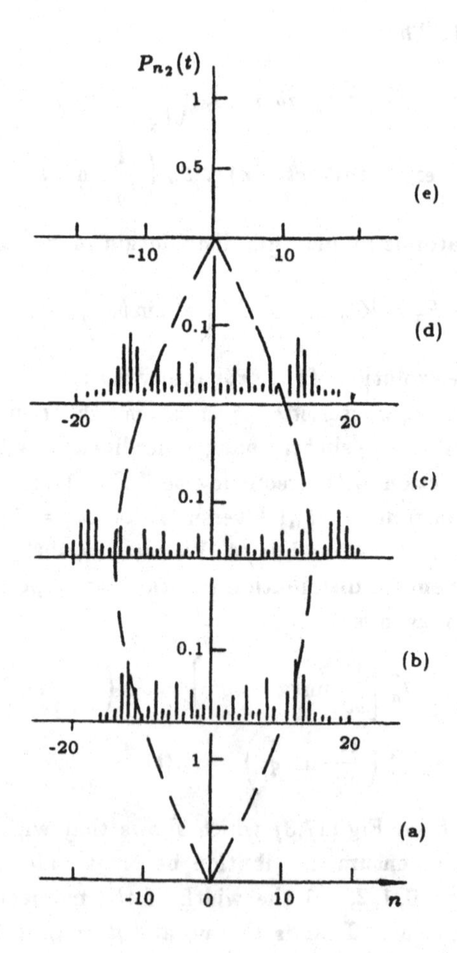

Figure 17.4: Distribution of the atomic momentum at different time when $\Delta = 0$ and $P_x \neq 0$, (a) $bqt = 0$, (b) $bqt = \pi/4$, (c) $bqt = \pi/2$, (d) $bqt = 3\pi/4$, and (e) $bqt = \pi$

$|n| \approx \Omega|\sin bqt|/(bq)$, that is to say, under the interaction of the standing-wave field, the alteration of the atomic momentum is related not only to the atomic Rabi frequency Ω, but also to the parameter bq associated with the Doppler shift, and this alteration is periodical with period $\pi/(bq)$. It is worthwhile to mention that all these above theoretical results are verified by recent experiments.

Next let us to discuss the influence of the detuning on the atomic diffraction. In order to simplify the calculation, we assume that the initial atomic momentum is perpendicular to the wave vector $\mathbf{k}$, so that the Doppler shift terms in (17.30) and (17.29) can be ignored. We also assume that the atomic recoil kinetic energy is sufficiently less than the detuning Δ and the Rabi frequency Ω, so that the term bn^2C_n can be dropped. Therefore, (17.29) and (17.30) are simplified as

$$i\frac{d}{dt}C_{2n} = \frac{\Delta C_{2n}}{2} + \frac{\Omega}{2}(C_{2n-1} + C_{2n+1}) \tag{17.49}$$

$$i\frac{d}{dt}C_{2n-1} = -\frac{\delta C_{2n-1}}{2} + \frac{\Omega}{2}(C_{2n} + C_{2n-1}) \tag{17.50}$$

If the detuning $|\Delta|$ is sufficiently larger than the atomic Rabi frequency Ω, then when the atom initial in its ground state goes to its excited state, it follows to its ground state immediately. Thus the time of the atom staying in its excited state is very short. In this case, adopting the adiabatic approximation to solve (17.49) and (17.50) is valid, that is,

$$\frac{d}{dt}C_{2n-1} = 0 = \frac{\Delta C_{2n-1}}{2} + \frac{\Omega}{2}(C_{2n} + C_{2n-1})$$

i.e.,

$$C_{2n-1} = \frac{\Omega}{\Delta}(C_{2n} + C_{2n-2}) \tag{17.51}$$

Substituting equation (17.51) and $C_{2n+1} = \Omega(C_{2n+2} + C_{2n})/\Delta$ into (17.49) we obtain

$$i\frac{d}{dt}C_{2n} = \left(\frac{\Delta}{2} + \frac{\Omega^2}{\Delta}\right) C_{2n} + \frac{\Omega^2(C_{2n+2} + C_{2n})}{2\Delta} \tag{17.52}$$

Assuming that

$$C_{2n}(t) = (-i)^n \exp[-i(\Delta/2 + \Omega^2/\Delta)t]Q_n(t) \tag{17.53}$$

then (17.52) becomes

$$\frac{d}{dt}Q_n = \frac{\Omega^2(Q_{n-1} - Q_{n+1})}{2\Delta} \tag{17.54}$$

By means of (17.35) $C_{2n}(t)$ reads

$$C_{2n}(t) = (-i)^n \exp[-i(\Delta/2 + \Omega^2/\Delta)t] A J_n(\Omega^2 t/\Delta) \tag{17.55}$$

here A is the normalization constant. From (17.37) we obtain

$$1 = \sum_n |C_{2n}|^2 + |C_{2n-1}|^2 = A^2 \sum_n \{J_n^2(\Omega^2 t/\Delta)$$
$$+(\Omega/\Delta)^2 |J_n(\Omega^2 t/\Delta) - iJ_{n-1}(\Omega^2 t/\Delta)|^2\} = A^2[1 + 2(\Omega/\Delta)^2]$$

That is

$$A = [1 + 2(\Omega/\Delta)^2]^{-1/2} \approx 1 - (\Omega/\Delta)^2 \tag{17.56}$$

here we have used $\Omega/|\Delta| \ll 1$ and only retained to the second-order correction of Ω/Δ. Therefore, the probability distribution function of the atomic momentum read as

$$P_{2n}(t) = |C_{2n}|^2 = [1 - 2(\Omega/\Delta)^2] J_n^2(\Omega^2 t/\Delta) \tag{17.57}$$
$$P_{2n-1}(t) = |C_{2n-1}|^2 \approx (\Omega/\Delta)^2 [J_n^2(\Omega^2 t/\Delta) + J_{n-1}^2(\Omega^2 t/\Delta)] \tag{17.58}$$

The functions P_{2n} and P_{2n-1} are the probabilities of finding the atomic momentum $2n\hbar k (n = 0, \pm 1, \pm 2, \cdots)$ and $(2n-1)\hbar k$ along the x-direction, respectively. Evidently, the odd probabilities P_{2n-1} are much smaller than the even ones P_{2n} as shown in Fig.(17.5). Inasmuch as the detuning is remarkably larger than the Rabi frequency, the atom initially in its ground state absorbs photons with momentum $\hbar\mathbf{k}_1$ from the first running-wave field to its excited state will return to its ground state immediately and simultaneously emits photons with momentum $\hbar\mathbf{k}_2$. In this processes, the atom transfer photons from the first running-wave field into the second ones, while the atomic momentum changes $2n\hbar k$. Therefore , the atomic momentum have clear distribution in $2n\hbar k$. During these transition processes, the lifetime of the atom staying in its excited state is very short, then the probabilities measuring the atomic momentum $(2n-1)\hbar k$ are very small. This phenomenon had been verified by the diffraction experiment of the Na atom as shown in Fig.(17.5), in which the vertical lines represent the theoretical calculations and the circles show the

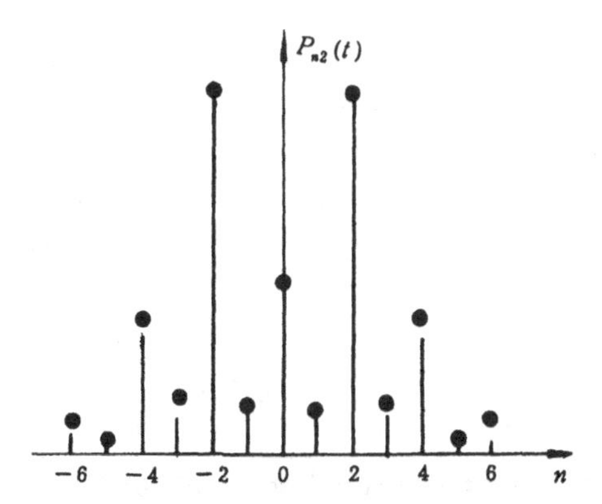

Figure 17.5: Distribution of the atomic momentum at nonresonant case. Here the vertical lines represent theoretical results and dots are experimental results

heights of the measurement results. It is worthwhile to mention that we have neglected the spontaneous emission effects in the above discussion. In fact, when the detuning Δ is larger than the spontaneous emission rate γ, neglecting the spontaneous emission effects is valid. The theoretical result shown in Fig.(17.5) is just the one obtained by neglecting the spontaneous emission effect.

So far we have analyzed the time evolution of the probability distribution of the atomic momentum due to the interaction of a standing-wave laser field, this corresponds to the phenomenon of the atomic diffraction. Next we discuss the time evolution of the atomic momentum, which corresponds to the phenomenon of the atomic deflection in the laser field. The atomic deflection phenomenon can be utilized to realize the isotope separation of the netural atom.

17.1.3 Deflection of the atom in a standing wave field

Suppose that the initial atomic momentum of the atom is perpendicular to the wave vector $\mathbf{k}$, i.e., q=0. In addition, the comparative weak recoil kinetic energy bn^2 is also ignored. In this case, $P_n(t) = P_{-n}(t) = J_n^2(\Omega t)$, the expectation value of the atomic momentum along the wave vector direction therefore gives

$$\langle P\rangle = \sum_{n=-\infty}^{\infty} n\hbar k P_n(t) = 0$$

This is because the atomic momentum exhibits the symmetric distribution in two sides of $n=0$, so that the expectation value $\langle P\rangle=0$. However, the expectation value $\langle|P|\rangle$ which represents the practical variation of the atomic momentum is not equal to zero, it reads

$$\langle|P|\rangle = \sum_{n=-\infty}^{\infty} |n|\hbar k J_n^2(\Omega t) = 2\hbar k \sum_{n=1}^{\infty} n J_n^2(\Omega t) \tag{17.59}$$

Using the summation formula of the Bessel function

$$\sum_{k=1}^{\infty} k J_k^2(x) = \frac{x^2}{2}[J_0^2(x) + J_1^2(x)] - \frac{x}{2} J_0(x) J_1(x) \tag{17.60}$$

we have

$$\langle|P|\rangle = \hbar k \Omega^2 t^2 [J_0^2(\Omega t) + J_1^2(\Omega t)] - \hbar k \Omega t J_0(\Omega t) J_1(\Omega t) \tag{17.61}$$

When $\Omega t \to \infty$, according to the asymptotic formula of the Bessel function

$$J_\mu(\Omega t) \approx \sqrt{\frac{2}{\pi \omega t}} \cos(\Omega t - \mu\pi/2 - \pi/4)$$

(17.61) becomes

$$\begin{aligned}\langle|P|\rangle &\approx \hbar k \Omega^2 t^2 \frac{2}{\pi \omega t}[\cos^2(\Omega t - \pi/4) + \cos^2(\Omega t - \pi/2 - \pi/4)] \\ &\quad - \hbar k \Omega t \frac{2}{\pi \Omega t} \cos(\Omega t - \pi/4)\cos(\Omega t - \pi/2 - \pi/4) \\ &\approx 2\hbar k \Omega t/\pi \approx 0.64 \hbar k \Omega t\end{aligned} \tag{17.62}$$

This means that for a long interaction time, the expectation value of the absolute of atomic momentum operator increases linearly with time.

The expectation value of the square of atomic momentum reads

$$\langle P^2\rangle = \sum_{n=-\infty}^{\infty} (n\hbar k)^2 J_n^2(\Omega t) = 2\hbar^2 k^2 (\Omega t/2)^2 = \frac{1}{2}\hbar^2 k^2 \Omega^2 t^2 \tag{17.63}$$

Here we have used the relation

$$\sum_{n=0}^{\infty} n^2 J_n^2(x) = (x/2)^2 \tag{17.64}$$

From (3.6.63), the square root of $\langle P^2 \rangle$ gives

$$\langle P^2 \rangle^{1/2} = \frac{\hbar k \Omega t}{\sqrt{2}} \tag{17.65}$$

That is to say, the spread of the atomic momentum increases linearly with time, i.e.,

$$(\Delta P) = \sqrt{\langle P^2 \rangle - \langle P \rangle^2} = \frac{\hbar k \Omega t}{\sqrt{2}} \tag{17.66}$$

Now we see that the atom whose momentum along **k** is initially equal to zero may gain momentum along **k** under the interaction of the standing-wave field, so the direction of atomic motion will diverge to **k** direction, that is, the motion of atom will deviate from its y-direction shown in Fig.(17.1). This phenomenon is called the atomic deflection. The average deflection angle is

$$\theta_{rms} = \langle P^2 \rangle^{1/2} / P_{y0} = \frac{\hbar \omega \Omega t}{\sqrt{2} c P_{y0}} \tag{17.67}$$

Next we analyze a typical case to estimate the value of deflection angle. If we assume that the atomic mass M is $1.6 \times 10^{-25} kg$, the length of the atom-field interaction region L equals to $2.6 \times 10^{-8} m$, the laser frequency ω is $3 \times 10^{15} s^{-1}$, the atomic Rabi frequency Ω equals to $5.6 \times 10^{11} s^{-1}$ (which corresponds to the laser power $I \approx 10^6 W/cm^2$), and the initial velocity of the atom along the y-direction v_{y0} is 420 m/s, which equals to the velocity of atomic thermal motion at room temperature, the interaction time $t = L/v_{y0} = 6.2 \times 10^{-9} s$, then from (17.67) we have

$$\theta_{rms} \approx 2.2^0$$

It is clear that the atom deviates from its original moving direction by 2.2^0, this result is able to be measured by experiments. In practice, the observation result in experiment can approach to 5^0. The atomic deflection phenomenon may be applied to realize the isotope separation technique of neutral atoms. The main idea of laser isotope separation is that by adjusting the laser frequencies to be resonant with the transition frequency of one of isotopes in atomic beam, the atoms belonging to this isotope will gain momentum along **k** direction due to

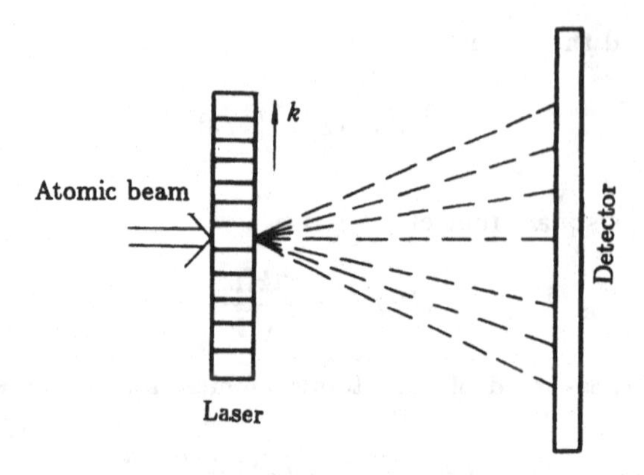

Figure 17.6: Deflection of an atom moving in the laser field

the resonant interaction of field, then these atoms will deviate from its original moving direction as shown in Fig.(17.6), therefore, these atoms are separated from the atomic beam. Since the laser frequency can be adjusted exactly, and the laser power can be high, this technique on the basis of the atomic deflection in the laser field is a very effective method to separate the isotope of neutral atoms.

17.2 Force on an atom exerted by the radiation field

In the previous discussions we know that when the atom passes through the laser field, the atomic diffraction and deflection phenomena will appear, that is to say, the mechanical motion of the atom is changed, which means that the radiation field exerts force on the atom. This force results from the transferring momentum of photons from radiation fields to the atoms. As we know, under the interaction of laser field, the atom can absorb one photon to reach its excited state. In this transition process, the atom not only absorbs the photon energy $\hbar\omega$, but also gains the photon momentum $\hbar\omega/c$ along the propagating direction of the laser beam. After time τ (τ is the lifetime of atom in its excited level), the atom spontaneously emit one photon with energy $\hbar\omega_0$ to go back its ground state, meanwhile, the atom also gain recoil momentum $\hbar\omega_0/c$ (ω_0 is the atomic transition frequency). Because the propagating directions of photons produced by atomic spontaneously emission are random and isotropy (shown in Fig.(17.7)), the accumulative effect of the photonic recoil momentum is zero, then the momentum gained by the atom is only $n\hbar\omega/c$ along the propagating direction of the light beam resulted from the pure absorption processes

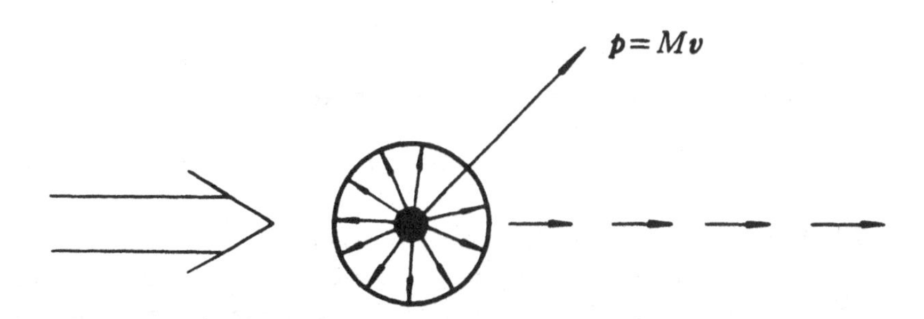

Figure 17.7: Momentum variation of a moving atom in absorption and spontaneous emission processes

of photons (here n is the sum number of the atomic absorption processes). The force on atom due to gaining this momentum is called the radiation force, usually it also named as scattering force. Since this force is associated with the spontaneous emission, it is also termed as the spontaneous emission force. Although there exist the stimulative emission processes in the atom-field coupling system, for the case that the space-distribution of the running-wave is homogenereous, the photons generated by the atomic induced emission processes have the same propagating direction as the light beam, the momentum gained by the absorption processes is therefore cancelled by the recoil momentum. In these absorption-emission processes, the atomic momentum is not changed and there does not exist force on the atom. However, if the space-distribution of the intensity of the light field is inhomogeneous (for example, the standing-wave field), the induced emission processes can also result in the force on atom. Because the atom can absorb photon with the wave vector $\mathbf{k}$ in this field, and stimulatively emit photon with the wave vector -$\mathbf{k}$. In this absorption-emission circle, the atomic momentum is changed with $2\hbar\omega/c$, thus the atom is suffered by a force, we call this force associated with induced transition as the radiation dipole force. Therefore, under the interaction of laser field, there exist two kinds of radiation forces on the atom, i.e., the spontaneous emission one and the dipole one. Next we discuss these two forces in detail by considering a system of a two-level atom interacting with the radiation field.

17.2.1 Quasi-classical description of the radiation force

We first analyze the properties of the spontaneous emission force and dipole force on the atom exerted by the radiation field by means of the quasi-classical theory.

The Hamiltonian for a two-level atom in a classical electromagnetic field may be taken in the form within the dipole approximation as

$$H = \frac{\mathbf{P}^2}{2M} + E_+|+\rangle\langle+| + E_-|-\rangle\langle-| - \mu \cdot \mathbf{E}(\mathbf{R}, t) \tag{17.68}$$

where $\mathbf{P}^2/2M$ is the kinetic energy associated with the center-of-mass momentum $\mathbf{P}$, E_+ (E_-) represents the energy of the atomic excited (ground) state, μ stands for the atomic dipole operator, and $\mathbf{E}(\mathbf{R}, t)$ is the electric field at the center-of-mass position $\mathbf{R}$. In the Heisenberg picture, the operators $\mathbf{R}$ and $\mathbf{P}$ of the atom obey the equations of motion

$$\frac{d}{dt}\mathbf{R} = \frac{1}{i\hbar}[\mathbf{R}, H] = \nabla_p H = \frac{\mathbf{P}}{M} \tag{17.69}$$

$$\frac{d}{dt}\mathbf{P} = \frac{1}{i\hbar}[\mathbf{P}, H] = -\nabla_R H = \nabla(\mu \cdot \mathbf{E}) \tag{17.70}$$

For simplicity to calculate, we assume the extension Δr of the atomic wave packet is sufficient small compared with the laser wavelength λ, i.e.,

$$\Delta r \ll \lambda, \quad \text{or} \quad \frac{1}{\Delta r} \gg \frac{1}{\lambda} \tag{17.71}$$

In this case, it is valid to replace the atomic position operator $\mathbf{R}$ by its expectation value $\mathbf{r} = \langle \mathbf{R} \rangle$ in treating the atomic motion. Therefore, combining (17.69) with (17.70), we obtain the force on the atom as

$$\mathbf{F} = M\frac{d^2}{dt^2}\mathbf{r} = \langle \nabla(\mu \cdot \mathbf{E}) \rangle \tag{17.72}$$

In order to simplify the calculation, we assume that the polarization vector ϵ of the electric field $\mathbf{E}(\mathbf{R}, t)$ is unrelated to $\mathbf{R}$ and t, i.e.,

$$\mathbf{E}(\mathbf{R}, t) = \epsilon\varepsilon(\mathbf{R}, t) = \frac{1}{2}\epsilon\varepsilon(\mathbf{E}) \exp\{i[\theta(\mathbf{R}) + \omega t]\} + \text{c.c.} \tag{17.73}$$

Inserting (17.73) into (17.72) we have

$$\mathbf{F} = \langle \mu \cdot \epsilon \nabla \varepsilon(\mathbf{R}, t) \rangle$$

Considering (17.71), the operator $\mathbf{R}$ can be replaced by $\mathbf{r}$ in the above equation, that is, the force on the atom exerted by radiation field becomes

$$\mathbf{F} = \langle \mu \cdot \epsilon \rangle \nabla \varepsilon(\mathbf{r}, t) \tag{17.74}$$

This is just a basic equation to discuss the force on the atom exerted by radiation field. In order to solve $\mathbf{F}$, we must obtain the time evolution of the atomic dipole moment $\langle \mu \cdot \epsilon \rangle$.

Here we only consider the velocity of atom is so small that in one spontaneous emission lifetime the distance of the atomic motion is less than the laser wavelength λ, i.e.,

$$v\Gamma^{-1} \ll \lambda \qquad or \quad kv/\Gamma \ll 1 \tag{17.75}$$

Noticing the Heisenberg uncertainty relation

$$M \Delta v \Delta r \geq \hbar \tag{17.76}$$

and (17.71), we obtain

$$\frac{\hbar k}{M \Delta v} \ll 1 \tag{17.77}$$

The above equation shows that the influence of absorpting or emitting one photon on the atomic momentum distribution is very weak. In order to make eqs. (17.75) and (17.77) to be valid simultaneously, there must exist

$$\frac{\hbar^2 k^2 / 2M}{\hbar \Gamma} \ll 1 \tag{17.78}$$

This means that the recoil momentum gained by the atom due to the interaction of laser field is small compared with the spontaneous emission linewidth Γ of the atomic excited state. That is to say, in one-photon absorption or emission process, the detuning between the atom and the laser field does not change clearly. Under this assumption condition, neglecting the effect of the kinetic energy term $\mathbf{P}^2/2M$ is reasonable when we discuss the atomic internal motion, therefore, the Hamiltonian of the atom-field coupling system may be rewritten as

$$H' = E_2|+\rangle\langle+| + E_1|-\rangle\langle-| - \mu \cdot \mathbf{E}(t) \tag{17.79}$$

here $\mathbf{E}(t) = \epsilon\varepsilon(\mathbf{r}(t), t)$.

The atomic internal state vector at time t can be described as

$$|\Psi(t)\rangle = C_2(t)|+\rangle + C_1(t)|-\rangle \tag{17.80}$$

From the Schrödinger equation it is easy to obtain that

$$i\hbar\frac{d}{dt}C_1 = E_1C_1 - \mu\cdot\varepsilon(t)C_2$$
$$i\hbar\frac{d}{dt}C_2 = E_2C_2 - \mu\cdot\varepsilon(t)C_1 \tag{17.81}$$

here $\mu = \langle 1|\mu\cdot\epsilon|2\rangle$ is the amplitude of the atomic dipole momentum, $\varepsilon(t) = \varepsilon(\mathbf{r}(t),t)$. $\varepsilon(t)$ in (17.81) contains the rapid oscillating term $\exp(\pm i\omega t)$. In order to solve (17.81), we make the following transforms

$$C_1(t) = D_1(t)\exp\{-iE_1t/\hbar + i[\Delta t + \theta(t)]/2\}$$
$$C_2(t) = D_2(t)\exp\{-iE_2t/\hbar - i[\Delta t + \theta(t)]/2\} \tag{17.82}$$

here Δ is the detuning, which obeys

$$\Delta = \omega - \omega_0 \qquad (\omega_0 = (E_2 - E_1)/\hbar) \tag{17.83}$$

and $\theta(t)$ is the phase of the laser field, it is defined by

$$\theta(t) = \theta(\mathbf{r}(t), t) \tag{17.84}$$

Therefore, (17.81) becomes

$$i\hbar\frac{d}{dt}D_1 = \frac{1}{2}\hbar\left[\Delta + \frac{d}{dt}\theta(t)\right]D_1 - \mu\varepsilon D_2\exp\{-i[\theta(t) + \omega t]\}$$
$$i\hbar\frac{d}{dt}D_2 = -\frac{1}{2}\hbar\left[\Delta + \frac{d}{dt}\theta(t)\right]D_2 - \mu\varepsilon D_1\exp\{i[\theta(t) + \omega t]\} \tag{17.85}$$

Substituting (17.73) into the above two equations and making the rotating - wave approximation, i.e., neglecting the terms containing $\exp[\pm i(\omega + \omega_0)t]$ in eq.(17.85) we have

$$i\hbar\frac{d}{dt}D_1 = \frac{1}{2}\hbar\left[\Delta + \frac{d}{dt}\theta(t)\right]D_1 - \frac{1}{2}\mu E(t)D_2$$
$$i\hbar\frac{d}{dt}D_2 = -\frac{1}{2}\hbar\left[\Delta + \frac{d}{dt}\theta(t)\right]D_2 - \frac{1}{2}\mu E(t)D_1 \tag{17.86}$$

From (17.80) we know that the density matrix elements representing the atomic internal motion can be defined by

$$\rho_{nm} = C_nC_m^* \tag{17.87}$$

Considering (17.82), then the slowly-varying elements of the density matrix

$$\sigma_{nm} = D_n D_m^* \tag{17.88}$$

and ρ_{nm} satisfy the relations

$$\begin{aligned} &\rho_{11} = \sigma_{11}, \quad \rho_{22} = \sigma_{22}, \quad \rho_{12} = \sigma_{12}\exp\{i[\theta(t) + \omega t]\} \\ &\rho_{21} = \sigma_{21}\exp\{-i[\theta(t) + \omega t]\} \end{aligned} \tag{17.89}$$

It is easy to obtain from (17.86) that σ_{nm} obey

$$\begin{aligned} \frac{d}{dt}\sigma_{11} &= -\frac{1}{2}i\Omega(\sigma_{12} - \sigma_{21}) \\ \frac{d}{dt}\sigma_{22} &= \frac{1}{2}i\Omega(\sigma_{12} - \sigma_{21}) \\ \frac{d}{dt}\sigma_{12} &= -i\left[\Delta + \frac{d}{dt}\theta(t)\right]\sigma_{12} + \frac{1}{2}i\Omega(\sigma_{22} - \sigma_{11}) \end{aligned} \tag{17.90}$$

here $\Omega(t) = \mu E(t)/\hbar$, it is just the on-resonance Rabi flopping frequency for a two-level atom in a field with amplitude E. As yet the atomic damping resulted from the interaction of the vacuum field is not included. According to Chapters 2, 5 and 8 we know that the time evolution equations of density matrix elements σ_{nm} included the spontaneous damping become

$$\begin{aligned} \frac{d}{dt}\sigma_{11} &= -\frac{1}{2}i\Omega(\sigma_{12} - \sigma_{21}) + \Gamma\sigma_{22} \\ \frac{d}{dt}\sigma_{22} &= \frac{1}{2}i\Omega(\sigma_{12} - \sigma_{21}) - \Gamma\sigma_{22} \\ \frac{d}{dt}\sigma_{12} &= -i\left[\Delta + \frac{d}{dt}\theta(t)\right]\sigma_{12} + \frac{1}{2}i\Omega(\sigma_{22} - \sigma_{11}) - \frac{1}{2}\Gamma\sigma_{12} \end{aligned} \tag{17.91}$$

The appearance of the last term in (17.91) is based on $\sigma_{11} + \sigma_{22} = 1$.

By use of the density matrix elements ρ_{nm}, the atomic dipole momentum $\langle\mu \cdot \epsilon\rangle$ can be expressed as

$$\langle\mu \cdot \epsilon\rangle = \mu(\rho_{12} + \rho_{21}) = \mu\{\sigma_{12}\exp\{i[\theta(t) + \omega t]\} + \text{c.c}\} \tag{17.92}$$

Inserting (17.73) and (17.92) into (17.74), and neglecting the fast oscillating terms $\exp(\pm 2i\omega t)$, and noticing

$$E(t) = E(\mathbf{r}(t)), \quad \frac{d}{dt}\theta(t) = \frac{d}{dt}\theta(\mathbf{r}(t)) = \frac{\partial}{\partial t}\mathbf{r}\cdot\nabla\theta(\mathbf{r}) = \nabla\theta(\mathbf{r})\cdot\frac{d}{dt}\mathbf{r} \tag{17.93}$$

we obtain

$$\mathbf{F} = \frac{1}{2}\mu\nabla E(\sigma_{12} + \sigma_{21}) - \frac{i}{2}\mu E\nabla\theta(\sigma_{12} - \sigma_{21}) \tag{17.94}$$

If let

$$U = \sigma_{12} + \sigma_{21}, \qquad V = -i(\sigma_{12} - \sigma_{21}), \qquad W = \sigma_{22} - \sigma_{11} \tag{17.95}$$

then (17.91) and (17.95) become

$$\mathbf{F} = \frac{1}{2}\hbar(U\nabla\Omega + V\Omega\nabla\theta) \tag{17.96}$$

and

$$\begin{aligned} \frac{d}{dt}U &= \left(\Delta + \frac{d}{dt}\theta\right)V - \frac{1}{2}\Gamma U \\ \frac{d}{dt}V &= -\left(\Delta + \frac{d}{dt}\theta\right)U + \Omega W - \frac{1}{2}\Gamma V \\ \frac{d}{dt}W &= -\Omega V - \Gamma(W+1) \end{aligned} \tag{17.97}$$

Equation (17.97) is just the optical Bloch equation. Solving (17.97) and inserting the solutions of U and V into (17.96) we can obtain the force on the two-level atom exerted by the laser field with amplitude $E(\mathbf{r})$ and phase $\theta(\mathbf{r})$.

From (17.95) we see that during one spontaneous emission lifetime $1/\Gamma$, the variations of the amplitude $E(t) = E(\mathbf{r}(t), t)$ and the phase derivative $\frac{d}{dt}\theta(t) = \nabla\theta(\mathbf{r}(t), t) \cdot \frac{d\mathbf{r}}{dt}$ of the laser field are very small, so U, V, and W are fast variables compared with $E(t)$ and $\frac{d}{dt}\theta(t)$. Thus U and V in (17.96) can be replaced by their steady solutions which may be resulted from the following equations

$$\begin{aligned} \frac{d}{dt}U &= \left(\Delta + \frac{d}{dt}\theta\right)V - \frac{1}{2}\Gamma U = 0 \\ \frac{d}{dt}V &= -\left(\Delta + \frac{d}{dt}\theta\right)U + \Omega W - \frac{1}{2}\Gamma W = 0 \\ \frac{d}{dt}W &= -\Omega V - \Gamma(W+1) = 0 \end{aligned} \tag{17.98}$$

Solving the above three equations we obtain

$$U = -\frac{\Omega(\Delta + \frac{d}{dt}\theta)}{4(\Delta + \frac{d}{dt}\theta)^2 + \Gamma^2 + 2\Omega^2}$$

$$V = -\frac{2\Gamma\Omega}{4(\Delta + \frac{d}{dt}\theta)^2 + \Gamma^2 + 2\Omega^2} \tag{17.99}$$

Therefore, the force $\mathbf{F}$ on the two-level atom exerted by the radiation field gives

$$\mathbf{F} = -\hbar\frac{\Gamma\Omega^2\nabla\theta + (\Delta + \frac{d}{dt}\theta)\nabla\Omega^2}{4(\Delta + \frac{d}{dt}\theta)^2 + \Gamma^2 + 2\Omega^2} \tag{17.100}$$

Suppose that the laser field is a running-wave field with amplitude ε_0 unrelated to the position of space

$$\varepsilon(\mathbf{r}(t)) = \varepsilon_0 \cos(\mathbf{k}\cdot\mathbf{r} - \omega t) \tag{17.101}$$

In this case, there exist

$$\Omega = \mu\varepsilon_0/\hbar = \text{constant}, \qquad \theta(\mathbf{r}) = -\mathbf{k}\cdot\mathbf{r} \tag{17.102}$$

then (17.100) becomes

$$\mathbf{F}_1 = \frac{\Gamma\Omega^2\hbar\mathbf{k}}{4(\Delta - \mathbf{k}\cdot\frac{d\mathbf{r}}{dt})^2 + \Gamma^2 + 2\Omega^2} \tag{17.103}$$

Now we analyze the properties of the force $\mathbf{F}_1$ on the atom exerted by the radiation field. Evidently, the direction of the force $\mathbf{F}_1$ is the same as the wave vector, its value is proportional to the spontaneous emission rate Γ of the atom. Since the stimulative emission of the atom in the running-wave field does not contribute to the force on the atom, the force $\mathbf{F}_1$ is just the spontaneous emission force or the scattering force. From (17.103), we can see that the force F_1 appears as a Lorentzian function of ω centered at $\omega_0 + \mathbf{k}\cdot\frac{d}{dt}\mathbf{r}$, here $\mathbf{k}\cdot\frac{d}{dt}\mathbf{r}$ is the Doppler frequency shift. Inasmuch as the direction of $\mathbf{F}_1$ is the same as the wave vector $\mathbf{k}$, the force $\mathbf{F}_1$ plays the acceleration role in the atomic motion when the atom moves along the direction of the wave vector $\mathbf{k}$. However, when the atom moves along the direction contrary to the wave vector $\mathbf{k}$, the role of $\mathbf{F}_1$ is to decelerate the atomic motion. In the view of physical mechanism, when the atom absorbs one photon with momentum $\hbar\mathbf{k}$, its momentum increases by $\hbar\mathbf{k}$, then the atom will spontaneously decay to its ground state and emit a photon with homogenous direction, so the accumulative effect of atomic recoil momentums average to be zero. Therefore, the atomic momentum alters $\Delta\mathbf{P}$ after a number of absorption processes. When the direction of the atomic velocity $\frac{d}{dt}\mathbf{r}$ is along to $\mathbf{k}$, the atomic velocity will increase, but if the direction of $\frac{d}{dt}\mathbf{r}$ is contrary to $\mathbf{k}$, the atomic velocity will decrease.

From (17.103) we find that $\mathbf{F}_1$ is associated with the detuning Δ and the Doppler frequency shift. If $\Delta < 0$, i.e., the frequency ω of the field is less than the atomic transition frequency ω_0, and the direction of the atomic velocity is the same as $\mathbf{k}$, then the Doppler frequency shift will be increased due to the acceleration effect of $\mathbf{F}_1$ on the atom, the spontaneous emission force therefore decreases gradually (It must be mentioned that when $\mathbf{k}\cdot\frac{d}{dt}\mathbf{r}\sim\Gamma$, (17.103) is invalid). If the direction of the atomic velocity is contrary to $\mathbf{k}$, i.e., $\mathbf{k}\cdot\frac{d}{dt}\mathbf{r}<0$, it is clear that $\mathbf{F}_1$ reaches its maximum value when $\Delta-\mathbf{k}\cdot\frac{d}{dt}\mathbf{r}=0$. However, when the atomic velocity and the wave vector have the same direction , i.e., $\mathbf{k}\cdot\frac{d}{dt}\mathbf{r}>0$, then $(\Delta-\mathbf{k}\cdot\frac{d}{dt}\mathbf{r})^2$ increases, so $\mathbf{F}_1$ decreases. In fact, this is because the increasing of the Doppler frequency shift, the detuning $-(|\Delta|+\mathbf{k}\cdot\frac{d}{dt}\mathbf{r})$ of the atom-field coupling system becomes large as shown in Fig.(17.8a), then the probability of finding the atom to be excited decreases, therefore, the force

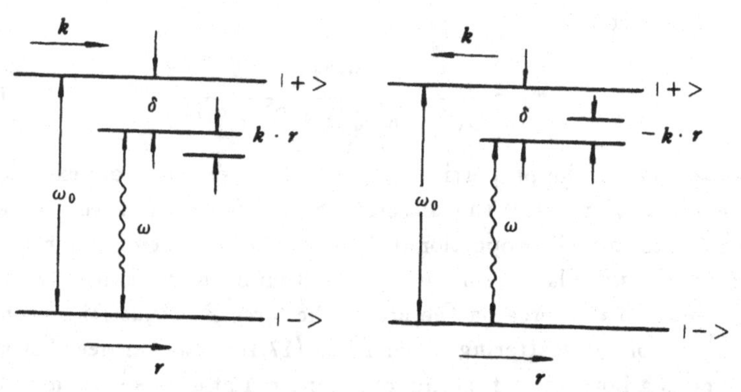

Figure 17.8: Variation of the atom-field detuning due to Doppler effects

$\mathbf{F}_1$ decreases. On the contrary, when $\mathbf{k}\cdot\frac{d}{dt}\mathbf{r}<0$, the detuning decreases because of the decreasing of Doppler frequency shift as shown in Fig(17.8b), then the probability of finding the atom to be excited increases. In this case, $\mathbf{F}_1$ increases. When $|\Delta|=|\mathbf{k}\cdot\frac{d}{dt}\mathbf{r}|$, F_1 reaches its maximum value, that is to say, the force F_1 on the atom exerted by radiation field reaches its maximum value. For the case of $\Delta<0$, the relation between $\mathbf{F}_1$ and $\mathbf{k}\cdot\frac{d}{dt}\mathbf{r}$ may be discussed similarly.

From (17.103) we can also see that $\mathbf{F}_1$ depends on the intensity of laser field, since the Rabi flopping frequency Ω is proportional to the amplitude ε_0 of the radiation field. For the weak field case, $\mathbf{F}_1$ is proportional to the intensity of laser field. With the increasing of the laser intensity, $\mathbf{F}_1$ also increases. When the field is so strong, i.e., $\Omega^2 \gg \Gamma^2 + 4(\Delta - \mathbf{k}\cdot\mathbf{r})^2$, the force F_1 reaches to a saturation value as

$$\mathbf{F}_1 = \frac{\hbar \mathbf{k} \Gamma}{2} \tag{17.104}$$

The reason is that for the strong field case, the atomic populations reach to their saturation values, then the photons absorbed by the atom at per time is saturated, and the photonic momentum gained by the atom is also saturated, consequently the radiation force achieves to its maximum saturation value. If we choose the atomic mass $M = 2.5 \times 10^{-25} kg$, the spontaneous emission linewidth $\Gamma = 10^8 Hz$, and the frequency of field $\omega = 3 \times 10^{15} Hz$, then according to (17.104) the acceleration of atom due to the interaction of the strong field is approximated to be $a_s = 10^6 m/s$, which is the 10^5 times than the acceleration of gravity. Thus the influence of radiation force on the atomic motion is very evident.

If the direction of the atomic velocity V_0 is contrary to the wave vector $\mathbf{k}$, then the spontaneous emission force leads to decelerate the velocity of the moving atom. For a moving atom with thermal velocity $v_0 = 10^2 m/s$, by how many times of the absorption and spontaneous emission processes can the atomic velocity decline to zero ? According to the kinematics law, we find

$$v_0 = a_s t = \frac{a_s N}{\Gamma}$$

so that

$$N = v_0 \Gamma / a_s = 10^2 \times 10^8 / 10^6 = 10^4$$

This means that the atom must carry out about ten thousand processes of one-photon absorption to decline its velocity to zero. In this case, the atomic thermal motion disappears nearly and the temperature of the atomic gas decreases to very low value. We will discuss this problem in detail in next chapter.

The above discussions show that there exists the spontaneous emission force acting on the atom in a running-wave field. This spontaneous emission force is associated with the spontaneous emission rate, the Doppler frequency shift, the intensity of field, and the detuning between the atom and the field. What is about the force acting on the atom in a standing-wave field ? Now we turn to discuss this question.

For a standing-wave field

$$\varepsilon(\mathbf{x},t) = E(\mathbf{x})\cos(\omega t) \tag{17.105}$$

According to (17.102) we have

$$\Omega(\mathbf{x}) = \mu E(\mathbf{x})/\hbar, \qquad \frac{d}{dt}\theta(\mathbf{x}) = 0 \tag{17.106}$$

Inserting (17.106) into (17.100) we obtain

$$\mathbf{F}_2 = -\hbar\Delta\frac{\nabla\Omega^2}{4\Delta^2+\Gamma^2+2\Omega^2} \tag{17.107}$$

It is clear that there does not exist the spontaneous emission force related to $\nabla\theta$, but there displays the force proportional to the gradient of light intensity $\nabla\Omega^2$, this one is usually called the radiative dipole force, which is associated with the stimulative transitions of the atom. The relation between the radiative dipole force $\mathbf{F}_2$ and the detuning Δ is displayed in Fig.(17.9). When $\omega > \omega_0$, i.e., the detuning is positive, the direction of $\mathbf{F}_2$ is along the negative $\nabla\Omega^2$. If

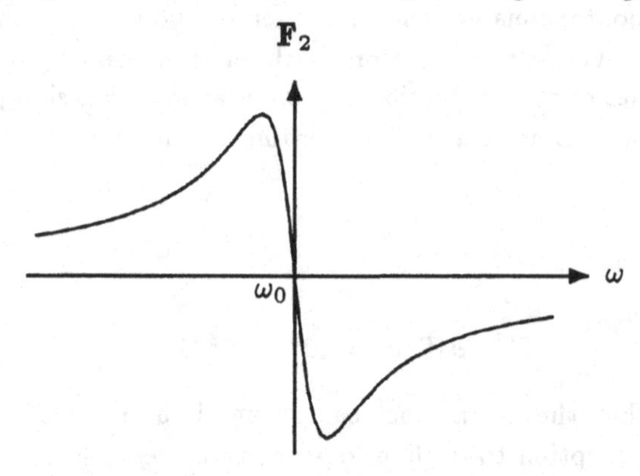

Figure 17.9: Dependence of radiation force $\mathbf{F}_2$ on the detuning Δ

the space-distribution of the field is inhomogenous, the radiative dipole force is in the direction of decreasing the intensity of the field, and the atom tends to be expelled by the field. When the field is tuned below resonance ($\omega < \omega_0$), the radiative dipole force is in the direction of increasing the intensity of the field, and the atom tends to be trapped by the radiation field. On the resonance case

($\Delta = 0$), the radiative dipole force vanishes. We also find that the radiative force $\mathbf{F}_2$ can not be saturated with the increasing of the field intensity, which is different from $\mathbf{F}_1$. The larger gradient of the light intensity $\nabla\Omega^2$ is, the larger the radiative dipole force $\mathbf{F}_2$. So the radiative dipole force is efficient to restrain the atomic motion. In experiments, this force is used to produce the laser for trapping atoms.

We can also obtain the radiative dipole force $\mathbf{F}_2$ through solving the optical Bloch equation (17.97). However, since the radiative dipole force is related to the stimulative transition of the atom, the physical mechanism of the radiative dipole force can not be revealed clearly compared with the spontaneous emission force. In order to realize the properties of the radiative dipole force acting on the atom in a laser field, next we adopt the dressed-state method to describe the radiative dipole force.

17.2.2 Description of the radiative dipole force by means of the dressed state method

Different from the previous quasi-classical method, here we adopt the quantum theory to discuss the radiative dipole force acting on the atom in a radiation field. The Hamiltonian to describe a two-level atom both in a laser field with frequency ω and in a vacuum radiation field is in the form

$$H = H_A + H_L + V_{A-L} + V_{A-R} \tag{17.108}$$

here H_A is the sum of the kinetic and the internal energies of the two-level atom

$$H_A = \frac{\mathbf{P}^2}{2M} + \hbar\omega_0 S_z \tag{17.109}$$

H_L is the free Hamiltonian of the laser field, i.e.,

$$H_L = \hbar\omega a_L^\dagger a_L \tag{17.110}$$

H_R represents the free Hamiltonian of vacuum field,

$$H_R = \hbar\sum_\lambda \omega_\lambda a_\lambda^\dagger a_\lambda \tag{17.111}$$

The interaction between the atom and the laser field is described by the Hamiltonian V_{A-L}, it is written in the rotating-wave approximation as

$$V_{A-L} = -\{\mathbf{d}\cdot\varepsilon_L(\mathbf{R})S_+ a_L + \mathbf{d}\cdot\varepsilon_L^*(\mathbf{R})S_- a_L^\dagger\} \tag{17.112}$$

where $\mathbf{d}$ and $\varepsilon_L(\mathbf{R})$ represent, respectively, the atomic dipole moment vector and the space distribution of laser field. The last term V_{A-R} stands for the interaction between the atom and the vacuum field

$$V_{A-R} = -\sum_{\lambda}[\mathbf{d}\cdot\varepsilon_\lambda(\mathbf{R})S_+a_\lambda + \mathbf{d}\cdot\varepsilon_\lambda^*(\mathbf{R})S_-a_\lambda^\dagger] \tag{17.113}$$

A. Density matrix equation describing the atomic internal property under the interaction of a laser field

Now we use the dressed-atom approach to obtain the equation of the density matrix to describe the atomic internal property. The basic idea of the dressed-atom approach used here is similar to that in Chapter 8. We first diagonalize the Hamiltonian of the atom-field coupling system and obtain the dressing states of the system. Then we consider the spontaneous emission effect due to the interaction of the vacuum field . Note that, when we derive the dressed states of the system, we also regard that the atomic motion satisfies (17.71) and (17.78), that is to say, the atomic kinetic energy term $\mathbf{P}^2/2M$ is omitted in H, and the atomic position operator $\mathbf{R}$ is replaced by its average value $\mathbf{r} = \langle\mathbf{R}\rangle$. Consequently, the Hamiltonian describing the interaction between the atom and the laser field may be rewritten as

$$H_{DA}(\mathbf{r}) = H_1(\mathbf{r}) + V_{A-L}(\mathbf{r}) \tag{17.114}$$

with

$$H_1(\mathbf{r}) = \hbar\omega_0 S_z + \hbar\omega_L a_L^\dagger a_L \tag{17.115}$$

$H_1(\mathbf{r})$ is the unperturbed Hamiltonian of the atom and the laser field, whose eigenstates are $\{|e,n\rangle\}$ and $\{|g,n+1\rangle\}$, and the corresponding eigenvalues are $(n+1)\hbar\omega_L - \hbar\delta/2$ and $(n+1)\hbar\omega_L + \hbar\delta/2$. Here δ is the detuning between the atom and the laser field, i.e., $\delta = \omega_L - \omega_0$. The atom-field coupling connecting the eigenstate $|e,n\rangle$ with $|g,n+1\rangle$ is characterized by matrix elements of the dipole transitions as

$$\frac{2}{\hbar}\langle e,n|V_{A-L}|g,n+1\rangle = -\frac{2}{\hbar}\sqrt{n+1}\mathbf{d}\cdot\varepsilon_L(\mathbf{r}) = \omega_1(\mathbf{r})\exp[i\phi(\mathbf{r})] \tag{17.116}$$

Here V_{A-L} is defined by (17.112), $\phi(\mathbf{r})$ represents the relative phase between the atomic dipole moment and the laser field, $\omega_1(\mathbf{r})$ is the resonant Rabi flopping frequency of the atom, it actually depends on the number n of photons. If the laser field we consider here is a strong coherent field with a sharp Poissonian distribution for number n of photons, and the width Δn is very small compared with the average number $\overline{n}$ of photons, i.e., $\Delta n \ll \overline{n}$, then $\omega_1(\mathbf{r})$ can be regard to be independent of n, and n can be replaced by the average photon number $\overline{n}$. At this approximation, similar to (17.20), the eigenstates and the corresponding eigenvalues of the Hamiltonian (17.114) are, respectively as

$$\begin{aligned} E_{1n}(\mathbf{r}) &= (n+1)\hbar\omega_L - \hbar\delta/2 + \hbar\Omega(\mathbf{r})/2, \\ |1,n,\mathbf{r}\rangle &= \exp[i\phi(\mathbf{r})/2]\cos\theta(\mathbf{r})|e,n\rangle \\ &\quad + \exp[-i\phi(\mathbf{r})/2]\sin\theta(\mathbf{r})|g,n+1\rangle \end{aligned} \tag{17.117}$$

$$\begin{aligned} E_{2n}(\mathbf{r}) &= (n+1)\hbar\omega_L - \hbar\delta/2 - \hbar\Omega(\mathbf{r})/2, \\ |2,n,\mathbf{r}\rangle &= -\exp[i\phi(\mathbf{r})/2]\sin\theta(\mathbf{r})|e,n\rangle \\ &\quad + \exp[-i\phi(\mathbf{r})/2]\cos\theta(\mathbf{r})|g,n+1\rangle \end{aligned} \tag{17.118}$$

with

$$\begin{aligned} \Omega(\mathbf{r}) &= [\omega_1^2(\mathbf{r}) + \delta^2]^{1/2}, \\ \cos 2\theta(\mathbf{r}) &= -\frac{\delta}{\Omega(\mathbf{r})}, \quad \sin 2\theta(\mathbf{r}) = \frac{\omega_1(\mathbf{r})}{\Omega(\mathbf{r})} \end{aligned} \tag{17.119}$$

As shown in Fig.(17.10), the gap between $E_{1n}(\mathbf{r})$ and $E_{2n}(\mathbf{r})$ depends on $\Omega_1(\mathbf{r})$. Since $\omega_1(\mathbf{r})$ is independent of n in the strong field case, the gaps for the different $E_{2n}(\mathbf{r})$ and $E_{1n}(\mathbf{r})$ can be approximated to be identical. Here the dressed states $|1,n,\mathbf{r}\rangle$ and $|2,n,\mathbf{r}\rangle$ and the corresponding eigenvalues $E_{1n}(\mathbf{r})$ and $E_{2n}(\mathbf{r})$ are associated with the position $\mathbf{r}$ of the atomic center-of-mass. Therefore, for an inhomogeneous laser beam (for example, a standing-wave field), these dressed states and eigenvalues will vary with the position $\mathbf{r}$.

We now take into account the effect of V_{A-R}, i.e., the effect of the atomic spontaneous emission. In the uncoupled basis $\{|e,n\rangle\}$ and $\{|g,n\rangle\}$, the spontaneous emission process occurs from $|e,n\rangle$ to $|g,n\rangle$. Because the dressed eigenstates $|j,n,\mathbf{r}\rangle$ and $|i,n-1,\mathbf{r}\rangle$ contain $|e,n\rangle$ and $|g,n\rangle$, respectively, the dipole matrix elements in the spontaneous emission processes are connected with the states $|j,n,\mathbf{r}\rangle$ and $|i,n-1,\mathbf{r}\rangle$, i.e.,

$$\mathbf{d}_{ij}(\mathbf{r}) = \langle i,n-1,\mathbf{r}|\mathbf{d}(S_+ + S-)|j,n,\mathbf{r}\rangle \tag{17.120}$$

From (17.118) and (17.119) we obtain

$$\mathbf{d}_{11} = -\mathbf{d}_{22} = \mathbf{d}\cos\theta\sin\theta\exp(i\phi), \tag{17.121}$$

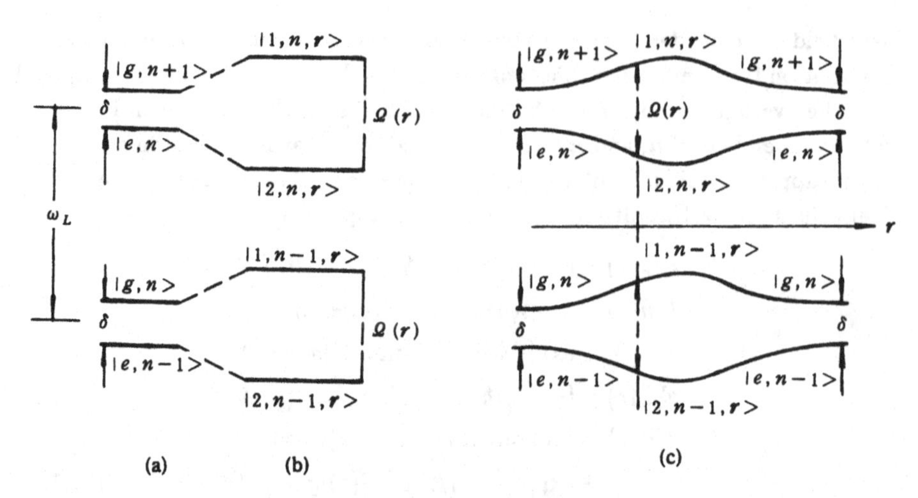

Figure 17.10: Diagram of eigenvectors for atom-field coupling system

$$\mathbf{d}_{12} = -\mathbf{d}\sin^2\theta\exp(i\phi), \quad \mathbf{d}_{21} = \mathbf{d}\cos^2\theta\exp(i\phi) \tag{17.122}$$

Identical to the triplet structure of the fluorescence spectrum, there display three kinds of fluorescent photons with three different frequencies (shown in Fig.(17.11)), $\omega_L + \Omega(\mathbf{r})$ for $|1,n,\mathbf{r}\rangle$ to $|2,n-1,\mathbf{r}\rangle$, $\omega_L - \Omega(\mathbf{r})$ for $|2,n,\mathbf{r}\rangle$ to $|1,n-1,\mathbf{r}\rangle$, and ω_L for $|1,n,\mathbf{r}\rangle$ to $|1,n-1,\mathbf{r}\rangle$ and $|2,n,\mathbf{r}\rangle$ to $|2,n-1,\mathbf{r}\rangle$.

Now we derive the master equation for the density matrix ρ of the dressed atom. For a given point $\mathbf{r}$, From Fig.(17.11) we can easily find that the population ($\pi_{1,n}(\mathbf{r}) = \langle 1,n,\mathbf{r}|\rho|1,n,\mathbf{r}\rangle$) of the dressed atom satisfies

$$\frac{d}{dt}\pi_{1,n}(\mathbf{r}) = -(\Gamma_{11}+\Gamma_{21})\pi_{1,n}(\mathbf{r}) + \Gamma_{11}\pi_{1,n+1}(\mathbf{r}) + \Gamma_{12}\pi_{2,n+1}(\mathbf{r}) \tag{17.123}$$

here Γ_{ij} are the transition rates, which obey

$$\Gamma_{ij} = |\mathbf{d}_{ij}|^2 \tag{17.124}$$

Inasmuch as ω_1 is independent of n for the intensive field, Γ_{ij} is also independent of n. From (17.121) we see that $\Gamma_{11} = \Gamma_{22} = d^2\sin^2\theta\cos^2\theta = \Gamma\cos^2\theta\sin^2\theta$, where Γ is the normal spontaneous emission rate. Eq.(17.124) shows that the decrease of the population $\pi_{1,n}(\mathbf{r})$ is due to the decay from the upper level $|1,n,\mathbf{r}\rangle$ to the two lower levels $|1,n-1,\mathbf{r}\rangle$ and $|2,n-1,\mathbf{r}\rangle$, on the contrary, the increase of $\pi_{1,n}(\mathbf{r})$ results from the transitions from the upper

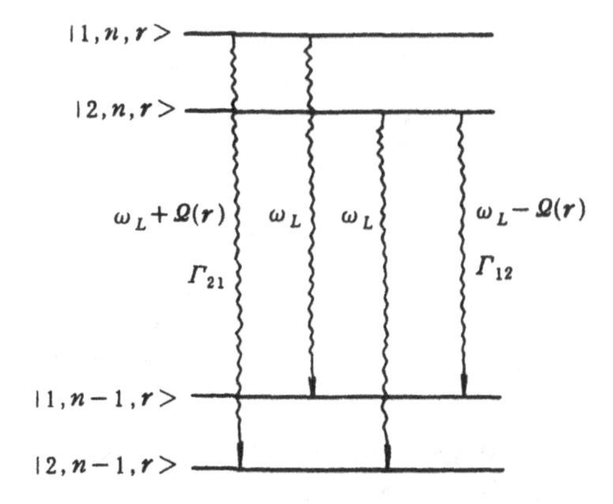

Figure 17.11: Diagram of fluorescence transitions

levels $|1, n+1, \mathbf{r}\rangle$ and $|2, n+1, \mathbf{r}\rangle$ to $|1, n, \mathbf{r}\rangle$. Similarly, the evolution of the population $\pi_{2,n}(\mathbf{r}) = \langle 2, n, \mathbf{r}|\rho|2, n, \mathbf{r}\rangle$ is written as

$$\frac{d}{dt}\pi_{2,n}(\mathbf{r}) = -(\Gamma_{12} + \Gamma_{22})\pi_{2,n}(\mathbf{r}) + \Gamma_{21}\pi_{1,n+1}(\mathbf{r}) + \Gamma_{22}\pi_{2,n+1}(\mathbf{r}) \quad (17.125)$$

Since the laser field we consider here is a strong coherent one, the photons are located at the neighbourhood of $n = \bar{n}$, the approximation $P_0(n) = P_0(n+1)$ is reasonable, which means that the photon distribution function $P_0(n)$ exhibits the periodical property in the width Δn. Therefore, $\pi_{i,n+1}(\mathbf{r})$ is approximated to be equal to $\pi_{i,n}(\mathbf{r})$, i.e.,

$$\pi_{i,n}(\mathbf{r}) \approx \pi_{i,n+1}(\mathbf{r}) \quad (17.126)$$

If we define the reduced populations of the dressed atom as

$$\pi_i(\mathbf{r}) = \sum_n \langle i, n, \mathbf{r}|\rho|i, n, \mathbf{r}\rangle \quad (17.127)$$

then from (17.123) and (17.125) one can obtain the evolution equation of $\pi_i(\mathbf{r})$

$$\begin{aligned}\frac{d}{dt}\pi_1(\mathbf{r}) &= -\Gamma_{21}\pi_1(\mathbf{r}) + \Gamma_{12}\pi_2(\mathbf{r}) \\ \frac{d}{dt}\pi_2(\mathbf{r}) &= -\Gamma_{12}\pi_2(\mathbf{r}) + \Gamma_{21}\pi_1(\mathbf{r})\end{aligned} \quad (17.128)$$

For the off-diagonal terms $\rho_{ij,n}(\mathbf{r}) = \langle i, n, \mathbf{r}|\rho|j, n, \mathbf{r}\rangle$ $(i \neq j)$, we first analyze the decay processes associated with $\rho_{ij,n}(\mathbf{r})$. As shown in Fig.(17.12), $\rho_{12,n}(\mathbf{r})$

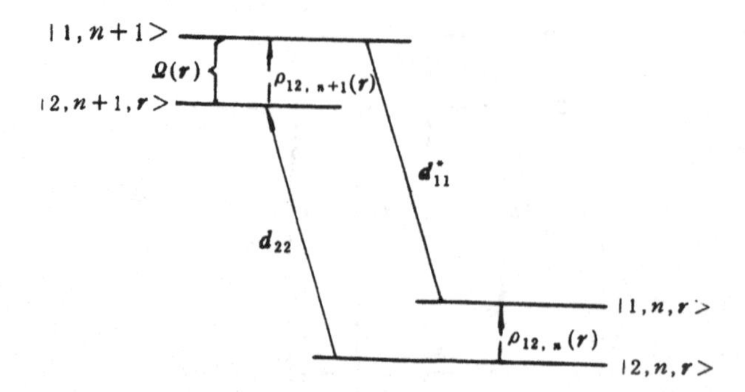

Figure 17.12: Schematic diagram of transition processes associated with non-diagonal elements of the density matrix

evolves freely with frequency $\Omega(\mathbf{r})$, so in the master equation of $\rho_{12,n}(\mathbf{r})$, there must exist the term $\Omega(\mathbf{r})\rho_{12,n}(\mathbf{r})$. Since the levels $|1,n,\mathbf{r}\rangle$ and $|2,n,\mathbf{r}\rangle$ are damped with the rates $\Gamma_{11}+\Gamma_{21}$ and $\Gamma_{22}+\Gamma_{12}$, respectively, the master equation of $\rho_{12,n}(\mathbf{r})$ must contain the damping term with decay rate $(\Gamma_{11}+\Gamma_{21}+\Gamma_{22}+\Gamma_{12})/2$, i.e.,$-(\Gamma_{11}+\Gamma_{21}+\Gamma_{22}+\Gamma_{12})\rho_{12,n}(\mathbf{r})/2$. Meantime, $\rho_{12,n}(\mathbf{r})$ and the other off-diagonal element $\rho_{12,n+1}(\mathbf{r})$ have the same free evolution frequency, there is a coupling between these two terms as shown in Fig.(17.12), the coupling constant is $\mathbf{d}_{11}^{*}\cdot\mathbf{d}_{22}$. So the evolution equation of $\rho_{12,n}(\mathbf{r})$ is

$$\begin{aligned}\frac{d}{dt}\rho_{12,n}(\mathbf{r}) = & -i\Omega(\mathbf{r})\rho_{12,n}(\mathbf{r}) - (\Gamma_{11}+\Gamma_{21}+\Gamma_{22}+\Gamma_{12})\rho_{12,n}(\mathbf{r})/2 \\ & +\mathbf{d}_{11}^{*}\cdot\mathbf{d}_{22}\rho_{12,n+1}(\mathbf{r})\end{aligned} \tag{17.129}$$

By use of the property of the strong laser field, $P_0(n)=P_0(n+1)$, we have

$$\rho_{12,n}(\mathbf{r}) \approx \rho_{12,n+1}(\mathbf{r}) \tag{17.130}$$

Defining the total relaxation rate

$$\begin{aligned}\Gamma_{coh}(\mathbf{r}) &= (\Gamma_{11}+\Gamma_{21}+\Gamma_{22}+\Gamma_{12})/2 - \mathbf{d}_{11}^{*}\cdot\mathbf{d}_{22} \\ &= \Gamma(\sin^4\theta+\cos^4\theta+2\sin^2\theta\cos^2\theta)/2+\Gamma\sin^2\theta\cos^2\theta \\ &= \Gamma(1/2+\sin^2\theta\cos^2\theta)\end{aligned} \tag{17.131}$$

then the off-diagonal elements of the reduced density matrix of the dressed atom

$$\rho_{ij}(\mathbf{r}) = \sum_n \rho_{ij,n}(\mathbf{r}), \qquad (i\neq j) \tag{17.132}$$

obey the following equations

$$\frac{d}{dt}\rho_{12}(\mathbf{r}) = -i\Omega(\mathbf{r})\rho_{12}(\mathbf{r}) - \Gamma_{coh}(\mathbf{r})\rho_{12}(\mathbf{r}) \tag{17.133}$$

and

$$\frac{d}{dt}\rho_{21}(\mathbf{r}) = i\Omega(\mathbf{r})\rho_{21}(\mathbf{r}) - \Gamma_{coh}(\mathbf{r})\rho_{21}(\mathbf{r}) \tag{17.134}$$

Up to now, we have derived the evolution equations (17.128, 17.129, 17.133, 17.134) of the reduced density matrix of the dressed atom. The steady-state solutions of these equations can be derived from the following equations

$$\begin{aligned}
\frac{d}{dt}\pi_1(\mathbf{r}) &= 0 = -\Gamma_{21}(\mathbf{r})\pi_1(\mathbf{r}) + \Gamma_{12}(\mathbf{r})\pi_2(\mathbf{r}) \\
\frac{d}{dt}\pi_2(\mathbf{r}) &= 0 = -\Gamma_{12}(\mathbf{r})\pi_2(\mathbf{r}) + \Gamma_{21}(\mathbf{r})\pi_1(\mathbf{r}) \\
\frac{d}{dt}\pi_{12}(\mathbf{r}) &= 0 = [-i\Omega(\mathbf{r}) - \Gamma_{coh}(\mathbf{r})]\rho_{12}(\mathbf{r}) \\
\frac{d}{dt}\pi_{21}(\mathbf{r}) &= 0 = [i\Omega(\mathbf{r}) - \Gamma_{coh}(\mathbf{r})]\rho_{21}(\mathbf{r})
\end{aligned}$$

Solving the above four equations we obtain

$$\pi_1^{st}(\mathbf{r}) = \frac{\Gamma_{12}(\mathbf{r})}{\Gamma_{21}(\mathbf{r}) + \Gamma_{12}(\mathbf{r})} = \frac{\sin^4\theta(\mathbf{r})}{\sin^4\theta(\mathbf{r}) + \cos^4\theta(\mathbf{r})} \tag{17.135}$$

$$\pi_2^{st}(\mathbf{r}) = \frac{\Gamma_{21}(\mathbf{r})}{\Gamma_{21}(\mathbf{r}) + \Gamma_{12}(\mathbf{r})} = \frac{\cos^4\theta(\mathbf{r})}{\sin^4\theta(\mathbf{r}) + \cos^4\theta(\mathbf{r})} \tag{17.136}$$

$$\rho_{12}^{st}(\mathbf{r}) = \rho_{21}^{st}(\mathbf{r}) = 0 \tag{17.137}$$

here we have used the relations

$$\Gamma_{12}(\mathbf{r}) = \Gamma\sin^4\theta(\mathbf{r}), \quad \Gamma_{21}(\mathbf{r}) = \Gamma\cos^4\theta(\mathbf{r}) \tag{17.138}$$

If defining

$$\Gamma_{pop}(\mathbf{r}) = \Gamma_{12}(\mathbf{r}) + \Gamma_{21}(\mathbf{r}) = \Gamma[\sin^4\theta(\mathbf{r}) + \cos^4\theta(\mathbf{r})] \tag{17.139}$$

which represents the contributions of the off-diagonal elements ρ_{ij} describing the atomic coherence on the pumping rate of populations, and using the relation $\pi_1 + \pi_2 = 1$, then (17.128) and (17.129) can be rewritten as

$$\frac{d}{dt}\pi_i(\mathbf{r}) = -\Gamma_{pop}(\mathbf{r})[\pi_i(\mathbf{r}) - \pi_1^{st}(\mathbf{r})] \quad (i = 1, 2) \tag{17.140}$$

Now we have only given the time evolution of the reduced density matrix of the atom in the dressed-state representation at a given point $\mathbf{r}$, the effect of the atomic motion is not considered. For the case of a moving atom, i.e., the point $\mathbf{r}$ varies with time t, the eigenvectors $|i, n, \mathbf{r}(t)\rangle$ and $|j, n, \mathbf{r}(t)\rangle$ depend on time, the master equation must be modified by taking into account the time dependence of $|1, n, \mathbf{r}(t)\rangle$ and $|j, n, \mathbf{r}(t)\rangle$, i.e.,

$$\frac{d}{dt}\pi_i(\mathbf{r}) = \sum_n \left\{ \langle i, n, \mathbf{r}|\frac{d}{dt}\rho|i, n, \mathbf{r}\rangle + \frac{d}{dt}\overline{\langle i, n, \mathbf{r}|}\rho|i, n, \mathbf{r}\rangle + \langle i, n, \mathbf{r}|\rho\frac{d}{dt}\overline{|i, n, \mathbf{r}\rangle} \right\} \tag{17.141}$$

Noticing

$$\frac{d}{dt}\overline{|i, n, \mathbf{r}\rangle} = \frac{d}{dt}\mathbf{r}(t) \cdot \nabla|i, n, \mathbf{r}\rangle = \mathbf{v} \cdot \nabla|i, n, \mathbf{r}\rangle \tag{17.142}$$

and

$$\nabla|1, n, \mathbf{r}\rangle = \nabla\theta(\mathbf{r})|2, n, \mathbf{r}\rangle + \frac{i}{2}\nabla\phi(\mathbf{r})[\cos 2\theta(\mathbf{r})|1, n, \mathbf{r}\rangle - \sin 2\theta(\mathbf{r})|2, n, \mathbf{r}\rangle] \tag{17.143}$$

$$\nabla|2, n, \mathbf{r}\rangle = -\nabla\theta(\mathbf{r})|1, n, \mathbf{r}\rangle - \frac{i}{2}\nabla\phi(\mathbf{r})[\cos 2\theta(\mathbf{r})|2, n, \mathbf{r}\rangle + \sin 2\theta(\mathbf{r})|1, n, \mathbf{r}\rangle] \tag{17.144}$$

and considering (17.127), (17.132) and (17.140), then (17.141) takes the form

$$\frac{d}{dt}\pi_1(\mathbf{r}) = -\Gamma_{pop}(\mathbf{r})[\pi_1(\mathbf{r}) - \pi_1^{st}(\mathbf{r})] + \mathbf{v} \cdot \nabla\theta(\rho_{12} + \rho_{21}) + i\mathbf{v} \cdot \nabla\phi \sin\theta\cos\theta(\rho_{21} - \rho_{12}) \tag{17.145}$$

Following the similar procedure, one can obtain

$$\frac{d}{dt}\rho_{12}(\mathbf{r}) = -\{i[\Omega(\mathbf{r}) + \mathbf{v} \cdot \nabla\phi\cos 2\theta] + \Gamma_{coh}(\mathbf{r})]\}\rho_{12} + [\mathbf{v} \cdot \nabla\theta + i\mathbf{v} \cdot \nabla\phi\sin\theta\cos\theta](\pi_2 - \pi_1) \tag{17.146}$$

It is clear that there display the terms associated with the velocity of atomic center-of-mass in the density matrix elements π_i and ρ_{12}, which different from eqs.(17.133), (17.134) and (17.140), these terms are the reflection of the inhomogenous space-distribution of the laser field.

Next we give an example of π_i to discuss the influence of the atomic motion on the atomic density matrix. From (17.145) we see that the variation of π_i during the time dt is

$$d\pi_i = \frac{d}{dt}\pi_i dt = (d\pi_i)_{Rad} + (d\pi_i)_{NA} \tag{17.147}$$

where

$$\begin{aligned}(d\pi_i)_{Rad} =& -\Gamma_{pop}(\pi_i - \pi_i^{st}) \\ (d\pi_i)_{NA} =& \mathbf{v}dt \cdot [\nabla\theta(\rho_{12} + \rho_{21}) + i\nabla\phi \sin\theta\cos\theta(\rho_{21} - \rho_{12})] \\ =& -(d\pi_2)_{NA}\end{aligned} \tag{17.148}$$

$(d\pi_i)_{Rad}$ corresponds to the modification of π_i that is due to the radiative relaxation through the spontaneous emission, which is identical to the case resulted from the atom being stationary. $(d\pi_i)_{NA}$ represents the contribution of two terms proportional to the atomic velocity during the time dt. This modification of the dressed populations, induced by the atomic motion, is due to the spatial variation of dressed energy levels, i.e., the spatial variation of the eigenvalues and the eigenvectors. We call $(d\pi_i)_{NA}$ as the nonadiabatic kinetic coupling term since it describes the possibility for a moving atom initially in a level $|i, n, \mathbf{r}(0)\rangle$ to reach the other level $|j, n, \mathbf{r}(t)\rangle$ in the absence of spontaneous emissions. For example, when an atom moves from the antinode to the node of a standing-wave field, the atomic dressed levels display obvious change. At time t the probability for the atomic transition from $|i, n, \mathbf{r}\rangle$ to $|j, n, \mathbf{r}\rangle$ due to the atomic motion is approximated as

$$P_{i\to j} \leq \frac{|\langle j, n, \mathbf{r}(t)|\frac{d}{dt}\overline{i, n, \mathbf{r}(t)\rangle}|^2}{|\omega_{ij}(t)|^2} = \frac{|\langle j, n, \mathbf{r}(t)|\frac{d}{dt}\overline{i, n, \mathbf{r}(t)\rangle}|^2}{\Omega_{ij}^2(t)} \tag{17.149}$$

Because the relative phase between the dipole moment and the standing-wave laser field is independent of time, that is, ϕ is a constant. It is easy to obtain from eqs.(17.142) and (17.143) that

$$\frac{d}{dt}\overline{|i, n, \mathbf{r}(t)\rangle} = \pm\mathbf{v}\cdot\nabla\theta|j, n, \mathbf{r}(t)\rangle \tag{17.150}$$

And eq.(17.149) gives for $\nabla\theta$

$$\nabla\theta = -\frac{\delta\nabla\omega_1}{2(\delta^2 + \omega_1^2)} \tag{17.151}$$

In a standing-wave laser field, $\omega_1(\mathbf{r})$ may be expressed as

$$\omega_1(\mathbf{r}) = \overline{\omega}_1 \cos(\mathbf{k}\cdot\mathbf{r}) \tag{17.152}$$

here $\overline{\omega}_1$ is a constant unrelated to $\mathbf{r}$. Substituting (17.152) into (17.151), we have

$$|\nabla\theta| = \left|\frac{k\delta\tilde{\omega}_1 \sin(\mathbf{k}\cdot\mathbf{r})}{2(\delta^2 + \tilde{\omega}_1^2\cos^2(\mathbf{k}\cdot\mathbf{r})}\right| \leq \frac{k}{2}\left|\frac{\tilde{\omega}_1}{\delta}\right| \tag{17.153}$$

The transition probability $P_{i\to j}$ therefore reads

$$P_{i\to j} \leq \left(\frac{kv}{2}\right)^2 \left|\frac{\delta\tilde{\omega}_1 \sin(\mathbf{k}\cdot\mathbf{r})}{\delta^2 + \tilde{\omega}_1^2\cos^2(\mathbf{k}\cdot\mathbf{r})}\right|^2 \leq \left|\frac{kv\tilde{\omega}_1}{2\delta^2}\right|^2 \tag{17.154}$$

The above equation shows that the effect of the nonadiabatic kinetic coupling is maximal at the node of the standing-wave field. Thus for a moving atom, the effect of the atomic motion must be considered.

B. Radiative dipole force acting on the atom by a radiation field

Now we have obtained the evolution equation of the atomic density matrix and analyzed the influence of atomic motion on the master equation. Next we discuss the relation between the force acting on the atom by a radiation field and the atomic motion.

Similar to (17.70) and (17.72), the force operator $\mathbf{F}$ is defined as

$$\mathbf{F} = \frac{d}{dt}\mathbf{P} = \frac{i}{\hbar}[H, \mathbf{P}] = -\nabla_R H = -\nabla_R(V_{A_L} + V_{A_R}) \tag{17.155}$$

As usual, we are interested in the expectation value of $\mathbf{F}$ over both field and internal atomic states, i.e.,

$$\mathbf{f}(\mathbf{r}) = \langle \mathbf{F}(\mathbf{R})\rangle \tag{17.156}$$

In analogy with (17.72), here the position operator $\mathbf{R}$ of the atomic center-of-mass is replaced by its expectation value $\mathbf{r} = \langle\mathbf{R}\rangle$. Substituting (17.155) into (17.156) and noticing V_{A-R} describing the interaction between the atom and the vacuum field, i.e., we have $\langle\nabla_R V_{A-R}\rangle = 0$, then obtain

$$\mathbf{f}(\mathbf{r}) = \langle S_+ a_L \nabla[\mathbf{d}\cdot\varepsilon_L(\mathbf{r}) + S_- a_L^\dagger \nabla[\mathbf{d}\cdot\varepsilon_L^*(\mathbf{r})]\rangle \tag{17.157}$$

Using (17.116), the above equation can be written as

$$\begin{aligned}\mathbf{f}(\mathbf{r}) &= \frac{\hbar}{2}\sum_n \{\langle e,n|\rho|g,n+1\rangle\nabla[\omega_1(\mathbf{r})\exp[-i\phi(\mathbf{r})]] \\ &\quad +\langle g,n+1|\rho|e,n\rangle\nabla[\omega_1(\mathbf{r})\exp[i\phi(\mathbf{r})]]\} \\ &= \frac{i\hbar\omega_1}{2}\nabla\phi(\mathbf{r})\{\rho_{eg}\exp[-i\phi(\mathbf{r})] - \rho_{ge}\exp[i\phi(\mathbf{r})]\} \\ &\quad -\frac{\hbar\nabla\omega_1}{2}\{\rho_{eg}\exp[-i\phi(\mathbf{r})] + \rho_{ge}\exp[i\phi(\mathbf{r})]\}\end{aligned} \tag{17.158}$$

where we have put

$$\rho_{eg} = \sum_n \langle e,n|\rho|g,n+1\rangle, \qquad \rho_{ge} = \sum_n \langle g,n+1|\rho|e,n\rangle \tag{17.159}$$

It is clear that (17.158) and (17.96) have the similar formation. From the discussion of (17.100)-(17.107) we know that the term containing $\nabla\phi(\mathbf{r})$ in (17.158) is just the spontaneous emission force, which has been discussed in the preceding section. Next we focus on the term containing $\nabla\omega_1(\mathbf{r})$ in $\mathbf{f}(\mathbf{r})$, which is interpreted as the radiative dipole force. Using (17.117) and (17.118), ρ_{eq} and ρ_{ge} can be expressed in terms of π_i and ρ_{ij} in the dressed representation as

$$\rho_{eg} = e^{i\phi}[\sin\theta\cos\theta\pi_1 - \sin\theta\cos\theta\pi_2 + \cos^2\theta\rho_{12} - \sin^2\theta\rho_{21}] \tag{17.160}$$

$$\rho_{ge} = e^{-i\phi}[\sin\theta\cos\theta\pi_1 - \sin\theta\cos\theta\pi_2 + \cos^2\theta\rho_{21} - \sin^2\theta\rho_{12}] \tag{17.161}$$

Then the dipole force acting on the atom can be written as

$$\mathbf{f}_{dip} = -\frac{\hbar\nabla\omega_1}{2}\sin 2\theta(\pi_1 - \pi_2) - \frac{\hbar\nabla\omega_1}{2}\cos 2\theta(\rho_{12}+\rho_{21}) \tag{17.162}$$

Note

$$\nabla\Omega = \nabla[\omega_1^2(\mathbf{r}) + \delta^2]^{1/2} = \frac{\omega_1(\mathbf{r})}{\Omega(\mathbf{r})}\nabla\omega_1 = \sin 2\theta\nabla\omega_1 \tag{17.163}$$

and

$$\nabla(\cos 2\theta) = -\nabla\left(\frac{\delta}{\Omega}\right)$$

i.e.,

$$(\nabla\omega_1)\cos 2\theta = 2\Omega\nabla\theta \tag{17.164}$$

$\mathbf{f}_{dip}(\mathbf{r})$ can be rewritten as

$$\mathbf{f}_{dip}(\mathbf{r}) = \frac{\hbar}{2}\nabla\Omega(\pi_2 - \pi_1) - \hbar\Omega\nabla\theta(\rho_{12} + \rho_{21}) \tag{17.165}$$

Now we have obtained the expression of the dipole force acting on the atom in terms of density matrix elements of the dressed atom, so we can use the evolution equations (17.133), (17.134) and (17.140) of π_i, ρ_{21} and ρ_{12} to discuss the properties of the dipole force $\mathbf{f}_{dip}(\mathbf{r})$.

In order to get some physical insight into the expression (17.165) of the dipole force $\mathbf{f}_{dip}(\mathbf{r})$, we now calculate the work dW that has to be provided for moving dressed atom by a quantity $d\mathbf{r}$

$$dW = -\mathbf{f}_{dip}(\mathbf{r}) \cdot d\mathbf{r} = -\hbar\frac{\nabla\Omega}{2} \cdot d\mathbf{r}(\pi_2 - \pi_1) + \hbar\Omega\nabla\theta \cdot d\mathbf{r}(\rho_{12} + \rho_{21}) \tag{17.166}$$

If we define

$$\begin{aligned} E_1(\mathbf{r}) &= E_{1n}(\mathbf{r}) - \frac{E_{1n}(\mathbf{r}) + E_{2n}(\mathbf{r})}{2} = \frac{\hbar\Omega(\mathbf{r})}{2} \\ E_2(\mathbf{r}) &= E_{2n}(\mathbf{r}) - \frac{E_{1n}(\mathbf{r}) + E_{2n}(\mathbf{r})}{2} = -\frac{\hbar\Omega(\mathbf{r})}{2} \end{aligned} \tag{17.167}$$

here $E_1(\mathbf{r})$ and $E_2(\mathbf{r})$ are the parts depended on the position vector $\mathbf{r}$, then the first term of (17.166) can be simply written as

$$-\frac{\hbar\nabla\Omega(\mathbf{r})}{2} \cdot d\mathbf{r}(\pi_2 - \pi_1) = -\frac{\hbar d\Omega}{2}(\pi_2 - \pi_1) = \sum_{i=1,2} \pi_i dE_i \tag{17.168}$$

where dE_i represents the energy change of the dressed level $|i, n, \mathbf{r}\rangle$ when the atom moves from $\mathbf{r}$ to $\mathbf{r} + d\mathbf{r}$. By use of (17.149) and $d\mathbf{r} = \mathbf{v}dt$, the second term in (17.166) can be expressed as

$$\begin{aligned} &\hbar\Omega\nabla\theta \cdot d\mathbf{r}(\rho_{12} + \rho_{21}) = \hbar\Omega\nabla\theta \cdot \mathbf{v}dt(\rho_{12} + \rho_{21}) = \hbar\Omega(d\pi_1)_{NA} \\ &= -\hbar\Omega(d\pi_2)_{NA} = \hbar\Omega[(d\pi_1)_{NA} - (d\pi_2)_{NA}] \sum_{i=1,2} E_i(d\pi_i)_{NA} \end{aligned} \tag{17.169}$$

Substituting (17.168) and (17.169) into (17.166), we have

$$dW = \sum_{i=1,2} [\pi_i dE_i + E_i(d\pi_i)_{NA}] \tag{17.170}$$

(17.170) shows that the work of the dipole force contributes to the non-adiabatic energy coupling terms which change the populations, that is, only $(d\pi_i)_{NA}$ appear in dW. It follows that, if the velocity v of the atomic center-of-mass is low enough, the terms containing $(d\pi_i)_{NA}$ can be omitted in (17.170), and we can keep only the first term of (17.170), i.e.,

$$dW = \sum_{i=1,2} \pi_i dE_i \quad \text{(when } v \text{ is low enough)} \tag{17.171}$$

Define

$$U_A = \sum_{i=1,2} \pi_i E_i \tag{17.172}$$

which can be interpreted as the mean potential energy of the atom in the field. Since $(d\pi_i)_{NA} \neq d\pi_i$, then the work dW provided by the dipole force is not equal to the change of the potential energy, i.e.,

$$dW \neq dU_A \tag{17.173}$$

In order to understand the physical meaning of the difference between dW and dU_A, we may rewrite (17.170) as

$$\begin{aligned} dW &= \sum_i \pi_i dE_i + \sum_i E_i[(d\pi_i)_{NA} + (d\pi_i)_{Rad}] - \sum_i E_i(d\pi_i)_{Rad} \\ &= \sum_i [\pi_i dE_i + E_i d\pi_i] - \sum_i E_i(d\pi_i)_{Rad} \\ &= dU_A - \sum_i E_i(d\pi_i)_{Rad} \end{aligned} \tag{17.174}$$

Inasmuch as $(d\pi_i)_{Rad}$ represents the change of populations due to the spontaneous emission of the atom moving a displacement $d\mathbf{r}$, then we may use (17.128) and (17.129) to replace $(d\pi_i)_{Rad}$ in (17.174), thus obtain

$$dW = dU_A + (\Gamma_{21}\pi_1\hbar\Omega - \Gamma_{12}\pi_2\hbar\Omega)dt \tag{17.175}$$

here (17.167) has been used. We call the term $(\Gamma_{21}\pi_1\hbar\Omega - \Gamma\pi_2\hbar\Omega)dt$ in (17.175) as the energy change of the radiation field (laser+fluorescence photons) during the time dt of the displacement $d\mathbf{r}$, i.e.,

$$dU_F = (\Gamma_{21}\pi_1\hbar\Omega - \Gamma_{12}\pi_2\hbar\Omega)dt \tag{17.176}$$

Such a result can be understood by considering that, during the time dt, the laser photons (energy $\hbar\omega_L$) were transferred to the fluorescent photons. In the transition processes from $|i,n,\mathbf{r}\rangle$ to $|i,n-1,\mathbf{r}\rangle$, the fluorescent photons and the laser photons have the same energy $\hbar\omega_L$, so the energy of the radiation field can not be changed in these fluorescence emission processes. By contrast, the fluorescence photons emitted by the transitions $|1,n,\mathbf{r}\rangle \rightarrow |2,n-1,\mathbf{r}\rangle$ or $|2,n,\mathbf{r}\rangle \rightarrow |1,n-1,\mathbf{r}\rangle$ have an energy $\hbar\omega_L + \hbar\Omega(\mathbf{r})$ or $\hbar\omega_L - \hbar\Omega(\mathbf{r})$, and the emission of such fluorescent photons changes the energy of the radiation field by a quantity $\hbar\Omega(\mathbf{r})$ or $-\hbar\Omega(\mathbf{r})$. So $\Gamma_{21}\pi_1 d\Omega dt$ in (17.176) represents the increase of the energy of the radiation field due to the transitions $|1,n,\mathbf{r}\rangle \rightarrow |2,n-1,\mathbf{r}\rangle$ during the time dt, on the contrary, $-\Gamma_{12}\pi_2\hbar\Omega dt$ stands for the decrease of the energy of the radiation field due to the transitions on $|2,n,\mathbf{r}\rangle \rightarrow |1,n-1,\mathbf{r}\rangle$ during the time dt. Therefore, the work done against the dipole force for moving the atom from $\mathbf{r}$ to $\mathbf{r}+d\mathbf{r}$ is transformed by the variation of the atomic mean potential energy and the field energy, i.e.,

$$dW = -\mathbf{f}_{dip}(\mathbf{r}) \cdot \mathbf{r} = dU_A + dU_F \tag{17.177}$$

Now we discuss the dipole force acting on the atom when it is at rest at a given point $\mathbf{r}$. In this case, we can ignore the effects of non-adiabatic kinetic energy coupling terms and replace the populations π_i in (17.171) by their steady-state values π_i^{st} (17.135) and (17.136) , which give

$$\mathbf{f}_{dip}^{st}(\mathbf{r}) = -\sum_i (\nabla E_i)\pi_i^{st} = -(\nabla E_1)\pi_1^{st} - (\nabla E_2)\pi_2^{st} \tag{17.178}$$

The physical meaning of this expression is that the mean dipole force $\mathbf{f}_{dip}^{st}$ is the average of the forces $-\nabla E_1$ and $-\nabla E_2$, which acted on the atom being the state $|1,n,\mathbf{r}\rangle$ and $|2,n,\mathbf{r}\rangle$, respectively, and weighted by the probabilities of occupation π_1^{st} and π_2^{st} of these two types of states. Using eqs.(17.135), (17.136) and (17.129), we can write (17.178) as

$$\mathbf{f}_{dip}^{st}(\mathbf{r}) = -\hbar\delta\frac{\omega_1\nabla\omega_1}{\omega_1^2+2\delta^2} = -\hbar\delta\frac{\omega_1^2}{\omega_1^2+2\delta^2}\alpha = -\nabla\left[\frac{\hbar\delta}{2}\ln\left(1+\frac{\omega_1^2}{2\delta^2}\right)\right] \tag{17.179}$$

with

$$\alpha = \frac{\nabla\omega_1}{\omega_1} = \frac{\Omega}{\omega_1^2}\nabla\Omega \tag{17.180}$$

For a standing-wave laser field, $\omega_1(\mathbf{r}) = \tilde{\omega_1}\cos(\mathbf{k}\cdot\mathbf{r})$, here $\tilde{\omega_1}$ is independent of $\mathbf{r}$. From (17.179) we obtain the steady-state dipole force acting on the two-level atom to be

$$\mathbf{f}_{dip}^{st}(\mathbf{r}) = -\hbar\delta \frac{\nabla\omega_1^2}{4\delta^2 + 2\omega_1^2}$$

Evidently, the above equation is coincident with (17.107) except that Γ^2 is absent in the denominator of the above equation.

Eq.(17.179) shows that the direction of the dipole force $\mathbf{f}_{dip}^{st}$ is clearly dependent on the sign of the detuning δ between the laser field and the atom. This phenomenon can be understood as follows: if the detuning δ is positive (as shown in Fig.(17.13a), the probability of the atom being the state $|1, n, \mathbf{r}\rangle$

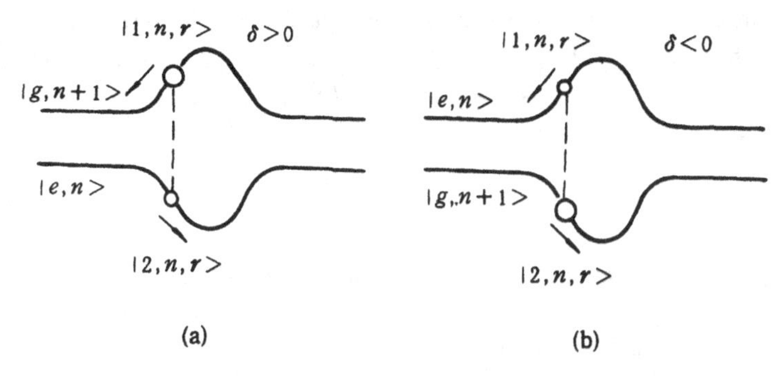

Figure 17.13: The dipole force direction depended on the detuning δ

with the factor of state $|e, n\rangle$ is less than that of being the state $|2, n, \mathbf{r}\rangle$, so that the probability of the spontaneous transitions from the state $|1, n, \mathbf{r}\rangle$ is less than those from $|2, n, \mathbf{r}\rangle$. This shows that the steady-state population π_1^{st} of the atom being the states $\{|1, n, \mathbf{r}\rangle\}$ are larger than the atom being the states $\{|2, n, \mathbf{r}\rangle\}$ $(\pi_1^{st}\rangle\pi_2^{st})$. The dipole force acting on the atom being the states $\{|1, n, \mathbf{r}\rangle\}$ is therefore dominant, and the atom is attracted toward the low intensity regions of the laser field. By contrast, if the detuning δ is negative (as shown in Fig.(17.13b), the steady-state populations of the atom being the states $\{|2, n, \mathbf{r}\rangle\}$ are larger than those in the states $\{|1, n, \mathbf{r}\rangle\}$, the dipole force acting on the atom being the states $\{|2, n, \mathbf{r}\rangle\}$ is then dominant, which leads to the atom to be expelled from the low intensity regions. Finally, if $\delta = 0$, the populations in both states $\{|1, n, \mathbf{r}\rangle\}$ and $\{|2, n, \mathbf{r}\rangle\}$ are equal, so the mean dipole force vanishes, i.e., $\mathbf{f}_{dip}^{st} = 0$.

We now consider the variation of the dipole force when the atom moves slowly. Suppose that the atomic velocity is so small that

$$kv/\Gamma \ll 1, \qquad i.e., \qquad v\Gamma^{-1} \ll \lambda \tag{17.181}$$

In this case the effects of the non-adiabatic kinetic energy coupling terms are still negligible. Therefore, we can also use (17.171) as our starting point, i.e.,

$$\mathbf{f}_{dip}(\mathbf{r}) = -\pi_1 \nabla E_1 - \pi_2 \nabla E_2 \tag{17.182}$$

Different from (3.6.178), the populations π_i here are no longer the steady-state values π_{dip}^{st} because the atom is in moving case. However (17.181) shows that the Doppler shift is small than the wave length of the light. And the distance of the atom travelled for a life time is much less than the laser wavelength λ. It follows that the populations $\pi_i(\mathbf{r})$ in (17.182) for the slowly moving atom are very close to the steady-state values $\pi_i^{st}(\mathbf{r})$ in (17.178), and the difference between $\pi_i(\mathbf{r})$ and $\pi_i^{st}(\mathbf{r})$ obeys

$$\pi_i - \pi_i^{st} \propto \frac{kv}{\Gamma} \tag{17.183}$$

Since $(\pi_i)_{NA}$ can be ignored, we can obtain from eq.(17.144)

$$\frac{d}{dt}\pi_i(\mathbf{r}) = \mathbf{v} \cdot \nabla \pi_i(\mathbf{r}) = -\Gamma_{pop}[\pi_i(\mathbf{r}) - \pi_i^{st}(\mathbf{r})]$$

i.e.,

$$\pi_i(\mathbf{r}) - \pi_i^{st}(\mathbf{r}) = -\tau_{pop}\mathbf{v} \cdot \nabla \pi_i(\mathbf{r}) \tag{17.184}$$

with

$$\tau_{pop} = \frac{1}{\Gamma_{pop}(\mathbf{r})} \tag{17.185}$$

Noticing $kv/\Gamma \ll 1$ and retaining only the first-order of kv/Γ , we can replace $\nabla\pi_i(\mathbf{r})$ by $\nabla\pi_i^{st}(\mathbf{r})$ in (17.184), i.e.,

$$\pi_i(\mathbf{r}) \approx \pi_I^{st}(\mathbf{r}) - \tau_{pop}\mathbf{v} \cdot \nabla \pi_i^{st}(\mathbf{r}) \tag{17.186}$$

Using the Taylor expansion expression

$$f(x + \Delta x) \approx f(x) + \Delta x f'(x) + \cdots \tag{17.187}$$

then (17.186) can be expressed as

$$\pi_i(\mathbf{r}) \approx \pi_i^{st}(\mathbf{r} - \mathbf{v}\tau_{pop}) \tag{17.188}$$

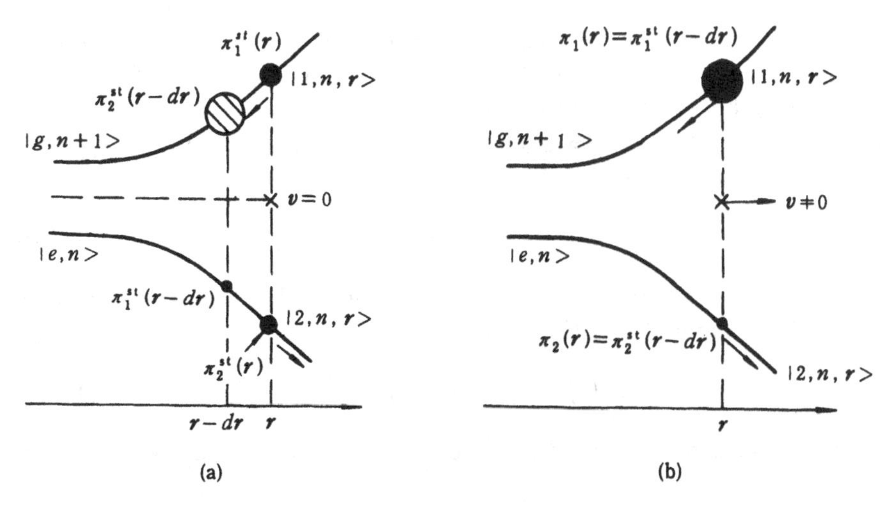

Figure 17.14: Dependence of dipole force direction on atomic velocity

The above equation has a clear physical meaning. The radiative relaxation between the two sets of states $\{|1,n,\mathbf{r}\rangle\}$ and $\{|2,n,\mathbf{r}\rangle\}$ take place with a certain time constant τ_{pop}. Since the atom is moving and the steady-state population $\pi_i^{st}(\mathbf{r})$ generally depends on $\mathbf{r}$, the radiative relaxation can not instantaneously adjust the population $\pi_i(\mathbf{r})$ to the steady-state value $\pi_i^{st}(\mathbf{r})$ which will be obtained if the atom were staying in $\mathbf{r}$. There is a certain response time, characterized by the time constant τ_{pop}, so that the population $\pi_i(\mathbf{r})$ for a moving atom in $\mathbf{r}$ at t is the steady-state population π_i^{st} corresponding to a previous time $t-\tau_{pop}$, i.e., corresponding to the position $\mathbf{r}-\mathbf{v}\tau_{pop}$, that is $\pi_i(\mathbf{r}) = \pi_i^{st}(\mathbf{r}-\mathbf{v}\tau_{pop}) \neq \pi_i^{st}(\mathbf{r})$. By substitution of (17.186) into (17.182) and using (17.119) and (17.164), the dipole force gives

$$\mathbf{f}_{dip}(\mathbf{r}) = \mathbf{f}_{dip}^{st}(\mathbf{r}) - \frac{2\hbar\delta}{\Gamma}\left[\frac{\omega_1^2(\mathbf{r})}{\omega_1^2(\mathbf{r})+2\delta^2}\right]^3 (\alpha\cdot\mathbf{v})\alpha \tag{17.189}$$

As we see, when the atomic motion is slow, there displays a velocity-dependent term in the dipole force, this can be analysed from Fig.(17.14). When the detuning δ is positive and $\mathbf{v}=0$, for the dressed atom, the probability of containing state $|g,n+1\rangle$ in the state $|1,n,\mathbf{r}\rangle$ is larger than that in the state $|2,n,\mathbf{r}\rangle$, therefore, the steady-state populations satisfy $\pi_1^{st}(\mathbf{r}) > \pi_2^{st}(\mathbf{r})$. The probability of the atom being $|e,n\rangle$ in the state $|1,n,\mathbf{r}\rangle$ increases when the Rabi frequency $\omega_1(\mathbf{r})$ increases. It follows that $\pi_1^{st}(\mathbf{r})$ decreases when $\mathbf{r}$ is shifted from $\mathbf{r}-d\mathbf{r}$ to $\mathbf{r}$ along the laser beam: i.e. , $\pi_1^{st}(\mathbf{r}-d\mathbf{r}) > \pi_1^{st}(\mathbf{r})$ (as shown in Fig.(17.14a)). Consider an atom moving with the velocity $\mathbf{v}$ (as shown in Fig.(17.14b)), according to (17.188), the population in $\mathbf{r}$ is $\pi_1(\mathbf{r})$ and not $\pi_1^{st}(\mathbf{r})$. When $\mathbf{r} = \mathbf{v}\tau_{pop}$, then

$$\pi_1(\mathbf{r}) = \pi_1^{st}(\mathbf{r}-\mathbf{v}\tau_{pop}) > \pi_1^{st}(\mathbf{r}) \tag{17.190}$$

as shown in Fig.(17.14b). The same argument gives that the population $\pi_2(\mathbf{r})$ for the dressed atom being the states $\{|2,n,\mathbf{r}\rangle\}$ satisfies

$$\pi_2(\mathbf{r}) = \pi_2^{st}(\mathbf{r}-\mathbf{v}\tau_{pop}) < \pi_2^{st}(\mathbf{r}) \tag{17.191}$$

Thus the atom is moving towards to the high-intensity region due to the friction force

$$\mathbf{f}_{dip}(\mathbf{r},\mathbf{v}) = -\frac{\hbar\nabla\Omega}{2}[\pi_1(\mathbf{r}) - \pi_2(\mathbf{r})] < \mathbf{f}_{dip}^{st}(\mathbf{r}) \tag{17.192}$$

Since $\pi_1(\mathbf{r})$ and $\pi_2(\mathbf{r})$ contain the velocity $\mathbf{v}$, there is a term containing $\mathbf{v}$ in (17.186). And $\mathbf{f}_{dip}(\mathbf{r},\mathbf{v}) < \mathbf{f}_{dip}^{st}(\mathbf{r})$, so the term containing $\mathbf{v}$ in (17.186) plays the role of damping force for $\delta > 0$. With a negative detuning, the conclusion will be reversed: the velocity-dependent dipole force will then be a heating force to accelerate the atomic motion.

References

1. A.Ashkin, *Phys.Rev.Lett.* **24** (1970) 156; **25**(1970) 1321.

2. R.J.Cook and A.F.Bernhardt, *Phys.Rev.* **A18** (1978) 2533.

3. V.S.Letokhov and V.G.Minogin, *Phys.Rep.* **73** (1981) 1.

4. A.P.Kazantsev, G.A.Ryabenko, G.I.Surdutovich and V.P.Yakovlev, *Phys. Rep.* **129** (1985) 75.

5. A.F.Bernhardt and B.W.Shore, *Phys.Rev.* **A23** (1981) 1298.

6. E.Arimondo and A.Bambini, *Opt.Commun.* **37** (1981) 103.

7. G.Compagno, J.S.Peng, and F.Persico, *Phys.Lett.* **A88** (1982) 285.

8. G.Compagno, J.S.Peng, and F.Persico, *Phys.Rev.* **A26** (1982) 2065.

9. P.E.Moskowitz, P.L.Gould, S.R.Atlas and D.E.Pritchard, *Phys. Rev. Lett.* **51** (1983) 370.

10. P.E.Moskowitz, P.L.Gould, and D.E.Pritchard, *J.Opt.Soc.Am.* **B2** (1985) 1784.

11. P.L.Gould, G.A.Ruff, and D.E.Pritchard, *Phys.Rev.Lett.* **56** (1986) 827.

12. P.J.Martin, B.G.Oldaker, A.H.Miklich and D.E.Pritchard, *Phys.Rev.Lett.* **60** (1988) 515.

13. M.J.Holland, D.F.Walls, and P.Zoller, *Phys.Rev.Lett.* **67** (1991) 1716.

14. A.M.Herkommer, V.M.Akulin, and W.P.Schleich, *Phys. Rev. Lett.* **69** (1992) 3298.

15. M.Lindberg, *Appl.Phys.* **B54** (1992) 476.

16. A.Ashkin, *Phys.Rev.Lett.* **40** (1979) 729.

17. R.J.Cook, *Phys.Rev.* **A204** (1979).

18. J.P.Gordon and A.Ashkin, *Phys.Rev.* **A21** (1980) 1606.

19. S.Stenholm, *Phys.Rev.* **A27** (1983) 2513.

20. J.Dalibard and C.Cohen-Tannoudji, *J.Opt.Soc.Am.* **B2** (1985) 1707.

21. P.R.Hemmber, M.G.Prentiss, M.S.Shahriar and N.P.Bigelow, *Opt. Commun.* **89** (1992) 335.

22. C.Cohen-Tannoudji and S.Reynaud *J.Phys.* **B.10** (1977) 345.

CHAPTER 18

LASER COOLING

In the previous chapter we have verified that an atom moving in a laser field will experience the radiation force described by (17.103). If the propagating direction of the laser beam is anti-parallel to the velocity $\mathbf{v}$ of the atomic motion, then the radiation force f will hinder the atomic moving, and lead to decelerate the atomic velocity. To decelerate the atomic moving by use of the laser field is an effective method to cool atom, this method is called the laser cooling.

If the atom is cooled to be at rest, then the Doppler frequency shift and the collision shift will disappear, the spectrum distribution of atom and the atomic parameters will be measured more exactly. So to realize the laser cooling is an important way for studying further the physical properties of the atom precisely, which may induce possible applications in laser spectroscopy, atomic clock and so on. In this chapter we first introduce how to decelerate the motion of the atomic beam, then discuss the quantum theoretical description of the laser cooling, finally we focus on discussing the limited temperature of the laser cooling.

18.1 Decelerating the motion of atoms by use of a laser field

How to decelerate the motion of atoms by use of a laser field is one of the elementary ideas in the laser cooling technology, here we give an outline of explanations. In order to realize the laser cooling efficiently, the first question to be necessarily overcomed is how to compensate continuously the variation of the Doppler frequency shift. For a two-level atom with transition frequency ω_0, when the laser frequency ω is chosen to obey

$$\omega = \omega_0 - \mathbf{k} \cdot \mathbf{v} \tag{18.1}$$

here $\mathbf{k} \cdot \mathbf{v}$ is the Doppler frequency shift, the effective resonant interaction between the atom and the laser field may occur. In this case the atomic velocity may be decelerating. But in the decelerated processes of the atom, the Doppler frequency shift changes with the variation of atomic velocity. If the laser frequency were not adjusted continuously to satisfy (18.1), then after a number of absorpting and emitting photons, the laser frequency will be deviated from the resonant relation (18.1), and the efficiency of decelerating processes decreases. Therefore, how to adjust the laser frequency continuously to compensate the variation of the Doppler frequency shift is very important in laser cooling. From the view of technology, there are different methods to compensate the Doppler frequency shift such as scanning the laser frequency continuously so that it can response the decreasing of the atomic velocity, and assures (18.1) to be valid correspondingly.

If the moving atom is in a standing-wave field, the forces acting on the atom by two running-wave fields are different due to the Doppler frequency shifts. The friction force undergone by the one whose propagating direction is opposite to the direction of the atomic motion becomes large, the Doppler frequency shift $-\mathbf{k} \cdot \mathbf{v}$ enables the atom to tend to be resonant with the field frequency. But the Doppler frequency shift induced by the field whose propagating direction is along to the atomic velocity increases the detuning between the atom and the field, so the force acting on the atom by this laser field becomes small. The difference of these two forces results in a friction force acting on the atom, whose direction is opposite to the atomic velocity and can be qualitatively expressed as

$$F = -\alpha v \tag{18.2}$$

here α is the friction coefficient, which depends on the laser power, the detuning between the atom and the laser field, the atomic decay rates and etc.. The expression of the coefficient α will be given in detail in Section 18.2.

If there arrange three pairs of laser beams which are orthogonal to each other in a three-dimensional space, then the moving atom can be confined in the overlapping range due to the interaction of these six beams of standing-wave laser fields. This arrangement of laser beams is interpreted as "optical molasses".

It is worthwhile to point out that even if the continuous compensation of the Doppler freuqnecy shift is solved prefectly in technology, the limit temperature of laser cooling is impossible to reach to zero Kelvin. There exists a limitation of cooling temperature, which is called the Doppler cooling limit temperature T'_D. On one hand in the processes of atomic absorption and spontaneous emssion of photons, the atom gains the photonic momentum to decelerate the atomic motion. On the other hand the atom gains the recoil mometum of spontaneous emission photons. This recoil momentum is fully random in time and direction, and leads to a fluctuation of the atomic momentum. This fluctuation process corresponds to a heating effect on the atoms by the laser field. The cooling process and the heating process on the atoms by the laser field reach to an equilibrium temperature, which is just the Doppler limitation temperature T'_D. However, the modern technology of the laser cooling enables the lowest cooling temperature to be below T'_D, which is predicted by the conventional theoretical treatment. That is to say, there are some difficulties in the conventional theory about the Doppler cooling mechanism. The difficulties result from that in the conventional theory, the atoms are assumed to be the two-level one, but in fact any atom is not so simple, even the ground state of atom usually has several Zeeman sublevels. Due to the effect of the optical pumping, every sublevel can be populated, those populations depend on the polarization of laser field. Meanwhile, under the interaction of laser field, the atomic ground level can be shifted, which is termed the optical shift. The different sublevels can be shifted by different values, which are associated with the polarization of laser field. In this chapter we focus our attention on the theoretical description of the laser cooling mechanism by considering the effects of the optical pumping and the optical shift.

In addition, from the view of technology, how to overcome the difficulty of the optical pumping is another key to realize the laser cooling efficiently. What is the optical puming? In briefly, after an atom in one of its ground sublevels absorbs a photon and jumps to its excited level, it may goes back to another ground sublevel due to spontaneous emission, the atom in this sublevel may be not resonant to the laser field, then the decelerating processes are paused. Therefore, the atom is accumulated in this sublevel, this process is called the optical pumping process. This optical pumping process is an important effect in disturbing the laser cooling. In order to cancel the optical pumping effect, some methods such as the circular transition by using the polarization lasers are adopted in technology. Here we do not intend to introduce the technology of the laser cooling in detail, the readers who are interested in this technology matter may refer to the references listed in the end of this chapter, here we will focus on the quantum description of the laser cooling.

18.2 Quantum theoretical description of the laser cooling

18.2.1 Hamiltonian describing the system of a polarization laser field interacting with a quasi-two-level atom

The system we consider here is consisted of a quasi-two-level atom with Zeeman sublevels and a polarization laser such as cicular or linear polarization one. We suppose that a quasi-two-level atom with angular momentums $J_g = 1/2$ and $J_e = 3/2$ moves in two laser plane waves which have same frequency ω_L and opposite propagating directions. Then in the rotating-wave-approximation, the interaction Hamiltonain of the laser-atom coupling system is written as

$$V = \mathbf{D}^+ \cdot \varepsilon^+(\mathbf{r}) \exp(-i\omega_L t) + \mathbf{D}^- \cdot \varepsilon^-(\mathbf{r}) \exp(i\omega_L t) \tag{18.3}$$

here $\mathbf{D}^+$ and $\mathbf{D}^-$ are the raising and lowering operators of the atomic electric-dipole operator, respectively, $\varepsilon^+(\mathbf{r})$ and $\varepsilon^-(\mathbf{r})$ are positive and negative-frequency components of the laser electric field., i.e.,

$$\varepsilon^+(\mathbf{r}) = \varepsilon_0 \mathbf{e}_1 \exp(i\mathbf{k} \cdot \mathbf{r}) + \varepsilon_0' \mathbf{e}_2 \exp(-i\mathbf{k} \cdot \mathbf{r}) \tag{18.4}$$

If the amplitudes of the two lasers are equal, i.e., $\varepsilon_0 = \varepsilon_0'$, the wave vector obeys $\mathbf{k} = k\mathbf{e}_z$, and the polarization unit vectors obey $\mathbf{e}_1 = \mathbf{e}_x$, $\mathbf{e}_2 = \mathbf{e}_y$, that is to say, these two laser beams are the running-wave fields which have the same amplitude and opposite propagating directions as well as the orthogonal linear polarizations, then the positive frequency part of laser fields becomes

$$\varepsilon^+(z) = \varepsilon_0 \sqrt{2} \left[\frac{\mathbf{e}_x + \mathbf{e}_y}{\sqrt{2}} \cos(kz) - i \frac{\mathbf{e}_y - \mathbf{e}_x}{\sqrt{2}} \sin(kz) \right] \tag{18.5}$$

As we see, the total field has the different polarization directions at different points. For example, when z=0, the laser field is a linear polarization one with the amplitude $\sqrt{2}\varepsilon_0$ and polarization vector $\frac{1}{\sqrt{2}}(\mathbf{e}_x + \mathbf{e}_y)$; for $z = \lambda/8$, (18.5) becomes

$$\varepsilon^+(\lambda/8) = \sqrt{2}\varepsilon_0 e^{i\pi/4} \frac{\mathbf{e}_x - i\mathbf{e}_y}{\sqrt{2}} = \sqrt{2}\varepsilon_0 e^{i\pi/4} \mathbf{e}_- \tag{18.6}$$

According to (3.25), we know that the photon described by (18.6) is a right-circular polarization one. And when $z = \lambda/4$, the laser field becomes a linear polarization one with the polarization vector $\frac{\mathbf{e}_y-\mathbf{e}_x}{\sqrt{2}}$. But for $z = 3\lambda/8$, equation (18.5) becomes

$$\varepsilon^+(3\lambda/8) = -\sqrt{2}\varepsilon_0 e^{i\pi/4}\frac{\mathbf{e}_x + \mathbf{e}_y}{\sqrt{2}} = \sqrt{2}\varepsilon_0 e^{i\pi/4}\mathbf{e}_+ \tag{18.7}$$

here $\mathbf{e}_+ = -(\mathbf{e}_x + i\mathbf{e}_y)/\sqrt{2}$ is the left-circular polarization unit vector. It is clear that the polarization directions of laser field vary with the spatial position as shown in Fig.(18.1). In order to simplify the discussion, we make a coordinate

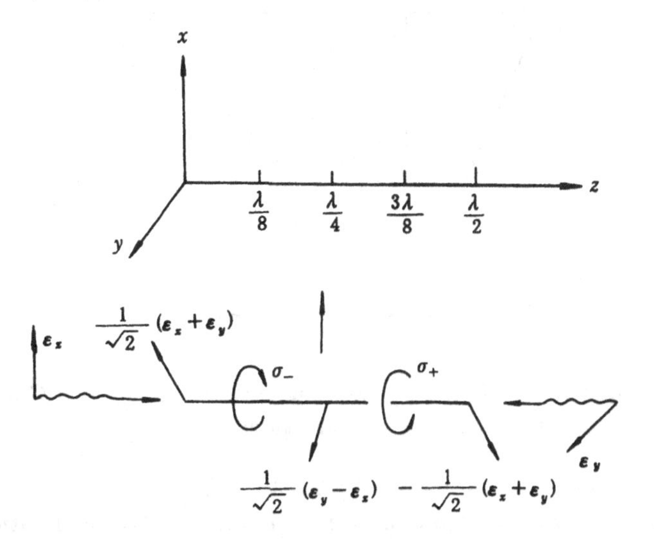

Figure 18.1: Polarization directions of laser field at different space points

displacement by $\lambda/8$ in (18.5), i.e., $z \to z + \lambda/8$, then (18.5) becomes

$$\begin{aligned}
\varepsilon^+(z) &= \sqrt{2}\varepsilon_0\left[\frac{\mathbf{e}_x + \mathbf{e}_y}{\sqrt{2}}\cos k(z+\lambda/8) - i\frac{\mathbf{e}_y - \mathbf{e}_x}{\sqrt{2}}\sin k(z+\lambda/8)\right]\\
&= \sqrt{2}\varepsilon_0\left[e^{i\pi/4}\cos kz\frac{\mathbf{e}_x - i\mathbf{e}_y}{\sqrt{2}} - e^{-i\pi/4}\sin kz\frac{\mathbf{e}_x + \mathbf{e}_y}{\sqrt{2}}\right]\\
&= \sqrt{2}\varepsilon_0[\mathbf{e}_- e^{i\pi/4}\cos kz + \mathbf{e}_+ e^{-i\pi/4}\sin kz]
\end{aligned} \tag{18.8}$$

In this case, the laser field can be equivalent to the superposition of two circular-polarization standing waves with polarization unit vectors $\mathbf{e}_+$ and $\mathbf{e}_-$, respectively.

As we know, for a quasi-two-level atom with angular quantum number of ground state $J_g = 1/2$ and that of excited state $J_e = 3/2$, its excited state has four Zeeman sublevels $e_{\pm 3/2}$ and $e_{\pm 1/2}$, and its ground state has two Zeeman sublevels $g_{\pm 1/2}$, as shown in Fig.(18.2). The Clebsch-Gordon coefficients of the various transitions $g_m \leftrightarrow e_{m'}$ are indicated in Fig.(18.2). The square of these coefficients give the probabilities of the corresponding transitions. In the transition processes of $g_{-1/2} \rightarrow e_{-3/2}$ and $g_{1/2} \rightarrow e_{-1/2}$, the magnetic quan-

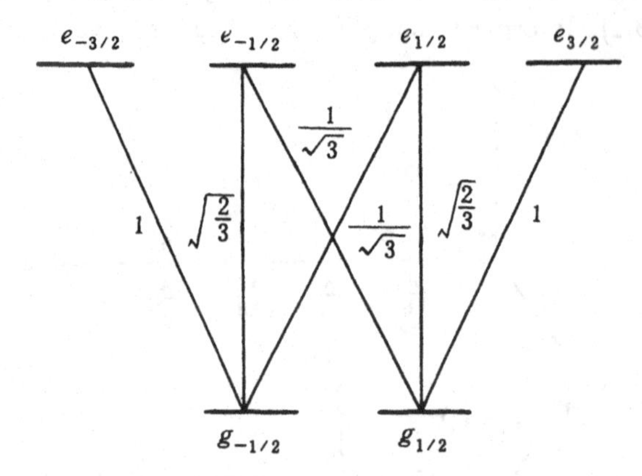

Figure 18.2: Diagram of the atomic level structure and Clebsch-Gordon coefficients representing the strengthes of transitions $g_m \leftrightarrow e_{m'}$

tum number of the atom decreases by 1. According to the conservation law of angular momentum, the atom must absorb one circular-polarization photon from the laser field described by (18.8), the angular quantum number of this kind photons is -1. Similarly, in the transition processes of $g_{-1/2} \rightarrow e_{1/2}$ and $g_{1/2} \rightarrow e_{3/2}$, the atom must absorb one left-polarization photon with angular quantum number 1. However, in the transition processes of $g_{-1/2} \rightarrow e_{-1/2}$ and $g_{1/2} \rightarrow e_{1/2}$, the magnetic quantum number does not change, so under the interaction of the laser field described by (18.8), these transition processes are forbidden. Therefore, after considering the conservation law of the angular momentum and the Clebsch-Gordon coefficients of various transitions $g_m \leftrightarrow e_{m'}$, the raising operator $\mathbf{D}^+$ of the atomic dipole moment in equation (18.3) can be written by

$$\mathbf{D}^+ = d\left[\mathbf{e}_- e^{-i\pi/4}\left(|e_{-3/2}\rangle\langle g_{-1/2}| + \sqrt{\frac{1}{3}}|e_{-1/2}\rangle\langle g_{1/2}|\right)\right.$$

$$+\mathbf{e}_+ e^{i\pi/4}\left(|e_{3/2}\rangle\langle g_{1/2}| + \sqrt{\frac{1}{3}}|e_{1/2}\rangle\langle g_{-1/2}|\right)\Bigg] \tag{18.9}$$

here d represents the strength of the dipole moment, the aim of introducing the phase factors $e^{i\pi/4}$ and $e^{-i\pi/4}$ is for making d to be real. Substituting (18.8) and (18.9) into (18.3), we obtain that the interaction Hamiltonian of the atom-field coupling system can be given by

$$\begin{aligned} V = & \frac{\hbar\Omega}{\sqrt{2}}\sin kz\left(|e_{3/2}\rangle\langle g_{1/2}| + +\sqrt{\frac{1}{3}}|e_{1/2}\rangle\langle g_{-1/2}|\right)\exp(-i\omega_L t) \\ & +\frac{\hbar\Omega}{\sqrt{2}}\cos kz\left(|e_{-3/2}\rangle\langle g_{-1/2}| + \sqrt{\frac{1}{3}}|e_{-1/2}\rangle\langle g_{1/2}|\right)\exp(-i\omega_L t) \\ & +\text{h.c.} \end{aligned} \tag{18.10}$$

where

$$\Omega = \frac{2d\varepsilon_0}{\hbar} \tag{18.11}$$

According to (17.72) and (18.10), the expectation value of the radiation force acting on the atom by laser field gives

$$\begin{aligned} f = & \langle -\frac{dV}{dz}\rangle \\ = & -\frac{\hbar k\Omega}{\sqrt{2}}\cos kz \mathrm{Tr}\left[\rho\left(|e_{3/2}\rangle\langle g_{1/2}| + \sqrt{\frac{1}{3}}|e_{1/2}\rangle\langle g_{-1/2}|\right)\right]\exp -i\omega_L t \\ & +\frac{\hbar k\Omega}{\sqrt{2}}\sin kz \mathrm{Tr}\left[\rho\left(|e_{-3/2}\rangle\langle g_{-1/2}| + \sqrt{\frac{1}{3}}|e_{-1/2}\rangle\langle g_{1/2}|\right)\right]\exp(-i\omega_L t) + \text{c.c.} \\ = & -\frac{\hbar k\Omega}{\sqrt{2}}\cos kz\left[\tilde{\rho}(g_{1/2}, e_{3/2}) + \sqrt{\frac{1}{3}}\tilde{\rho}(g_{-1/2}, e_{1/2}) + \text{c.c.}\right] \\ & +\frac{\hbar k\Omega}{\sqrt{2}}\sin kz\left[\tilde{\rho}(g_{-1/2}, e_{-3/2}) + \sqrt{\frac{1}{3}}\tilde{\rho}(g_{1/2}, e_{-1/2}) + \text{c.c.}\right] \end{aligned} \tag{18.12}$$

here ρ stands for the steady-state density matrix describing the atomic internal motion, and $\tilde{\rho}(g_i, e_j)$ is defined by

$$\tilde{\rho}(g_i, e_j) = \langle g_i|\rho|e_j\rangle\exp(-i\omega_L t) \tag{18.13}$$

From (18.12) we see that we need to obtain the density matrix elements such as $\tilde{\rho}(g_{1/2}, e_{3/2})$ and $\tilde{\rho}(g_{-1/2}, e_{1/2})$ for solving the radiation force acting on the atom under the interaction of laser field. So we rewrite the internal Hamiltonian of the atom-field coupling system in the Hilbert space which is formed by the following atomic internal states

$$|e_{3/2}\rangle = |1\rangle = \begin{pmatrix} 1 \\ 0 \\ 0 \\ 0 \\ 0 \\ 0 \end{pmatrix}, \qquad |e_{1/2}\rangle = |2\rangle = \begin{pmatrix} 0 \\ 1 \\ 0 \\ 0 \\ 0 \\ 0 \end{pmatrix},$$

$$|e_{-1/2}\rangle = |3\rangle = \begin{pmatrix} 0 \\ 0 \\ 1 \\ 0 \\ 0 \\ 0 \end{pmatrix}, \qquad |e_{-3/2}\rangle = |4\rangle = \begin{pmatrix} 0 \\ 0 \\ 0 \\ 1 \\ 0 \\ 0 \end{pmatrix},$$

$$|g_{1/2}\rangle = |5\rangle = \begin{pmatrix} 0 \\ 0 \\ 0 \\ 0 \\ 1 \\ 0 \end{pmatrix}, \qquad |g_{-1/2}\rangle = |6\rangle = \begin{pmatrix} 0 \\ 0 \\ 0 \\ 0 \\ 0 \\ 1 \end{pmatrix} \tag{18.14}$$

i.e.,

$$\begin{aligned} H_A &= \frac{\mathbf{P}^2}{2m} + \sum_{i=1}^{4} \hbar\omega_A |i\rangle\langle i| \\ &= \frac{\mathbf{P}^2}{2m} + \begin{pmatrix} \hbar\omega_A & 0 & 0 & 0 & 0 & 0 \\ 0 & \hbar\omega_A & 0 & 0 & 0 & 0 \\ 0 & 0 & \hbar\omega_A & 0 & 0 & 0 \\ 0 & 0 & 0 & \hbar\omega_A & 0 & 0 \\ 0 & 0 & 0 & 0 & 0 & 0 \\ 0 & 0 & 0 & 0 & 0 & 0 \end{pmatrix} \end{aligned} \tag{18.15}$$

where $\frac{\mathbf{P}^2}{2m}$ represents the atomic kinetic energy, $\sum_{i=1}^{4} \hbar\omega_A |i\rangle\langle i|$ stands for the unperturbed Hamiltonian of the atomic internal states, here we have assumed that $E_{g_{\pm 1/2}} = 0$. Therefore, the interaction Hamiltonian V of the atom-field coupling system is expressed as

$$V = \begin{pmatrix} 0 & 0 & 0 & 0 & x_1^* & 0 \\ 0 & 0 & 0 & 0 & 0 & \frac{1}{\sqrt{3}}x_1^* \\ 0 & 0 & 0 & 0 & \frac{1}{\sqrt{3}}x_2^* & 0 \\ 0 & 0 & 0 & 0 & 0 & x_2^* \\ x_1 & 0 & \frac{1}{\sqrt{3}}x_2 & 0 & 0 & 0 \\ 0 & \frac{1}{\sqrt{3}}x_1 & 0 & x_2 & 0 & 0 \end{pmatrix} \tag{18.16}$$

with

$$x_1 = \frac{\hbar\Omega}{\sqrt{2}} \sin kz \exp(i\omega_L t), \qquad x_2 = \frac{\hbar\Omega}{\sqrt{2}} \cos kz \exp(i\omega_L t) \tag{18.17}$$

Now we can obtain the density matrix of the system by solving the Schrödinger equation

$$i\hbar\frac{\partial}{\partial t}\rho = [H_A + V, \rho] \tag{18.18}$$

18.2.2 Time evolution of the density matrix elements of the atomic internal states

Similar to eqs.(17.75) and (17.78), here we also assume that the atom moves so slowly that the recoil kinetic energy gained by the atom due to the interaction of laser field is sufficiently less than the linewidth Γ of the spontaneous emission. In this assumption, the kinetic energy term $\frac{\mathbf{P}^2}{2m}$ in (18.15) can be neglected when we discuss the evolution of atomic internal states.

In (18.18) we let

$$\tilde{\rho}_{ij} = \langle i|\rho|j\rangle e^{-i\omega_L t} = \rho_{ij}e^{-i\omega_L t} \quad (i = 5, 6, \quad j = 1, 2, 3, 4) \tag{18.19}$$

and substitute (18.15) and (18.16) into (18.18), then the time evolution of density matrix elements obeys

$$i\hbar \begin{pmatrix} A_1 & A_2 \\ A_3 & A_4 \end{pmatrix} = \begin{pmatrix} B_1 & B_2 \\ B_3 & B_4 \end{pmatrix} \tag{18.20}$$

where

$$A_1 = \begin{pmatrix} \dot{\rho}_{11} & \dot{\rho}_{12} & \dot{\rho}_{13} \\ \dot{\rho}_{21} & \dot{\rho}_{22} & \dot{\rho}_{23} \\ \dot{\rho}_{31} & \dot{\rho}_{32} & \dot{\rho}_{33} \end{pmatrix}$$

$$A_2 = \begin{pmatrix} \dot{\rho}_{14} & \dot{\tilde{\rho}}_{15}\exp(-i\omega_L t) - i\omega_L\rho_{15} & \dot{\tilde{\rho}}_{16}\exp(-i\omega_L t) - i\omega_L\rho_{16} \\ \dot{\rho}_{24} & \dot{\tilde{\rho}}_{25}\exp(-i\omega_L t) - i\omega_L\rho_{25} & \dot{\tilde{\rho}}_{26}\exp(-i\omega_L t) - i\omega_L\rho_{26} \\ \dot{\rho}_{34} & \dot{\tilde{\rho}}_{35}\exp(-i\omega_L t) - i\omega_L\rho_{35} & \dot{\tilde{\rho}}_{36}\exp(-i\omega_L t) - i\omega_L\rho_{36} \end{pmatrix}$$

$$A_3 = \begin{pmatrix} \dot{\rho}_{41} & \dot{\rho}_{42} & \dot{\rho}_{43} \\ \dot{\tilde{\rho}}_{51}e^{i\omega_L t} + i\omega_L\rho_{51} & \dot{\tilde{\rho}}_{52}e^{i\omega_L t} + i\omega_L\rho_{52} & \dot{\tilde{\rho}}_{53}e^{i\omega_L t} + i\omega_L\rho_{53} \\ \dot{\tilde{\rho}}_{61}e^{i\omega_L t} + i\omega_L\rho_{61} & \dot{\tilde{\rho}}_{62}e^{i\omega_L t} + i\omega_L\rho_{62} & \dot{\tilde{\rho}}_{63}e^{i\omega_L t} + i\omega_L\rho_{63} \end{pmatrix}$$

$$A_4 = \begin{pmatrix} \dot{\rho}_{44} & \dot{\tilde{\rho}}_{45}e^{-i\omega_L t} - i\omega_L\rho_{45} & \dot{\tilde{\rho}}_{46}e^{-i\omega_L t} - i\omega_l\rho_{46} \\ \dot{\tilde{\rho}}_{54}e^{i\omega_L t} + i\omega_L\rho_{54} & \dot{\rho}_{55} & \dot{\rho}_{56} \\ \dot{\tilde{\rho}}_{64}e^{i\omega_L t} + i\omega_L\rho_{64} & \dot{\rho}_{56} & \dot{\rho}_{66} \end{pmatrix}$$

$$B_1 = \begin{pmatrix} a(\tilde{\rho}_{51} - \tilde{\rho}_{15}) & a\tilde{\rho}_{52} - a_1\tilde{\rho}_{16} & a\tilde{\rho}_{53} - b_1\tilde{\rho}_{15} \\ a_1\tilde{\rho}_{61} - a\tilde{\rho}_{25} & a_1(\tilde{\rho}_{62} - \tilde{\rho}_{26}) & a_1\tilde{\rho}_{63} - b_1\tilde{\rho}_{25} \\ b\tilde{\rho}_{51} - a\tilde{\rho}_{35} & b_1\tilde{\rho}_{52} - a_1\tilde{\rho}_{36} & b_1(\tilde{\rho}_{53} - \tilde{\rho}_{35}) \end{pmatrix}$$

$$B_2 = \begin{pmatrix} a\tilde{\rho}_{54} - b\tilde{\rho}_{16} & \begin{matrix}(a\rho_{55} + \omega_A\tilde{\rho}_{15} - a\rho_{11} \\ -b_1\rho_{12})\exp(-i\omega_L t)\end{matrix} & \begin{matrix}(\omega_A\tilde{\rho}_{16} + a\rho_{56} - b_1\rho_{12} \\ -b_1\rho_{14})\exp(-i\omega_L t)\end{matrix} \\ a_1\tilde{\rho}_{64} - b\tilde{\rho}_{26} & \begin{matrix}(a_1\rho_{65} + \omega_A\tilde{\rho}_{25} - a\rho_{21} \\ -b_1\rho_{23})\exp(-i\omega_L t)\end{matrix} & \begin{matrix}\omega_A\tilde{\rho}_{26} + a_1\rho_{66} - a_1\rho_{22} \\ -b\rho_{24})\exp(-i\omega_L t)\end{matrix} \\ b_1\tilde{\rho}_{54} - b\tilde{\rho}_{36} & \begin{matrix}(b_1\rho_{55} + \omega_A\tilde{\rho}_{35} - a\rho_{31} \\ -b_1\rho_{33})\exp(-i\omega_L t)\end{matrix} & \begin{matrix}\omega_A\tilde{\rho}_{36} + b_1\rho_{56} - a_1\rho_{13} \\ -b\rho_{24})\exp(-i\omega_L t)\end{matrix} \end{pmatrix}$$

$$B_3 = \begin{pmatrix} b\tilde{\rho}_{61} - a\tilde{\rho}_{45} & b\tilde{\rho}_{62} - a_1\tilde{\rho}_{46} & b\tilde{\rho}_{63} - b_1\tilde{\rho}_{45} \\ \begin{matrix}(b_1\rho_{31} - \omega_A\tilde{\rho}_{51} \\ -a\rho_{55} + a\rho_{11})e^{i\omega_L t}\end{matrix} & \begin{matrix}(b_1\rho_{32} - \omega_A\tilde{\rho}_{52} \\ -a_1\rho_{56} - a\rho_{12})e^{i\omega_L t}\end{matrix} & \begin{matrix}(b_1\rho_{33} - \omega_A\tilde{\rho}_{53} \\ -b_1\rho_{55} - a_1\rho_{13})e^{i\omega_L t}\end{matrix} \\ \begin{matrix}(a_1\rho_{21} - \omega_A\tilde{\rho}_{61} \\ -a\rho_{65} + b\rho_{41})e^{i\omega_L t}\end{matrix} & \begin{matrix}(a_1\rho_{22} - \omega_A\tilde{\rho}_{62} \\ -a_1\rho_{66} + b\rho_{42})e^{i\omega_L t}\end{matrix} & \begin{matrix}(a_1\rho_{23} - \omega_A\tilde{\rho}_{63} \\ -b_1\rho_{65} + b\rho_{43})e^{i\omega_L t}\end{matrix} \end{pmatrix}$$

$$B_4 = \begin{pmatrix} b(\tilde{\rho}_{64} - \tilde{\rho}_{46}) & \begin{matrix}(\omega_A\tilde{\rho}_{45} - b_1\rho_{43} \\ +b\rho_{65} - a\rho_{41})e^{-i\omega_L t}\end{matrix} & \begin{matrix}(\omega_A\tilde{\rho}_{46} - a_1\rho_{42} \\ b\rho_{66} - b\rho_{44})e^{-i\omega_L t}\end{matrix} \\ \begin{matrix}(b_1\rho_{34} - \omega_A\tilde{\rho}_{54} \\ -b\rho_{56} + a\rho_{14})e^{-i\omega_L t}\end{matrix} & \begin{matrix}(a\tilde{\rho}_{15} + b_1\tilde{\rho}_{35} \\ -a\tilde{\rho}_{51} - b_1\tilde{\rho}_{53})\end{matrix} & \begin{matrix}(a\tilde{\rho}_{16} + b_1\tilde{\rho}_{36} \\ -a_1\tilde{\rho}_{52} - b\tilde{\rho}_{54})\end{matrix} \\ \begin{matrix}(a_1\rho_{24} + -\omega_A\tilde{\rho}_{64} \\ -b\rho_{66} + b\rho_{44})e^{i\omega_L t}\end{matrix} & \begin{matrix}(a_1\tilde{\rho}_{25} + b\tilde{\rho}_{45} \\ -a\tilde{\rho}_{61} - b_1\tilde{\rho}_{63})\end{matrix} & \begin{matrix}(a_1\tilde{\rho}_{26} + b\tilde{\rho}_{46} \\ -a_1\tilde{\rho}_{62} - b_1\tilde{\rho}_{46})\end{matrix} \end{pmatrix}$$

with

$$a = \frac{\hbar\Omega\sin kz}{\sqrt{2}}, \quad a_1 = \frac{a}{\sqrt{3}}, \quad b = \frac{\hbar\Omega\cos kz}{\sqrt{2}}, \quad b_1 = \frac{b}{\sqrt{3}} \tag{18.21}$$

It is worthwhile to point out that the influences of atomic spontaneous emission effects are not included in (18.18). When the spontaneous emission is taken into account, we can obtain from (18.20) the evolution equation of atomic internal density matrix elements, which is usually termed as the optical Bloch equation. For example, from (18.20) we have

$$\begin{aligned}&\dot{\tilde{\rho}}_{51}\exp(i\omega_L t)+i\omega_L\tilde{\rho}_{51}\exp(i\omega_L t)\\&=\frac{1}{i\hbar}(a\rho_{11}+b_1\rho_{31}-\omega_A\tilde{\rho}_{51}-a\rho_{55})\exp(i\omega_L t)-\frac{1}{2}\Gamma\tilde{\rho}_{51}\exp(i\omega_L t)\end{aligned}$$

Substituting (18.21) into the above equation and assuming the detuning

$$\delta=\omega_L-\omega_A \tag{18.22}$$

then obtain the evolution equation of $\tilde{\rho}_{51}$ as

$$\dot{\tilde{\rho}}_{51}=-\left(i\delta+\frac{\Gamma}{2}\right)\tilde{\rho}_{51}-\frac{i\Omega}{\sqrt{6}}\cos kz\rho_{31}+i\frac{\Omega}{\sqrt{2}}\sin kz(\rho_{55}-\rho_{11}) \tag{18.23}$$

This equation is valid for any laser power and for any atomic velocity , but it is difficult to evaluate. In order to simply the calculation, we only discuss the case in the low-power and low-velocity domains. If the intensity of the laser field is so weak that $\Omega\ll\Gamma$, then the probabilities from ground states $|g_{-1/2}\rangle$ and $|g_{1/2}\rangle$ to the excited states $|e_j\rangle$ ($j=\pm1/2,\pm3/2$) are very small. Consequently, comparing with the populations ρ_{55}, we can neglect ρ_{11} and ρ_{31} when we calculate ρ_{51} to first order in Ω/Γ. (18.23) may be therefore simplified to

$$\dot{\tilde{\rho}}_{51}=-\left(i\delta+\frac{\Gamma}{2}\right)\tilde{\rho}_{51}+\frac{i\Omega\sin kz\rho_{55}}{\sqrt{2}} \tag{18.24}$$

We also assume that the atomic velocity is so low that the Doppler frequency shift $kv\ll\Gamma$. Then during the relaxation time $2\Gamma^{-1}$ the variation of $\sin kz=\sin kvt$ in (18.24) is slight. In other words, $\tilde{\rho}_{51}$ is a fast decay variable compared with $\rho_{55}\sin kz$, then we have

$$\begin{aligned}\tilde{\rho}_{51}(t)&=\rho_{51}(0)\exp\left[-\left(i\delta+\frac{\Gamma}{2}\right)t\right]\\&\quad+i\frac{\Omega}{\sqrt{2}}\int_0^t\sin kz\rho_{55}\exp\left[-\left(i\delta+\frac{\Gamma}{2}\right)(t-t')\right]dt'\\&=\rho_{51}(0)\exp\left[-\left(i\delta+\frac{\Gamma}{2}\right)t\right]\\&\quad+\frac{i\Omega\sin kz}{\sqrt{2}(i\delta+\Gamma/2)}\rho_{55}\left\{1-\exp\left[-\left(i\delta+\frac{\Gamma}{2}\right)t\right]\right\}\end{aligned}$$

When $t \gg \Gamma/2$, the exponential term in the above equation tends to zero rapidly, in this case $\tilde{\rho}_{51}(t)$ and ρ_{55} satisfy the following relation

$$\tilde{\rho}_{51}(t) = \frac{i\Omega \sin kz}{i\delta + \Gamma/2}; \qquad \rho_{55} = \frac{\Omega \sin kz}{\delta - i\Gamma/2}\rho_{55} \tag{18.25}$$

In as much as $\tilde{\rho}_{51}(t)$ is a variable with rapid decay rate, we can directly let $\dot{\tilde{\rho}}_{51} = 0$ in (18.24), that is we can adopt the adiabatic elimination approximation, i.e.,

$$\dot{\tilde{\rho}}_{51} = 0 = -\left(i\delta + \frac{\Gamma}{2}\right)\tilde{\rho}_{51} + \frac{i\Omega \sin kz\rho_{55}}{\sqrt{2}}$$

It is apparent that we can obtain eq.(18.25) from the above equation by a straightforward way. Now we need to calculate ρ_{55}. From (18.20) we have

$$\dot{\rho}_{11} = -\Gamma\rho_{11} + \frac{i\Omega}{\sqrt{2}}\sin kz(\tilde{\rho}_{15} - \tilde{\rho}_{51}) \tag{18.26}$$

Here the term $-\Gamma\rho_{11}$ describes the effect of the spontaneous emission. Substituting (18.25) into (18.26), we obtain ρ_{11} in the adiabatic elimination approximation to be

$$\rho_{11} = s_0 \sin^2 kz\rho_{55} \tag{18.27}$$

where s_0 is the detuning-dependent saturation parameter

$$s_0 = \frac{\Omega^2/2}{\delta^2 + \Gamma^2/4} \tag{18.28}$$

Eq.(18.27) reflects the relation between the populations of the atom being in the excited state $|e_{3/2}\rangle$ and the ground state $|g_{1/2}\rangle$, apparently $\rho_{11}/\rho_{55} \propto \Omega^2/\Gamma^2$. Similarly, we can write the optical Bloch equations for the coherent terms $\dot{\tilde{\rho}}_{35}$ and $\dot{\tilde{\rho}}_{62}$ as

$$\dot{\tilde{\rho}}_{35} = \left(i\delta - \frac{\Gamma}{2}\right)\tilde{\rho}_{35} + i\Omega\frac{\sin kz}{\sqrt{2}}\rho_{31} + \frac{i\Omega\cos kz(\rho_{33} - \rho_{55})}{\sqrt{6}} \tag{18.29}$$

$$\dot{\tilde{\rho}}_{62} = -\left(i\delta + \frac{\Gamma}{2}\right)\tilde{\rho}_{62} + i\Omega\frac{\sin kz}{\sqrt{6}}\rho_{66} - \frac{i\Omega\sin kz\rho_{22}}{\sqrt{6}} - \frac{i\Omega\cos kz\rho_{42}}{\sqrt{2}} \tag{18.30}$$

Following the same procedure of calculating $\tilde{\rho}_{51}$, i.e., neglecting the populations ρ_{33} and ρ_{22} of the excited states and the coherence terms ρ_{31} and ρ_{42}, whose magnitudes are proportional to the second-order infinitesimal terms of Ω/Γ. Taking the adiabatic elimination approximation, one obtain

$$\tilde{\rho}_{35} = \frac{\Omega/\sqrt{6}}{\delta + i\Gamma/2} \sin kz \rho_{55} \tag{18.31}$$

$$\tilde{\rho}_{62} = \frac{\Omega/\sqrt{6}}{\delta - i\Gamma/2} \sin kz \rho_{66} \tag{18.32}$$

We finally calculate the populations in excited states $|e_{1/2}\rangle$ and $|e_{-1/2}\rangle$ associated with the density matrix elements $\tilde{\rho}_{62}$ and $\tilde{\rho}_{35}$. Using (18.20) and considering the spontaneous emission effects we have

$$\dot{\rho}_{22} = -\Gamma\rho_{22} + \frac{i\Omega}{\sqrt{6}} \sin kz(\tilde{\rho}_{26} - \tilde{\rho}_{62}) \tag{18.33}$$

$$\dot{\rho}_{33} = -\Gamma\rho_{33} + \frac{i\Omega}{\sqrt{6}} \cos kz(\tilde{\rho}_{35} - \tilde{\rho}_{53} \tag{18.34}$$

Following the same procedure of calculating ρ_{11}, we can obtain ρ_{22} and ρ_{33} as

$$\rho_{22} = \frac{1}{2} s_0 \sin^2 kz \rho_{66} \tag{18.35}$$

$$\rho_{33} = \frac{1}{3} s_0 \cos^2 kz \rho_{55} \tag{18.36}$$

Now we have obtained the nondiagonal matrix elements $\tilde{\rho}_{51}$ and $\tilde{\rho}_{35}$ associated with the state $|g_{1/2}\rangle$ undergoing the interaction of the laser field, and the populations $\rho_{ii}(i = 1, 2, 3)$ associated with $|g_{1/2}\rangle$ undergoing the spontaneous emission processes. It is also easy to obtain the time-dependent equation of the population ρ_{55} in state $|g_{1/2}\rangle$ from (18.20) to be

$$\dot{\rho}_{55} = \Gamma(\rho_{11} + 2\rho_{22}/3 + \rho_{33}/3) - \left[\frac{i\Omega}{\sqrt{2}} \sin kz \tilde{\rho}_{15} + \frac{i\Omega}{\sqrt{6}} \cos kz \tilde{\rho}_{35} - \frac{i\Omega}{\sqrt{2}} \sin kz \tilde{\rho}_{51} - \frac{i\Omega}{\sqrt{6}} \cos kz \tilde{\rho}_{53}\right] \tag{18.37}$$

The meaning of the terms $\Gamma\rho_{ii}(i = 1, 2, 3)$ stand for the contribution of the atomic decay from the excited states $|e_{3/2}\rangle$, $|e_{1/2}\rangle$ and $|e_{-1/2}\rangle$ to the ground state $|g_{1/2}\rangle$ due to the interaction of the vacuum field, these decay processes obey the electric dipole selection rule $\Delta m = 0, \pm 1$. Substituting (18.25), (18.31) and their conjugations, and eqs.(18.27), (18.35) and (18.36) into (18.37), we obtain

$$\dot{\rho}_{55} = -\frac{2}{9}s_0\Gamma(\cos^2 kz\rho_{55} - \sin^2 kz\rho_{66}) \tag{18.38}$$

Similarly, the density matrix elements $\tilde{\rho}_{64}$, ρ_{44} and ρ_{66} associated with $|g_{-1/2}\rangle$ satisfy the following equations

$$\tilde{\rho}_{64} = \frac{\Omega\sqrt{2}}{\delta - i\Gamma/2}\cos kz\rho_{66} \tag{18.39}$$

$$\rho_{44} = s_0\cos^2 kz\rho_{66} \tag{18.40}$$

$$\begin{aligned}\dot{\rho}_{66} &= \Gamma\left(\rho_{44} + \frac{2\rho_{33}}{3} + \rho_{22}\right) + \frac{i\Omega}{\sqrt{6}}\sin kz(\tilde{\rho}_{62} - \tilde{\rho}_{26}) \\ &\quad + \frac{i\Omega}{\sqrt{2}}\cos kz(\tilde{\rho}_{64} - \tilde{\rho}_{46}) \\ &= -\frac{2}{9}\Gamma s_0(\sin^2 kz\rho_{66} - \cos^2 kz\rho_{55})\end{aligned} \tag{18.41}$$

Considering the normalization condition

$$1 = \sum_{i=1}^{6}\rho_{ii} \approx \rho_{55} + \rho_{66} \tag{18.42}$$

and using (18.38) and (18.41), the steady-state populations in the ground states $|g_{-1/2}\rangle$ and $|g_{1/2}\rangle$ give

$$\rho_{55}^{st} = \sin^2 kz \tag{18.43}$$

$$\rho_{66}^{st} = \cos^2 kz \tag{18.44}$$

Substituting (18.42)-(18.44) into (18.38) and (18.41) we have

$$\dot{\rho}_{55} = -[\cos^2 kz\rho_{55} - sin^2 kz(1 - \rho_{55})]/\tau_p = -(\rho_{55} - \rho_{55}^{st})/\tau_p \tag{18.45}$$

$$\dot{\rho}_{66} = -(\rho_{66} - \rho_{66}^{st})/\tau_p \tag{18.46}$$

with

$$\frac{1}{\tau_p} = \frac{2}{9}\Gamma s_0 \tag{18.47}$$

In analogy with (17.139), $1/\tau_p$ is called the pumping rate. Eqs.(18.45) and (18.46) show that the populations in states $|g_{-1/2}\rangle$ and $|g_{1/2}\rangle$ of a rest atom at point z reach their steady-state values after time τ_p. Clearly τ_p is proportional to the laser intensity $I \propto \Omega^2$. And eqs.(18.43) and (18.44) imply that the steady-state populations ρ_{55}^{st} and ρ_{66}^{st} depend on the atomic position in the laser field. After having the expressions of density matrix elements for the atomic internal states, we can now take into account the radiation force acting on the atom by the laser field from (18.12).

18.2.3 Radiation force acting on the atom by the laser field

In order to calculate the radiation force acting on the atom by the laser field, we substitute $\tilde{\rho}_{51}$, $\tilde{\rho}_{62}$, $\tilde{\rho}_{64}$ and $\tilde{\rho}_{53}$ and their complex conjugations into equation (18.12)

$$\begin{aligned}
f = & -\frac{\hbar k\Omega}{\sqrt{2}}\cos kz\left\{\frac{\Omega}{2}\sin kz\rho_{55}\left(\frac{1}{\delta - i\Gamma/2}+\frac{1}{\delta + i\Gamma/2}\right)\right. \\
& \left.+\frac{\Omega}{3\sqrt{2}}\sin kz\rho_{66}\left(\frac{1}{\delta - i\Gamma/2}+\frac{1}{\delta + i\Gamma/2}\right)\right\} \\
& +\frac{\hbar k\Omega}{\sqrt{2}}\sin kz\left\{\frac{\Omega}{\sqrt{2}}\cos kz\rho_{66}\left(\frac{1}{\delta - i\Gamma/2}+\frac{1}{\delta + i\Gamma/2}\right)\right. \\
& \left.+\frac{\Omega}{3\sqrt{2}}\cos kz\rho_{55}\left(\frac{1}{\delta - i\Gamma/2}+\frac{1}{\delta + i\Gamma/2}\right)\right\} \\
= & -\frac{2}{3}\hbar k s_0\delta(\rho_{55}-\rho_{66})\sin 2kz
\end{aligned} \tag{18.48}$$

The above equation shows that under the interaction of the weak laser field, the radiation force f acting on the atom depends on the difference of populations in ground states $|g_{1/2}\rangle$ and $|g_{-1/2}\rangle$. That is to say, f is determined by the property of the atom in its two ground sublevels. In order to understand the physical meaning of this phenomenon, we discuss the variation of the property of the ground levels.

If the laser field is very weak, then the probability of the atom jumping from the ground state to the excited state $|e_j\rangle$ is very small. For the atom initially in the state $|g_i\rangle$ under the interaction of a weak laser field, these transition processes can be discussed by the perturbation theory. Inasmuch as the atom has two possible ground states $|g_{-1/2}\rangle$ and $|g_{1/2}\rangle$, the atom-field coupling system can be treated as two isolated sub-systems. For the atom initially in the state $|g_{1/2}, n\rangle$, according to the steady-state perturbation theory, the dressed state vector of the atom interacting with the laser field may be written as

$$|\Psi_{1/2}\rangle = A(|g_{1/2}, n\rangle + \tilde{\rho}_{15}|e_{3/2}, n-1\rangle + \tilde{\rho}_{35}|e_{-1/2}, n-1\rangle) \tag{18.49}$$

where A is the normalization constant. Replacing ρ_{55} by 1 in (18.25) and (18.31), then (18.49) becomes

$$|\Psi_{1/2}\rangle = A\left(|g_{1/2}, n\rangle + \frac{\Omega/\sqrt{2}}{\delta + i\Gamma/2}\sin kz|e_{3/2}, n-1\rangle + \frac{\Omega/\sqrt{6}}{\delta + i\Gamma/2}\cos kz|e_{-1/2}, n-1\rangle\right) \tag{18.50}$$

Considering the normalization condition

$$\langle\Psi_{1/2}|\Psi_{1/2}\rangle = 1$$

we have

$$A = \left(1 + s_0\sin^2 kz + \frac{s_0}{6}\cos^2 kz\right)^{-1/2} \tag{18.51}$$

The dressed state vector of the atom initially in $|g_{1/2}, n\rangle$ is therefore written as

$$|\Psi_{1/2}\rangle = \left(1 + s_0\sin^2 kz + \frac{s_0}{6}\cos^2 kz\right)^{-1/2} \left(|g_{1/2}, n\rangle + \frac{\Omega/\sqrt{2}}{\delta + i\Gamma/2}\sin kz|e_{3/2}, n-1\rangle + \frac{\Omega/\sqrt{6}}{\delta + i\Gamma/2}\cos kz|e_{-1/2}, n-1\rangle\right) \tag{18.52}$$

Following the similar method to calculate the Lamb shift of the Hydrogen atom (11.32), the modification of the ground state energy due to the effect of the laser field is

$$\begin{aligned}
\Delta E_{1/2} = &-\left\langle \Psi_{1/2}\left|\left(\sum_{i=1}^{4}\hbar\omega_A|i\rangle\langle i|\right)\right|\Psi_{1/2}\right\rangle - \langle\Psi_{1/2}|\hbar\omega_L a^+ a|\Psi_{1/2}\rangle \\
&+\left\langle g_{1/2}, n\left|\left(\sum_{i=1}^{4}\hbar\omega_A|i\rangle\langle i|\right)\right|g_{1/2}, n\right\rangle + \langle g_{1/2}, n|\hbar\omega_L a^+ a|g_{1/2}, n\rangle \\
= &-\hbar n\omega_L + \hbar\omega_L s_0\left(\sin^2 kz + \frac{\cos^2 kz}{3}\right)\left(1 + s_0\sin^2 kz + \frac{s_0}{3}\cos^2 kz\right)^{-1} \\
&-\hbar\omega_A s_0\left(\sin^2 kz + \frac{\cos^2 kz}{3}\right)\left(1 + s_0\sin^2 kz + \frac{s_0}{3}\cos^2 kz\right)^{-1} + \hbar n\omega_L
\end{aligned}$$

$$\approx \hbar\delta s_0 \left(\frac{\cos^2 kz}{3} + sin^2 kz\right) \tag{18.53}$$

$$= E_0 - \frac{\hbar\delta \cos 2kz}{3} \tag{18.54}$$

where

$$E_0 = \frac{2\hbar\delta s_0}{3} \tag{18.55}$$

Eq.(18.54) shows that the ground sublevel $|g_{1/2}\rangle$ of the atom is shifted due to the interaction of laser field, the value of this shift is $E_0 - \hbar\delta s_0 \cos 2kz/3$. This shift is called the light shift as mentioned in the preceding. Using the same method, we can also obtain the light shift for another ground sublevel $|g_{-1/2}\rangle$ as

$$\Delta E_{-1/2} = E_0 + \frac{\hbar\delta s_0 \cos 2kz}{3} \tag{18.56}$$

Apparently, the light shifts of the two ground sublevels are different. This results from that these two ground sublevels are coupled to the excited states with different coupling strengths. From (18.10) we see that the ground state $|g_{1/2}\rangle$ is coupled to the excited states $|e_{3/2}\rangle$ and $|e_{-1/2}\rangle$ with strengths $\frac{\hbar\Omega}{\sqrt{2}} \sin kz$ and $\frac{\hbar\Omega}{\sqrt{6}} \cos kz$, respectively, and the state $|g_{-1/2}\rangle$ is coupled to the states $|e_{-3/2}\rangle$ and $|e_{1/2}\rangle$ with strengths $\frac{\hbar\Omega}{\sqrt{2}} \cos kz$ and $\frac{\hbar\Omega}{\sqrt{6}} \sin kz$, respectively. Since the intensity of the laser field is spatial-dependent, the light shifts of two sublevels have different values, which depend on the spatial distribution of laser field. Fig.(18.3) gives the light shifts $\Delta E_{\pm 1/2}$ oscillate in space with a period $\lambda/2$.

Since the light shifts of two ground states vary with the space positions, the forces experienced by the atom in the states $|g_{1/2}\rangle$ and $|g_{-1/2}\rangle$ are respectively written as

$$f_{1/2} = -\frac{d}{dt}\Delta E_{1/2} = -\frac{2}{3}\hbar k\delta s_0 \sin 2kz \tag{18.57}$$

$$f_{-1/2} = \frac{2}{3}\hbar k\delta s_0 \sin 2kz \tag{18.58}$$

In terms of (18.57) and (18.58), (18.48) becomes

$$f = f_{1/2}\rho_{55} + f_{-1/2}\rho_{66} \tag{18.59}$$

This force f is then just the average of two state-dependent forces $f_{\pm 1/2}$ weighted by the populations of these states. Such an expression is similar to (17.182) obtained for a two-level atom in intensive laser light in which the

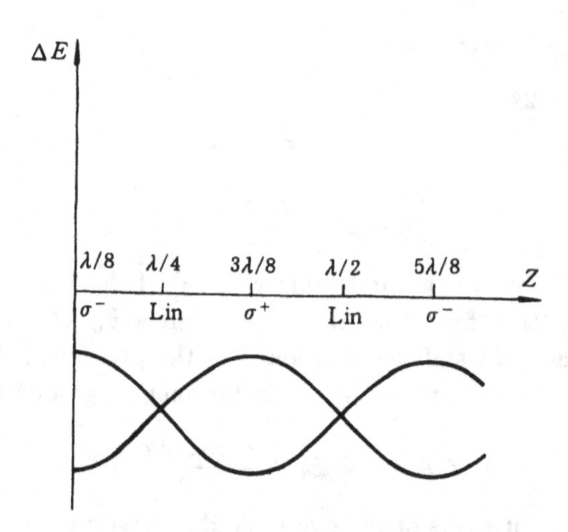

Figure 18.3: Variation versus z for light shifts $\Delta E_{\pm 1/2}$ of two ground states sublevels

radiation force is also the average of forces in two dressed levels with corresponding weights. However, an important difference arises here, because each of the two levels involved in (18.59) is essentially a ground-state sublevel with a long residence time τ_p, but in (17.182) the two dressed states are strongly contaminated by the excited state, so that their lifetime is approximated to be Γ^{-1}.

Now we turn to calculate the populations ρ_{55} and ρ_{66} for evaluating the force acting on the atom. First we assume that the atom is at rest. In this case the atomic velocity v=0 and the functions $\sin kz$ and $\cos kz$ are time-independent. According to (17.178) we know that the atomic populations ρ_{55}^{st} and ρ_{66}^{st} in states $|g_{\pm 1/2}\rangle$ are the stationary ones. Then the force acting on the atom is given by

$$f(z, v=0) = f_{1/2}\rho_{55}^{st} + f_{-1/2}\rho_{66}^{st} = \frac{2}{3}\hbar k \delta s_0 \sin 2kz \cos 2kz \tag{18.60}$$

Next we consider the case of the atom with a very slow velocity. Since eqs.(18.45) and (18.46) have the same expressions as (17.140), when the Doppler frequency shift kv is smaller than the pumping rate τ_p^{-1}, i.e., $kv\tau_p \ll 1$, then from (17.189) we have

$$\rho_{55}(z,v) \approx \rho_{55}^{st}(z) - v\tau_p \frac{d}{dz}\rho_{55}^{st} + \cdots = \sin^2 kz - kv\tau_p \sin 2kz \quad (18.61)$$

$$\rho_{66}(z,v) \approx \rho_{66}^{st}(z) - v\tau_p \frac{d}{dz}\rho_{66}^{st} + \cdots = \cos^2 kz + kv\tau_p \sin 2kz \quad (18.62)$$

Substituting the above two equations into (18.48), we obtain the force acting on the very slow moving atom by the laser field to be

$$f(z,v) = f(z, v=0) + \frac{4}{3}\hbar k^2 \delta s_0 v\tau_p \sin^2(2kz) \quad (18.63)$$

We now calculate the force acting on the atom during the atom moving for one wavelength, that is to say, to average the force f over a spatial period. Evidently, from (18.60) and (18.63) we have

$$\overline{f}(v=0) = \frac{1}{\lambda}\int_0^\lambda f(z, v=0)dz = \frac{2}{3\lambda}\hbar k\delta s_0 \int_0^\lambda \sin 2kz dz = 0 \quad (18.64)$$

$$\overline{f}(v) = \frac{1}{\lambda}\int_0^\lambda f(z,v)dz = -\alpha v \quad (18.65)$$

This is just the force acting on the atom by the standing-wave laser field described by (18.2). Here the coefficient α is given by

$$\alpha = -\frac{2}{3}\hbar k^2 \delta s_0 \tau_p = -\frac{3\hbar k^2 \delta}{\Gamma} \quad (18.66)$$

The coefficient α depends on the detuning between the frequency ω_L of laser field and the atomic transition frequency ω_A. For $\delta > 0$, there exists $\alpha < 0$, so the spatial average force $\overline{f}(v) = |\alpha| v$ acting on the atom plays an acceleration role in the atomic motion. On the contrary for $\delta < 0$, i.e., $\alpha > 0$, the force $\overline{f}(v) = -|\alpha| v$ decelerates the atomic motion. α (18.66) is just the friction coefficient named in (18.2). The hinderance effect of the radiation force eq.(18.65) on the atomic motion enables the atomic velocity to decrease gradually, then the atomic kinetic energy E_k also decrease. According to the thermodynamics law $E_k = k_B T/2$, with the decreasing of the atomic kinetic energy, the temperature decreases gradually, so the force (18.65) can realizes the aim of cooling atom by use of the laser field. Next we discuss the physical mechanism of the laser cooling in more detail.

18.2.4 Physical mechanism of the laser cooling

The coordinate has been assumed to be shifted by $\lambda/8$ when we deduce (18.12), so at z=0, the atom is acted by the right circular polarization field. In this moment the atom placed at z=0 can absorb a σ^- photon and jump to $|e_{-1/2}\rangle$ from its initial state $|g_{1/2}\rangle$. Furthermore, the atom will decay to the ground state $|g_{-1/2}\rangle$ due to the vacuum fluctuations. (If the atom decays to the ground state $|g_{1/2}\rangle$ under the interaction of the vacuum field, it can absorb another σ^- photon to jump to $|e_{-1/2}\rangle$, then the atom may have another chance to transit to the state $|g_{-1/2}\rangle$). On the other hand, if the atom initially in state $|g_{-1/2}\rangle$ absorbs a σ^- photon and jump to the state $|e_{-3/2}\rangle$, then the atom in $|e_{-3/2}\rangle$ will decay to the ground state $|g_{-1/2}\rangle$ due to the effect of the vacuum field. Thus in the steady-state case, the atom will be pumped into state $|g_{-1/2}\rangle$. According to (18.43) and (18.44), the steady-state atomic populations in the two ground sublevels at z=0 may be written by

$$\rho_{55}^{st}(z=0)=0, \qquad \rho_{66}^{st}(z=0)=1 \tag{18.67}$$

In addition, from (18.12) we also know that the σ^- transition starting from $|g_{1/2}\rangle$ is three times as intense as the σ^- transition starting from $|g_{-1/2}\rangle$, so the light shift $\Delta E_{-1/2}$ of $|g_{-1/2}\rangle$ is three times larger than $\Delta E_{1/2}$ of $|g_{1/2}\rangle$. Obviously, from (18.54) and (18.56) we obtain

$$\Delta E_{1/2}(z=0)=\frac{\hbar\delta s_0}{3}, \qquad \Delta E_{-1/2}(z=0)=\hbar\delta s_0 \tag{18.68}$$

If the atom is at $z=\lambda/4$, where the polarization of light is σ^+, thus the previous conclusions are reversed, i.e., the atom will be pumped into state $|g_{1/2}\rangle$ in the steady-state case. The steady-state populations are

$$\rho_{55}^{st}(z=\lambda/4)=1, \qquad \rho_{66}^{st}(z=\lambda/4)=0 \tag{18.69}$$

and the light shifts of $|g_{1/2}\rangle$ and $|g_{-1/2}\rangle$ become

$$\Delta E_{1/2}(z=\lambda/4)=\hbar\delta s_0, \qquad \Delta E_{-1/2}(z=\lambda/4)=\frac{\hbar\delta s_0}{3} \tag{18.70}$$

Finally, if the atom is located in a place where the light polarization is linear, for example, at $z=\lambda/8,\ 3\lambda/8,\cdots$, thus the laser field can be treated as two polarization fields σ^+ and σ^- with equal amplitudes. From the symmetry we know that both sublevels are equally populated and undergo the same light shift, which can be confirmed by eqs.(18.43), (18.44), (18.54), and (18.56). Fig.(18.4) is the illustration of these results: the light shifts are represented by

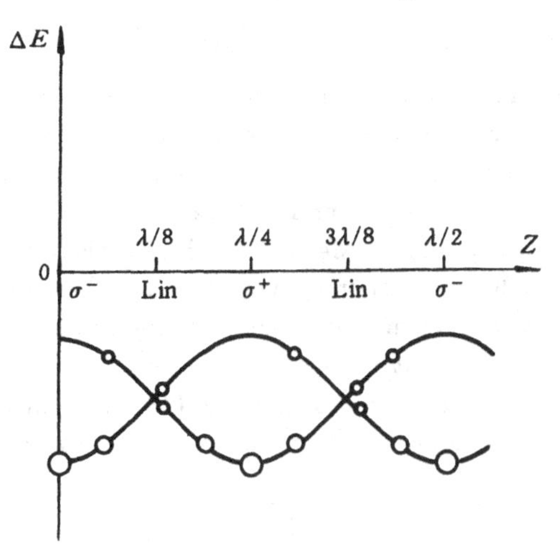

Figure 18.4: Light shifts and populations versus z for two sublevels of the ground state

two curves of cosine functions, the size of black circles is proportional to the steady-state populations of the corresponding sublevels.

If the light shift curves are treated as the potential hill that the atom will climb, the atomic populations are symmetrically distributed in the neighbour valleys (i.e., z=0, $\lambda/4, \cdots$) or in the neighbour tops (i.e., $z = \lambda/8,\ 3\lambda/8, \cdots$), so the spatial average of the force acting on the atom is equal to zero as shown in (18.64). However, when the atom moves, this symmetry is broken, meanwhile the average friction force $\overline{f}$ (18.65) appears. According to the Taylor expansion formula

$$f(x+\Delta x) \approx f(x) + \Delta x f'(x)$$

we may rewrite eqs.(18.61) and (18.62) which are valid under the condition $kv\tau_p \ll 1$ as

$$\rho_{55} \approx \rho_{55}^{st}(z - v\tau_p), \qquad \rho_{66} \approx \rho_{66}^{st}(z - \tau_p) \tag{18.71}$$

then we find that the radiation force acting on the atom at z is decided by the steady-state populations at $z - v\tau_p$. Suppose that the atom is started from

z=0, where the atom is populated in the potential valley due to the pumping effect (as shown in Fig.(18.5)), and follows a moving to the right. If the atomic velocity v is such that the atom travels over a distance $\lambda/4$ during τ_p, i.e., $v = \lambda/(4\tau_p) = \pi/(2k\tau_p)$, from (18.71) we have

$$\rho_{66}(z = \lambda/4) = \rho_{66}^{st}(z = 0) = 1 \tag{18.72}$$

So that the atom is remained in the sublevel $|g_{-1/2}\rangle$ after travelling a distance $\lambda/4$. However, the atom has passed a process of climbing the potential hill, so the atomic kinetic energy is reduced and transformed into the potential energy partially. On the top of the potential hill ($z = \lambda/4$), the atom can absorb a σ^+ photon and jump to $|e_{1/2}\rangle$. Meanwhile, due to the effect of vacuum field, the atom in $|e_{1/2}\rangle$ will emit a fluorescent photon and decay to the ground state $|g_{1/2}\rangle$, that is to say, the atom falls in another potential valley and the atomic potential energy radiates out through fluorescent photons. From there, the same process can be repeated as shown in Fig.(18.5). It thus displays that, like Sisphus in Greek mythology, the atom always seems to be climbing potential

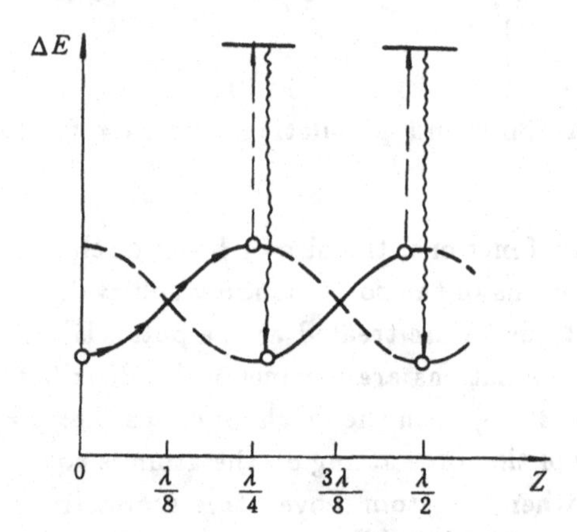

Figure 18.5: Sisphus effect of he atom under interaction of laser field

hills, transforming part of its kinetic energy into potential energy.

The previous physical picture shows clearly that this new cooling mechanism is most effective when the atom travels a distance of the order $\lambda/4$ during the pumping time τ_p. Thus the velocity-capture range may be defined by

$$v_p \approx \frac{\lambda}{4\tau_p} \qquad or \qquad kv_p\tau_p \approx \frac{1}{4} \tag{18.73}$$

Because the pumping rate $1/\tau_p$ is proportional to the saturation constant s_0, then v_p is also proportional to s_0. When $s_0 \ll 1$, i.e., $\Omega \ll \Gamma$, v_p tends to zero, which corresponds to a very low temperature of the laser cooling.

The above discussion shows that the effective laser cooling occurs at $v = v_p$. But it must be mentioned that this result is obtained on the basis of eqs.(18.61) and (18.62) which are only valid under the condition $kv \ll 1/\tau_p$, so the spatial average force $\overline{f}$ is invalid for the case $v = 1/(4\tau_p)$. Now we come to the calculation of the force for $kv \ll \Gamma$, which is the starting point to derive the optical Bloch equation without considering $\mathbf{P}^2/(2M)$. Since the laser field is very weak, i.e., $\Omega \ll \Gamma$, $s_0 \ll 1$, then the velocity range defined by $kv \ll \Gamma$ is wider than that $kv \ll 1/\tau_p$, that is to say, there is no restriction on the relative values of kv and $1/\tau_p$ under the condition $kv \ll \Gamma$. In this case, it is necessary to solve ρ_{55} and ρ_{66} starting from (18.45) and (18.46). Noticing that

$$\frac{d}{dt}\rho_{55} = v\frac{d}{dz}\rho_{55} = -\frac{1}{\tau_p}(\rho_{55} - \sin^2 kz) \tag{18.74}$$

$$\frac{d}{dt}\rho_{66} = v\frac{d}{dz}\rho_{66} = -\frac{1}{\tau_p}(\rho_{66} - \cos^2 kz) \tag{18.75}$$

and using the integral formulas

$$\int dx e^{ax}\sin^2 bx = \frac{e^{ax}}{2a} - \frac{e^{ax}}{a^2+4b^2}\left(\frac{a}{2}\cos 2bx + b\sin 2bx\right)$$

$$\int dx e^{ax}\cos^2 bx = \frac{e^{ax}}{2a} + \frac{e^{ax}}{a^2+4b^2}\left(\frac{a}{2}\cos 2bx + b\sin 2bx\right)$$

we have

$$\rho_{55}(z,v) = \frac{1}{2} - \frac{\cos 2kz + 2kv\tau_p \sin 2kz}{2(1+4k^2v^2\tau_p^2)} \tag{18.76}$$

$$\rho_{66}(z,v) = \frac{1}{2} + \frac{\cos 2kz + 2kv\tau_p \sin 2kz}{2(1+4k^2v^2\tau_p^2)} \tag{18.77}$$

Substituting (18.76) and (18.77) into (18.48), we finally obtain the radiation force to be

$$f(z,v) = \frac{2\hbar k\delta s_0 \sin 2kz \cos 2kz + 4\hbar k^2 v\delta s_0\tau_p \sin^2 2kz}{3(1+4k^2v^2\tau_p^2)} \tag{18.78}$$

Obviously, (18.78) is identified with (18.63) for $kv\tau_p \ll 1$. Averaging $f(z,v)$

over a wavelength, the spatial average force acting on the atom is written as

$$\overline{f}(v) = -\frac{\alpha v^2}{1 + v^2/v_c^2} \tag{18.79}$$

with

$$v_c = \frac{1}{2k\tau_p} = \frac{\lambda}{4\pi\tau_p} \tag{18.80}$$

Eq.(18.79) predicts that $\overline{f}(v)$ reaches to its maximum value when $v = v_c$

$$\overline{f}_{max}(v = v_c) = -\frac{\alpha v_c}{2} \tag{18.81}$$

Since the atom moving with velocity $v = v_c$ experiences a maximum friction force ($\delta < 0$) in the laser field, we call v_c as the critical velocity. During a pumping time, the atom travels a distance

$$z = v_c \tau_p = \frac{1}{2k\tau_p}\tau_p = \frac{\lambda}{4\pi} \sim \frac{\lambda}{4} \tag{18.82}$$

Evidently, v_c and v_p have the same order of magnitude, so the previous discussion about the physical mechanism of the laser cooling is still valid.

18.3 Limited temperature of laser cooling

With the development in the technique of the laser cooling, it is expected that the atoms may be cooled to be very close to absolute zero. Although one may predict theoretically the limited temperature of the laser cooling, the correctness is determined by whether the theoretical predications coincided with the experimental results. Recently, the magnitude of the limited temperature of the laser cooling has decreased gradually with the development of technology, which also promotes the theoretical investigation furthermore. In this section, we will focus on discussing the limited temperature of the laser cooling by means of the above quantum theory treatment.

18.3.1 Atomic momentum diffusion in a laser field

As yet we have discussed the laser cooling processes. However, in the interaction processes between the atom and the field, the atomic momentums be diffused, so there exist the laser heating processes. For the atom-field coupling

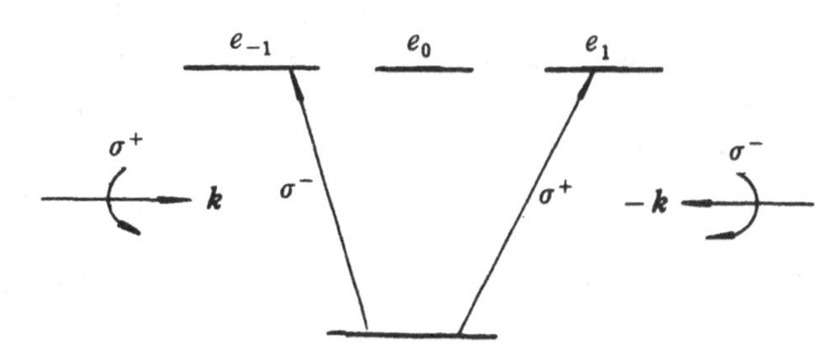

Figure 18.6: Atomic transition processes from $J_g = 0 \leftrightarrow J_e = 1$

system under the consideration here, there are three aspects leading to the atomic momentum diffusion: (1). Since fluorescence photons emitted by the atom have stochastic property due to the effect of the vacuum field, there are fluctuations of the momentum carried away by fluorescence photons, which is responsible for the atomic momentum diffusion. (2). There are fluctuations in the difference among the number of photons absorbed in each of two laser fields, which leading to the atomic momentum diffusion. (3). Because spontaneous emission processes are involved in the optical pumping process, there are fluctuations of the instantaneous dipole force f oscillating back and forth between $f_{1/2}(z)$ and $f_{-1/2}(z)$ at a rate $1/\tau_p$, which results in the atomic momentum diffusion. Now we turn to discuss the influences of these three processes on the atomic momentum diffusion coefficient

$$D_p = \frac{1}{2}\frac{d}{dt}(\langle \mathbf{P}^2 \rangle - \langle \mathbf{P} \rangle^2) = \frac{1}{2}\frac{d}{dt}(\Delta \mathbf{P})^2 \tag{18.83}$$

In order to analyze the influences of the first and the second fluctuation processes on the atomic momentum diffusion coefficient D_p, we first evaluate the contributions of these two processes to D_p in a relative simple model to estimate the value of D_p in the atom with $J_g = 1/2$ and $J_e = 3/2$ under the interaction of two linear polarization fields.

For the sake of simplicity, we choose an atomic transition from $J_g = 0$ (ground level) to $J_e = 1$ (excited level) (Fig.18.6) under the interaction of two counter-propagating running waves with equal amplitude and orthogonal polarizations σ^+ and σ^-, respectively. After the atom absorbs one photon and transits to an excited sublevel from the ground state, for example, if the atom in $|g_0\rangle$ absorbs a σ^+ photon to jump to $|e_1\rangle$, subsequently, the atom can stimulately emit a σ^+ photon and returns back to $|g_0\rangle$. In this absorption and emission process, the atomic momentum is unchanged. Meanwhile, the atom in $|e_1\rangle$ can also emit a fluorescence photon with momentum $\hbar\mathbf{k}'$ arisen in the spontaneous emission. In this absorption and emission process, the atomic momentum varies. Suppose that during the time ΔT, the atom transfers N_+ σ^+ photons to N_+ fluorescence photons, and also transfers N_- σ^- photons to N_- fluorescence photons, then the variation of the atomic momentum is

$$\Delta\mathbf{P} = (N_+ - N_-)\hbar\mathbf{k} - \sum_{i=1}^{N_+ + N_-} \hbar\mathbf{k}_i \tag{18.84}$$

The first term represents the variation of the atomic momentum after the atom absorbs the net N_+ σ^+ photons and N_- σ^- photons . The second term stands for the atomic momentum variation resulting from the emission of $N_+ + N_-$ fluorescence photons. Since the fluorescence photons are homogenous in the momentum-space, the average of $\mathbf{k}_i$ is equal to zero. Because the amplitudes of two laser fields are equal, the average values of N_+ and N_- are equal, i.e., $\overline{N}_+ = \overline{N}_-$. The average value of $\Delta\mathbf{P}$ therefore gives

$$\overline{\Delta\mathbf{P}} = (\overline{N}_+ - \overline{N}_-)\hbar\mathbf{k} - \sum_{i=1}^{N_+ + N_-} \hbar\mathbf{k}_i = 0 \tag{18.85}$$

That is to say, the average of the atomic momentum variation is zero. However, during the time ΔT, the fluctuation of the atomic momentum is

$$(\Delta\mathbf{P})^2 = \left[(N_+ - N_-)\hbar\mathbf{k} - \sum_{i=1}^{N_+ + N_-} \hbar\mathbf{k}_i\right]^2 = (N_+ - N_-)^2\hbar^2 k^2 + \left(\sum_{i=1}^{N_+ + N_-} \hbar\mathbf{k}_i\right)^2 - 2(N_+ - N_-)\hbar\mathbf{k} \cdot \sum_{i=1}^{N_+ + N_-} \hbar\mathbf{k}_i \tag{18.86}$$

Averaging the above expression, then

$$\overline{(\Delta\mathbf{P})}^2 = \overline{(N_+ - N_-)^2\hbar^2 k^2} + \overline{\left(\sum_{i=1}^{N^+ + N_-} \hbar\mathbf{k}\right)^2} \tag{18.87}$$

Notice that the momentum $\hbar\mathbf{k}_i$ of a given fluorescence photon is not correlated to the momentum of other fluorescence photons, nor to the numbers N_+ and

N_-, the last term in (18.87) reduces to

$$\overline{\left(\sum_{i=1}^{N_+ + N_-} \hbar \mathbf{k}_i\right)^2} = \overline{\sum_{i=1}^{N_+ + N_-} \hbar^2 k^2} + 2 \overline{\sum_{i \neq j}^{N_+ + N_-} \hbar^2 \mathbf{k}_i \cdot \mathbf{k}_j} = \overline{(N_+ + N_-)} \hbar^2 k^2 \tag{18.88}$$

So (18.87) becomes

$$\overline{(\Delta \mathbf{P})^2} = \overline{(N_+ - N_-)^2} \hbar^2 k^2 + \overline{(N_+ + N_-)} \hbar^2 k^2 \tag{18.89}$$

Therefore, the momentum diffusion coefficient is

$$D'_p = \frac{\overline{(\Delta \mathbf{P})^2}}{2\Delta T} = \frac{1}{2} \hbar^2 k^2 \frac{\overline{N_+ + N_-}}{\Delta T} + \frac{1}{2} \hbar^2 k^2 \frac{\overline{(N_+ - N_-)^2}}{\Delta T} \tag{18.90}$$

The factor $\overline{(N_+ + N_-)}/\Delta T$ in the first term in (18.90) represents the emission rate R of fluorescence photons. In Chapter 8, it is defined as

$$R = \Gamma s_0 \tag{18.91}$$

So the first term in (18.90) describes the contribution of fluctuations of the momentum carried away by the fluorescence photons to the momentum diffusion coefficient, i.e.,

$$D_{pvac} = \frac{1}{2} \hbar^2 k^2 \Gamma s_0 \tag{18.92}$$

It is necessary to point out that here we did not consider the influence of the polarization distribution of fluorescence photons. It can be verified when the influence of the polarization distribution of fluorescence photons is considered, the diffusion coefficient D_{pvac} becomes

$$D_{pvac} = \frac{1}{5} \hbar^2 k^2 \Gamma s_0 \tag{18.93}$$

We now discuss the contribution of the second term in (18.90) to D_p. This term describes the fluctuations of the difference $N_+ - N_-$ between the number of photons absorbed in each wave. Remark now that there is no correlation between the polarizations of two laser fields, the amplitudes of two laser fields are equal, and the transition strengths for $|g_0\rangle \leftrightarrow |e_1\rangle$ and $|g_0\rangle \leftrightarrow |e_{-1}\rangle$ are also equal , so the atom in $|g_0\rangle$ has the same probability of absorbing either a σ^+ photon or a σ^- one. Therefore, in the processes of the atom totally absorbing

$N_+ + N_-$ laser photons and transforming them into $N_+ + N_-$ fluorescence photons, the probability $P(N_+, N_-)$ for finding the atom absorbing N_+ σ^+ photons and N_- σ^- photons is simply related to the probability $P(N_+ + N_-)$ for finding a total number $N_+ + N_-$ laser photons absorbed by atom

$$P(N_+, N_-) = \frac{(N_+ + N_-)!}{2^{N_+ + N_-} N_+! N_-!} P(N_+ + N_-) \tag{18.94}$$

Using the above equation and letting $N = N_+ + N_-$, then

$$\begin{aligned}\overline{(N_+ - N_-)^2} &= \sum_{N_+=0}^{\infty} \sum_{N_-=0}^{\infty} (N_+ - N_-)^2 P(N_+, N_-) \\ &= \sum_{N=0}^{\infty} \sum_{N_-=0}^{N} (N - 2N_-)^2 \frac{N!}{2^N (N - N_-)! N_-!} P(N) \\ &= \sum_{N=0}^{\infty} [N^2 - 2N^2 + N(N-1) + 2N] P(N) = \overline{N_+ + N_-}\end{aligned} \tag{18.95}$$

here we have used the binomial theorem

$$(a+b)^n = \sum_{k=0}^{n} \frac{a^k b^{n-k} n!}{k!(n-k)!}$$

Inserting (18.95) into the second term of (18.90), the contribution of fluctuations of the difference $N_+ - N_-$ between the photons absorbed in two laser fields to the atomic momentum diffusion is

$$D_{pdif} = \frac{1}{2} \hbar^2 k^2 \Gamma s_0 \tag{18.96}$$

Consequently, the total atomic momentum diffusion of the first two diffusion processes gives

$$D_p' = D_{pdif} + D_{pvac} = \frac{7}{10} \hbar^2 k^2 \Gamma s_0 \tag{18.97}$$

It is necessary to mention that we need to know the atomic momentum diffusion coefficient for the atom with $J_g = 1/2$ and $J_e = 3/2$. However, this quasi-two-level atom can be approximately treated as the superposition of two atoms with $J_g = 0$ and $J_e = 1$, so the first two diffusion processes contribute to the momentum diffusion for the quasi-two-level atom with $J_g = \frac{1}{2}$ and $J_e = \frac{3}{2}$ have the order of magnitude of D_p' (18.97). As a reasonable approximation, we also treat D'_p as the first two contributions to atomic momentum diffusion coefficient for the quasi-two-level atom with $J_g = 1/2$ and $J_e = 3/2$.

Now we turn to evaluate the contribution of the third diffusion process, i.e., the stochastic property of the instantaneous dipole force f results in the atomic momentum diffusion. First we analyze the random property of the instantaneous dipole force. For the atom initially at point z, if the atom is in the state $|g_{1/2}\rangle$, it undergoes a force $-\nabla E_{1/2}$ as shown in Fig.(18.7). Because of the spontaneous emission process involving in the optical pumping processes,

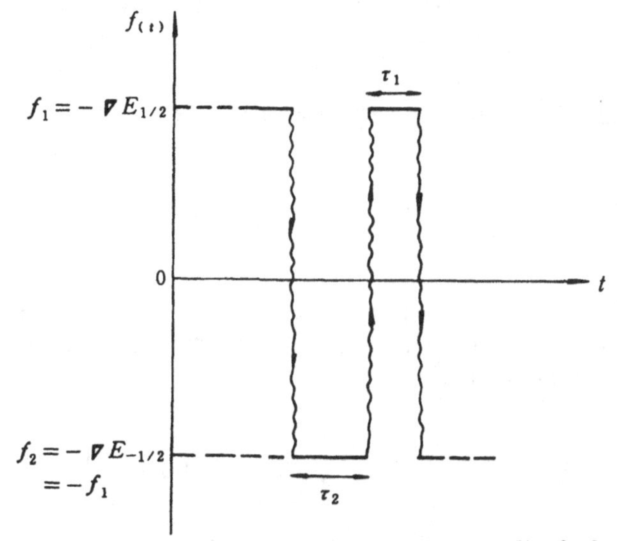

Figure 18.7: Stochastic property of instaneous dipole force **f**

the atom absorbs a σ^+ photon and at a random time jumps into $|g_{-1/2}\rangle$ by spontaneous emitting a fluorescence photon. In this sublevel $|g_{-1/2}\rangle$, the atom undergoes a force $-\nabla E_{-1/2} = \nabla E_{1/2}$. Subsequently, the atom will absorb a σ^- photon and be put back into $|g_{1/2}\rangle$ at a random time by spontaneous emission of a fluorescence photon, and so on. It thus displays that the instantaneous force experienced by the atom switches back and forth between $-\nabla E_{1/2}$ and $-\nabla E_{-1/2}$ after random intervals τ_1 and $\tau_2, \cdots$. If $\overline{\tau}_1$ and $\overline{\tau}_2$ are the mean values of these successive time intervals spent by the atom in $|g_{1/2}\rangle$ and $|g_{-1/2}\rangle$, then the mean force experienced by the atom is

$$\mathbf{f} = -\nabla E_{1/2}\frac{\overline{\tau}_1}{\overline{\tau}_1 + \overline{\tau}_2} - \nabla E_{-1/2}\frac{\overline{\tau}_2}{\overline{\tau}_1 + \overline{\tau}_2} \tag{18.98}$$

Clearly, $\frac{\overline{\tau}_1}{\overline{\tau}_1+\overline{\tau}_2}$ and $\frac{\overline{\tau}_2}{\overline{\tau}_1+\overline{\tau}_2}$ are actually just the steady-state atomic populations in $|g_{1/2}\rangle$ and $|g_{-1/2}\rangle$, respectively.

Second we calculate the contribution of fluctuations of the instantaneous force experienced by the atom to the momentum diffusion coefficient. According to (18.83), i.e., $\mathbf{f} = \frac{d}{dt}\mathbf{P}$ we have

$$D_p'' = \langle \mathbf{P} \cdot \mathbf{f} \rangle - \langle \mathbf{P} \rangle \cdot \langle \mathbf{f} \rangle$$
$$= \int_{-\infty}^{0} [\langle \mathbf{f}(t) \cdot \mathbf{f}(0) \rangle - \langle \mathbf{f}(t) \rangle \cdot \langle \mathbf{f}(0) \rangle] dt$$

Here $\mathbf{P}(0) = \int_{-\infty}^{0} \mathbf{f}(t)dt$ has been used. Rewriting (18.98), we have

$$D_p'' = \int_0^\infty [\langle f(0)f(t) \rangle - \langle f \rangle^2] dt = \int_0^\infty d\tau [\langle f(\tau)f(t+\tau) \rangle - \langle f \rangle^2] \tag{18.99}$$

In order to obtain D_p'', we must calculate the expectation value of the two-time correlation function $f(t)f(t+\tau)$. By use of eq.(18.59), the two-time correlation function at z may be written as

$$\langle f(t)f(t+\tau) \rangle = \sum_{i,j=-1/2}^{1/2} f_i f_j P(i,t;j,t+\tau) \tag{18.100}$$

That is to say, the two-time correlation function is equal to the product of the two instantaneous forces f_i in $|i\rangle$ and f_j in $|j\rangle$ (i,j=-1/2,1/2) weighted by the probability $P(i,t;j,t+\tau)$ to be in a state $|i\rangle$ at time t and in a state $|j\rangle$ at time $t+\tau$ and summed over i and j. In the steady-state case, $P(i,t;j,t+\tau)$ depends only on τ and can be written as

$$P(i,t;j,t+\tau) = P_i P(j,\tau|i,0) \tag{18.101}$$

where P_i is the steady-state probability of the atom being in the state $|i\rangle$, which is

$$P_{1/2} = \rho_{55}^{st}, \qquad P_{-1/2} = \rho_{66}^{st} \tag{18.102}$$

and $P(j,\tau|i,0)$ is the conditional probability of the atom being in the state $|j\rangle$ at τ, when it is initially in $|i\rangle$. From eqs.(18.45) and (18.46) we know when the initial conditions are $\rho_{55}(0) = 1$ and $\rho_{66}(0) = 1$, then

$$\rho_{55}(\tau) = \rho_{55}^{st} + \rho_{66}^{st} \exp\left(-\frac{\tau}{\tau_p}\right) \tag{18.103}$$

$$\rho_{66}(\tau) = \rho_{66}^{st} + \rho_{55}^{st} \exp\left(-\frac{\tau}{\tau_p}\right) \tag{18.104}$$

Thus the conditional probabilities of finding the atom in $|j\rangle$ at time τ which initially in $|i\rangle$ at time t=0 are written as

$$P\left(\frac{1}{2},\tau|\frac{1}{2},0\right) = \rho_{55}(\tau) \tag{18.105}$$

$$\begin{aligned} P\left(-\frac{1}{2},\tau|\frac{1}{2},0\right) &= 1 - P\left(\frac{1}{2},\tau|\frac{1}{2},0\right) = 1 - \rho_{55}(\tau) \\ &= \rho_{66}^{st} - \rho_{66}^{st} \exp\left(-\frac{\tau}{\tau_p}\right) \end{aligned} \tag{18.106}$$

$$P\left(-\frac{1}{2},\tau|-\frac{1}{2},0\right) = \rho_{66}(\tau) \tag{18.107}$$

$$\begin{aligned} P\left(\frac{1}{2},\tau|-\frac{1}{2},0\right) &= 1 - P\left(-\frac{1}{2},\tau|-\frac{1}{2},0\right) = 1 - \rho_{66}(\tau) \\ &= \rho_{55}^{st} - \rho_{55}^{st} \exp\left(-\frac{\tau}{\tau_p}\right) \end{aligned} \tag{18.108}$$

Inserting eqs.(18.102) and (18.105)-(18.108) into (18.100), then we have

$$\langle f(t)f(t+\tau)\rangle = (f^{st})^2 + 4(f_{1/2})^2 \rho_{55}^{st}\rho_{66}^{st} \exp\left(-\frac{\tau}{\tau_p}\right) \tag{18.109}$$

Substituting (18.109) into (18.99) and using eqs.(18.57), (18.43) and (18.44) , the contribution of fluctuations of the intantaneous force to the atomic momentum diffusion coefficient yields

$$\begin{aligned} D_P'' &= \int_0^\infty 4(f_{1/2})^2 \rho_{55}^{st}\rho_{66}^{st} \exp\left(-\frac{\tau}{\tau_p}\right) d\tau \\ &= 2\hbar^2 k^2 \delta^2 s_0 \sin^4(2kz)/\Gamma \end{aligned} \tag{18.110}$$

Once this is averaged over a wavelength, it gives

$$\overline{D_p''} = \frac{1}{\lambda}\int_0^\lambda D_p'' dz = \frac{3}{4}\hbar^2 k^2 \frac{\delta^2}{\Gamma} s_0 \tag{18.111}$$

Therefore, the total momentum diffusion coefficient D_p is

$$D_p = D'_p + \overline{D''_p} = \frac{7}{10}\hbar^2 k^2 \Gamma s_0 + \frac{3}{4}\hbar^2 k^2 \frac{\delta^2}{\Gamma} s_0 \tag{18.112}$$

If $|\delta| \gg \Gamma$, then $D'_p \ll \overline{D''_p}$. In this case, the dominant contribution to the atomic momentum diffusion is $\overline{D''_p}$, i.e.,

$$D_p \approx \overline{D''_p} = \frac{3}{4}\hbar^2 k^2 \frac{\delta^2}{\Gamma} s_0 \tag{18.113}$$

18.3.2 Equilibrium temperature of the laser cooling

We now turn to calculate the equilibrium temperature in this laser cooling scheme. Because the laser fields can cool the atom and also may heat the atom due to the atomic momentum diffusion, then the question is that in what situation these two mechanisms may reach to the equilibrium. In other words, what is the equilibrium temperature ?

As we know, the friction force $f = -\alpha v$ results in the decreasing of the atomic kinetic energy, the decreasing rate is

$$\left(\frac{d}{dt}E\right)_{cool} = fv = -\alpha v^2 \tag{18.114}$$

On the other hand, the momentum diffusion also leads to the increasing of atomic kinetic energy, the increasing rate is

$$\left(\frac{d}{dt}E\right)_{heat} = \frac{1}{2M}\frac{d}{dt}(\Delta P)^2 = \frac{D_p}{M} \tag{18.115}$$

here we have used the relation $\frac{d}{dt}\langle \mathbf{P} \rangle^2 = 0$ in the steady-state case. After sufficiently long time interaction between the laser fields and the atom, the laser cooling and heating processes reach to equilibrium. The changing rate of atomic kinetic energy is therefore equal to zero, i.e.,

$$\left(\frac{d}{dt}E\right)_{cool} + \left(\frac{d}{dt}E\right)_{heat} = -\alpha v^2 + \frac{D_p}{M} = 0$$

Then we obtain

$$\frac{1}{2}Mv^2 = \frac{D_p}{\alpha} \tag{18.116}$$

Noticing that

$$E_k = \frac{Mv^2}{2} = \frac{k_B T}{2} \tag{18.117}$$

the equilibrium temperature, i.e., the lowest temperature of laser cooling atom obeys

$$k_B T_D = \frac{D_p}{\alpha} \tag{18.118}$$

When $|\delta| \gg \Gamma$, this temperature becomes

$$T_D \approx \frac{\hbar|\delta|s_0}{4k_B} \approx \frac{\hbar\Omega^2}{8|\delta|k_B} \tag{18.119}$$

The above equation indicates that the temperature of the laser cooling depends on the intensity of laser field and the detuning between the laser field and the atom. At a given laser intensity (Ω fixed), the temperature decreases when the detuning increases. At a given detuning, the temperature decreases when the intensity of laser field decreases. These results have been confirmed in experiments.

Finally, we discuss the achievable lowest temperature in this laser cooling scheme. Eq.(18.119) suggests that an arbitrarily low temperature can be reached, for instance, by decreasing the laser power. Actually, that is not true. Indeed, for $v > v_c$ the cooling force is nonlinear for v, but (18.119) is deduced from that f is linear for v, which demands that the average velocity of the atom in thermal equilibrium is sufficiently below v_c. We therefore get

$$v_c \gg v_{rms} \Longrightarrow \Omega \gg \sqrt{\frac{\hbar^2 k^2 |\delta|^3}{M\Gamma^2}} \tag{18.120}$$

which puts a lower bound on the laser power. This gives a lower bound on the achievable average velocity

$$V_{rms} = \sqrt{\frac{\hbar\Omega^2}{8M|\delta|}} \gg \frac{\hbar k|\delta|}{2\sqrt{2}M\Gamma} \tag{18.121}$$

Consequently, for $|\delta| \gg \Gamma$, the lowest achievable average velocity in this system under consideration here remains larger than the recoil velocity $\hbar k/M$, which gives a restriction of the lowest temperature T_D of laser cooling. However, the temperature T_D defined by (18.119) and (18.121) is lower than the Doppler limit temperature T_D' mentioned in Section 1. Eqs.(18.119) and (18.121) are in a good agreement with the results of experiments. For example, the observation of the lowest cooling temperature of C_e atom is $2.5 \pm 0.6 \mu K$, the corresponding thermal velocity at this temperature is a few times faster than recoil velocity of single-photon emission.

18.3.3 Laser cooling below the one-photon recoil energy by the velocity-selective coherent population trapping

From the above discussion we see that the lowest temperature T_D is limited by the atomic momentum diffusion resulting from the random recoil momentum due to the spontaneous emissions, so the lowest thermal equilibrium velocity of the atom is impossible to be less than one-photon recoil momentum $\hbar k/M$. However, is it possible to enable the atom in a state in which the atomic spontaneous emission disappears ? From the theoretical point of view, if the atom can be excited to the coherent population trapping state mentioned in Chapter 14, then the interaction between the atom and the laser field can only lead to the atomic oscillating between (or among) superposition levels of coherent population trapping states. Once the atom is pumped into this kind of coherent population trapping state, the spontaneous emission is in absence, then there does not exist random recoil momentum due to spontaneously emitted photons, therefore, the atomic velocity may be cooled to be below the one-photon recoil momentum.

In 1988, a new laser cooling mechanism based on the above idea, that is, the velocity-selective coherent population trapping was proposed to overcome the recoil limit by Cohen-Tannoudji and his co-workers, and their experimental result showed that the equilibrium temperature may be below $2\mu K$, which is corresponding to below the one-photon recoil momentum. The basic idea of this mechanism is as follows. Only the atom whose velocity v is zero can be in coherent population trapping state, but the atom with $v \neq 0$ can also absorb and emit photons. The larger v is, the higher the absorption rate. When the atomic velocity is cooled to be very low, the atom with $v \neq 0$ can also absorb photons, but after it spontaneously emit fluorescence photons, the atomic momentum is redistributed due to the recoil effect, so there is a certain probability for an atom initially in an absorbing velocity ($v \neq 0$) to be optically pumped into the v=0 nonabsorbing trapping state. When this happens, the atom is "hidden" from the light and also protected from the random recoils. After a interaction time interval Θ, there are a number of atoms to pile up in a narrow velocity interval δv around v=0. Atoms for which v is not exactly 0 are not perfectly trapped. As a result the width δv of the interval around v=0 is determined by the interaction time Θ. For a given Θ only the atoms which can remain to be trapped are those for which the absorption rate times Θ is smaller than 1. Since the absorption rate is proportional to v, the larger Θ, the smaller v for the remaining atoms. So that there is no lower limit to the atomic velocity. The recent experimental result is reported that the lower-temperature of the laser cooling may reach to $10^{-12}K$, and the mean velocity of the atom approaches to zero nearly.

In the previous discussions, we only gave a qualitative statement to the method of the laser cooling atom by the velocity-selective coherent population trapping. It is necessary to mention that the development in laser cooling is widely extended recently, several theoretical and technological treatments of the laser cooling have been suggested, such as cooling mechanism by "magnetic induction" and cooling mechanism by "velocity-selective of Raman transition". Considering the physical mechanism of those treatments are coincide with our previous statement, here we do not give any more discussions, the reader who is interested in this problems may find some references listed in the last of this chapter.

References

1. D.Wineland and W.Itano, *Phys.Rev.* **A20** (1979) 1521.

2. J.Gordon and A.Ashkin, *Phys.Rev.* **A21** (1980) 1606.

3. V.S.Letokov and V.G.Minogin, *Phys.Rep.* **73** (1981) 1.

4. Steven Chu, *Optics and Photonic News* **1** (1990) 40,

5. K.Gibble and S.Chu, *Phys.Rev.Lett.* **70** (1993) 1771

6. J.Dalibard, S.Reynaud, and C.Cohen-Tannoudji, *J.Phys.* **B.17** (1984) 4577.

7. J.Dalibard and C.Cohen-Tannoudji, *J.Phys.* **B18** (1985) 1661.

8. C.Cohen-Tannoudji, N.Claude, *Physics Today*,**43** (1990) 33.

9. S.Stenholm, *Rev.Mod.Phys.* **58** (1986) 699.

10. A.Aspect, J.Dalibard, A.Heidmann, C.Salomon and C. Cohen-Tannoudji, *Phys. Rev. Lett.* **57** (1986) 1688.

11. P.Lett, R.Watts, C.Westbrook, W.D.Phillips and H.Metcalf, *Phys. Rev. Lett.* **61** (1988)169.

12. A.Aspect, E.Arimodo, R.Kaiser, et al, *Phys. Rev. Lett.* **61** (1988) 826.

13. J.Dalibard and C.Cohen-Tannoudji, *J.Opt.Soc.Am.* **B6** (1989) 2023.

14. P.J.Ungar, D.S.Weiss, E.Piis and S.Chu, *J.Opt.Soc.Am.* **B6** (1989) 2058.

15. Y.Castin, H.Wallis and J.Dalibard, *J.Opt.Soc.Am.* **B6** (1989) 2046.

16. A.Aspect, E.Arimondo, et al, *J. Opt. Soc. Am.* **B6** (1989) 2112.

17. Y.Castin and K.Molmer, *J.Phys.* **B23** (1990) 4101.

18. Y.Shevy, *Phys.Rev.Lett.* **64** (1990) 2905.

19. C.Cohen-Tannoudji and W.D.Phillips, *Phys.Today* (Oct.1990) 33.

20. C.Salomon, J.Dalibard, W.Philips , A.Clairon and S.Gurllati, *Europhys. Lett.* **12** (1990) 683.

21. P.R.Berman, *Phys.Rev.* **A43** (1991) 1470.

22. S.Q.Shang, B.Sheey, H.Metcalf, P.van der Straten and G. Nienhuis, *Phys. Rev. Lett.* **67** (1991) 1094.

23. T.W.Mossberg, M.Lewenstein, and D.J.Gauthier, *Phys. Rev. Lett.* **67** (1991) 1723.

24. A.Aspect, *Phys.Rep.* **219** (1992) 141.

25. C.Cohen-Tannoudji, *Phys.Rep.* **219** (1992) 153.

26. J.Chen, J.G.Story, J.J.Tollett, and R.G.Hulet, *Phys. Rev. Lett.* **69** (1992) 1344.

27. W.Ketterle, A.Martin, M.A.Joffe and D.E.Pitchard, *Phys. Pev. Lett.* **69** (1992) 2483.

28. A.M.Steane, G.Hillenbrand, and C.J.Font, *J. Phys.* **B25** (1992) 4721.

29. V.Finkelstein, J.Guo, and P.R.Berman, *Phys.Rev.* **A46** (1992) 7108.

30. J.Y.Courtois and G.Grynberg, *Phys.Rev.* **A46** (1992) 7060.

31. J.Dalibard, Y.Castin, and K.Molmer, *Phys. Rev. Lett.* **68** (1992) 580.

32. E.Korsunsky, D.Kosachiov, B.Matisov, Yu.Rozhdestvensky, L. Windholz and C. Neureiter, *Phys.Rev.* **A48** (1993) 1419.

33. J.Guo and P.R.Berman, *Phys.Rev.* **A48** (1993) 3225.

34. M.Doery, M.Widmer, J.Bellance, E.Vredenbregt, T. Bergeman, and H. Metcalf, *Phys. Rev. Lett.* **72** (1994) 2546.

35. H.Metcalf and P.van der Straten, *Phys.Rep.* **224** (1994) 203.

36. J.Lawall, F.Bardou, B.Saubamea, K.Shimizu, M.Leduc, A. Aspect, and C. Cohen -Tannoudji, *Phys. Rev. Lett.* **73** (1915) 1915.

37. M.R.Doery, E.J.D.Vredenbregt, and T.Bergeman, *Phys.Rev.* **A51** (1995) 4881.

38. M.R.Doery, M.T.Widmer, M.J.Bellance, W.F.Buell, T.H.Bergeman, H. Metcalf and E. J. D. Vredenbregt, *Phys. Rev.* **A52** (1995) 2295.

Index

N

O

P

Q

R